T0235877

Open-Channel Flow

Third Edition

M. Hanif Chaudhry

Open-Channel Flow

Third Edition

 Springer

Dr. M. Hanif Chaudhry
University of South Carolina
Department of Civil and Environmental Engineering
Columbia, SC, USA

This book includes lecturer materials at sn.pub/lecturer-material

Originally published by Prentice-Hall, Inc., Englewood Cliffs, 1993

ISBN 978-3-030-96449-8   ISBN 978-3-030-96447-4   (eBook)
https://doi.org/10.1007/978-3-030-96447-4

To Shamim

Preface

The flow in natural channels, such as rivers and streams, or in built water-conveyance structures, such as canals, is referred to as open-channel or free-surface flow. The analysis of these flows is needed for the planning, design, and operation of water resource projects. Although empirical approaches have been used for this purpose in the past, the availability of efficient computational procedures during the last 50 years or so have made it possible to analyze large complex systems in addition to providing more accurate results that can be used with confidence. This book covers an introduction of these flows, presents modern numerical methods and computational procedures for analyses, and provides up-to-date information on the topic. The book is suitable as a text for senior-level undergraduate and graduate courses and as a reference for researchers and practicing engineers. Strong emphasis is given to the application of efficient solution techniques and numerical methods suitable for computer analysis. In addition, the coverage of unsteady flow is as detailed as that of steady flow. Except for Chapter 17 and parts of Chapters 9 and 18, the material is related to channels with rigid boundaries.

The book is divided into two parts: Chapters 1 through 10 cover steady flow and Chapter 11 through 18 cover unsteady flow. Chapter 1 summarizes basic flow concepts, and Chapter 2 presents the conservation laws of mass, momentum, and energy and their applications. Critical and uniform flows are discussed in Chapters 3 and 4, respectively. A qualitative discussion of gradually varied flows and methods for the computation of these flows are presented in Chapters 5 and 6, respectively. Chapter 7 deals with rapidly varied flow following mainly an empirical approach while modern numerical methods for the computation of these flows are outlined in Chapter 8. A number of procedures for the design of channels are presented in Chapter 9 and a number of special topics are discussed in Chapter 10. Unsteady flow is introduced in Chapter 11 and the governing equations for unsteady flow are derived in Chapter 12. Numerical integration of these equations and the initial and boundary conditions are discussed in Chapter 13. A number of explicit and implicit finite-difference methods are presented in Chapters 14 and 15 for one- and two-dimensional flows, respectively. Modeling of levee breach is discussed in Chapter 16, sediment transport in Chapter 17, and special topics related to unsteady flows in Chapter 18.

The text is based on the lecture notes for a course on open-channel flow for senior-level undergraduate and graduate students and for an advanced level graduate course on unsteady flow at Old Dominion University, at Washington State University, and at the University of South Carolina. In this edition, a new section on hydraulic models is added in Chapter 1, a new section on velocity measurements is added in Chapter 10, and a new chapter, Chapter 16, is added on the modeling of levee breach. Suggestions and comments of students, instructors, and reviewers are incorporated as appropriate. References are updated and additional problems are included, some of which are suitable for take-home tests. To facilitate and enhance learning, photographs are used extensively and short computer programs in Python related to different chapters are made available at the publisher's website.

In recent years, the author has used Chapters 1 through 6, 9, and 10 and parts of Chapter 7 in a three-semester-hour course for senior-level undergraduate and graduate students in water resources and Chapters 11 through 15 and Chapter 18 in an advanced three-credit class on unsteady flow. Students develop in each course three or four computer codes using any modern programming language, such as Python and MATLAB. For an introductory class on open-channel flow, some instructors may prefer a reduced coverage of Chapters 6 and 7 and instead include parts of Chapters 11 through 13. Parts of different chapters may be utilized in a course on computational hydraulics and hydraulic structures or to cover special topics.

Thanks are extended to anonymous reviewers for their suggestions for the clarity of presentation. I am thankful to my former postdoctoral fellows and graduate students for the inclusion of their jointly published contributions in the field. The assistance of Dr. M. Elkholy, Alexandria University, Egypt, for the preparation of the manuscript and for updating references and other material; Dr. Melih Calamak, Middle East Technical University, for reviewing the entire manuscript and making suggestions for its improvement; and my assistant, Katherine Tse, for proofreading, are thankfully acknowledged. The patience and understanding of my family, especially our grandchildren, Aryan, Amira, Rohan and Zain, during many hours spent in the preparation of this book is appreciated.

Columbia, SC, USA   M. Hanif Chaudhry

Contents

1

BASIC CONCEPTS

Areal photo of the Mississippi River Basin Model; with Atchafalaya outlet to the left, Ohio River to the right, Sioux City to top right (Courtesy, US Army Corps of Engineers)

Supplementary Information The online version contains supplementary material available at (https://doi.org/10.1007/978-3-030-96447-4_1).

1-1 Introduction

Liquids are transported from one location to another using natural or constructed conveyance structures. The cross section of these structures may be open or closed at the top. The structures with closed tops are referred to as *closed conduits* and those with the top open are called *open channels*. For example, tunnels and pipes are closed conduits whereas rivers, streams, estuaries etc. are open channels. The flow in an open channel or in a closed conduit having a free surface is referred to as *free-surface flow* or *open-channel flow*. The properties and the analyses of these flows are discussed in this book.

In this chapter, commonly used terms are first defined. The classification of flows is then discussed, and the terminology and the properties of a channel section are presented. Expressions are then derived for the energy and momentum coefficients to account for nonuniform velocity distribution at a channel section. The chapter concludes with a discussion of the pressure distribution in a channel section.

1-2 Definitions

The terms open-channel flow or free-surface flow (Fig. 1-1) are used synonymously in this book. The free surface is usually subjected to atmospheric pressure. Groundwater or subsurface flows are excluded from the present discussions. If there is no free surface and the conduit is flowing full, then the flow is called *pipe flow,* or *pressurized flow* (Fig. 1-2).

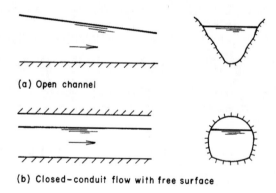

(a) Open channel

(b) Closed-conduit flow with free surface

Fig. 1-1   Free-surface flow

In a closed conduit, it is possible to have both free-surface flow and pressurized flow at different times. It is also possible to have these flows at a given time in different reaches of a conduit. For example, the flow in a storm sewer may be free-surface flow at a certain time. Then, due to large inflows produced

Fig. 1-2   Pipe or pressurized flow

by a sudden storm, the sewer may flow full and pressurize it. Similarly, the flow in a closed conduit may be free flow in part of the length and pipe flow in the remaining length. This type of combined free-surface, pressurized flow usually occurs in a closed conduit when the downstream end of the conduit is submerged (Fig. 1-3).

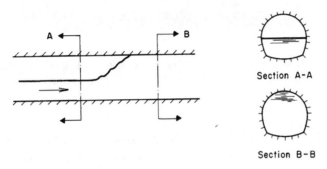

Section A-A

Section B-B

Fig. 1-3   Combined free-surface and pressurized flow

The photographs of Fig. 1-4 show unsteady flow in the 1:84-scale hydraulic model of the tailrace tunnel of Mica Power Plant, located on the Columbia River in Canada. The flow in the two unlined, horseshoe tailrace tunnels, each 18.3 m high and 14.6 m wide, is normally free-surface flow. However, during periods of high tailwater levels, the tunnels may be pressurized following major load changes on the turbogenerators that produce large changes in the inflow to the tunnels. The transient flow conditions shown in Fig. 1-4 are produced by increasing or decreasing in 9 seconds the discharge of three turbines on tunnel no. 2 while the discharge from the three turbines on tunnel no. 1 remains constant. The discharge increase in Fig. 1-4a is from zero to $850\,\text{m}^3/\text{s}$ and the discharge reduction in Fig. 1-4b is from $850\,\text{m}^3/\text{s}$ to zero. The free-surface and pressurized flows in a laboratory experimental setup are shown in Fig. 1-5. The initial steady state flow is from left to right and thus the upstream end is located on the left-hand side of the photographs.

The height to which liquid rises in a small-diameter piezometer inserted in a channel or a closed conduit depends upon the pressure at the location of the piezometer. A line joining the top of the liquid surface in the piezometers is

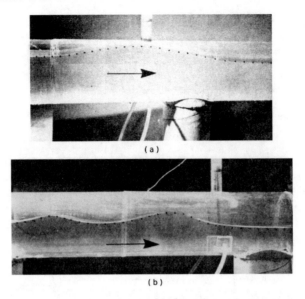

Fig. 1-4   Transient flow in the hydraulic model of Mica Tail-race Tunnel (Courtesy, British Columbia Hydro and Power Authority, Canada)

called the *hydraulic-grade line* (Fig. 1-6). In pipe flow, the height of hydraulic-grade line above a specified datum is called the *piezometric head* at that location. In free-surface flow, the hydraulic grade line usually, but not always, coincides with the free surface (see Section 1-6). If the velocity head, $V^2/(2g)$, in which V = mean flow velocity for the channel cross section, and g = acceleration due to gravity, is added to the top of the hydraulic grade line and the resulting points are joined by a line, then this line is called the *energy-grade line*. This line represents the total head at different sections of a channel.

1-3 Classification of Flows

Free-surface flows may be classified into different types (Fig. 1-7), as discussed in the following paragraphs.

Steady and Unsteady Flows

If the flow velocity at a given point does not change with respect to time, then the flow is called *steady flow*. However, if the velocity at a given location changes with respect to time, then the flow is called *unsteady flow*.

Note that this classification is based on the time variation of flow velocity v at a specified location. Thus, the local acceleration, $\partial v/\partial t$, is zero in steady

(a) Positive surge from downstream

(b) Positive surge from upstream

(c) Negative surge from downstream

(d) Negative surge from upstream

Fig. 1-5 **Free-surface and pressurized flows** (Courtesy, Professor C. S. Song [1984])

flows. In two- or three-dimensional steady flows, the time variation of all components of flow velocity is zero.

It is possible in some situations to transform unsteady flow into steady flow by having coordinates with respect to a moving reference. This simplification is helpful in the visualization of flow and in the derivation of governing equations.

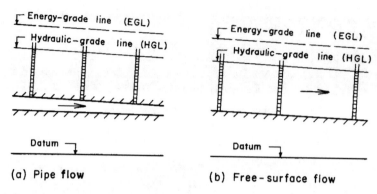

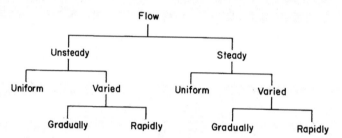

Fig. 1-7   Classification of flows

However, a transformation is possible only if the wave shape does not change as the wave propagates. For example, the shape of a surge wave propagating in a smooth channel does not change and consequently the propagation of a surge wave in an otherwise unsteady flow may be converted into steady flow by moving the reference coordinates at the absolute surge velocity. This is equivalent to an observer traveling beside the surge wave so that the surge wave appears to the observer to be stationary; thus the flow may be considered as steady. If the wave shape changes as it propagates, then it is not possible to transform such a wave motion into steady flow. A typical example of such a situation is the movement of a flood wave in a natural channel, where the shape of the wave is modified as it propagates in the channel.

Uniform and Nonuniform flows

If the flow velocity at an instant of time does not vary within a given length of channel, then the flow is called *uniform flow*. However, if the flow velocity at a time varies with respect to distance, then the flow is called *nonuniform flow*, or *varied flow*.

This classification is based on the variation of flow velocity with respect to space at a specified instant of time. Thus, the convective acceleration in uniform flow is zero. In mathematical terms, the partial derivatives of the velocity components with respect to x, y, and z direction are all zero. However, many times this strict restriction is somewhat relaxed by allowing a nonuniform velocity distribution at a channel section. In other words, a flow is considered uniform as long as the flow velocity in the direction of flow at different locations along a channel remains the same.

Depending upon the rate of variation with respect to distance, flows may be classified as *gradually varied flow* or *rapidly varied flow*. As the name implies, the flow is called gradually varied flow, if the flow depth varies at a slow rate with respect to distance, whereas the flow is called rapidly varied flow if the flow depth varies significantly in a short distance.

Note that the steady and unsteady flows are characterized by the variation of flow variables with respect to time at a given location, whereas uniform or varied flows are characterized by the variation at a given instant of time with respect to distance. Thus, in a steady, uniform flow, the total derivative $dV/dt = 0$. In one-dimensional flow, this means that $\partial v/\partial t = 0$, and $\partial v/\partial x = 0$. In two- and three-dimensional flow, the partial derivatives of the velocity components in the other two coordinate directions with respect to time and space are also zero.

Laminar and Turbulent Flows

The flow is called *laminar flow* if the liquid particles appear to move in definite smooth paths and the flow appears to be as a movement of thin layers on top of each other. In *turbulent flow,* the liquid particles move in irregular paths which are not fixed with respect to either time or space.

The relative magnitude of viscous and inertial forces determines whether the flow is laminar or turbulent: The flow is laminar if the viscous forces dominate, and the flow is turbulent if the inertial forces dominate.

The ratio of viscous and inertial forces is defined as the *Reynolds number,*

$$\mathbf{R}_e = \frac{VL}{\nu} \tag{1-1}$$

in which $\mathbf{R}_e$ = Reynolds number; V = mean flow velocity; L = a characteristic length; and ν = kinematic viscosity of the liquid. Unlike pipe flow in which the pipe diameter is usually used for the characteristic length, either hydraulic depth or hydraulic radius may be used as the characteristic length in free-surface flows. *Hydraulic depth* is defined as the flow area divided by the top water-surface width and the *hydraulic radius* is defined as the flow area divided by the wetted perimeter. The transition from laminar to turbulent flow in free-surface flows occurs for R_e of about 600, in which R_e is based on the hydraulic radius as the characteristic length.

In real-life applications, laminar free-surface flows are extremely rare. A smooth and glassy flow surface may be due to surface velocity being less than that required to form capillary waves and may not necessarily be due to the fact that the flow is laminar. Care should be taken while selecting geometrical scales for the hydraulic model studies so that the flow depth on the model is not very small. Very small depth may produce laminar flow on the model even though the prototype flow to be modeled is turbulent. The results of such a model are not reliable because energy losses are not simulated properly.

Subcritical, Supercritical, and Critical Flows

A flow is called *critical* if the flow velocity is equal to the velocity of a gravity wave having small amplitude. A gravity wave may be produced by a change in the flow depth. The flow is called *subcritical flow*, if the flow velocity is less than the critical velocity, and the flow is called *supercritical flow* if the flow velocity is greater than the critical velocity. The *Froude number*, $\mathbf{F}_r$, is equal to the ratio of inertial and gravitational forces and, for a rectangular channel, it is defined as

$$\mathbf{F}_r = \frac{V}{\sqrt{gy}} \tag{1-2}$$

in which $y =$ flow depth. General expressions for $\mathbf{F}_r$ are presented in Section 3-2. Depending upon the value of $\mathbf{F}_r$, flow is classified as *subcritical* if $\mathbf{F}_r < 1$; *critical* if $\mathbf{F}_r = 1$; and *supercritical* if $\mathbf{F}_r > 1$.

1-4 Terminology

Channels may be natural or artificial. Various names have been used for the artificial channels: A long channel having mild slope usually excavated in the ground is called a *canal*. A channel supported above ground and built of wood, metal, or concrete is called a *flume*. A *chute* is a channel having very steep bottom slope and almost vertical sides. A *tunnel* is a channel excavated through a hill or a mountain. A short channel flowing partly full is referred to as a *culvert*.

A channel having the same cross section and bottom slope throughout is referred to as a *prismatic channel*, whereas a channel having varying cross section and/or bottom slope is called a *non-prismatic channel*. A long channel may be comprised of several prismatic channels. A cross section taken *normal* to the direction of flow (e.g., Section BB in Fig. 1-8) is called a *channel section*. The depth of flow, y, at a section is the *vertical* distance of the lowest point of the channel section from the free surface. The *depth of flow section*, d, is the depth of flow *normal* to the direction of flow. The *stage*, Z, is the elevation or vertical distance of free surface above a specified datum (Fig. 1-8). The *top width*, B, is the width of channel section at the free surface. The *flow area*, A, is the cross-sectional area of flow *normal* to the direction of flow. The *wetted perimeter*, P is defined as the length of line of intersection of channel wetted surface with a cross-sectional plane normal to the flow direction. The *hydraulic radius*, R, and *hydraulic depth*, D, are defined as

$$R = \frac{A}{P}$$

$$D = \frac{A}{B} \qquad (1\text{-}3)$$

Expressions for A, P, D and R for typical channel cross sections are presented in Table 1-1.

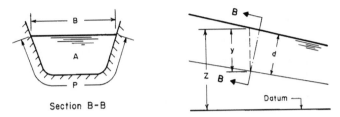

Section B-B

Fig. 1-8 **Definition sketch**

1-5 Velocity Distribution

The flow velocity in a channel section may vary from one point to another. This is due to shear stress at the bottom and at the sides of the channel and due to the presence of free surface. Fig. 1-9 shows typical velocity distributions in typical channel sections.

The flow velocity may have components in all three Cartesian coordinate directions. However, the components of velocity in the vertical and transverse directions are usually small and may be neglected. Therefore, only the flow velocity in the direction of flow needs to be considered. This velocity component varies with depth from the free surface. A typical variation of velocity with depth is shown in Fig. 1-10.

Energy Coefficient

As discussed in the previous paragraphs, the flow velocity in a channel section usually varies from one point to another. Therefore, the mean velocity head in a channel section, $(V^2/2g)_m$, is not the same as the velocity head, $V_m^2/(2g)$, computed by using the mean flow velocity, V_m, in which the subscript m refers to the mean values. This difference may be taken into consideration by introducing an *energy coefficient*, α, which is also referred to as the *velocity-head*, or *Coriolis coefficient*. An expression for this coefficient is derived in the following paragraphs.

Referring to Fig. 1-11, the mass of liquid flowing through area ΔA per unit time $= \rho V \Delta A$, in which $\rho = $ mass density of the liquid. Since the kinetic energy of mass m traveling at velocity V is $(1/2)mV^2$, we can write

Table 1-1 Properties of typical channel cross sections

Section	Area, A	Wetted Perimeter, P	Hydraulic radius, R	Top width, B	Hydraulic depth, D	
Rectangular	$B_o y$	$B_o + 2y$	$\dfrac{B_o y}{B_o + 2y}$	B_o	y	
Trapezoidal	$(B_o + sy)y$	$B_o + 2y\sqrt{1+s^2}$	$\dfrac{(B_o + sy)y}{B_o + 2y\sqrt{1+s^2}}$	$B_o + 2sy$	$\dfrac{(B_o + sy)y}{B_o + 2sy}$	
Triangular	sy^2	$2y\sqrt{1+s^2}$	$\dfrac{sy}{2\sqrt{1+s^2}}$	$2sy$	$0.5y$	
Circular	$\frac{1}{8}(\theta - \sin\theta)D_o^2$	$\frac{1}{2}\theta D_o$	$\frac{1}{4}\left(1 - \dfrac{\sin\theta}{\theta}\right)D_o$	$D_o \sin\frac{1}{2}\theta$	$\left(\dfrac{\theta - \sin\theta}{\sin\frac{1}{2}\theta}\right)\dfrac{D_o}{8}$	

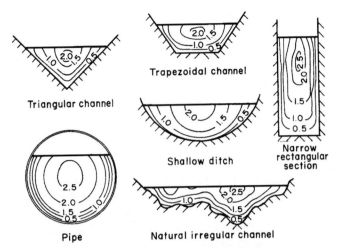

Fig. 1-9   Velocity distribution in different channel sections (After Chow [1959])

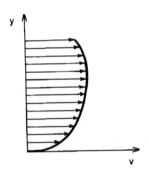

Fig. 1-10   Typical variation of velocity with depth

Kinetic energy transfer through area ΔA per unit time

$$= \frac{1}{2}\rho V \Delta A V^2$$

$$= \frac{1}{2}\rho V^3 \Delta A \qquad (1\text{-}4)$$

Hence,

Kinetic energy transfer through area A per unit time

$$= \frac{1}{2}\rho \int V^3\, dA \qquad (1\text{-}5)$$

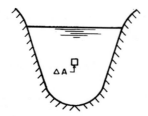

Fig. 1-11   Definition sketch

It follows from Eq. 1-4 that the kinetic energy transfer through area ΔA per unit time may be written as $(\gamma V \Delta A)V^2/(2g)$ = weight of liquid passing through area ΔA per unit time $\times$ velocity head, in which γ = specific weight of the liquid. Now, if V_m is the mean flow velocity for the channel section, then the weight of liquid passing through total area per unit time = $\gamma V_m \int dA$; and the velocity head for the channel section = $\alpha V_m^2/(2g)$, in which α = velocity-head coefficient. Therefore, we can write

Kinetic energy transfer through area A per unit time

$$= \rho \alpha V_m \frac{V_m^2}{2} \int dA \qquad (1\text{-}6)$$

Hence, it follows from Eqs. 1-5 and 1-6 that

$$\alpha = \frac{\int V^3 dA}{V_m^3 \int dA} \qquad (1\text{-}7)$$

Figure 1-12 shows a typical cross section of a natural river comprising of the main river channel and the flood plain on each side of the main channel. The flow velocity in the floodplain is usually very low as compared to that in the main section. In addition, the variation of flow velocity in each subsection is small. Therefore, each subsection may be assumed to have the same flow velocity throughout. In such a case, the integration of various terms of Eq. 1-7 may be replaced by summation as follows:

$$\alpha = \frac{V_1^3 A_1 + V_2^3 A_2 + V_3^3 A_3}{V_m^3 (A_1 + A_2 + A_3)} \qquad (1\text{-}8)$$

in which

$$V_m = \frac{V_1 A_1 + V_2 A_2 + V_3 A_3}{A_1 + A_2 + A_3} \qquad (1\text{-}9)$$

By substituting Eq. 1-9 into Eq. 1-8 and simplifying, we obtain

$$\alpha = \frac{(V_1^3 A_1 + V_2^3 A_2 + V_3^3 A_3)(A_1 + A_2 + A_3)^2}{(V_1 A_1 + V_2 A_2 + V_3 A_3)^3} \qquad (1\text{-}10)$$

Note that Eq. 1-10 is written for a cross section which may be divided into three subsections each having uniform velocity distribution. For a general case in which total area A may be subdivided into N such subareas each having uniform velocity, an equation similar to Eq. 1-10 may be written as

$$\alpha = \frac{\sum_{i=1}^{N}(V_i^3 A_i) \cdot \left(\sum A_i\right)^2}{\left(\sum V_i A_i\right)^3} \tag{1-11}$$

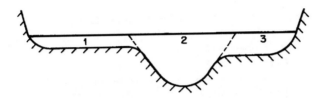

Fig. 1-12   Typical river cross section

Momentum Coefficient

Similar to the energy coefficient, a coefficient for the momentum transfer through a channel section may be introduced to account for nonuniform velocity distribution. This coefficient, also called *Boussinesq coefficient*, is denoted by β. An expression for this coefficient may be obtained as follows.

The mass of liquid passing through area ΔA per unit time $= \rho V \Delta A$. Therefore, the momentum passing through area ΔA per unit time $= (\rho V \Delta A)V = \rho V^2 \Delta A$. By integrating this expression over the total area, we get

Momentum transfer through area A per unit time

$$= \rho \int V^2 \, dA \tag{1-12}$$

By introducing the momentum coefficient, β, we may write the momentum transfer through area A in terms of the mean flow velocity, V_m, for the channel section, i.e.,

Momentum transfer through area A per unit time $\quad = \beta \rho V_m^2 \int dA \quad$ (1-13)

Hence, it follows from Eqs. 1-12 and 1-13 that

$$\beta = \frac{\int V^2 \, dA}{V_m^2 \int dA} \tag{1-14}$$

Theoretical values for α and β can be derived by using the power law and the logarithmic law for velocity distribution in wide channels. Chen (1992) derived the theoretical values of α and β using the power law distribution.

The values of α and β for typical channel sections [Temple, 1986; Watts et al., 1967; Chow, 1959] are listed in Table 1-2. For turbulent flow in a straight channel having a rectangular, trapezoidal, or circular cross section, α is usually less than 1.15 [Henderson, 1966]. Therefore, it may be neglected in the computations since its value is not precisely known and it is nearly equal to unity.

Table 1-2   Values of α and β for typical sections*

Channel section	α	β
Regular channels	1.10–1.20	1.03–1.07
Natural channels	1.15–1.50	1.05–1.17
Rivers under ice cover	1.20–2.00	1.07–1.33
River valleys, over-flooded	1.50–2.00	1.17–1.33

* Compiled from data given by Chow [1959]

Example 1-1

The velocity distribution in a channel section may be approximated by the equation, $V = V_o(y/y_o)^n$, in which V is the flow velocity at depth y; V_o is the flow velocity at depth y_o, and n = a constant. Derive expressions for the energy and momentum coefficients.

Solution:

Let us consider a unit width of the channel. Then, we can replace area A in the equations for the energy and momentum coefficients by the flow depth y. Now,

$$V_m = \frac{\int V \, dA}{\int dA}$$

For a unit width, this equation becomes

$$V_m = \frac{\int V \, dy}{\int dy}$$

By substituting the expression for V into this equation, we obtain

$$V_m = \frac{\int_0^{y_o} V_o(\frac{y}{y_o})^n \, dy}{\int_0^{y_o} dy}$$

$$= \frac{V_o}{y_o^n} \frac{y^{n+1}}{n+1} \Big|_0^{y_o} \frac{1}{y_o}$$

$$= \frac{V_o}{n+1}$$

By substituting $V = V_o(y/y_o)^n$, $V_m = V_o/(n+1)$, and $dA = dy$ into Eq. 1-7, we obtain

$$\alpha = \frac{\int_0^{y_o} V_o^3 (y/y_o)^{3n} dy}{[V_o/(n+1)]^3 \int_0^{y_o} dy}$$

$$= \frac{(V_o^3/y_o^{3n})[y^{3n+1}/(3n+1)]}{y_o[V_o/(n+1)]^3}$$

$$= \frac{(n+1)^3}{3n+1}$$

Substitution of $V = V_o(y/y_o)^n$ and $V_m = V_o/(n+1)$ into Eq. 1-14 yields

$$\beta = \frac{\int_0^{y_o} V_o^2 (y/y_o)^{2n} \, dy}{[V_o/(n+1)]^2 \int_0^{y_o} dy}$$

$$= \frac{(V_o^2 y_o)/(2n+1)}{[V_o/(n+1)]^2 y_o}$$

$$= \frac{(n+1)^2}{2n+1}$$

1-6 Pressure Distribution

The distribution of pressure in a channel section depends upon the flow conditions. Let us discuss a number of possible cases, starting with the simplest and then proceeding progressively to more complex. In this discussion, we will consider pressures as above the atmospheric pressure.

Static Conditions

Let us consider a column of liquid having cross-sectional area ΔA, as shown in Fig. 1-13. The horizontal and vertical components of the resultant force acting on the liquid column are zero, since the liquid is stationary. If $p =$ pressure intensity at the bottom of the liquid column, then the force due to pressure at the bottom of the column acting vertically upwards $= p\Delta A$. The weight of the liquid column acting vertically downwards $= \rho g y \Delta A$. Since the vertical component of the resultant force is zero, we can write

$$p\Delta A = \rho gy \Delta A$$

or

$$p = \rho gy \qquad (1\text{-}15)$$

In other words, the pressure intensity is directly proportional to the depth below the free surface. Since ρ is constant for typical engineering applications, the relationship between the pressure intensity and depth plots as a straight line, and the liquid rises to the level of the free surface in a piezometer, as shown in Fig. 1-13. The linear relationship, based on the assumption that ρ is constant, is usually valid except at very large depths, where large pressures result in increased density.

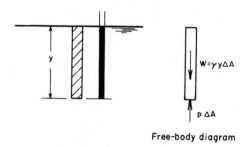

Free-body diagram

Fig. 1-13   Pressure in stationary fluid

Horizontal, Parallel Flow

Let us now consider the forces acting on a vertical column of liquid flowing in a horizontal, frictionless channel (Fig. 1-14). Let us assume that there is no acceleration in the direction of flow and the flow velocity is parallel to the channel bottom and is uniform over the channel section. Thus the streamlines are parallel to the channel bottom. Since there is no acceleration in the direction of flow, the component of the resultant force in this direction is zero. Referring to the free-body diagram shown in Fig. 1-14 and noting that the vertical component of the resultant force acting on the column of liquid is zero, we may write

$$\rho gy \Delta A = p\Delta A$$

or

$$p = \rho gy = \gamma y \qquad (1\text{-}16)$$

in which $\gamma = \rho g$ = specific weight of the liquid. Note that this pressure distribution is the same as if the liquid were stationary; it is, therefore, referred to as the *hydrostatic pressure distribution*.

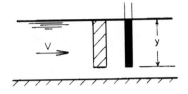
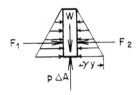

Free-body diagram

Fig. 1-14   Horizontal, parallel flow

Parallel Flow in Sloping Channels

Let us now consider the flow conditions in a sloping channel such that there is
no acceleration in the flow direction, the flow velocity is uniform at a channel
cross section and is parallel to the channel bottom; i.e., the streamlines are
parallel to the channel bottom. Figure 1-15 shows the free-body diagram of a
column of liquid normal to the channel bottom. The cross-sectional area of the
column is ΔA. If $\theta =$ slope of the channel bottom, then the component of the
weight of column acting along the column is $\rho g d \Delta A \cos \theta$ and the force acting
at the bottom of the column is $p \Delta A$. There is no acceleration in a direction
along the column length, since the flow velocity is parallel to the channel
bottom. Hence, we can write $p \Delta A = \rho g d \Delta A \cos \theta$, or $p = \rho g d \cos \theta = \gamma d \cos \theta$.
By substituting $d = y \cos \theta$ into this equation ($y =$ flow depth measured
vertically, as shown in Fig. 1-15), we obtain

$$p = \gamma y \cos^2 \theta \tag{1-17}$$

Note that, in this case, the pressure distribution is not hydrostatic in spite

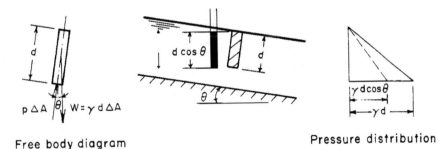

Free body diagram   Pressure distribution

Fig. 1-15   Parallel flow in a sloping channel

of the fact that we have parallel flow with no acceleration in the direction of
flow. However, if the slope of the channel bottom is small, then $\cos \theta \simeq 1$ and
$d \simeq y$. Hence,

$$p \simeq \rho g d \simeq \rho g y \qquad (1\text{-}18)$$

In several derivations in the subsequent chapters, we assume that the slope of the channel bottom is small. With this assumption, the pressure distribution may be assumed to be hydrostatic if the streamlines are almost parallel and straight, and the flow depths measured vertically or measured normal to the channel bottom are approximately the same.

Curvilinear Flow

In the previous three cases, the streamlines were straight and parallel to the channel bottom. However, in several real-life situations, the streamlines have pronounced curvature. To determine the pressure distribution in these flows, let us consider the forces acting in the vertical direction on a column of liquid with cross-sectional area ΔA, as shown in Fig. 1-16.

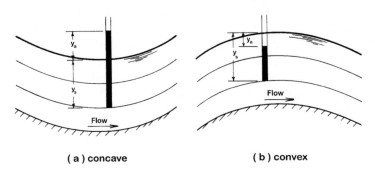

(a) concave   (b) convex

Fig. 1-16   Curvilinear flow

$$\text{Mass of the liquid column} = \rho y_s \Delta A \qquad (1\text{-}19)$$

If r = radius of curvature of the streamline and V is the flow velocity at the point under consideration, then

$$\text{Centrifugal acceleration} = \frac{V^2}{r} \qquad (1\text{-}20)$$

and

$$\text{Centrifugal force} = \rho y_s \Delta A \frac{V^2}{r} \qquad (1\text{-}21)$$

Dividing the centrifugal force by the area of the column and converting the pressure to pressure head, we obtain the following expression for the pressure head, y_a, acting at the bottom of the liquid column due to centrifugal acceleration

$$y_a = \frac{1}{g} y_s \frac{V^2}{r} \qquad (1\text{-}22)$$

The pressure due to centrifugal force is in the same direction as the weight of column if the curvature is concave, as shown in Fig. 1-16a, and it is in a direction opposite to the weight if the curvature is convex (Fig. 1-16b). Therefore, the total pressure head acting at the bottom of the column is an algebraic sum of the pressure due to centrifugal action and the weight of the liquid column, i.e.,

$$\text{Total pressure head} = y_s(1 \pm \frac{1}{g}\frac{V^2}{r}) \qquad (1\text{-}23)$$

A positive sign is used if the streamline is concave, and a negative sign is used if the streamline is convex. Note that the first term in Eq. 1-23 is the pressure head due to static conditions while the second term is the pressure head due to centrifugal action. Thus, the liquid in a piezometer inserted into the flow rises above the water surface, as shown in Fig. 1-16a. In other words, pressure increases due to centrifugal action in concave flows and decreases in convex flows (Fig. 1-16b).

Boussinesq derived an equation for flows with small water surface curvatures. Detailed derivations are presented in Subramanya [1991] and Jaeger [1957].

1-7 Reynolds Transport Theorem

The Reynolds transport theorem relates the flow variables for a specified fluid mass to that of a specified flow region. We will utilize it in later chapters to derive the governing equations for steady and unsteady flow conditions. To simplify the presentation of its application, we include a brief description in this section; for details, see Roberson and Crowe [1997].

We will call a specified fluid mass the *system* and a specified region, the *control volume*. The boundaries of a system separate it from its *surroundings* and the boundaries of a control volume are referred to as the *control surface*. The three well-known conservation laws of mass, momentum, and energy describe the interaction between a system and its surroundings. However, in hydraulic engineering, we are usually interested in the flow in a region as compared to following the motion of a fluid particle or the motion of a fluid mass. The Reynolds transport theorem relates the flow variables in a control volume to those of a system.

Let the *extensive property* of a system be B and the corresponding *intensive* property be β. The intensive property of a system is defined as the amount of B per unit mass, m, i.e.,

$$\beta = \lim_{\Delta m \to 0} \frac{\Delta B}{\Delta m} \qquad (1\text{-}24)$$

Thus, the total amount of B in a control volume

$$B_{cv} = \int_{cv} \beta \rho d\mathcal{V} \tag{1-25}$$

in which $\rho =$ mass density and $d\mathcal{V} =$ differential volume of the fluid, and the integration is over the control volume.

We will consider mainly one-dimensional flows in this book. The control volume will be fixed in space and will not change its shape with respect to time, i.e., it will not stretch or contract. For such a control volume for one-dimensional flow, the following equation relates the system properties to those in the control volume:

$$\frac{dB_{sys}}{dt} = \frac{d}{dt} \int_{cv} \beta \rho d\mathcal{V} + (\beta \rho AV)_{out} - (\beta \rho AV)_{in} \tag{1-26}$$

in which the subscripts *in* and *out* refer to the quantities for the inflow and outflow from the control volume and $V =$ flow velocity. The system is assumed to occupy the entire control volume, i.e., the system boundaries coincide with the control surface.

Let us now discuss the application of this equation to a control volume. As an example, the time rate of change of momentum of a system is equal to the sum of the forces exerted on the system by its surroundings (Newton's second law of motion). To use this equation to describe the conservation of momentum of the water of mass of fluid, m, in a control volume, the extensive property B is the momentum of fluid $= mV$ and the corresponding intensive property, $\beta = \lim_{\Delta m \to 0} V(\Delta m / \Delta m) = V$. To describe the conservation of mass, B is the mass of fluid and the corresponding intensive property $\beta = \lim_{\Delta m \to 0}(\Delta m / \Delta m) = 1$.

1-8 Hydraulic Models

For open-channel flow applications or for the design of hydraulic structures, scale models (e.g., on a 1:20 scale model, 1 m on the model represents 20 m on the prototype) have been extensively used for over 100 years. This is necessitated due to the complexity of flow for which analytical or closed form solutions are not available. These models have geometrical, kinematic and dynamic similarity to the prototype. Geometrical similarity is obtained if all the solid boundaries on the prototype and on the model are similar. Kinematic similarity requires similar flow patterns, i.e., ratio of the corresponding flow velocities as well as the directions on the prototype and on the model are the same. For dynamic similarity, theoretically all corresponding similitude relationships should be the same on the model and on the prototype. This is almost impossible. Therefore, typically only one dominant similitude relationship is satisfied and the other similitude relationships are ignored, as discussed in the following paragraphs.

For the model results to be valid, it is necessary that the flow on the model is turbulent if it is turbulent on the prototype, which is usually the case in most of practical problems. Depending on the model scale, the flow depth on the model might be too small to produce turbulent flow. Therefore, the flow depth on the model should be deep enough to produce turbulent flows on the model if the prototype has turbulent flow. For this purpose, a minimum flow depth of 2 to 3 cm is required on the model. Many times, to meet this requirement, the vertical scale is selected that is different from the horizontal scale. These models on which vertical and horizontal scales are different are called *distorted* scale models. Similarly, sometimes a different fluid, such as air, may be used on the model instead of water to satisfy similitude relationship.

For dynamic similarity, the dominant similitude numbers on the model and on the prototype should be same. In open-channel flows, there is a free-surface and thus gravity affects the flow and Froude similitude relationship is used to predict prototype behavior from the model results. Let us indicate quantities for the prototype by subscript "p" and for the model, by subscript "m". For a model scale of $L_r = L_p/L_m$, the scale ratios for the other quantities for Froude similitude are:

Time, T_r: $L_r^{1/2}$
Velocity, V_r: $L_r^{1/2}$
Discharge, Q_r: $L_r^{2.5}$
Force, F_r: $\rho_r L_r^3$
Pressure, p_r: $\rho_r L_r$
Mass, M_r: $\rho_r L_r^3$

Dimensional analysis shows that there are four different dimensionless numbers for different types of flow. Depending upon the dominant force in a particular flow, one of these numbers is dominant and may be used to predict corresponding prototype quantities from the measurements on the model. For example, flows having a free surface, gravitational force is dominant and Froude similitude is employed. If viscous forces are predominant, then Reynolds similitude is used and for surface tension or compressibility, Weber and Cauchy similitude are used. Expressions for these relationships are:

Froude Number, $F_r = \frac{V}{\sqrt{gL}}$
Reynolds number, $R_e = \frac{\rho L V}{\nu}$
Weber number, $We = \frac{V^3 \rho L}{\sigma}$
Cauchy number, $C_a = \frac{V^2 \rho}{E_c}$

In the above expressions, L is significant length (e.g., diameter for circular pipes, hydraulic radius, R for other cross sections), σ is surface tension, ρ is mass density of the fluid, and E_c is bulk modulus of elasticity. Almost universally water is used in the model studies of water resource projects.

Thus $\rho_r = 1$. However, if air is used on the model, then ρ_r is the ratio of the mass density of water to that of air.

Sometimes more than one similitude relationship may be important. For example, to study the formation of vortices at the power and pump intakes, viscous force play a part. To handle this on Froude similitude models, flow velocity equal to that on the prototype have been used during model tests. It was found that a small disturbance on the model could be taken as possibility of formation of small vortices on the prototype and small vortices on the model, formation of air-entraining vortices on the prototype. However, some caution is in order because using significantly higher velocities than required by the Froude relationship may result in different flow patterns.

The chapter opener photograph shows an areal view of the US Army Corps of Engineers' Mississippi River Basin Model. The entire Mississippi River basin was reproduced on the model (horizontal scale 1:2000 and vertical scale 1:100) and operated according to Froude similitude. Such models on which vertical scale is larger to reduce surface tension and to reproduce turbulence better are called *distorted scale models*. The model construction started in 1943 and continued in phases until 1966. The model covered an area of 200 acres. Tests on individual problems were conducted from 1949 through 1973 and the historic floods of 1937, 1943, 1945 and 1952 as well as hypothetical floods were reproduced [Wikipedia, 2021].

Figures 1-17, 1-18, 1-19, and 1-20 show the photographs of a number of scale-hydraulic models.

Fig. 1-17   **1:50 scale model of the Shawinigan hydroelectric complex on the Saint-Maurice River, Quebec, two power plants and two spillways maximum flow 8065 m^3/s (Courtesy, LaSalle/NHC, PQ, Canada)**

Fig. 1-18   Scale Model of Kohala Project, 58.5-m high concrete gravity dam and spillway, located on the Jhelum River, Pakistan; mean daily discharge 312 m³/s (Courtesy, Eric J Lesleighter)

1-9 Summary

In this chapter, commonly used terms were defined, classification of flows using several different criteria was outlined, and the properties of a channel section were presented. The distribution of velocity and pressure in a channel section was discussed and two coefficients were introduced to account for the nonuniform velocity distribution. A brief description of the Reynolds transport theorem was presented to facilitate its application in later chapters and hydraulic models were discussed.

Problems

1-1 In the following situations, is the flow steady or unsteady?

i. Flow in a storm sewer during a large storm;

Fig. 1-19   **Geometric scale model of morning glory spillway, Waller Creek Tunnel Project. Prototype flows evaluated up to 312 m³/s; Light reflecting on the distorted water surface as flow passes through the bar racks and approaches the spillway** (Courtesy, Alden Research Laboratory, Holden Mass)

   ii. Flow in a canal following closing of control gates at the downstream end;
  iii. Flow in a canal with fully open control gates;
  iv. Flow in an estuary during a tide;
   v. Flow downstream of breached dam.

1-2 Mark true or false:

   i. Flow in a partially full tunnel is open-channel flow;
  ii. Hydraulic jump is formed when supercritical flow changes to subcritical flow.

1-3 In the following situations, is the flow uniform or nonuniform?

   i. Flow in a channel contraction or expansion;
  ii. Flow at a channel entrance;
  iii. Flow in the vicinity of a bridge pier;
  iv. Flow at the end of a long prismatic channel;
   v. Flow downstream of a channel contraction.

Fig. 1-20   **1:55 scale model of Low-Sill Control Structure, located on the Mississippi River, with eleven 13.4-m wide sluice gates, design flow 9911 m³/s** (Courtesy, ERDC, US Army Corps of Engineers, Vicksburg, MS)

1-4 Derive expressions for the flow area, A, wetted perimeter, P, hydraulic radius, R, top-water surface width, B, and hydraulic depth, D, for the following channel cross sections:

    i. Rectangular (bottom width $=B_o$);
   ii. Trapezoidal (bottom width $= B_o$, side slopes $=$ s H : 1 V);
  iii. Triangular (side slopes $=$ s H : 1 V);
  iv. Partially-full circular (diameter $= D$);
   v. Standard horseshoe (Fig. 1-21).

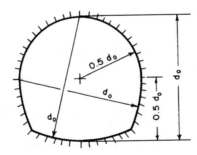

Fig. 1-21   **Horseshoe section**

1-5 The discharge in a channel is proportional to $AR^{2/3}$ if the flow is uniform. For a circular conduit having an inside diameter D, prove that the discharge is maximum when the flow depth is $0.94D$.

1-6 Compute $(R/R_f)^{2/3}$ and $AR^{2/3}/(AR^{2/3})_f$ for different values of y/D for a circular conduit flowing partially full, in which y = flow depth; D = conduit diameter; and the subscript 'f' refers to the values for the full section. At what values of ratio y/D do the curves have maximum values?

1-7 Determine the energy and momentum coefficients for the velocity distribution, $V = 5.75V_o \log(30y/k)$, in which V_o = flow velocity at the free surface; y_o = flow depth, and k = height of surface roughness. Assume the channel is very wide and rectangular.

1-8 The flow velocities measured at different flow depths in a wide rectangular flume are listed in Table 1-3. Write a computer program to determine the values of α and β. Use Simpson's rule for the numerical integration.

Table 1-3   Flow velocities at different depths

y (m)	0.0	0.2	0.4	0.6	0.8	1.0	1.2
V (m/s)	0.0	3.87	4.27	4.53	4.72	4.87	5.0

1-9 At a bridge crossing, the mean flow velocities (in m/s) were measured at the midpoints of different subareas, as shown in Fig. 1-22. Compute the values of α and β for the cross section.

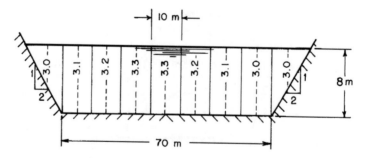

Fig. 1-22   Velocities at bridge crossing

1-10 Write a computer program to compute α and β for the flow in a channel having a general cross section. By using this program, compute α and β for the velocity distribution shown in Fig. 1-23.

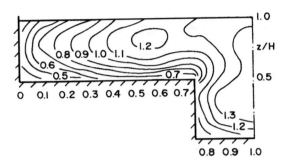

Fig. 1-23 **Dimensionless isovels** (After Knight and Hamed [1984])

1-11 Fig. 1-24 shows the velocity distribution measured on the scale model of a canal. By using the computer program of Problem 1-7, compute the energy and momentum coefficients.

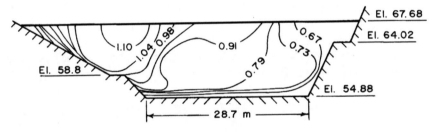

Fig. 1-24 **Velocity distribution** (After Babb and Amorocho [1965])

1-12 While computing the bending moment and the shear force acting on the side walls of the spillway chute of Fig. 1-25, a structural engineer assumed that the water pressure varies linearly from zero at the free surface to $\rho g y$ at the invert of the chute, in which $y=$ flow depth measured vertically. What are his computed values for the bending moment and the shear force at the invert level? Are the computed results correct? If not, compute the percentage error.

1-13 A spillway flip bucket has a radius of 20 m (Fig. 1-26). If the flow velocity at section BB is 20 m/s and the flow depth is 5 m, compute the pressure intensity at point C.

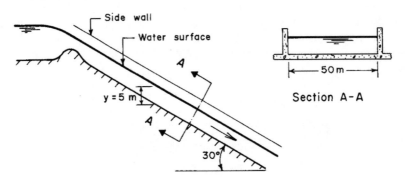

Fig. 1-25   Spillway chute

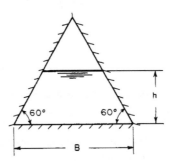

Fig. 1-26   Flip bucket

1-14 In a partially full channel having a triangular cross section (Fig. 1-27), the rate of discharge $Q = kAR^{2/3}$, in which $k =$ a constant; $A =$ flow area, and $R =$ hydraulic radius. Determine the depth at which the discharge is maximum. For the triangular channel section shown, $A = [B - (h/\sqrt{3})]h$, and $P = B + (4h/\sqrt{3})$.

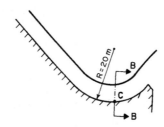

Fig. 1-27   Triangular channel cross section

1-15 In the following situations, is the flow uniform or nonuniform?

i. Flow in a channel contraction or expansion;

  ii. Flow at a channel entrance;

  iii. Flow in the vicinity of a bridge pier;

  iv. Flow at the end of a long prismatic channel.

1-16 In the following situations, is the flow steady or unsteady?

  i. Flow in a storm sewer during a large storm;

  ii. Flow in a power canal following shutting down of turbines;

  iii. Flow in a power canal when the turbines have been producing constant power;

  iv. Flow in an estuary during a tide.

1-17 In the following cases, is the flow laminar or turbulent?

  i. Flow in a wide rectangular channel at a flow velocity of $1\,\mathrm{m/s}$ at $1\,\mathrm{m}$ flow depth;

  ii. Flow in a wide rectangular channel at a flow velocity of $0.1\,\mathrm{m/s}$ at $2\,\mathrm{mm}$ flow depth.

1-18 Is it possible to have uniform flow in a frictionless sloping channel? Give reasons for your answer.

1-19 Is it possible to have uniform flow in a horizontal channel? Justify your answer.

1-20 If the angle between the flow surface and horizontal axis is ϕ and the angle between the channel bottom and horizontal is ϕ, prove that the pressure intensity at the channel bottom is

$$p = \frac{1}{1 + \tan\theta\tan\phi}\rho g y$$

in which $y = $ flow depth measured vertically.

1-21 For velocity distribution, $V = 5.75\log(30y/k)$, prove that $\alpha = 1 + 3r^2 - 2r^3$ and $\beta = 1 + r^2$, in which, $r = V_{max}/\bar{V} - 1$; $\bar{V} = $ mean velocity; and $V_{max} = $ maximum velocity.

1-22 Show that the bending moment on the side walls of a steep channel with a bottom slope θ for a flow depth of y is $\frac{1}{6}\gamma y^3\cos^4\theta$. Derive an expression for the shear force.

References

Babb, A. F. and Amorocho, J., 1965, "Flow Conveyance Efficiency of Transitions and Check Structures in a Trapezoidal Channel," Dept. of Irrigation, University of California, Davis.

Chen, C.L., 1992, "Momentum and Energy Coefficients Based on Power-Law Velocity Profile." *Jour, Hydraulic Engineering,* Amer. Soc. of Civil Engrs., vol. 118, no. 11, 1571-1584.

Chow, V. T., 1959, *Open-Channel Hydraulics,* McGraw-Hill Book Co., New York, NY.

Henderson, F. M., 1966, *Open Channel Flow,* MacMillan Publishing Co, New York, NY.

Jaeger, C., 1957, *Engineering Fluid Mechanics,* Blackie and Son, London.

Knight, D. W., and Hamed, M. E., 1984, "Boundary Shear in Symmetrical Compound Channels," *Jour. Hydraulic Engineering,* Amer. Soc. Civil Engrs., Oct.

Roberson, J.A. and Crowe, C.T. 1997, *Engineering fluid mechanics,* 6th ed., John Wiley & Sons, New York, NY.

Song, C. C. S., 1984, "Modeling of Mixed-Transient Flows," *Proc.,* Southeastern Conference on Theoretical and Applied Mechanics, SECTAM XII, vol. I, May, pp. 431-435.

Subramanya, K., 1991, *Flow in Open Channels,* Tata McGraw-Hill Publishing Co. Ltd., New Delhi, India.

Temple, D. M., 1986, "Velocity Distribution Coefficients for Grass-lined Channels," *Jour. of Hydraulic Engineering,* Amer. Soc. Civil Engrs., vol. 112, no. 3, pp. 193-205.

Watts, F. J., Simons, D. B., and Richardson, E. V., 1967, "Variation of α and β values in Lined Open Channels," *Jour., Hydraulics Div.,* Amer. Soc. Civil Engrs., vol 93, HY6, pp. 217-234 (see also Discussions: vol. 94, 1968, HY3, pp. 834-837; HY6, pp. 1560-1564; and vol. 95, 1969, HY3, p. 1059)

Wikipedia, 2021, "Mississippi River Basin Model," Last modified August 29, 2021. https://en.wikipedia.org/wiki/Mississippi River Basin Model

2

STEADY FLOW CONSERVATION LAWS

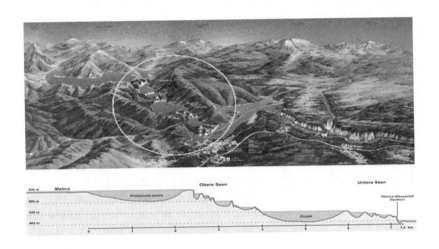

Water drops through a train of 16 lakes and 92 waterfalls, Plitvice Lakes, Croatia (Courtesy, Prof. Gary Parker)

© Springer Nature Switzerland AG 2022
M. H. Chaudhry, *Open-Channel Flow*,
https://doi.org/10.1007/978-3-030-96447-4_2

2-1 Introduction

Three conservation laws – conservation of mass, conservation of momentum, and conservation of energy– describe steady, free-surface flows. In this chapter, equations describing these laws for steady flows are derived and their application for the analysis of these flows is demonstrated.

For simplicity, only one-dimensional, steady flows are considered in this chapter. The governing equations are derived first using the basic principles of mechanics. Then it is shown that the same equations may be derived by applying the Reynolds Transport Theorem.* The flow velocity in a one-dimensional flow is only in the direction of flow and the components of the flow velocity in the transverse or lateral direction and in the vertical direction are both zero. In most flows, the lateral and vertical flow velocities are negligible. Therefore assuming such flows as one-dimensional is a valid assumption that simplifies the analysis considerably. As discussed in Chapter 1, the flow velocity may vary with the flow depth at a channel cross section. This variation in the flow velocity may be taken into consideration by using the mean flow velocity at the cross section. Such simulation models are referred to as one-dimensional, depth-averaged flow models. For example, the widely used HEC-2 computer model developed by the US Army Corps of Engineers in the 1960s is such a model.

2-2 Conservation of Mass

Civil engineers deal primarily with the flow of incompressible liquids, usually water, i.e., the mass density of the liquid is constant. Therefore, the law of conservation of mass between two channel cross sections with no lateral inflows or outflows implies conservation of the volume between the two sections and thus the volumetric flow rates at these sections are equal.

Let us consider the flow of an incompressible liquid in a channel, as shown in Fig. 2-1, with no inflow or outflow across the channel boundaries. Let the flow be steady. Let us denote the instantaneous flow velocity at a point normal to the flow area A by v, the flow depth by y, the mass density by ρ, top water-surface width by B, and use subscripts 1 and 2 to designate quantities for sections 1 and 2 respectively. Then, we may write

$$\text{Rate of mass inflow through area } dA_1 \text{ at section } 1 = \rho_1 v_1 dA_1 \qquad (2\text{-}1)$$

$$\text{Rate of mass outflow through area } dA_2 \text{ at section } 2 = \rho_2 v_2 dA_2 \qquad (2\text{-}2)$$

According to the law of conservation of mass, the rate of mass inflow at section 1 must be equal to the rate of mass outflow at section 2, since

---

* The derivation of the governing equations for unsteady flows using Reynolds Transport Theorem is presented in Chapter 11.

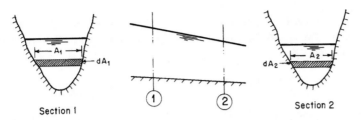

Fig. 2-1   Notation for the continuity equation

the volume of liquid stored in the channel between sections 1 and 2 remains unchanged, i.e.,

$$\int \rho_1 v_1 dA_1 = \int \rho_2 v_2 dA_2 \qquad (2\text{-}3)$$

Since the liquid is assumed incompressible, $\rho_1 = \rho_2$. Therefore,

$$\int v_1 \, dA_1 = \int v_2 \, dA_2 \qquad (2\text{-}4)$$

If the flow velocity is assumed uniform at each section, then Eq. 2-4 may be written as

$$V_1 \int dA_1 = V_2 \int dA_2 \qquad (2\text{-}5)$$

or

$$V_1 A_1 = V_2 A_2 \qquad (2\text{-}6)$$

Note that Eq. 2-6 is valid for nonuniform velocity distribution provided V_1 and V_2 are the mean flow velocities at sections 1 and 2, respectively. In terms of volumetric flow rate, Q, this equation becomes

$$Q_1 = Q_2 \qquad (2\text{-}7)$$

In hydraulic engineering, this equation is usually referred to as the *continuity equation*.

We may derive this equation by applying the Reynolds Transport Theorem to the control volume between the channel bottom and the top water surface, and between cross sections 1 and 2 (Fig. 2-1). Since the flow is assumed steady, the first term on the right-hand side of Eq. 1-26 is zero. The extensive property, B, is the mass of water in the control volume and the corresponding intensive property, $\beta = \lim_{\Delta m \to 0}(\Delta m / \Delta m) = 1$. Hence, Eqs. 2-6 and 2-7 follow from Eq. 1-26.

2-3 Conservation of Momentum

To derive an equation describing the conservation of momentum, let us consider the steady flow of an incompressible liquid in a channel, as shown in

Fig. 2-2. The channel is prismatic and there is no lateral inflow or outflow. Referring to this figure and using subscripts 1 and 2 to designate quantities for section 1 and 2

$$\text{Time rate of mass inflow at section } 1 = \frac{\gamma}{g}Q \qquad (2\text{-}8)$$

in which γ = specific weight of liquid. If V_1 is the mean flow velocity at section 1, then

$$\text{Time rate of momentum inflow at section } 1 = \frac{\gamma}{g}\beta_1 Q V_1 \qquad (2\text{-}9)$$

in which β_1 = momentum or Boussinesq coefficient introduced to account for the nonuniform velocity distribution. Similarly, we can write for section 2 that

$$\text{Time rate of momentum outflow} = \frac{\gamma}{g}\beta_2 Q V_2 \qquad (2\text{-}10)$$

Hence, it follows from Eqs. 2-9 and 2-10 that for the liquid volume between sections 1 and 2

$$\text{Time rate of increase of momentum} = \frac{\gamma}{g}Q(\beta_2 V_2 - \beta_1 V_1) \qquad (2\text{-}11)$$

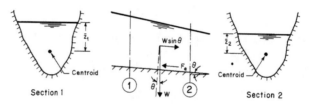

Fig. 2-2   Notation for the momentum equation

The following forces are acting on the volume of liquid between sections 1 and 2.

$$\text{Pressure force at section } 1, \ P_1 = \gamma \bar{z}_1 A_1 \qquad (2\text{-}12)$$

$$\text{Pressure force at section } 2, \ P_2 = \gamma \bar{z}_2 A_2 \qquad (2\text{-}13)$$

in which $\bar{z}$ = depth of the centroid of flow area A.

$$\text{Component of the weight of liquid between sections 1 and 2} = W \sin\theta \qquad (2\text{-}14)$$

in which W = weight of the volume of liquid between sections 1 and 2; and θ = slope of the channel bottom. Note that the weight component is acting in the downstream direction. Let us neglect the shear stress at the free surface between air and liquid and let us designate the external force due to shearing force between the liquid and the channel sides by F_e. Then the resultant force, F_r, acting on the volume of liquid in the downstream direction is

$$F_r = \gamma A_1 \bar{z}_1 - \gamma A_2 \bar{z}_2 + W \sin\theta - F_e \qquad (2\text{-}15)$$

According to the Newton's second law of motion, the time rate of change of momentum of the liquid volume is equal to the resultant of the external forces acting on the liquid volume. Hence, noting that $\gamma = \rho g$, we obtain from Eqs. 2-11 and 2-15

$$\rho \beta_2 Q_2 V_2 - \rho \beta_1 Q_1 V_1 = \rho g A_1 \bar{z}_1 - \rho g A_2 \bar{z}_2 + W \sin\theta - F_e \qquad (2\text{-}16)$$

This is the Momentum equation for the volume of liquid between sections 1 and 2 of Fig. 2-2.

We can derive this equation by applying the Reynolds Transport Theorem as follows. For the steady flow we are considering, the first term on the right-hand side of Eq. 1-26 is zero. Since momentum mV is the extensive property, B, the intensive property, $\beta = \lim_{\Delta m \to 0} V(\Delta m/\Delta m) = V$ and dB_{sys}/dt is equal to the resultant force acting on the control volume between the channel bottom, the channel cross sections 1 and 2 and the top water surface. (To be consistent with the general literature, we are using β to designate both the Boussenesq coefficient and the intensive property.) The resultant force is given by the right-hand side of Eq. 2-16. And the right-hand side of Eq. 1-26 is the same as the left-hand side of Eq. 2-16 which is equal to resultant force acting on the control volume given by the right-hand side of Eq. 2-16. Hence we obtain the momentum equation.

Note that F_e in the momentum equation is the external shearing force acting on the volume of liquid and does not depend upon the internal forces and the losses occurring inside the control volume.

Equation 2-16 is obtained by a general application of the momentum principle. Let us discuss how this is simplified for a special case. For a prismatic channel with horizontal bottom, the component of the weight of liquid in the downstream direction is zero. If we assume that the channel bottom and sides are smooth, then the shearing force is zero. If the flow velocity is uniform at sections 1 and 2, then $\beta_1 = \beta_2 = 1$. With these simplifications, Eq. 2-16 for a smooth, horizontal channel becomes

$$Q_2 V_2 - Q_1 V_1 = g A_1 \bar{z}_1 - g A_2 \bar{z}_2 \qquad (2\text{-}17)$$

From the continuity equation, $Q_1 = Q_2 = Q$ (say). Then, Eq. 2-17 may be written as

$$\frac{Q^2}{g A_1} + \bar{z}_1 A_1 = \frac{Q^2}{g A_2} + \bar{z}_2 A_2 \qquad (2\text{-}18)$$

Note that each side of this equation is similar except for the subscripts designating quantities for sections 1 and 2, respectively. Let us define

$$F_s = \frac{Q^2}{gA} + \bar{z}A \qquad (2\text{-}19)$$

in which F_s is referred to as the *specific force* or *momentum function*. Since each term on the right-hand side of Eq. 2-19 represents force/unit weight, we will herein refer to F_s as the specific force. The concept of specific force is very helpful in the application of the momentum equation; we will illustrate this in later sections.

2-4 Equation of Motion

Let us consider a rectangular fluid element along a streamline in a nonviscous fluid, as shown in Fig. 2-3. Let the length of the fluid element along the streamline be Δs, normal to the streamline be Δn, and the thickness of the fluid element perpendicular to the plane of paper be unity. Since the fluid is assumed nonviscous, there are no frictional forces acting on the fluid element.

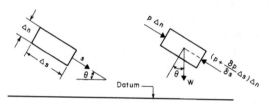

Fig. 2-3   Forces acting on fluid element

If p = pressure intensity at section 1, then the pressure at section 2 is $p + (\partial p/\partial s)\Delta s$. Hence,

$$\text{Pressure force acting on the upstream face} = p\Delta n \qquad (2\text{-}20)$$

and

$$\text{Pressure force acting on the downstream face} = (p + \frac{\partial p}{\partial s}\Delta s)\Delta n \qquad (2\text{-}21)$$

If ρ = mass density of the fluid, then

$$\text{Weight of the fluid element} = \rho g \Delta s \Delta n \qquad (2\text{-}22)$$

$$\text{Component of this weight in the s-direction} = \rho g \Delta s \Delta n \sin\theta \qquad (2\text{-}23)$$

in which $\sin\theta = -(\partial z/\partial s)$, and z = height above the datum, measured positive in the upward direction. Thus, the resultant force acting on the element in the downstream direction,

$$F_r = p\Delta n - (p + \frac{\partial p}{\partial s}\Delta s)\Delta n - \rho g\Delta s\Delta n \frac{\partial z}{\partial s} \tag{2-24}$$

This equation may be simplified as

$$F_r = -\frac{\partial p}{\partial s}\Delta s\Delta n - \rho g\Delta s\Delta n \frac{\partial z}{\partial s} \tag{2-25}$$

According to the Newton's second law of motion, the resultant force is equal to the mass of the fluid element times the acceleration of the fluid element, a_s; i.e.,

$$\rho\Delta s\Delta n a_s = -\frac{\partial p}{\partial s}\Delta s\Delta n - \rho g\Delta s\Delta n \frac{\partial z}{\partial s} \tag{2-26}$$

This equation may be simplified as

$$\rho a_s = -\frac{\partial}{\partial s}(p + \gamma z) \tag{2-27}$$

Since the flow velocity, $V = V(s,t)$, acceleration, a_s, in the s-direction may be written as

$$a_s = \frac{dV_s}{dt} = \frac{\partial V_s}{\partial t} + \frac{\partial V_s}{\partial s}\frac{ds}{dt} = \frac{\partial V_s}{\partial t} + V_s\frac{\partial V_s}{\partial s} \tag{2-28}$$

The first term on the right-hand side of this equation, $\partial V_s/\partial t$, represents the *local* acceleration; and the second term, $V_s(\partial V_s/\partial s)$, represents the *convective* acceleration. Substitution of Eq. 2-28 into Eq. 2-27 yields

$$\rho\left(\frac{\partial V_s}{\partial t} + V_s\frac{\partial V_s}{\partial s}\right) + \frac{\partial}{\partial s}(p + \gamma z) = 0 \tag{2-29}$$

This equation is called the Euler's equation of motion. Note that we only assumed that the fluid is nonviscous; otherwise, the equation is valid along a streamline in unsteady, nonuniform flows.

Let us now discuss how this equation may be simplified for a number of special cases.

Steady Flow

The local acceleration in steady flow is zero, i.e., $(\partial V_s/\partial t) = 0$. Hence, Eq. 2-29 becomes

$$\rho V_s\frac{dV_s}{ds} + \frac{d}{ds}(p + \gamma z) = 0 \tag{2-30}$$

Note that we have the total derivatives in Eq. 2-30 instead of the partial derivatives since both p and V_s for steady flow are now functions of s only. By multiplying throughout by ds and integrating the resulting equation, we obtain

$$\frac{1}{2}\rho V_s^2 + p + \gamma z = \text{constant} \tag{2-31}$$

Dividing by γ, this equation becomes

$$z + \frac{p}{\gamma} + \frac{V_s^2}{2g} = H = \text{constant} \tag{2-32}$$

The constant of integration H is referred to as the *total head*. Equation 2-32 is referred to as the *Bernoulli equation*. This equation is valid along a streamline. However, if the flow is irrotational, then it can be shown that this equation is valid throughout the flow field [Roberson and Crowe, 1997]. Recall that the Euler equation is valid for nonviscous fluid and we assumed in the derivation that ρ is constant. Therefore, we may say that the Bernoulli equation is valid for steady, irrotational, incompressible, and nonviscous flow.

Each term of Eq. 2-32 represents energy/unit weight and has the dimensions of length. In addition, note that the total head comprises three parts: datum head, z; pressure head, p/γ; and velocity head, $V_s^2/(2g)$. The datum head represents the potential energy, whereas the velocity head represents the kinetic energy.

Steady, Uniform Flow

Both the local and convective accelerations in steady-uniform flow are zero. Hence, Eq. 2-29 becomes

$$\frac{d}{ds}(p + \gamma z) = 0 \tag{2-33}$$

By integrating this equation, we obtain

$$\frac{p}{\gamma} + z = \text{constant} \tag{2-34}$$

The term $(p/\gamma + z)$ is referred to as the *piezometric head*. Note that this equation represents hydrostatic pressure distribution.

Unsteady, Nonuniform Flow

In unsteady, nonuniform flow, neither the local nor convective acceleration is zero. By multiplying Eq. 2-29 throughout by ds and integrating the resulting equation, we obtain

$$\rho \int \frac{\partial V_s}{\partial t} \, ds + \rho \int V_s \, dV_s + (p + \gamma z) = \text{Constant} \tag{2-35}$$

Dividing by ρg, this equation becomes

$$\frac{1}{g} \int \frac{\partial V_s}{\partial t} \, ds + \frac{V_s^2}{2g} + \frac{p}{\gamma} + z = \text{Constant} \tag{2-36}$$

A comparison of Eqs. 2-32 and 2-36 shows that an additional term, $\frac{1}{g}\int\frac{\partial V_s}{\partial t}\,ds$, is introduced due to unsteadiness. To evaluate the integral of this term, we need an expression for the variation of V_s with respect to time. Such an expression is not usually known. Therefore, this equation is not useful for a general analysis.

The preceding derivation shows us that the Bernoulli equation is valid only for steady flows. We will discuss its applications in the following sections.

2-5 Specific Energy

The concept of specific energy was introduced by Bakhmeteff in 1912 [Bakhmeteff, 1932]. As we shall see in the following sections, it is very useful for the application of the Bernoulli equation.

Let us now consider each term of the Bernoulli equation (Eq. 2-32) in detail. The flow velocity at a channel cross section may vary from point to point. However, as we discussed in Chapter 1, we may use the mean velocity at the section to calculate the velocity head by introducing the energy coefficient, α. The sum of the other two terms, $(z+p/\gamma)$, represents the piezometric head at a point. The piezometric head is constant at a section if the pressure distribution is hydrostatic. Assuming that the velocity distribution is uniform (i.e., $\alpha = 1$) and the pressure distribution is hydrostatic, (i.e., $p = \gamma y$), Eq. 2-32 may be written as

$$z + y + \frac{V^2}{2g} = H \tag{2-37}$$

Now, let us use the channel bottom as the datum. Then $z = 0$, and Eq. 2-37 simplifies to

$$y + \frac{V^2}{2g} = E \tag{2-38}$$

in which E is referred to as the *specific energy*. Note that E is the total head above the channel bottom.

To facilitate understanding the concept of specific energy, let us first consider it for a rectangular cross section having uniform velocity distribution, i.e., $\alpha = 1$. Let the channel width be B and the channel discharge be Q. Then, the discharge per unit width, q (hereinafter called the *unit discharge*), is $q = Q/B$, and $V = q/y$. Eq. 2-38 may now be written as

$$E = y + \frac{q^2}{2gy^2} \tag{2-39}$$

or

$$(E - y)y^2 = \frac{q^2}{2g} \tag{2-40}$$

For a specified unit discharge, q, the right-hand side of Eq. 2-40 is constant. Hence, we may write this equation as

$$Ey^2 - y^3 = \text{Constant} \tag{2-41}$$

This equation describes the relationship between E and y for a specified q. The E-y curve represented by this equation is plotted in Fig. 2-4. (The significance of the dotted curve is discussed later.)

Mathematically, we can prove that the E-y curve has two asymptotes: $(E - y) = 0$ and $y = 0$. The first asymptote represents a straight line passing through the origin and inclined at $45°$ to the horizontal axis; and the second asymptote is the horizontal axis. From physical considerations, we may explain the existence of these asymptotes as follows. It follows from Eq. 2-38 that the specific energy, E, comprises two parts: the flow depth, y, and the velocity head, $V^2/(2g)$. The value of V decreases to pass the same amount of q as y increases, thereby decreasing the velocity head. Hence, referring to Fig. 2-4, the upper limb of the curve approaches the straight line, $E = y$, as the velocity head becomes very small for very large values of y. In a similar manner, the value of V increases to pass the specified q as the value of y decreases, thereby increasing the velocity head. Hence, as y tends to zero, the velocity head tends to infinity, and the lower limb of the curve approaches the horizontal axis.

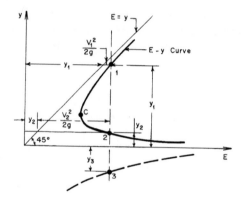

Fig. 2-4   Specific energy for a given unit discharge

Equation 2-41 is a cubic equation in y for a given E. This equation may have three distinct roots. One of these roots is always negative. However, since it is not physically possible to have a negative depth, there can be only two different values of y for a given value of E. These two depths, say y_1 and y_2, are called the *alternate depths*. As a special case, it is possible that $y_1 = y_2$ – i.e., at point C in Fig. 2-4. Such a depth is called the *critical depth, y_c,* and the corresponding flow is called the *critical flow*. The critical flow has a number of characteristics which are discussed in the next chapter. A flow having depth greater than the critical depth is called *subcritical flow* and a flow having depth less than the critical depth is called *supercritical flow* (Chapter 1).

The E-y curve for a specified value of q is presented in Fig. 2-4. To show the existence of three roots for a given value of E, the negative depths are plotted by a dotted curve.

The preceding discussion was for the specific energy curve for a specified value of q. Let us now discuss how curves for the other values of q will plot relative to that for q. Referring to Eq. 2-39, E increases as q increases for a given value of y. In other words, if we draw a line parallel to the E-axis for any given y, then the E-y curve for q_1 intersects it to the left of that for q if $q_1 < q$, and the E-y curve for q_2 intersects it to the right of that for q if $q_2 > q$. This is clear from the curves shown in Fig. 2-5.

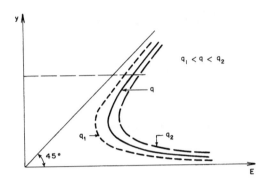

Fig. 2-5   Specific energy curves for different unit discharges

In terms of discharge, Q, the Bernoulli equation for a general channel section may be written as

$$H = z + \frac{p}{\gamma} + \frac{\alpha Q^2}{2gA^2} \tag{2-42}$$

Now, let us consider channels having steep bottom slopes. As we discussed in Chapter 1, as a general case, $p = \gamma d \cos\theta$, in which d = depth of flow normal to the channel bottom, and θ = angle between the channel bottom and the horizontal axis (Fig. 2-6). In addition, because of steep bottom slope, the flow depth, d, measured normal to the channel bottom is different from the flow depth, y, measured vertically. Let us use the channel bottom as the datum. Then, noting that the total head above the channel bottom is referred to as the specific energy, we may write Eq. 2-42 as

$$E = d\cos\theta + \frac{\alpha Q^2}{2gA^2} \tag{2-43}$$

Since $d = y\cos\theta$, Eq. 2-43 becomes

$$E = y \cos^2 \theta + \frac{\alpha Q^2}{2gA^2} \qquad\qquad (2\text{-}44)$$

Figure 2-7 shows the E-y curves for a channel having a steep bottom slope for three rates of discharges, $Q_1 < Q < Q_2$. In this case, note that the angle between the horizontal axis and the straight line to which the upper limbs of the E-y curves are asymptotes is not $45°$; this angle depends upon the slope of the channel bottom (see Prob. 2-1).

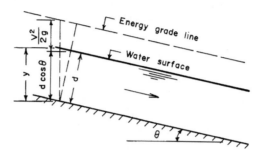

Fig. 2-6   **Definition sketch for steep channel**

2-6 Application of Momentum and Energy Equations

The momentum and energy equations should yield the same results if properly applied to any flow problem. However, which of these equations should be preferred for a particular situation depends upon the problem under consideration. Although it is difficult to set rigid guidelines, the following discussion of the advantages and limitations of each equation should be helpful for such a selection.

The energy equation provides computational ease and conceptual simplicity, since energy is a scalar quantity, as compared to the momentum equation, in which different terms are vector quantities. Therefore, in the latter case, the magnitude as well as the direction of each term should be known. This may make the analysis more difficult and cumbersome.

The head losses to be included in the energy equation are the internal losses that occur in the volume of liquid. The losses to be considered in the momentum equation are those due to the external shear stress acting on the boundaries of the control volume. The local losses, such as those in a bend or in a hydraulic jump, occur in a short length of the channel. In such short lengths, the losses due to shear at the boundaries are very small and may be neglected. Therefore, the momentum equation is preferable in such situations, since the energy equation cannot be used directly because the amount of internal losses is not known. However, it is advantageous to use the energy equation first if

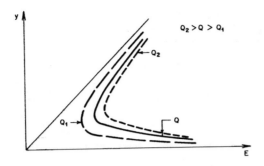

Fig. 2-7   Specific energy diagram for general channel cross section

there are some unknown external forces, e. g., forces on the sides in a channel expansion or contraction, provided the losses in the transition are negligible, and then to use the momentum equation to determine the magnitude of these external forces.

The energy and momentum equations may thus be used either alone or in sequence to solve a particular problem. The concepts of specific force (Sec. 2-3) and specific energy (Sec. 2-5) are very useful for such applications. The discussion of a channel transition, hydraulic jump, and hydraulic jump at a sluice gate outlet presented in the following sections should help in understanding the application of these concepts.

2-7 Channel Transition

A channel transition may be defined as a change in the channel cross section, e. g., change in the channel width and/or channel bottom slope. Such a geometrical change may be over a distance or it may be sudden. A channel transition is usually designed so that the losses in the transition are small. Thus, the energy losses in the transition may be neglected; and consequently the energy equation is more appropriate for the analysis.

For illustration purposes, we will consider a constant-width rectangular channel having a bottom step, as shown in Fig. 2-8. We want to determine whether the water surface rises or drops downstream of the transition for a specified flow depth and flow velocity upstream of the transition.

Since the channel width is constant, the unit discharge, q, is the same on both sides of the transition and the same specific energy curve is applicable to the upstream and downstream sides. Because the energy losses in the transition are assumed to be negligible, the total head H_1 is equal to H_2, in which the subscripts 1 and 2 refer to the quantities for the upstream and downstream sides of the transition respectively. Referring to Fig. 2-8a, $E_1 = H_1$, and $E_2 = H_2 - \Delta z$. Hence, $E_2 = E_1 - \Delta z$.

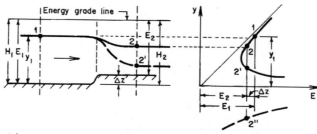

(a) Possible channel depths downstream of transition

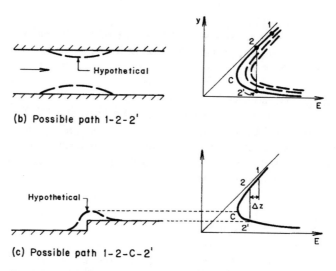

(b) Possible path 1-2-2'

(c) Possible path 1-2-C-2'

Fig. 2-8   Constant-width channel transition

On the specific energy diagram of Fig. 2-8, the point corresponding to flow conditions at section 1 is marked as 1. To determine the point corresponding to section 2, a vertical line is drawn such that $E = E_2$ as shown in Fig. 2-8a. The flow depths corresponding to the points where this line intersects the specific energy curve are the possible downstream depths. In this case, there are three such points, marked as 2, 2', and 2''. Point 2'' corresponds to a negative depth which is not physically possible. Hence, we shall not consider this point any further in our discussion. Of the other two points 2 and 2', let us determine which one is actually feasible. We see no particular problem in going from point 1 to 2 along the specific energy curve (Fig. 2-8a). However, to go from 2 to 2', two different paths [Henderson, 1966] may be followed, as shown in Fig. 2-8b and 2-8c. For the path along the vertical line 2-2' (Fig. 2-8b), we have to move off the specified specific energy curve and pass through the curves corresponding to higher unit discharges which is possible only if the channel width is reduced at the transition, as shown by a hypothetical channel in this figure. However, since the channel width is constant, there is no such contraction and consequently this path is not feasible. Similarly, a decrease in E is necessary to follow the second path, 2-C-2', as shown in Fig. 2-8c. E decreases only if the channel bottom rises, as shown by a hypothetical channel in this figure, i.e., the channel bottom rises until $E = E_c$ and then drops again until $E = E_2$. There is no such rise or drop in the bottom of the channel under consideration. Hence, the second path, 2-C-2', is not feasible. Therefore, only one depth is possible which corresponds to point 2 of Fig. 2-8a. In other words, subcritical flow remains subcritical downstream of the transition.

Following a similar argument, we can show that if the flow upstream of the transition is supercritical, then the possible flow depth downstream of the transition is that corresponding to point 2 and not that corresponding to point 2' (Fig. 2-9), i.e., flow downstream of the transition remains supercritical as it was upstream of the transition.

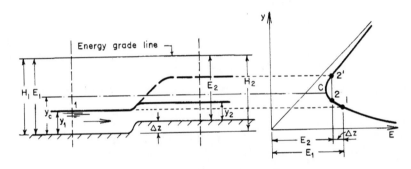

Fig. 2-9   Supercritical flow upstream of transition

From the preceding discussion, we conclude that for a step rise in the channel bottom, the flow depth decreases downstream of the step if the flow upstream of the transition is subcritical and the flow depth increases if the upstream flow is supercritical. These conclusions were drawn by considering all possible paths on a specific energy diagram. Let us now derive them mathematically in a more rigorous manner.

If the pressure distribution is hydrostatic and $\alpha = 1$, then the total head, H, at a channel section may be written as

$$H = z + y + \frac{V^2}{2g} \tag{2-45}$$

or

$$H = z + y + \frac{Q^2}{2gA^2} \tag{2-46}$$

We are interested in determining the sign of variation of y with a variation in the elevation of channel bottom, z. Assuming the downstream flow direction as positive for distance x measured along the channel bottom, flow depth increases if dy/dx is positive and it decreases if dy/dx is negative. By differentiating Eq. 2-46 with respect to x, we obtain

$$\frac{dH}{dx} = \frac{dz}{dx} + \frac{dy}{dx} + \frac{Q^2}{2g}\frac{d}{dx}\left(\frac{1}{A^2}\right) \tag{2-47}$$

Now

$$\frac{d}{dx}\left(\frac{1}{A^2}\right) = \frac{-2}{A^3}\frac{dA}{dx} \tag{2-48}$$

and

$$\frac{dA}{dx} = \frac{dA}{dy}\frac{dy}{dx} \tag{2-49}$$

For a small change in the flow depth, Δy, change in the flow area, $\Delta A \simeq B\Delta y$, in which $B = $ top water surface width. In the limit, as $\Delta y \to 0$, we may write $dA = Bdy$. Hence, Eq. 2-49 becomes

$$\frac{dA}{dx} = B\frac{dy}{dx} \tag{2-50}$$

In addition, we define the Froude number as

$$\mathbf{F}_r^2 = \frac{V^2}{gA/B} = \frac{BQ^2}{gA^3} \tag{2-51}$$

On the basis of Eqs. 2-50 and 2-51, Eq. 2-47 may be written as

$$\frac{dH}{dx} = \frac{dz}{dx} + (1 - \mathbf{F}_r^2)\frac{dy}{dx} \tag{2-52}$$

Note that this equation is valid only if the pressure distribution is hydrostatic. If there are no losses, then $dH/dx = 0$, and Eq. 2-52 becomes

$$\frac{dz}{dx} = (\mathbf{F}_r^2 - 1)\frac{dy}{dx} \tag{2-53}$$

This equation describes the variation of the flow depth for any variation in the bottom elevation. If there is a step rise in the channel bottom, then $dz/dx > 0$. For the right-hand side of Eq. 2-53 to be positive, two possible conditions are either $(\mathbf{F}_r^2 - 1)$ and dy/dx are both positive or both are negative. The first condition implies that if $\mathbf{F}_r > 1$ (i.e., flow is supercritical), then $dy/dx > 0$; i.e., flow depth increases at the step. Similarly, the second condition implies that if $\mathbf{F}_r < 1$ (i.e., flow is subcritical), then $dy/dx < 0$, i.e., flow depth decreases at the step. Following a similar argument, it can be shown from Eq. 2-53 that for a drop in the channel bottom, the flow depth decreases if the flow upstream of the step is supercritical, and it increases if the upstream flow is subcritical.

Example 2-1

A 4-m wide rectangular channel is carrying $10\,m^3/s$ at a depth of $2.5\,m$. There is a step rise of $0.2\,m$ in the channel bottom. Assuming there are no losses at the transition, determine the flow depth downstream of the bottom step. Does the water surface rise or fall at the step?

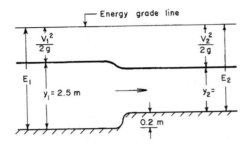

Fig. 2-10   Definition sketch for bottom step

Given:

$Q = 10\,\mathrm{m^3/s}$;
$B = 4\,\mathrm{m}$;
$y_1 = 2.5\,\mathrm{m}$;
$\Delta z = 0.2\,\mathrm{m}$;
No head losses at the transition.

Determine:

$y_2 = ?$
Change in water-surface level $= ?$

Solution:

The flow velocity and the specific energy at section 1 are

$$V_1 = \frac{Q}{A_1}$$
$$= \frac{10}{4 \times 2.5}$$
$$= 1\,\text{m/s}$$
$$E_1 = y_1 + \frac{V_1^2}{2g}$$
$$= 2.5 + \frac{1}{2 \times 9.81}$$
$$= 2.55\,\text{m}$$

Since there are no losses, referring to Fig. 2-10

$$E_2 = E_1 - 0.2$$
$$= 2.55 - 0.2$$
$$= 2.35\,\text{m}$$

Now

$$E_2 = y_2 + \frac{Q^2}{2gA_2^2}$$

By substituting values of E_2, Q, and $A_2 = 4y_2$ into this equation and simplifying, we obtain

$$y_2^3 - 2.35y_2^2 + 0.32 = 0$$

Solution of this equation by trial and error yields three roots: 2.29, 0.405, and -0.345 m. The third root is physically impossible because of the negative depth. In addition, only the first root is possible, since the upstream flow is subcritical, i.e., $\mathbf{F}_r < 1$; the second root requires that the flow has to pass through the critical depth at the step. Hence, $y_2 = 2.29$ m is the only possible downstream depth.

Let us use the channel bottom upstream of the transition as the datum. Then

$$\text{Water level downstream of the transition} = 0.2 + 2.29$$
$$= 2.49\,\text{m}$$

Thus, the water-surface level drops by $2.5 - 2.49 = 0.01$ m.

2-8 Hydraulic Jump

A hydraulic jump is formed in a channel whenever supercritical flow changes to subcritical flow. At the jump, there is a sharp discontinuity in the water surface and considerable amount of energy is dissipated due to turbulence.

A detailed description of the hydraulic jump is presented in Chapter 7; at present, we are interested only in developing a relationship between the flow depths and the flow velocities upstream and downstream of the jump. The flow depths upstream and downstream of the jump are called *sequent depths*, or *conjugate depths*.

To simplify the derivation, we will consider a rectangular, horizontal channel. Since the amount of energy loss in a jump is not known *a priori*, we cannot apply the energy equation directly. However, since the length of the jump is usually short, the losses due to shear at the channel bottom and sides are small as compared to the pressure forces and may be neglected. In addition, since the channel is horizontal, the component of the weight of water in the downstream direction is zero. Therefore, referring to Fig. 2-11, specific force, F_s, at section 1 is equal to that at section 2, i.e.,

$$\frac{Q^2}{gA_1} + \bar{z}_1 A_1 = \frac{Q^2}{gA_2} + \bar{z}_2 A_2 \tag{2-54}$$

Rearranging the terms of this equation, we obtain

$$\frac{Q^2}{g}\frac{A_2 - A_1}{A_1 A_2} = \bar{z}_2 A_2 - \bar{z}_1 A_1 \tag{2-55}$$

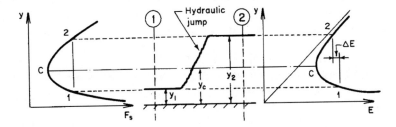

Fig. 2-11   Hydraulic jump

For a rectangular channel, $A = By$ and $\bar{z} = \frac{1}{2}y$. By substituting these relationships into Eq. 2-55 and simplifying, we obtain

$$\frac{Q^2}{g}(y_2 - y_1) = \frac{1}{2}B^2 y_1 y_2 (y_2^2 - y_1^2) \tag{2-56}$$

or

$$\frac{Q^2}{g}(y_2 - y_1) = \frac{1}{2}B^2 y_1 y_2 (y_2 + y_1)(y_2 - y_1) \tag{2-57}$$

Substituting $Q = By_1 V_1$ and simplifying, Eq. 2-57 becomes

$$\frac{y_1 V_1^2}{g} = \frac{1}{2}y_2 (y_2 + y_1) \tag{2-58}$$

Dividing throughout by y_1^2 yields

$$\frac{2V_1^2}{gy_1} = \frac{y_2}{y_1}\left(\frac{y_2}{y_1} + 1\right) \tag{2-59}$$

Now, Froude number, $\mathbf{F}_{r1} = V_1/\sqrt{gy_1}$. Hence, Eq. 2-59 may be written as

$$\left(\frac{y_2}{y_1}\right)^2 + \frac{y_2}{y_1} - 2\mathbf{F}_{r1}^2 = 0 \tag{2-60}$$

Solution of this equation yields

$$\frac{y_2}{y_1} = \frac{1}{2}\left(-1 + \sqrt{1 + 8\mathbf{F}_{r1}^2}\right) \tag{2-61}$$

Note that the negative sign with the radical term is neglected because it gives a negative ratio, which is physically impossible. This equation specifies a relationship between the depths upstream and downstream of the jump in terms of $\mathbf{F}_{r1}$. Proceeding similarly, we can derive the following equation in terms of $\mathbf{F}_{r2}$.

$$\frac{y_1}{y_2} = \frac{1}{2}\left(-1 + \sqrt{1 + 8\mathbf{F}_{r2}^2}\right) \tag{2-62}$$

Thus if the flow depth and flow velocity on one side of the jump are known, then their values on the other side can be determined by using Eq. 2-61 or 2-62 and the continuity equation. The energy losses can then be computed from the energy equation.

It is easier to visualize the relationships between E, F_s, flow depths on the upstream and downstream sides and the energy loss in the jump by plotting the specific energy and specific force diagrams side by side, as shown in Fig. 2-11.

Example 2-2

The reservoir level upstream of a 30-m wide spillway for a flow of $800\,m^3/s$ is at El. 200 m. The downstream river level for this flow is at El. 100 m. Determine the invert level of a stilling basin having the same width as the spillway so that a hydraulic jump is formed in the basin. Assume the losses in the spillway are negligible.

Given:

$Q = 800\,\text{m}^3/\text{s};$
$B = 30\,\text{m};$
Upstream water level $=$ El 200 m;
Downstream water level $=$ El 100 m.

Determine:

Stilling basin invert elevation to form the jump?

Solution:

Let z be the invert elevation of the stilling basin. Then, referring to Fig. 2-12, $y_2 = 100 - z$. Since the losses on the spillway face are negligible and assuming y_1 to be small,

$$V_1 = \sqrt{2g(200 - z)}$$

Now, $Q = BV_1 y_1$. Hence

$$y_1 = \frac{800}{30 \times \sqrt{19.62(200 - z)}}$$

$$= \frac{6.02}{\sqrt{200 - z}}$$

Substituting expressions for y_1 and V_1

$$\mathbf{F}_{r1}^2 = \frac{V_1^2}{gy_1}$$

$$= \frac{19.62(200 - z)}{9.81 \times 6.02/\sqrt{200 - z}}$$

$$= 0.332(200 - z)^{1.5}$$

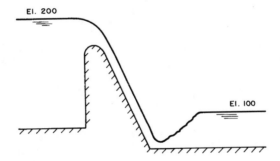

Fig. 2-12   Sketch for Example 2-2

Substitution of expressions for y_1, y_2, and $\mathbf{F}_{r1}^2$ into Eq. 2-61 yields

$$\frac{100 - z}{6.02\sqrt{200 - z}} = \frac{1}{2}\left(-1 + \sqrt{1 + 8 \times 0.332(200 - z)^{1.5}}\right)$$

Simplifying this equation, we obtain

$$(100 - z)\sqrt{200 - z} = -3.01 + 3.01\sqrt{1 + 2.656(200 - z)^{1.5}}$$

Solving this equation by trial and error

$$z = 84.18\,\text{m}$$

Thus, the stilling basin invert should be at El. 84.18 to form the jump.

2-9 Hydraulic Jump at Sluice Gate Outlet

Figure 2-13 shows the flow conditions downstream of a sluice gate used to control outflow from the reservoir. A hydraulic jump is formed in this case just downstream of the gate. A combined use of the specific-energy and specific-force diagrams, as shown in this figure, illustrates the usefulness of these concepts.

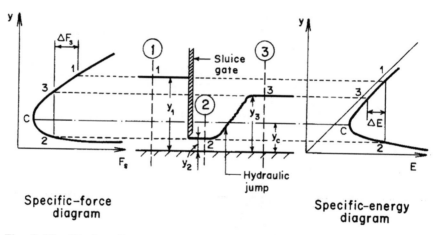

Fig. 2-13   Hydraulic jump at sluice-gate outlet

Assuming there are no losses at the gate, $E_1 = E_2$. However, because of an additional external force between sections 1 and 2, (i.e., thrust on the gate, F_g), F_{s1} is not equal to F_{s2}. There is loss of energy in the hydraulic jump between sections 2 and 3. Hence, E_2 is not equal to E_3. Since the losses due to shear at the channel bottom and sides between sections 2 and 3 are small and can be neglected, $F_{s3} = F_{s2}$ provided the channel bottom slope is either zero or can be assumed as zero. Referring to Fig. 2-13 , the thrust on the gate, $F_g = \gamma(F_{s1} - F_{s2})$; and the energy losses in the jump $= E_2 - E_3$.

The following example illustrates the application of the specific energy and specific force diagrams.

Example 2-3

A hydraulic jump is formed in a 5-m wide outlet at a short distance down-stream of a control gate (Fig. 2-13). If the flow depths just downstream of the gate is 2 m and the outlet discharge is 150 m³/s, determine

i. *Flow depth downstream of the jump;*
ii. *Thrust on the gate; and*
iii. *Head losses in the jump.*

Assume there are no losses in the flow through the gate.

Given:

$Q = 150 \, \text{m}^3/\text{s}$;
$B = 5 \, \text{m}$;
$y_2 = 2 \, \text{m}$.

Determine:

$y_3 = ?$
Thrust on the gate $= ?$
Head losses in the jump $= ?$

Solution:

The unit discharge in the outlet,

$$q = 150/5 = 30 \text{ m}^3/\text{s/m}$$

Referring to Fig. 2-13

$$V_2 = \frac{q}{y_2}$$
$$= 30/2 = 15 \text{ m/s}$$

$$\mathbf{F}_{r2}^2 = \frac{V_2^2}{gy_2}$$
$$= \frac{(15)^2}{9.81 \times 2}$$
$$= 11.47$$

Depth Downstream of Jump

Note that in this example, section 2 is upstream of the jump and section 3 is downstream of the jump. Hence, substituting this value of $\mathbf{F}_{r2}^2$ into Eq. 2-61, we obtain

$$y_3 = \frac{1}{2}y_2(\sqrt{1 + 8\mathbf{F}_{r2}^2} - 1)$$
$$= \frac{1}{2}2(\sqrt{1 + 8 \times 11.47} - 1)$$
$$= 8.63 \text{ m}$$

Head Loss in the Jump

$$E_2 = E_3 + H_l$$

or

$$H_l = E_2 - E_3$$

or

$$H_l = (y_2 + \frac{q^2}{2gy_2^2}) - (y_3 + \frac{q^2}{2gy_3^2})$$

Substituting the values for y_2, y_3 and q

$$H_l = 2 + \frac{(30)^2}{2 \times 9.81(2)^2} - 8.63 - \frac{(30)^2}{2 \times 9.81(8.63)^2}$$
$$= 4.21 \text{ m}$$

Thrust on the Gate

The depth upstream of the gate can be determined by applying the energy equation between sections 1 and 2. If we use the channel bottom as the datum and neglect the losses at the gate which are negligible, then

$$y_1 + \frac{V_1^2}{2g} = y_2 + \frac{V_2^2}{2g}$$

By substituting the values for y_2 and V_2 and noting that V_1 is almost zero, we obtain $y_1 = 13.468$ m.

Referring to Fig. 2-13

$$P_f = F_{s_1} - F_{s_2}$$
$$= (\frac{y_1^2}{2} + \frac{q^2}{gy_1}) - (\frac{y_2^2}{2} + \frac{q^2}{gy_2})$$
$$= \left[\frac{(13.68)^2}{2} + \frac{(30)^2}{9.81 \times 13.468}\right] - \left[\frac{(2)^2}{2} + \frac{(30)^2}{9.81 \times 2}\right]$$
$$= 49.634 \text{ m}^2$$

Therefore,

$$\text{Thrust on the gate} = \gamma P_f \times \text{gate width}$$
$$= 9.81\,(\text{kN/m}^3) \times 49.634(\text{m}^2) \times 5(\text{m})$$
$$= 2434.5\,\text{kN}$$

2-10 Summary

In this chapter, the continuity and momentum equations describing the steady, free-surface flows were derived. The concepts of specific energy and specific force were introduced and their applications to analyze different problems were illustrated.

Problems

2-1 Derive the expression for the slope of the straight line to which the upper limb of the specific energy curve is an asymptote for a channel having a bottom slope of θ.

2-2 Plot the specific energy versus depth curves for $Q = 400$, 600, and $800\,\text{m}^3/\text{s}$ in a trapezoidal channel having a bottom width of $20\,\text{m}$ and side slopes of 2H : 1V. Assume the bottom slope is small. From these curves, determine the critical depth for each discharge.

2-3 Derive an expression for the specific force for a rectangular channel.

2-4 The flow depth and the flow velocity upstream of a 0.2-m sudden step rise in the bottom of a 5-m wide rectangular channel are $5\,\text{m}$ and $4\,\text{m/s}$, respectively. Assuming there are no losses in the transition, determine:

  i. The flow depth downstream of the step and the change in the water level;
  ii. The flow depth and the water level downstream of the step if the channel bottom has a 0.2-m drop instead of the rise, as in (i).

2-5 A 10-m wide rectangular channel is carrying a discharge of $200\,\text{m}^3/\text{s}$ at a flow depth of $5\,\text{m}$.

  i. If the channel bottom has a sudden rise of $0.3\,\text{m}$, determine the depth of flow downstream of the step. Does the water surface rise or drop at the step?
  ii. Compute the flow depth and the water surface level downstream of the bottom step if the step is a drop of $0.2\,\text{m}$;
  In both cases, assume that there are no losses at the bottom step.

2-6 If the discharge in the channel of Prob. 2-5 is $400\,\mathrm{m^3/s}$, determine the flow depth downstream of the step if the step is: (i) a rise, and (ii) a drop. Does the water surface rise or drop downstream of the step in each case?

2-7 An 8-m wide rectangular channel carries a flow of $96\,\mathrm{m^3/s}$ at a flow depth of 4 m. The channel width is constricted to 6 m in a length of 5 m. Assuming the channel transition has straight and vertical sides, and there are no losses, plot the water-surface profile in the transition.

2-8 A hydraulic jump is formed in a 4-m wide outlet just downstream of the control gate, which is located at the upstream end of the outlet. The flow depth upstream of the gate is 20 m. If the outlet discharge is $100\,\mathrm{m^3/s}$, determine

  i. Flow depth downstream of the jump;
 ii. Thrust on the gate; and
iii. Energy losses in the jump.

Assume there are no losses in the flow through the gate.

2-9 On the specific-energy diagram for a rectangular channel, prove that the slope of the straight line joining the critical depth for different unit discharges is 2/3.

2-10 An 8-m wide rectangular channel has a flow velocity and flow depth of $4\,\mathrm{m/s}$ and 4 m, respectively. The channel bottom is at El. 700. Assuming no losses, design a transition so that the water level downstream of the transition is at El. 703.54, if

  i. The channel width remains constant;
 ii. The channel bottom level downstream of the transition is at El. 700.2 m and the channel width is variable.

2-11 Figure 2-14 shows a step rise in the channel bottom. If the channel width is constant and there are no losses, determine:

  i. Flow depth downstream of the transition;
 ii. The maximum height of the step so that the upstream water levels are not affected.

2-12 If there is no step in the channel bottom of Fig. 2-14 (i.e., channel bottom is horizontal), determine the minimum width of channel such that the upstream water level remains unchanged. Assume there are no losses in the channel transition.

2-13 A 5-ft wide rectangular channel carries a flow of $80\,\mathrm{ft^3/sec}$ at a depth of 8 ft. The channel width is reduced to 4 ft in a length of 10 ft. By assuming that there are no losses in the contraction,

  i. Compute the water surface profile for horizontal channel bottom;

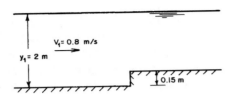

Fig. 2-14   Channel of Prob. 2-11

ii. Determine the variation of bottom elevation so that the water level remains constant.

2-14 For the sequent depths, y_1 and y_2 of an hydraulic jump in a horizontal, frictionless, rectangular channel, prove that

$$\frac{y_1}{y_2} = \frac{1}{2}\left(-1 + \sqrt{1 + 8\mathbf{F_{r2}}^2}\right)$$

2-15 A hydraulic jump is formed in a 4-m wide outlet just downstream of the control gate, which is located at the upstream end of the outlet. The flow depth upstream of the gate is 20 m. If the outlet discharge is 40 m^3/s, determine

1. i Flow depth downstream of the gate;
2. ii Flow depth downstream of the jump;
3. iii Energy losses in the jump; and
4. iv Thrust on the gate.

2-16 A trapezoidal channel (bottom width = 10 m; side slope 2H : 1V) carrying a flow of 30 m³/s ends in a free overfall at the downstream end. Determine the flow depth immediately upstream of the fall.

2-17 Two flow depths in a channel with the same specific energy and discharge are called alternate depths. If the flow depth in a 10-ft wide rectangular channel for a discharge of 100 cu ft per sec is 5 ft, determine the alternate depth.

2-18 The discharge in a 20-m wide, rectangular, horizontal channel is 80 m³/s at a flow depth of 0.5 m upstream of a hydraulic jump. Determine the flow depth downstream of the jump and the head losses in the jump.

2-19 An 8-m wide rectangular channel is carrying a flow of 54 m³/s at a flow depth of 0.6 m. A sluice gate located at the downstream end of the channel controls the flow depth y_2. Determine y_2 so that a hydraulic jump is formed upstream of the gate.

2-20 The channel bottom at the junction of two channels is raised by 0.1 m. The upstream channel is 10 m wide and the downstream channel is 8 m wide. If the channel discharge is $10\,\text{m}^3/\text{s}$ and the depth in the upstream channel is 1.5 m, determine the flow depth downstream of the junction assuming the losses at the junction to be $0.2V_2^2/(2g)$, where $V_2 =$ flow velocity in the downstream channel.

2-21 The flow depth in an 8-m wide rectangular channel upstream of a 0.15-m sudden drop is 2 m and the flow velocity is 3 m/s. Assuming no head losses, compute the flow depth downstream of the drop.

2-22 The flow velocity in a rectangular channel is 3 m/s at a depth of 3 m. If the channel bottom has a step rise of 0.3 m, determine the flow depth downstream of the step assuming no losses at the step.

2-23 Determine the required depth in the river downstream of a 120-m wide spillway to form a hydraulic jump at its toe for the following data. The upstream reservoir level is at El. 160 m, the spillway discharge is $1200\,\text{m}^3/\text{s}$ and the river bottom level is at El. 120. Assume the losses on the spillway face are negligible and the stilling basin walls are vertical. Compute the energy losses in the jump.

2-24 A 6-m wide rectangular channel is carrying a flow of $18\,\text{m}^3/\text{s}$. Plot a diagram between the flow depth and the specific energy for these conditions. What are the alternate and sequent depths corresponding to $y = 0.3\,\text{m}$? Determine the head losses in the jump.

2-25 The flow depth and velocity in a 6-ft wide rectangular channel are 5 ft and 3 ft/sec, respectively. The channel width is reduced to 5 ft in a distance of 15 ft.

  i. Compute and plot the flow depth in the contraction assuming the bottom is horizontal and the head losses are negligible;
  ii. In order to keep the water surface in the contraction horizontal, determine the variation of the bottom elevation.

2-26 The width of a rectangular channel is reduced from 5 ft to 4 ft and the channel bottom has a step rise. If the flow depths upstream and downstream of the contraction are 3 ft and 2.5 ft respectively for a flow of $50\,\text{ft}^3/\text{sec}$, determine the height of the step. Assume the losses in the transition are negligible.

2-27 The flow depth upstream of the hydraulic jump in a rectangular channel is 1.5 ft and the Froude number is 5. Compute the flow depth downstream of the jump and the head losses in the jump.

2-28 A hydraulic jump is formed just downstream of a sluice gate located at the entrance of a channel. There is a constant-level lake upstream of the sluice gate. The flow depth and velocity in the channel downstream of the jump are 5.2 ft and 4.3 ft/sec, respectively. Determine the water level in the lake. Assume the losses for flow through the gate are negligible.

2-29 Plot a family of curves between E and y for a trapezoidal channel with bottom width of 8 m and side slopes of 2H : 1V for $Q = 10, 20, 40,$ and $50 \, \text{m}^3/\text{s}$. Draw the locus of the critical depths on these curves.

2-30 Prove that the Froude number, $\mathbf{F}_r$, for a channel having steep bottom slope is $\mathbf{F}_r = V/\sqrt{gD \cos \theta/\alpha}$, where $D =$ hydraulic depth and $\alpha =$ velocity-head coefficient.

2-31 Two different flow depths, y_1 and y_2, (called alternate depths) are possible in a channel for a specified discharge Q and specific energy, E. If y_1 is known for a given value of Q in a trapezoidal channel, determine y_2 by using the bisection method. Write the governing equation, computational steps you will use, and a computer program based on this procedure.

2-32 The reservoir level upstream of an overflow spillway is at El 400 ft. The downstream water level for the design flow of 80,000 ft^3/sec is at El 220 ft. If the spillway width at the entrance to the stilling basin is 200 ft, determine the invert level of the basin so that a hydraulic jump is formed in the basin at design flow. No baffle piers, chute blocks, or end sill are to be provided.

2-33 A 2.5-m wide rectangular channel with a bottom-step rise of 0.1 m has a flow velocity of 2.5 m/s at a depth of 4 m upstream of the step. If the channel width is constant and there are no losses at the step, determine:

  i. Flow depth downstream of the step;
  ii. Maximum height of the step so that the upstream water levels are not affected.

2-34 A 4-m wide, horizontal rectangular channel is carrying a discharge of $20 \, \text{m}^3/\text{s}$ at a depth of 2.5 m. If the channel width is reduced linearly from 4 m to 3 m in a length of 5 m, determine the flow depth at the location where the channel width is 3.5 m. Assume there are no losses in the contraction.

References

Bakhmeteff, B. A., 1932, *Hydraulics of Open Channels*, McGraw-Hill Book Co., New York, NY.

Henderson, F. M., 1966, *Open Channel Flow*, MacMillan Publishing Co., New York, NY.

Roberson, J. A. and Crowe, C. T., 1997, *Engineering Fluid Mechanics,* 6th ed., John Wiley & Sons, New York, NY.

3

CRITICAL FLOW

Jinjia Falls on the Oshira River, Japan (Courtesy, Prof. Gary Parker)

Supplementary Information The online version contains supplementary material available at (https://doi.org/10.1007/978-3-030-96447-4_3).

3-1 Introduction

In Chapter 2, the depth at which the specific energy was minimum for a given discharge was called the *critical depth*; the corresponding flow as the *critical flow* which has a number of special properties. We discuss these properties and show how they can be used for engineering applications. Then, a procedure for computing the critical depth in a channel is presented and we show that there may be more than one critical depth for a specified discharge in a compound channel.* The chapter concludes by discussing the possible number of critical depths in a compound channel and how to determine their values.

3-2 Rectangular Channel

For simplicity, we will first consider properties of critical flow in a channel having a rectangular cross section and then channels having a non-rectangular cross section.

Specific Energy

As we discussed in Chapter 2, the specific energy for a rectangular channel having hydrostatic pressure distribution and uniform velocity distribution may be written as

$$E = y + \frac{V^2}{2g} \tag{3-1}$$

or

$$E = y + \frac{q^2}{2gy^2} \tag{3-2}$$

in which y = flow depth and q = discharge per unit width. We know from calculus that $dE/dy = 0$ for E to be minimum or maximum for a given q. Hence, differentiating Eq. 3-2 with respect to y and equating the resulting expression to zero, we obtain

$$\frac{dE}{dy} = 1 - \frac{q^2}{gy^3} = 0 \tag{3-3}$$

According to the above definition, the depth at which E is minimum is called the critical depth, y_c. In Section 3-7 we discuss that this definition is not sufficient to determine the critical depths in a compound channel.

It follows from Eq. 3-3 that

---

* A channel having a cross section comprising the main flow section and one or two overbank flow sections (e.g., a stream with flood plains) is called a compound channel.

$$y_c = \sqrt[3]{\frac{q^2}{g}} \tag{3-4}$$

When $dE/dy = 0$, E may be minimum or maximum. For E to be minimum, d^2E/dy^2 is positive at that depth. Let us prove that this is the case when $y = y_c$. By differentiating Eq. 3-3 with respect to y, we obtain

$$\frac{d^2E}{dy^2} = \frac{3q^2}{gy^4} \tag{3-5}$$

On the basis of Eq. 3-4, this equation becomes

$$\frac{d^2E}{dy^2} = \frac{3}{y_c} \tag{3-6}$$

The right-hand side of Eq. 3-6 is always positive. Hence, E is minimum at $y = y_c$ for a given value of q.

We will derive from Eq. 3-4 in the following paragraphs three important properties of critical flows:

1. It follows from Eq. 3-4 that

$$q^2 = gy_c^3 \tag{3-7}$$

By designating V_c as the flow velocity at critical flow, Eq. 3-7 may be written as

$$\frac{V_c^2}{2g} = \frac{1}{2}y_c \tag{3-8}$$

Hence, the velocity head in critical flow is *one-half* of the critical depth.

2. By substituting Eq. 3-8 into Eq. 3-1, we obtain

$$E = y_c + \frac{1}{2}y_c$$

or

$$y_c = \frac{2}{3}E \tag{3-9}$$

i.e., the critical depth is equal to *two-thirds* of the specific energy.

3. It follows from Eq. 3-8 that

$$\frac{V_c}{\sqrt{gy_c}} = 1$$

or the Froude number

$$\mathbf{F}_r = \frac{V_c}{\sqrt{gy_c}} = 1 \tag{3-10}$$

This equation shows that the Froude number, $\mathbf{F}_r = 1$ at critical flow.

Unit discharge

To determine the variation of unit discharge q with y for a specified value of E, let us re-write Eq. 3-2 as

$$q^2 = 2gEy^2 - 2gy^3 \tag{3-11}$$

It is clear from this equation that $q = 0$ when $y = 0$, and also when $y = E$. Thus we have two points on the q-y curve for a given E. To study the shape of this curve, let us determine the locations of the maxima and minima of this curve and the value of q at these points. For q to be maximum or minimum, $dq/dy = 0$. Hence, differentiating Eq. 3-11 with respect to y and simplifying, we obtain

$$q\frac{dq}{dy} = gy(2E - 3y) \tag{3-12}$$

Equating the derivative to zero and simplifying, we obtain

$$y(2E - 3y) = 0 \tag{3-13}$$

Equation 3-13 has two roots: $y = 0$ and $y = \frac{2}{3}E$. We discussed previously that $q = 0$ when $y = 0$. Hence, we will not gain any more information by studying this root further. The second root yields the same depth as the critical depth (see Eq. 3-9).

To verify whether the flow is maximum or minimum at this depth, we have to determine the sign of d^2q/dy^2. Differentiating Eq. 3-12 with respect to y, we get

$$q\frac{d^2q}{dy^2} + \left(\frac{dq}{dy}\right)^2 = 2gE - 6gy \tag{3-14}$$

Substitution of $dq/dy = 0$ and $y = \frac{2}{3}E$ into this equation yields

$$\frac{d^2q}{dy^2} = -\frac{2gE}{q} \tag{3-15}$$

It is clear from this equation that the second derivative of q with respect to y is always negative. Hence, for a given E, the unit discharge, q, is maximum at critical depth, y_c. An expression for the maximum discharge may be obtained by substituting $y = \frac{2}{3}E$ into Eq. 3-11 and then simplifying the resulting expression. This procedure yields

$$q^2_{max} = \frac{8}{27}gE^3 \tag{3-16}$$

Based on the preceding information, a typical q-y curve for a specified E may be plotted as shown in Fig. 3-1. The q-y curves for two other values of specific energy, such that $E_1 < E < E_2$, are also shown in this figure.

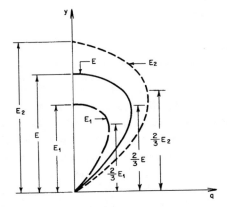

Fig. 3-1   Variation of unit discharge

Specific force

As we discussed in Chapter 2, the expression for specific force, F_s, for a rectangular channel is

$$F_s = \frac{q^2}{gy} + \frac{1}{2}y^2 \qquad (3\text{-}17)$$

The maxima and minima for the F_s–y curve may be determined as follows:

By differentiating Eq. 3-17 with respect to y and equating the resulting expression to zero, we get

$$\frac{dF_s}{dy} = -\frac{q^2}{gy^2} + y = 0 \qquad (3\text{-}18)$$

Noting that $V = q/y$, this equation may be written as

$$\frac{V^2}{2g} = \frac{1}{2}y \qquad (3\text{-}19)$$

This equation is the same as Eq. 3-8, which is valid when the flow is critical. To determine whether F_s is maximum or minimum at critical depth, let us differentiate Eq. 3-18 with respect to y again and substitute $y = y_c$. This procedure yields

$$\frac{d^2F_s}{dy^2} = 1 + \frac{2q^2}{gy_c^3} \qquad (3\text{-}20)$$

Since the right-hand side of this equation is always positive, the specific force is *minimum* at critical depth.

Wave Celerity

So far our discussion of critical flows has been in terms of the specific energy and the specific force. Another parameter of great importance in free-surface flows is the celerity of a small wave. The *celerity* is defined as the wave velocity with respect to the velocity of the medium in which the wave is traveling.

To derive an expression for the wave celerity, let us consider a small wave in a horizontal, frictionless channel, as shown in Fig. 3-2a. The wave is considered to be small if $|\delta y| \ll y$. Let us assume the wave shape does not change as it travels in the downstream direction with absolute wave velocity, V_w and that, as a result of the wave motion, the flow velocity is changed from V to $V + \delta V$. By superimposing a constant velocity V_w in the upstream direction, we may transform the unsteady flow of Fig. 3-2a to the steady flow of Fig. 3-2b. Let the thickness of the control volume perpendicular to the paper be unity.

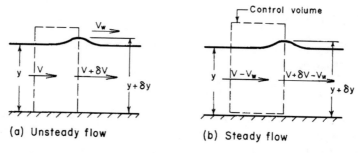

(a) Unsteady flow   (b) Steady flow

Fig. 3-2   Definition sketch for wave propagation

Since the channel is horizontal, the component of the weight of water in the control volume in the downstream direction is zero. Similarly, there is no shear force acting on the channel boundary, since the channel is assumed to be frictionless. Thus, the pressure force, F_1, acting on the upstream side of the control volume, and the pressure force, F_2, acting on the downstream side are the only forces acting on the control volume in the flow direction. Expressions for these forces are

$$F_1 = \frac{1}{2}\rho g y^2 \tag{3-21}$$

$$F_2 = \frac{1}{2}\rho g (y + \delta y)^2 \tag{3-22}$$

in which ρ = mass density of water. Hence, the resultant force, F_r, acting on the control volume is

$$F_r = F_1 - F_2$$
$$= \frac{1}{2}\rho g \left[(y^2 - (y + \delta y)^2\right] \tag{3-23}$$

Now, the time rate of change of momentum

$$= \rho y(V - V_w)\left[(V + \delta V - V_w) - (V - V_w)\right] \tag{3-24}$$

By equating the resultant force to the time rate of change of momentum and dividing throughout by ρg, it follows from Eqs. 3-23 and 3-24 that

$$\frac{y}{g}(V - V_w)\left[(V + \delta V - V_w) - (V - V_w)\right] = \frac{1}{2}[y^2 - (y + \delta y)^2] \tag{3-25}$$

By neglecting the higher-order terms, this equation may be simplified as

$$(V - V_w)\delta V = -g\delta y \tag{3-26}$$

The continuity equation for Fig. 3-2b may be written as

$$y(V - V_w) = (y + \delta y)(V + \delta V - V_w) \tag{3-27}$$

Neglecting the higher-order terms and simplifying, this equation becomes

$$y\delta V = -(V - V_w)\delta y \tag{3-28}$$

Combining Eqs. 3-26 and 3-28, we obtain

$$(V - V_w)^2 = gy$$

or

$$V_w = V \pm \sqrt{gy} \tag{3-29}$$

As we defined previously, celerity, c, is the wave velocity relative to the medium in which the wave is traveling, i.e., $V_w = V \pm c$. Hence, it follows from Eq. 3-29 that

$$c = \sqrt{gy} \tag{3-30}$$

We proved in a previous section that the Froude number $\mathbf{F}_r = 1$ when the flow is critical. By substituting the expression for $\mathbf{F}_r$ and using subscript c to denote various quantities for critical flow, we obtain

$$\frac{V_c}{\sqrt{gy_c}} = 1$$

or

$$V_c = \sqrt{gy_c} \tag{3-31}$$

Hence, on the basis of Eqs. 3-30 and 3-31, we may write

$$V_c = c \tag{3-32}$$

Thus, the celerity of a small wave is equal to the flow velocity when the flow is critical. Since, $V < V_c$ in subcritical flows, it follows that $V < c$ in these flows. Similarly, we may prove that $V > c$ in the supercritical flows.

Three different flow situations for the propagation of a disturbance are possible depending upon the relative magnitudes of V and c, i.e., whether the flow is subcritical, critical, or supercritical. These three cases are shown in Fig. 3-3. In subcritical flow, the wave travels in the upstream and downstream directions at velocities $(V - c)$ and $(V + c)$ respectively, since $V < c$, as shown in Fig. 3-3a. In critical flow, since $c = V$, the upper end of the wave remains stationary, and only the downstream end travels in the downstream direction at velocity $V + c$ (Fig. 3-3b). In supercritical flow, since $V > c$, the upstream and the downstream ends travel in the downstream direction at velocities $(V - c)$ and $(V + c)$, respectively (Fig. 3-3c). In other words, supercritical flow carries the wave downstream and the wave does not travel in the upstream direction.

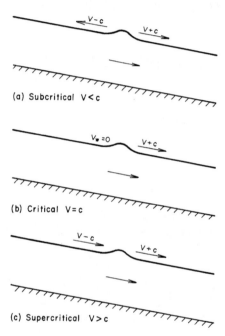

(a) Subcritical $V < c$

(b) Critical $V = c$

(c) Supercritical $V > c$

Fig. 3-3   Wave Propagation

This property, which defines the direction of travel of a disturbance, may be utilized in the field to determine the type of flow by producing a small disturbance on the flow surface by throwing a small object and noting the directions in which this disturbance travels. If the disturbance travels both in the upstream and in the downstream directions, then the flow is subcritical. However, if the flow carries the disturbance in the downstream direction, then the flow is supercritical.

Whether the disturbance in a flow travels in the upstream direction or not has some practical significance. For example, since a disturbance does not travel in the upstream direction in supercritical flow, it means that the flow upstream of a specified location does not 'know' what is happening on the downstream side of that location. In other words, to change the flow conditions at a section, flow conditions must be changed at an upstream location. In the hydraulic-engineering jargon, this condition is referred to as the supercritical flows upstream *control*. Following a similar argument, we can show that subcritical flows have downstream control.

The question might be asked, Can we keep on changing flow conditions at a downstream location in supercritical flow without affecting the flow upstream of that location? The answer to this question depends upon what happens to the flow upstream of that location: If the flow changes to subcritical flow, then the upstream flow is affected. However, if the flow conditions remain supercritical, then the flow conditions do not change upstream of that location.

3-3 Non-Rectangular Channel

Let us now develop relationships for critical flow in a prismatic channel having non-rectangular cross section (e.g., trapezoidal, triangular, circular, parabolic etc.). We call a cross section regular if the top water-surface width, B, is a continuous function of y and there are no floodplains or overbanks. Channels with floodplains are considered in Section 3-7.

Specific Energy

To simply the derivation, let us assume that the pressure distribution is hydrostatic and the velocity is uniform. Then, the specific energy,

$$E = y + \frac{Q^2}{2gA^2} \tag{3-33}$$

For E to be minimum, $dE/dy = 0$. Hence, differentiating Eq. 3-33 with respect to y and equating it to zero, we obtain

$$\frac{dE}{dy} = 1 - \frac{Q^2}{gA^3}\frac{dA}{dy} = 0 \tag{3-34}$$

Since $dA/dy = B$ for a regular cross section, Eq. 3-34 may be written as

$$1 - \frac{BQ^2}{gA^3} = 0$$

or

$$\frac{V^2}{2g} = \frac{D}{2} \tag{3-35}$$

in which $D = A/B$ is defined as the *hydraulic depth*. If we differentiate Eq. 3-34 with respect to y again, we can show that d^2E/dy^2 is positive provided $3B^2/A > dB/dy$, a condition which is usually satisfied. Therefore, E is minimum at the depth when $dE/dy = 0$. We refer to this depth as the *critical depth*. It is clear from Eq. 3-35 that the velocity head is one-half the hydraulic depth when the flow is critical. Recall that the velocity head was one-half the flow depth in a rectangular cross section.

For critical flow, $\mathbf{F}_r = 1$. Hence, we can derive the following expression for $\mathbf{F}_r$ from Eq. 3-35:

$$\mathbf{F}_r = \frac{V}{\sqrt{gD}} \tag{3-36}$$

For flows in a steep channel having nonuniform velocity, the following expression for $\mathbf{F}_r$ may be derived by introducing the velocity-head coefficient, α and the slope of the channel bottom:

$$\mathbf{F}_r = \frac{V}{\sqrt{gD\cos\theta/\alpha}} \tag{3-37}$$

Specific Force

Now, let us prove that the specific force, F_s is minimum when the flow is critical. As we discussed in Chapter 2,

$$F_s = \frac{Q^2}{gA} + \bar{z}A \tag{3-38}$$

For F_s to be minimum, $dF_s/dy = 0$ and $d^2F_s/dy^2 > 0$. By differentiating Eq. 3-38 with respect to y, we obtain

$$\frac{dF_s}{dy} = \frac{d}{dy}\left(\frac{Q^2}{gA}\right) + \frac{d}{dy}(\bar{z}A) = 0 \tag{3-39}$$

Let us now consider each term of this equation one by one. Since Q is constant

$$\frac{d}{dy}\left(\frac{Q^2}{gA}\right) = -\frac{Q^2}{gA^2}\frac{dA}{dy}$$
$$= -\frac{BQ^2}{gA^2} \tag{3-40}$$

The derivative of the second term may be evaluated as follows. The moment of area A about the top water surface is $\bar{z}A$. Referring to Fig. 3-4, the change in the moment of area A due to a small change in the flow depth, $\varDelta y$ about the top water surface is

$$\varDelta(\bar{z}A) = \left[A(\bar{z} + \varDelta y) + (B\varDelta y)\frac{1}{2}\varDelta y\right] - \bar{z}A \tag{3-41}$$

By neglecting terms of higher order, this equation becomes

$$\Delta(\bar{z}A) = A\Delta y \qquad (3\text{-}42)$$

In the limit, as $\Delta y \to 0$, this equation may be written as

$$d(\bar{z}A) = Ady \qquad (3\text{-}43)$$

By substituting Eqs. 3-40 and 3-43 into Eq. 3-39, and simplifying the resulting equation, we obtain

$$\frac{V^2}{2g} = \frac{D}{2} \qquad (3\text{-}44)$$

This condition is satisfied when the flow is critical. Hence, the specific force is minimum at critical depth. Note that we have not proved the necessary condition for F_s to be minimum, i.e., that d^2F_s/dy^2 is positive when $y = y_c$. This is left as an exercise for the reader.

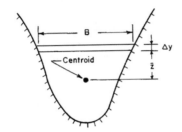

Fig. 3-4   **Definition sketch for** $\Delta(\bar{z}A)$

3-4 Application of Critical Flow

It is clear from the preceding properties that the discharge and flow depth have a unique relationship when the flow is critical (Eqs. 3-4 and 3-35). Utilizing this unique relationship, several flow-measuring devices have been developed. These devices are called *critical-flow meters*. [†] As we discussed in Section 3-2, the discharge is maximum when the flow is critical for a given specific energy. Therefore, the length of bridges and other structures on a channel may be reduced by producing critical flow at that section (see Example 3-1).

Critical flow may be produced in a channel by raising the channel bottom, by reducing the channel width, or by a combination of these measures. Let us now discuss how to determine the size of bottom step or the magnitude of channel contraction to produce critical flow.

[†] Critical-flow meters are discussed in Chapters 7 and 10.

Constant-width Channel with Bottom Step

Figure 3-5 shows the variation of water surface due to a step rise in the channel bottom. We showed in Chapter 2 that the water level at the step rises if the flow upstream of the step is supercritical, and it drops if the flow is subcritical. We may ask the question, Is there an upper limit on the size of the step such that the upstream water levels are not affected? A casual look at the specific-energy diagram indicates that there is a limit. As we raise the channel bottom, the point on the specific energy curve moves towards point C, which corresponds to critical flow. Thus, if we had subcritical flow at section 1, then the maximum height of this step, $(\Delta z)_{max}$, is equal to $E_1 - E_c$, as shown in Fig. 3-5. In this expression, subscripts 1 and c refer to section 1 and critical flow, respectively. Raising the bottom elevation further requires additional reduction in the specific energy. However, that is not possible, since E is minimum when the flow is critical. Therefore, if we raise the bottom level more than this maximum amount, either the unit discharge is reduced if the upstream water level is constant or the upstream water level is raised to increase the specific energy to produce the specified discharge. Similarly, if the flow is supercritical upstream of the step, then there is an upper limit on the height of the step that does not affect the upstream water level or the channel discharge. Referring to Fig. 3-5, this limiting height is again $(\Delta z)_{max}$ if the flow depth at section 1 is y_2.

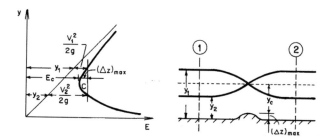

Fig. 3-5   Water-surface variation for a bottom step

Horizontal, Variable-width Channel

Figure 3-6 shows the variation of water level produced by a reduction in the channel width with the channel bottom remaining horizontal. The water depth decreases when the width decreases if the upstream flow is subcritical and it increases if the flow upstream of the constriction is supercritical. Similar to a step rise discussed previously, there is an upper limit by which the channel width may be contracted. We may reduce the channel width until critical flow is produced at the constriction. A further reduction in the channel width either reduces the unit discharge or raises the upstream water level.

The constriction in the channel width or a step rise in the channel bottom, or a combination of these two, such that the upstream water level for a specified discharge is influenced, is called a *choke*.

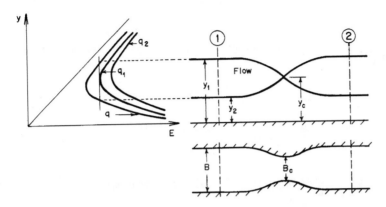

Fig. 3-6   **Water-level variation in a variable-width, horizontal channel**

Example 3-1

A bridge is planned on a 50-m wide rectangular channel carrying a flow of $200\,m^3/s$ at a flow depth of $4.0\,m$. For reducing the length of the bridge, what is the minimum channel width such that the upstream water level is not influenced for this discharge?

Given:

$Q = 200\,\mathrm{m^3/s}$;
$B = 50\,\mathrm{m}$;
$y = 4.0\,\mathrm{m}$.

Determine:

Minimum channel width at the bridge site = ?

Solution:

The flow velocity and specific energy are

$$V = \frac{200}{50 \times 4}$$
$$= 1.0\,\text{m/s}$$

$$E = 4. + \frac{(1)^2}{2 \times 9.81}$$
$$= 4.05\,\text{m}$$

For the discharge to be maximum at the bridge site for a given upstream specific energy of 4.05 m, the flow should be critical. Hence, it follows from Eq. 3-9 that

$$y_c = \frac{2}{3}E$$
$$= \frac{2}{3} \times 4.05$$
$$= 2.7\,\text{m}$$

The unit discharge, q, corresponding to this critical depth may be computed from Eq. 3-4, i.e.,

$$q = \sqrt{gy_c^3}$$
$$= \sqrt{9.81 \times (2.7)^3}$$
$$= 13.9\,\text{m}^3/\text{s/m}$$

The width needed for this unit discharge is

$$B_c = \frac{200}{13.9}$$
$$= 14.4\,\text{m}$$

Therefore, the channel width may be reduced from 50 to 14.4 m without affecting the upstream level for a flow of 200 m^3/s.

3-5 Location of Critical Flow

In the previous section, we showed that critical flow may be produced in a channel by raising the channel bottom, and/or by decreasing the channel width.[‡] However, we did not discuss where the critical depth will occur. For this purpose, we will consider a rectangular channel: first, a constant-width channel having a variable bottom level, followed by a horizontal channel having a variable width.

[‡] The occurrence of critical flow at changes in bottom slope or at a free overfall is discussed in Chapter 5.

By neglecting the losses in a transition, we derived the following equation in Chapter 2 for a *rectangular channel* having constant width (Eq. 2-53) and variable bottom level:

$$\frac{dz}{dx} = (\mathbf{F}_r^2 - 1)\frac{dy}{dx} \tag{3-45}$$

It is clear from this equation that the right-hand side is equal to zero when the flow is critical, i.e., when $\mathbf{F}_r = 1$ or when $dy/dx = 0$. Hence, the flow is critical at a point where $dz/dx = 0$, i.e., at the highest point of the step. The proof that it is not at the lowest point of the step is left as an exercise for the reader (Hint: The second derivative of z with respect to x must be negative at the highest point).

For a horizontal rectangular channel having a variable width, the total head,

$$H = z + y + \frac{1}{2g}\left(\frac{Q}{By}\right)^2 \tag{3-46}$$

By differentiating this equation with respect to x, assuming there are no losses (i. e., $dH/dx = 0$), and there is no lateral inflow or outflow, and noting that $B = B(x)$, we obtain

$$\frac{dy}{dx} + \frac{Q^2}{2g}\frac{d}{dx}\left(\frac{1}{(By)^2}\right) = 0 \tag{3-47}$$

Upon expansion of the second term, this equation becomes

$$\frac{dy}{dx} - \frac{Q^2}{gB^2y^3}\frac{dy}{dx} - \frac{Q^2}{gy^2B^3}\frac{dB}{dx} = 0 \tag{3-48}$$

By definition $\mathbf{F}_r^2 = Q^2/(gB^2y^3)$. Hence, we may write this equation as

$$(1 - \mathbf{F}_r^2)\frac{dy}{dx} - \mathbf{F}_r^2\frac{y}{B}\frac{dB}{dx} = 0 \tag{3-49}$$

For critical flow, $\mathbf{F}_r = 1$. Hence, it follows from Eq. 3-49 that $dB/dx = 0$. In other words, critical flow occurs at the point of minimum channel width. The reader may prove that it does not occur where the channel width is maximum.

3-6 Computation of Critical Depth

For the analysis and design of open channels, it is necessary to know the critical depth. Procedures for computing the critical depth in a channel having a regular cross section are discussed in this section, and a procedure for determining the critical depths in a compound channel is presented in Section 3-7.

The critical depth for a specified discharge may be computed from the equation $\mathbf{F}_r = 1$. The effects of nonuniform velocity may be considered by including the velocity-head coefficient, α. The channel bottom slope may be large. Then, based on Eq. 3-37 this equation becomes

$$\frac{V}{\sqrt{gD\cos\theta/\alpha}} = 1 \qquad (3\text{-}50)$$

Since, $Q = VA$, Eq. 3-50 becomes

$$\frac{Q/A}{\sqrt{gD\cos\theta/\alpha}} = 1 \qquad (3\text{-}51)$$

Let us define the section factor, $Z = A\sqrt{D}$. Then, this equation may be written as

$$Z = A\sqrt{D} = \frac{Q/\sqrt{\cos\theta}}{\sqrt{g/\alpha}} \qquad (3\text{-}52)$$

The left-hand side of this equation is a function of the properties of the channel cross section and the value of y_c. Thus, there is only one critical depth for a specified discharge in a given channel if $A\sqrt{D}$ for the channel cross section increases monotonically with y. Or, critical flow at a given value of y_c in a channel is possible only for one value of discharge. In Section 3-7, we discuss how multiple critical depths are possible for a specified discharge in a compound channel.

An explicit relationship may be derived (see Problem 3-11) to determine the critical depth in a rectangular, triangular, or parabolic channel. For general applications, however, the critical depth may be determined by using the design curves [Chow, 1959], by solving Eq. 3-52 by a trial-and-error procedure or by using numerical methods for the solution of a nonlinear algebraic equation. These procedures are discussed in the following paragraphs.

Design curves

Design curves are presented in Fig. 3-7. Let $Z_c = A\sqrt{D}$, where Z_c = section factor for the critical depth. If we want to determine the critical depth for a specified discharge, then we know the values of Q, θ, and α. Therefore, we can compute the left-hand side of Eq. 3-52. Let us divide this computed value by $B_o^{2.5}$ for a trapezoidal cross section and by $D_o^{2.5}$ for a circular section (B_o = channel bottom width, and D_o = conduit diameter). The resulting value is then equal to $Z_c/B_o^{2.5}$ or $Z_c/D_o^{2.5}$ depending upon the cross section. Now, y_c/B_o or y_c/D_o may be read directly from Fig. 3-7 corresponding to this value of $Z_c/B_o^{2.5}$ or $Z_c/D_o^{2.5}$.

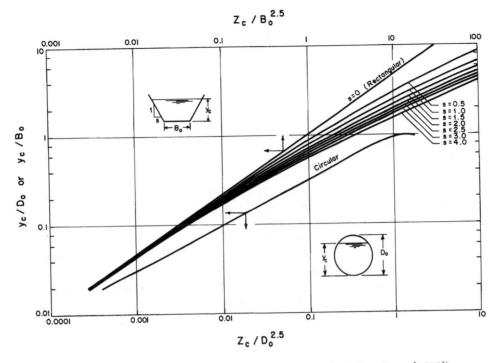

Fig. 3-7   **Curves for determining the critical depth** (After Chow [1959])

Trial-and-Error Procedure

In the trial-and-error procedure, we substitute expressions for flow area, A, and hydraulic depth, D, for the channel cross section into Eq. 3-52 and then solve the resulting equation by trial and error.

Numerical Methods

Several numerical methods are available for the solution of a nonlinear algebraic equation [Chapra and Canale, 2006; McCracken and Dorn, 1964], e.g., bisection method, Newton method, secant method, and the method of successive approximations. Of these methods, we present only the application of the Newton method, also called the Newton-Raphson method.

To determine the roots of an algebraic equation by the Newton method, we write the equation as

$$F(y) = 0 \tag{3-53}$$

Let us substitute $D = A/B$ into Eq. 3-52 and re-write it as

$$F(y) = A^{3/2}B^{-1/2} - \frac{Q/\sqrt{\cos\theta}}{\sqrt{g/\alpha}} = 0 \tag{3-54}$$

To solve Eq. 3-54 by the Newton method, we need the expression for dF/dy. An expression for this derivative may be obtained by differentiating Eq. 3-54 with respect to y, and noting that $dA/dy = B$; i.e.,

$$\frac{dF}{dy} = \frac{3}{2}A^{1/2}BB^{-1/2} - \frac{1}{2}A^{3/2}B^{-3/2}\frac{dB}{dy} \qquad (3\text{-}55)$$

This equation may be simplified as

$$\frac{dF}{dy} = \frac{3}{2}A^{1/2}B^{1/2} - \frac{1}{2}\left(\frac{A}{B}\right)^{3/2}\frac{dB}{dy} \qquad (3\text{-}56)$$

For a trapezoidal section having side slopes of 1 vertical to s horizontal, $dB/dy = 2s$. For any other channel section, an expression for dB/dy may be obtained similarly.

To start the iterative procedure, we need an initial estimate for y_c. The number of iterations are reduced considerably if this estimate is close to the actual value of y_c. For such an initial estimate, Eq. 3-52 may be approximately solved by assuming the channel as rectangular. For example, an approximate value for the initial estimate for a trapezoidal channel is

$$y_c = \left(\frac{Q^2}{gB_o^2}\right)^{1/3} \qquad (3\text{-}57)$$

in which $B = $ bottom width.

The following example illustrates the above procedures. First, the design curves are used; then a trial-and-error procedure is employed.

Example 3-2

Compute the critical depth in a trapezoidal channel for a flow of 30 m^3/s. The channel bottom width is 10.0 m, side slopes are 2H : 1V, the bottom slope is negligible, and $\alpha = 1$.

Given:

$B_o = 10.0\,\text{m}$;
$s = 2$;
$\theta = 0.0$;
$Q = 30\,\text{m}^3/\text{s}$;
$\alpha = 1$.

Determine:

$y_c = ?$

Solution:

Design Curves

Substituting the values of Q, θ, g, and α into the left-hand side of Eq. 3-52, we obtain

$$Z_c = \frac{Q/\sqrt{\cos\theta}}{\sqrt{g/\alpha}}$$

$$= \frac{30/\sqrt{\cos 0}}{\sqrt{9.81/1.}}$$

$$= 9.58$$

Now,

$$\frac{Z_c}{B_o^{2.5}} = \frac{9.58}{(10)^{2.5}} = 0.030$$

On Fig. 3-7, corresponding to $Z_c/B_o^{2.5} = 0.030$ on the abscissa and for $s = 2$, we read from the ordinate, $y_c/B_o = 0.09$. Therefore,

$$y_c = 0.09 \times 10 = 0.9\,\text{m}$$

Trial-and-Error Procedure

For critical flow, it follows from Eq. (3-52),

$$\frac{Q/\sqrt{\cos\theta}}{\sqrt{g/\alpha}} = A\sqrt{D}$$

By substituting the values of Q, g, θ and α into this equation, we obtain

$$A\sqrt{D} = \frac{30}{\sqrt{9.81}} = 9.58$$

Now, we have to determine the flow depth, y_c, for which $A\sqrt{D}$ for the specified channel cross section is 9.58. By substituting the specified values into the expressions for the channel properties of a trapezoidal section, we obtain

$$A = \frac{1}{2}(10.0 + 10.0 + 4.0y_c)y_c$$

$$= (10.0 + 2.0y_c)y_c$$

$$B = 10.0 + 4.0y_c$$

$$D = \frac{A}{B}$$

$$= \frac{(10.0 + 2.0y_c)y_c}{10.0 + 4.0y_c}$$

Substituting these expressions for A and D into Eq. 3-52 and simplifying, we obtain

$$A\sqrt{D} = (10.0 + 2.0y_c)y_c\sqrt{\frac{(10.0 + 2.0y_c)y_c}{10.0 + 4.0y_c}} = 9.58$$

Upon simplification, this equation becomes

$$[(10.0 + 2.0y_c)y_c]^{3/2} - 9.58\sqrt{10.0 + 4.0y_c} = 0$$

By solving this equation by trial and error, we obtain

$$y_c = 0.91\,\text{m}$$

Numerical Methods

Computer programs are developed for computing the critical depth in a trapezoidal channel using the Newton method or the bisection method [Chapra and Canale, 2006]. The initial estimate of $YI = 1.0\,\text{m}$ for the Newton method is determined from Eq. 3-57. The initial estimated values for YR and YL for the bisection method are arbitrarily assigned as 0.5 and 5 m. The critical depth computed by these methods is 0.91 m.

3-7 Critical Depths in Compound Channels

It is possible to have critical flow at more than one depth in a channel with a compound cross section. In this section, we will discuss how to determine the total number of critical depths for a given discharge in a compound channel and how to compute their values one by one.

General Remarks

We proved in Section 3-2 that the specific energy at critical depth is minimum for a given discharge in channels having regular cross sections (rectangular, trapezoidal, circular, etc.). This characteristics of critical flow has been widely used to determine the critical depth in a channel. For regular channel cross sections, the specific energy for a specified discharge is minimum only at one depth; hence, these procedures give only one critical depth. However, several investigators [French, 1985; Knight et al., 1984; Blalock and Sturm, 1981 and 1983; Konemann, 1982; Black, 1982; Rajaratnam and Ahmadi, 1979 Petryk and Grant, 1978; Myers and Elsawy, 1975; Wright and Carstens, 1970] have analytically and experimentally shown that there may be more than one critical depth for channels with overbank or floodplain flow (Fig. 3-8). Schoellhamer, Peters and Larock [1985] studied the problem using separate Froude numbers for the main-channel and floodplain flows. It was shown that the Froude number for the main-channel flow for their experimental section was equal to one at two different depths, thereby indicating that there is more than one critical depth. Because of the possibility of multiple critical depths, it is necessary to determine their values to compute correctly the water-surface flow profiles. Blalock and Sturm [1981] outlined difficulties associated with several available methods for critical-depth computations presented by Petryk and Grant [1978], Soil Conservation Service [1976], U. S. Army Corps of Engineers [1982], and U. S. Geological Survey [1976]. They defined a compound-channel Froude number for the entire section. The flow depths at which specific energy was minimum were called critical depths. Their Froude number correctly locates the minima on the specific energy diagram. However, the flow may be critical even when the specific energy is not minimum, as shown by Chaudhry and Bhallamudi [1988].

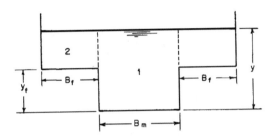

Fig. 3-8 Compound channel cross section

Only one critical depth is determined in a compound channel section by the algorithm presented by the U. S. Army Corps of Engineers [1982]. Same is true for the other procedures used by Soil Conservation Service [1976] and Davidian [1984]. Blalock and Sturm [1981] defined a Froude number which correctly locates the points of minimum specific energy although they did not present procedure for determining the critical depths. We present herein an algorithm which first determines the possible number of critical depths in a cross section for a given discharge and then computes their values one by one in an efficient manner.

In this section, we present the necessary expressions and outline a computational procedure to determine the critical depths in a compound channel. These expressions are derived for a symmetric cross section by assuming small bottom slope and hydrostatic pressure distribution even near the critical depth (For the detailed derivation of these expressions, the reader should see Chaudhry and Bhallamudi [1988]).

We first determine the characteristic directions of the governing equations (continuity and momentum) and write an expression for the compound channel Froude number, $\mathbf{F}_{cr}$. For critical flow, $\mathbf{F}_{cr} = 1$. By substituting expressions for the momentum coefficient in terms of the parameters for the main channel and for the floodplains into this equation and introducing the following non-dimensional parameters

$$y_r = \frac{y}{y_f}$$

$$b_r = \frac{B_f}{B_m}$$

$$b_f = \frac{B_f}{y_f}$$

$$n_r = \frac{n_m}{n_f} \tag{3-58}$$

we obtain

$$\frac{gB_m^2 y_f^3}{Q^2} = C \tag{3-59}$$

in which

$$C = \frac{1}{y_r + 2b_r(y_r - 1)} \left[\left(\frac{m}{y_r}\right)^2 + \left(\frac{1-m}{y_r - 1}\right)^2 \frac{1}{2b_r} \right]$$

$$+ \frac{2m(1-m)}{3[y_r + 2b_r(y_r - 1)]} \left(\frac{5}{y_r(y_r - 1)} - \frac{2}{b_f + y_r - 1} \right)$$

$$\times \left[\left(\frac{m}{y_r}\right) - \left(\frac{1-m}{y_r - 1}\right)\frac{1}{2b_r} \right] \tag{3-60}$$

and

$$m = \frac{1}{1 + 2n_r \left(A_2/A_1\right)^{\frac{5}{3}} \left(P_1/P_2\right)^{\frac{2}{3}}} \tag{3-61}$$

It is clear from Eq. 3-60 that C is a function of y_r, n_r, b_r and b_f. For any channel cross section, C-y_r relationship may be plotted as shown in Fig. 3-9. Those values of y_r at which $C = k = (gB_m^2 y_f^3)/Q^2$ are the critical depths for flows over the floodplain; i.e., the abscissa of the intersection point of this C-y_r curve with the horizontal line $C = k$ gives the value of y_r corresponding to the critical depth.

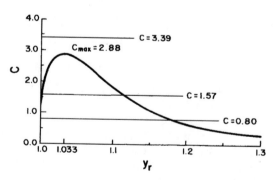

Fig. 3-9 C-y_r **curve for** $b_m = 1.$, $n_r = 0.9$, **and** $b_f = 3$

A procedure for solving Eq. 3-59 to determine the critical depths is illustrated by the following example.

Example 3-3

Determine the critical depths for $Q = 1.7$, 2.5, and 3.5 m^3/s in a channel having the following dimensions (Fig. 3-8): $B_m = 1.0\,m$; $B_f = 3.0\,m$; $y_f = 1.0\,m$; and $n_r = 0.9$.

Given:

$B_m = 1\,\text{m}$;
$B_f = 3\,\text{m}$;
$y_f = 1\,\text{m}$;
$n_r = 0.9$.

Determine:

Critical depths for three rates of discharge $= ?$

Solution:

The C-y_r curve for this cross section is plotted in Fig. 3-9 for $b_m = B_m/y_f = 1.0$, $b_f = 3.0$ and $n_r = 0.9$. When $y_r \to 1$, $m \to 1$ and $C \to 1$ and when $y_1 \to \infty$, $m \to 0$ and $C \to 0$. These properties of the C-y_r curve are clear from Fig. 3-9. In addition, this curve has a maximum value, C_{max}, equal to 2.88, which occurs at $y_r = 1.033$. Depending upon the value of discharge (and hence k), three typical cases are possible. These are discussed in the following paragraphs.

$Q = 1.7\,\text{m}^3/\text{s}$

For this discharge, $k = 3.39$. The horizontal line corresponding to $k = 3.39$ does not intersect the C-y_r curve (Fig. 3-9); therefore, there is no solution for Eq. 3-59 – i.e., critical flow cannot occur for this discharge if the flow depth $y_r > y_f$. However, there is a critical depth for flow in the main channel only, i.e. for $y < y_f$. This critical depth, y_c, may be determined from the equation

$$y_c = \left(\frac{Q^2}{gB_m^2}\right)^{\frac{1}{3}} \tag{3-62}$$

For $Q = 1.7\ m^3/s$ and for this channel cross section, $y_c = 0.67\,\text{m}$. This corresponds to the minimum specific energy, i.e., point C on the curve for $Q = 1.7\,\text{m}^3/\text{s}$ of Fig. 3-10.

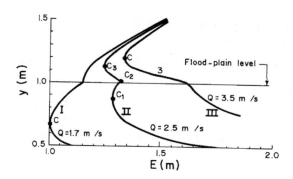

Fig. 3-10   **Specific energy vs depth curve for cross section of Fig. 3-8**

$Q = 2.5\,\text{m}^3/\text{s}$

For this discharge $k = 1.57$. Thus k is greater than 1.0 but less than C_{max} ($= 2.88$). The horizontal line corresponding to $k = 1.57$ intersects the C-y_r curve twice (Fig. 3-10), giving two critical depths : $y_{c_2} = 1.002\,\text{m}$, $y_{c_3} = 1.12\,\text{m}$. Since $k > 1$, critical flow also occurs when the flow is only in the main channel, at depth $y_{c_1} = [Q^2/(gB_m^2)]^{\frac{1}{3}} = 0.86\,\text{m}$. Hence, for this discharge, there are three critical depths, $y_{c_1} = 0.86\,\text{m}$, $y_{c_2} = 1.002\,\text{m}$ and $y_{c_3} = 1.12\,\text{m}$. Specific energy is locally minimum at two of these depths, i.e. at points C_1 and C_3, and locally maximum at the third, i.e. at point C_2, as shown by the curve for $Q = 2.5\,\text{m}^3/\text{s}$ in Fig. 3-10. In other words, one of the three possible critical depths would not have been computed if only the minimum specific energy had been used as the criteria for determining the critical depths. The importance of this conclusion will become apparent while plotting the water-surface profiles in Chapter 5 (Examples 5-4 and 5-5).

$Q = 3.5\ m^3/s$

For this discharge, $k = 0.8$. Since $k < 1$, critical depth cannot occur when the flow is only in the main channel. However, the horizontal line corresponding to $k = 0.8$ intersects the C-y_r curve at $y_r = 1.18$ (Fig. 3-10), giving critical depth $y_c = 1.18\,\text{m}$. This is illustrated by the single minimum specific energy point C of the curve for $Q = 3.5\,\text{m}^3/\text{s}$ in Fig. 3-10 occurring when the flow is over the banks.

Algorithm for Computing the Critical Depths

The following conclusions may be drawn from the typical cases of the preceding example:

1. For $k > C_{max}$, there is only one critical depth, and it is less than the depth of floodplains, y_f;

2. For $1 \leq k < C_{max}$, three critical depths are possible. Two of these critical depths are for the flow over the banks and the third occurs when the flow is only in the main channel;
3. For $k < 1$, there is only one critical depth, and it is greater than the depth of floodplains, y_f.

These conclusions may be utilized as follows to formulate an algorithm for computing the critical depths.

To solve Eq. 3-59 for y_r by an iterative procedure, we may write this equation as

$$
\begin{aligned}
y^* = &\frac{2b_r}{2b_r + 1} + \frac{1}{C(2b_r + 1)} \left[\left(\frac{m}{y}\right)^2 + \left(\frac{1-m}{y_r - 1}\right)^2 \frac{1}{2b_r} \right] \\
&+ \frac{2m(1-m)}{3C(2b_r + 1)} \left[\frac{5}{y_r(y_r - 1)} - \frac{2}{b_f + y_r - 1} \right] \left[\frac{m}{y_r} - (\frac{1-m}{y_r - 1}) \frac{1}{2b_r} \right]
\end{aligned}
$$

$$(3\text{-}63)$$

First, we estimate a value of y_r, and then compute y^* from Eq. 3-63. If $|y*-y_r|$ is less than a specified tolerance, then $y*$ is the correct value of y_r; otherwise y_r is set equal to $y*$ and the iterations are repeated until a solution is obtained.

The procedure for computing the critical depths for a specified discharge in a given channel (i.e. Q, B_m, B_f, y_f, n_m, and n_f are known) may be summarized as follows:

1. Calculate k.
2. If k is less than 1, solve Eq. 3-63 for y_r by using an iterative procedure, and then compute $y_c = y_r y_f$;
3. If k is greater than or equal to 1, then follow steps 4 to 9;
4. Compute C_{max}. To do this, C is computed from Eq. 3-60 for different values of y_r, starting with an initial y_r value of approximately 1.0 and continuing until a maximum value of C is reached;
5. If k is greater than C_{max}, then $y_c = [Q^2/(gB_m^2)]^{\frac{1}{3}}$;
6. For k less than C_{max}, follow steps 7 to 9;
7. Compute y_{c_1} from the equation: $y_{c_1} = [Q^2/(gB_m^2)]^{\frac{1}{3}}$;
8. Solve Eq. 3-63 for y_r using an iterative procedure. Then, $y_{c_3} = y_r y_f$;
9. Solve for y_{c_2}. To do this, C is computed for different values of y_r from Eq. 3-60, starting with an initial value of y_r close to 1.0 and continuing until the computed C value is equal to k. The value of y_r at which C is equal to k gives y_{c_2}/y_f. Then, $y_{c_2} = y_r y_f$.

3-8 Summary

In this chapter, several properties of critical flow were discussed and a number of expressions were derived for a rectangular cross section. Then, regular non-rectangular cross sections were considered. A number of engineering applications of critical flows were discussed. It was shown mathematically where critical depth occurs in a channel of varying width or bottom elevation. Three procedures for computing the critical depth in a regular cross section were presented. The chapter concluded with a discussion of the possibility of multiple critical depths in a compound channel and how to determine their values.

Problems

3-1 Derive an expression for the Froude number, $\mathbf{F}_r$, by assuming that the velocity-head coefficient, α is a function of the flow depth.

3-2 Write a general-purpose computer program for computing the critical depth in a regular prismatic channel by using the: (i) bisection method; (ii) Newton method. Of these two methods, which method do you prefer and why?

Use this program to compute the critical depth in a circular conduit having a diameter of 4 m and carrying a discharge of $16.0\,\mathrm{m}^3/\mathrm{s}$. Assume $\alpha = 1$.

3-3 A trapezoidal channel having a bottom width of 20 m and side slopes of 2H : 1V is carrying $60\,\mathrm{m}^3/\mathrm{s}$. Assuming $\alpha = 1.1$, determine the critical depth.

3-4 For the channel section shown in Fig. 3-11, determine the critical depth for a flow of $80\,\mathrm{m}^3/\mathrm{s}$.

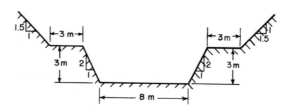

Fig. 3-11   Cross section for Prob. 3-4

3-5 For a discharge of $850\,\mathrm{m}^3/\mathrm{s}$, compute the critical depth in a tunnel having a standard horseshoe section (Fig. 3-12). The flow area, A, top water-surface width, B, and hydraulic radius, R, at different flow depths, y, are listed as

3-6 If y_1 and y_2 are the alternate depths in a rectangular channel, prove that

$$y_c = \sqrt[3]{\frac{2(y_1 y_2)^2}{y_1 + y_2}}$$

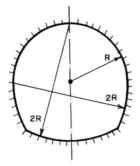

Fig. 3-12   Horseshoe section

y (m)	A (m²)	R (m)	B (m)
0	0.0	0.00	0.0
2	23.4	1.34	16.6
4	58.3	2.68	18.2
6	95.7	3.70	19.2
8	134.8	4.50	19.8
10	174.6	5.15	20.0
12	214.4	5.65	19.6
14	252.5	5.99	18.3
16	287.0	6.13	16.0
18	315.4	6.01	12.0
20	331.7	5.08	0.0

3-7 Derive expressions for the critical depth in a prismatic channel having the following cross sections and assuming in each case that the slope of the channel bottom is small:

   i. Trapezoidal;
  ii. Triangular;
 iii. Circular.

3-8 A 50-m wide rectangular channel is carrying a flow of $250\,\mathrm{m^3/s}$ at a flow depth of 5 m. To produce critical flow in this channel, determine:

   i. The height of the step in the channel bottom if the width remains constant;
  ii. The reduction in the channel width if the channel-bottom level remains unchanged;
 iii. A combination of the width reduction and the bottom step.

3-9 The drainage canal shown in Fig. 3-13 has a flow of $96\,\mathrm{m^3/s}$. If the flow depth at Section 1 is 4.22 m, what is the depth at Section 2? Assume there are no losses in the transition.

Determine the flow depth at the downstream end if the canal ends in a free overfall. Assume that critical depth occurs at the overfall.

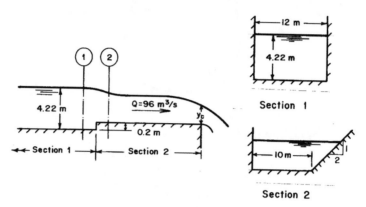

Fig. 3-13 **Drainage canal of Prob. 3-9**

3-10 Write a general-purpose computer program to determine the critical depth in a channel for a specified discharge and for either of the following channel cross sections:

   i. Circular;
  ii. Trapezoidal;
 iii. Triangular; and
 iv. Horseshoe.

3-11 Show that the critical depth in a channel having a triangular, rectangular, or parabolic cross section may be determined from the following explicit equation:

$$y_c = \left[\frac{Q^2}{g} \frac{(m+1)^3}{4k^2} \right]^{1/(2m+3)}$$

in which the x and y coordinates of the sides of the half-section may be defined as $x = ky^m$. For a triangular cross section, $k = s$, $m = 1$; for a rectangular cross section, $k = \frac{1}{2}B_o$, $m = 0$; and for a parabolic cross section, $k = (1/a)^m$, and $m = 1/n$ with the equation for the parabola being $y = ax^n$.

3-12 A mountain creek has a parabolic cross section with a top water surface width of 9 ft at a depth of 3 ft. Determine the critical depth for a flow of $50\,\text{ft}^3/\text{s}$.

3-13 An 8-ft diameter concrete-lined sewer is laid at a bottom slope of 1 ft/mile. Compute the critical depth for a discharge of 100 cfs.

3-14 A trapezoidal irrigation channel is 10 ft wide at the bottom and has side slopes of 2H : 1V. For a flow of 100 ft^3/s, determine the critical depth.

3-15 A 5-ft dia circular culvert carries a flow of 15 ft^3/s. Determine the critical depth.

3-16 For a horseshoe tunnel shown in Fig. 1-17 ($d_o = 30$ ft), determine the critical depth for a discharge of 300 cfs.

3-17 A 4-ft diameter culvert barrel carries a flow of 10 ft^3/sec and discharges free into a lake. What is the depth just upstream of the free fall?

3-18 A high-level rectangular outlet at a dam is 10 ft wide, 15 ft high and 500 ft long. The invert level at the entrance is at El. 122 and the bottom slope is 0.001. A sluice gate at the entrance controls the flow through the outlet. If the coefficient of discharge, C_d, is 0.7 ($Q = C_d A \sqrt{2gH}$), determine the thrust on the gate for the reservoir level of El 182. The water level in the river into which the outlet discharges is at El. 131 for this flow.

3-19 In order to reduce the flow velocity at a section, a fisheries biologist tied a 6-in diameter tree log at the bottom of a stream. The flow velocity and the flow depth prior to the installation of the log were 2 ft/sec and 4 ft, respectively. Determine the change in the flow velocity and flow depth just downstream of the log.

3-20 For the channel of Problem 3-8, what is the minimum channel width without affecting the upstream water level?

3-21 The reservoir level upstream of an overflow spillway is at El. 400 ft. The downstream water level for the design flow of 80,000 cfs is at El. 220 ft. If the spillway width at the entrance to the stilling basin is 200 ft, determine the invert level of the basin so that a hydraulic jump is formed in the basin at design flow. No baffle piers, chute blocks, or end sill is to be provided.

3-22 Write a computer program to determine the critical depth in a channel with circular cross section. Use the bisection and the Newton methods.

3-23 The flow velocity and flow depth in a 5-m wide rectangular channel are 1.5 m/s and 4 m, respectively. Design a converging transition so that the flow is critical in the transition. Assume the channel bottom to be horizontal, and losses in the transition to be negligible.

3-24 Let x be the number of the first letter of your last name in the English alphabet (e.g., $x = 10$ for Johnson) and k is your month of birth (e.g., $k = 4$ for April 20) and e is the last two digits of your year of birth (e.g., for 2018, $e = 18$.)
If $x < 12$, the channel is circular with a diameter in m of $D = k$;
If $x > 12$, the channel is trapezoidal with bottom width of k m and side slope of 1H:1V.
For a discharge of e in m^3/s, compute and plot the critical depth, starting with the current year and until you are 25 years old.

References

Black, R. G., 1982, *Discussion of* Blalock and Sturm [1981], *Jour. Hyd. Div.,* Amer. Soc. Civ. Engrs., vol. 108, pp. 798-800.

Blalock, M. E. and Sturm, T. W., 1981, "Minimum Specific Energy in Compound Channel," *Jour. Hyd. Div.,* Amer. Soc. Civ. Engrs., vol. 107, pp. 699-717 (see also closure, vol. 109, 1983, pp. 483-486).

Blalock, M. E., and Sturm, T. W. 1983. Closure to "Minimum Specific Energy in Compound Open Channel" by Merritt E. Blalock and Terry W. Sturm (June, 1981). *Jour. Hydraulic Engineering,* vol. 109, no. 3, pp. 483–486.

Chapra, S. C. and Canale, R. P., 2006, *Numerical Methods for Engineers,* 5th ed., McGraw Hill, New York, NY.

Chaudhry, M. H., and Bhallamudi, S. M., 1988, "Computation of Critical Depth in Compound Channels," *Jour., Hydraulic Research,* Inter. Assoc. for Hydraulic Research., vol. 26, no. 4, pp. 377-395.

Chow, V. T., 1959, *Open-Channel Hydraulics,* McGraw-Hill Book Company, New York, N. Y.

Davidian, J., 1984, "Computation of Water-Surface Profiles in Channels," *Techniques of Water-Resources Investigations,* United States Geological Survey, Book 3, Chapter A15.

French, R. H., 1985, *Open-Channel Hydraulics,* McGraw-Hill Book Co., New York, NY.

Konemann, N., 1982, *Discussion of* Blalock and Sturm [1981], *Jour. Hydraulics Division,* Amer. Soc. Civ. Engrs., vol. 108, pp.462-464

Knight, D. W., Demetriou, J. D. and Hamed, M. E., 1984, "Stage Discharge Relationships for Compound channels," *Channels and Channel Control Structures,* K. V. H. Smith (ed.), Springer-Verlag, pp. 4.21-4.35.

McCracken, D. D. and Dorn, W. S. 1964, *Numerical methods and FORTRAN programming,* John Wiley, New York, NY.

Myers, B. C. and Elsawy, E. M., 1975, "Boundary Shear in Channel With Flood Plains" *Jour. Hyd. Div.,* Amer. Soc. Civ. Engrs., vol. 101, pp. 933-946.

Petryk, S., and Grant, E. U., 1978, "Critical Flow in Rivers with Flood Plains," *Jour. Hyd. Div.,* Amer. Soc. Civ. Engrs., vol. 104, pp. 583-594.

Rajaratnam, N., and Ahmadi, R. M., 1979, "Interaction Between Main Channel and Flood Plain Flows," *Jour. Hyd. Div.,* Amer. Soc. Civ. Engrs., vol. 105, pp. 573-588.

Schoellhamer, D. H., Peters, J. C. and Larock, B. E., 1985, "Subdivision Froude number," *Jour. Hyd. Div.,* Amer. Soc. Civ. Engrs., vol. 111, pp. 1099-1104.

Soil Conservation Service., 1976, "WSP-2 Computer program," *Technical Release No.61.*

United States Army Corps of Engineers., 1982, "HEC-2; Water Surface Profile", *User's Manual,* Hydrologic Engineering Center.

United States Geological Survey., 1976, "Computer Applications for Step-Backwater and Floodway Analysis," *Open File Report 76-499.*

Wright, R. R. and Carstens, M. R., 1970, "Linear Momentum Flux to Overbank Sections," *Jour. Hyd. Div.,* Amer. Soc. Civ. Engrs., vol. 96, pp. 1781-1793.

4

UNIFORM FLOW

Part of 390-km long Colorado River Aqueduct. The trapezoidal canal is 6 m wide at the bottom, 16.8 m wide at the top and is 3.4 m deep. The maximum discharge is 51 m^3/s (Courtesy, Metropolitan Water District of Southern California)

Supplementary Information The online version contains supplementary material available at (https://doi.org/10.1007/978-3-030-96447-4_4).

© Springer Nature Switzerland AG 2022
M. H. Chaudhry, *Open-Channel Flow*,
https://doi.org/10.1007/978-3-030-96447-4_4

4-1 Introduction

In free-surface flow, the component of the weight of water in the downstream direction causes acceleration of flow (it causes deceleration if the bottom slope is negative), whereas the shear stress at the channel bottom and sides offers resistance to flow. Depending upon the relative magnitude of these accelerating and decelerating forces, the flow may accelerate or decelerate. For example, if the resistive force is more than the component of the weight, then the flow velocity decreases and, to satisfy the continuity equation, the flow depth increases. The converse is true if the component of the weight is more than the resistive force. However, if the channel is long and prismatic (i.e., channel cross section and bottom slope do not change with distance), then the flow accelerates or decelerates for a distance until the accelerating and resistive forces are equal. From that point on, the flow velocity and flow depth remain constant (Fig. 4-1). Such a flow, in which the flow depth does not change with distance, is called *uniform flow*, and the corresponding flow depth is called the *normal depth*.

Uniform flow is discussed in this chapter. An equation relating the bottom shear stress to different flow variables is first derived. Various empirical resistance formulas used for the free-surface flows are then presented. A procedure for computing the normal depth for a specified discharge in a channel of known properties is outlined.

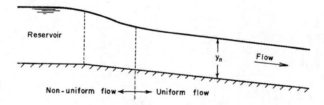

Fig. 4-1   **Uniform and non-uniform flows**

4-2 Flow Resistance

Leonardo da Vinci described the resistance offered by the channel bottom and sides to free-surface flows and its effects on the velocity distribution in an excellent manner as follows [Rouse and Ince, 1963]:

> "The water of straight rivers is the swifter the farther away it is from the walls, because of resistance.
> Water has higher speed on the surface than at the bottom. This happens because water on the surface borders on air which is of little

resistance, because lighter than water, and the water at the bottom is touching the earth which is of higher resistance, because heavier than water and not moving. From this follows that the part which is more distant from the bottom has less resistance than that below."

Because of the variation in resistance along the wetted perimeter and because of the shape of the channel cross section, secondary currents are usually set up in free-surface flows even if the channel is straight. In addition, the shear resistance offered to flow at the channel boundaries is not uniform. However, to simplify the analysis, we will assume that the flow is one-dimensional – i.e., there are no secondary currents in the flow and the shear resistance to flow at the boundaries is uniform.

4-3 Flow Resistance Equations

In this section, we present several equations relating the channel resistance to various flow variables. For a general derivation, we first derive an equation for nonuniform flow and then simplify it for uniform flow as a special case of nonuniform flow.

Chezy Equation

To derive the Chezy equation, we make the following *assumptions*: The flow is steady; the slope of the channel bottom is small; and the channel is prismatic.

Let us consider a control volume of length Δx, as shown in Fig. 4-2. At the upstream side of this control volume, let the distance be x, flow velocity be V, and the flow depth be y. Then, the values of these variables at the downstream side are $x + \Delta x$, $V + (dV/dx)\Delta x$, and $y + (dy/dx)\Delta x$.

The following forces are acting on the control volume: pressure force on the upstream side, F_1; pressure forces on the downstream side, F_2 and F_3; a component of the weight of water in the control volume in the downstream direction, W_x; and the shear force, F_f, acting on the channel bottom and the sides. Referring to Fig. 4-2, the expression for these forces may be written as follows:

$$\text{Pressure force, } F_1 = \gamma A \bar{z} \tag{4-1a}$$

in which $\bar{z}$ = depth of the centroid of flow area A below the water surface and γ = specific weight of water. The component of the weight of water in the downstream direction,

$$W_x = \gamma A \Delta x \sin \theta \tag{4-1b}$$

in which θ = angle between the channel bottom and the horizontal axis. Since the channel-bottom slope is assumed to be small, $\sin \theta \simeq \tan \theta \simeq -dz/dx$. Note that the negative sign is due to the fact that z decreases as x increases. Hence, we may write Eq. 4-1b as

$$W_x = -\gamma A \frac{dz}{dx} \Delta x \qquad (4\text{-}1c)$$

The pressure force acting on the downstream side of the control volume may be divided into two parts, as shown in Fig. 4-2. F_2 is the pressure force due to flow depth, y, and F_3 is the pressure force for the increase in depth in distance, Δx. The expressions for F_2 and F_3 are

$$F_2 = \gamma A \bar{z}$$

and

$$F_3 = \gamma A \frac{dy}{dx} \Delta x \qquad (4\text{-}1d)$$

Note that in the expression for F_3, we have neglected the higher-order term, which corresponds to the small triangle at the top. If the average shear stress

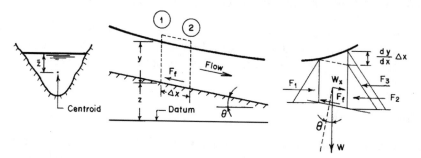

Fig. 4-2   Definition sketch

acting on the channel bottom and sides is τ_o, then the shearing force,

$$F_f = \tau_o P \Delta x \qquad (4\text{-}1e)$$

in which P = wetted perimeter. Referring to Fig. 4-2, the resultant force, F_r, acting on the control volume in the downstream direction is

$$F_r = \Sigma F = F_1 - (F_2 + F_3) + W_x - F_f \qquad (4\text{-}2)$$

Substituting Eqs. 4-1a through 4-1e into Eq. 4-2, and simplifying, we obtain

$$F_r = -\gamma A \Delta x \left(\frac{dy}{dx} + \frac{dz}{dx} + \frac{P \tau_o}{\gamma A} \right) \qquad (4\text{-}3)$$

To apply the Reynolds transport theorem, * the intensive property, $\beta = V$. Therefore,

---

* We may derive the same equation by applying Newton's second law to the liquid in the control volume, i.e., the resultant force F_r is equal to the time rate of change of momentum. The time rate of change of momentum $= \rho A V [(V + \frac{dV}{dx} \Delta x) - V]$. By equating this expression to that for F_r from Eq. 4-3 and simplifying, we obtain Eq. 4-5.

$$\sum F = \frac{\partial}{\partial t} \int_{x}^{x+\Delta x} \rho V A dx + (\rho A V^2)_2 - (\rho A V^2)_1 \tag{4-4}$$

Since the flow is assumed to be steady, the first term on the right-hand side of this equation is zero. By substituting for F_r from Eq. 4-3 into Eq. 4-4, dividing both sides by $\gamma A \Delta x$, and applying the mean value theorem to the right-hand side, we obtain

$$\frac{V}{g}\frac{dV}{dx} = -\left(\frac{dy}{dx} + \frac{dz}{dx} + \frac{P}{A}\frac{\tau_o}{\gamma}\right) \tag{4-5}$$

It follows from this equation that

$$\tau_o = -\gamma R\left(\frac{dy}{dx} + \frac{dz}{dx} + \frac{V}{g}\frac{dV}{dx}\right) \tag{4-6}$$

in which $R = A/P$ = hydraulic radius. This equation may be simplified as

$$\tau_o = -\gamma R\frac{d}{dx}\left(y + z + \frac{V^2}{2g}\right)$$
$$= -\gamma R\frac{dH}{dx}$$
$$= \gamma R S_f \tag{4-7}$$

in which S_f = slope of the energy grade line = $-dH/dx$. Note that we are using a negative sign since H decreases as x increases.

If the flow is steady and uniform, then by definition $dV/dx = 0$ and $dy/dx = 0$. In addition, since $S_o = -dz/dx$, Eq. 4-7 may be written as

$$\tau_o = \gamma R S_o \tag{4-8}$$

Based on dimensional analysis, we may write

$$\tau_o = k\rho V^2 \tag{4-9}$$

in which k is a dimensionless constant that depends upon the Reynolds number, roughness of the channel bottom and sides, etc. It follows from Eqs. 4-7 and 4-9 that

$$V = \sqrt{\frac{g}{k}R S_f} \tag{4-10}$$

This equation may be written as

$$V = C\sqrt{R S_f} \tag{4-11}$$

in which C = Chezy constant. This equation was introduced by a French engineer named Antoine Chezy, in 1768 while designing a canal for the water-supply system of Paris. [Henderson 1966; Chow 1959] Note that Eq. 4-11 is valid for nonuniform, steady flow. For uniform flow we use Eqs. 4-8 and 4-9 instead of Eqs. 4-7 and 4-9. As a result we obtain the following equation instead of Eq. 4-11:

$$V = C\sqrt{RS_o} \qquad (4\text{-}12)$$

Note that Eq. 4-11 for nonuniform flow and Eq. 4-12 for uniform flow are similar except that we use the slope of the energy grade line, S_f, for nonuniform flow, but we use the slope of the channel bottom, S_o, (which has the same value as the slope of the energy grade line or the slope of the water surface) for the uniform flow.

It is clear from Eq. 4-11 or 4-12 that C has dimensions of $\sqrt{\text{length}}/\text{time}$ as compared to the Darcy Weisbach friction factor, f, which is dimensionless. However, like f, C depends upon the channel roughness and the Reynolds number, $\mathbf{R}_e$. In addition, it may also depend upon the channel cross-sectional shape, although this dependence appears to be small [Anonymous, 1963] and may be neglected. Because the channel roughness may vary over a wide range, its effect on C has not been as thoroughly investigated as that on f.

Let us now compare the Chezy equation, Eq. 4-11, for open channels with the Darcy-Weisbach friction formula for pipes,

$$h_f = \frac{fL}{D}\frac{V^2}{2g} \qquad (4\text{-}13)$$

in which h_f = head loss in a pipe of diameter D and length L. The slope of the energy grade line, $S = h_f/L$. Therefore, we may write this equation as

$$V = \sqrt{\frac{2gDS}{f}} \qquad (4\text{-}14)$$

Noting that the hydraulic radius, R, for a pipe is equal to $D/4$, Eq. 4-11 becomes

$$V = C\sqrt{\frac{DS}{4}} \qquad (4\text{-}15)$$

It follows from the above two equations that $C = \sqrt{8g/f}$.

Figure 4-3 shows the Moody diagram plotted with C as the ordinate instead of f [Henderson, 1966]. This diagram is divided into three regions: hydraulically smooth, transition, and fully rough. A flow may be considered *hydraulically smooth* even though the channel surface is rough, provided the projections of the surface roughness are covered by the laminar sublayer. As the Reynolds number increases, the thickness of this layer decreases and the effect of roughness projections on flow becomes important. Then, the flow is in the *transition* region. However, when the roughness projections are not covered by the viscous sublayer and dominate the flow because losses are due to form drag, flow may be classified as *fully rough*. These flow types may be classified based on the value of a dimensionless number, $R_s = kV^*/\nu$. In this expression, ν is the kinematic viscosity of the liquid; k is a characteristic length parameter for the size of the channel-surface roughness; and, V^* is the *shear velocity*, which is defined as

$$V^* = \sqrt{\frac{\tau_o}{\rho}} = \sqrt{gRS_f} \qquad (4\text{-}16)$$

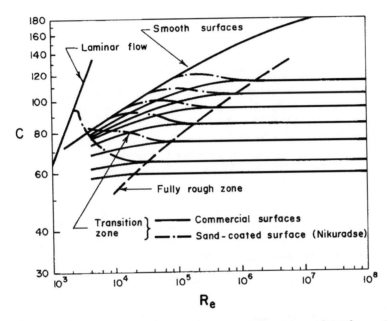

Fig. 4-3   Modified moody diagram (After Henderson [1966]; reprinted by permission of Pearson, Inc., Upper Saddle River, NJ)

The flow is considered *smooth* if $R_s < 4$; *transition* if $4 < R_s < 100$; and *fully rough* if $R_s > 100$. The expressions for C for smooth and rough flows derived from the experimental data on flow through pipes are [Henderson, 1966]:

Smooth Flows

$$C = 28.6\mathbf{R}_e^{1/8} \qquad \text{if } \mathbf{R}_e < 10^5 \qquad (4\text{-}17)$$

and

$$C = 4\sqrt{2g}\log_{10}\left(\frac{\mathbf{R}_e\sqrt{8g}}{2.51C}\right) \qquad \text{if } \mathbf{R}_e > 10^5 \qquad (4\text{-}18)$$

Rough Flows

$$C = -2\sqrt{8g}\log_{10}\left(\frac{k_s}{12R} + \frac{2.5}{\mathbf{R}_e\sqrt{f}}\right) \qquad (4\text{-}19)$$

The preceding equations are valid only for small channels with fairly smooth surfaces since these are based on pipe data. Empirical relationships and field observations should be employed for large channels with rough flow surfaces.

Manning Equation

Since the derivation of the Chezy equation in 1768, several researchers have tried to develop a rational procedure for estimating the value of Chezy constant, C. However, unlike the Darcy-Weisbach friction factor for the closed conduits, these attempts have not been very successful, because C depends upon several parameters in addition to the channel roughness. Based on the field observations, Ganguillet and Kutter [Chow, 1959] proposed a complex formula for C. Later, Gauckler and Hagen independently showed that

$$C \propto R^{1/6} \qquad (4\text{-}20)$$

According to Henderson [1966], a French engineer named A. Flamant incorrectly attributed the above equation to an Irishman, R. Manning, and expressed it in the following form in 1891

$$V = \frac{1}{n}R^{2/3}S_f^{\frac{1}{2}} \qquad (4\text{-}21)$$

in which n = Manning coefficient. This is the Manning equation, which has been very widely used in the English-speaking countries.

Again note that n is not a dimensionless constant and has the dimensions of time/(length)$^{1/3}$. Therefore, we convert this equation so that the value of n is the same in both SI and English units.

Equation 4-21 is valid for SI units, i.e., V is in m/s, and R is in m. In the foot-pound-second units, this equation becomes

$$V = \frac{1.49}{n}R^{2/3}S_f^{\frac{1}{2}} \qquad (4\text{-}22)$$

in which V is in ft/sec and R is in ft. As a result of this conversion, the value of n is the same in both system of units.

The value of n depends mainly upon the surface roughness, amount of vegetation, and channel irregularity, and, to a lesser degree, upon stage, scour and deposition, and channel alignment.

Tables 4-1 and 4-2 list the typical values of n for different materials. These tables are compiled from the maximum, minimum, and normal values of n presented by Chow [1959], and by Chen and Cotton [1988].

Table 4-1   Typical values* of Manning n

Material	n
Metals	
Steel	0.012
Cast iron	0.013
Corrugated metal	0.025
Non-metals	
Lucite	0.009
Glass	0.010
Cement	0.011
Concrete	0.013
Wood	0.012
Clay	0.013
Brickwork	0.013
Gunite	0.019
Masonary	0.025
Rock cuts	0.035
Natural streams	
Clean and straight	0.030
Bottom: gravel, cobbles and boulders	0.040
Bottom: cobbles with large boulders	0.050

*Compiled from tables presented by Chow [1959].

In addition, the photographs published by the United States Geological Survey [Barnes, 1967] and the Department of Agriculture [Ramser, 1929; Scobey, 1939] are excellent sources of reference for the selection of n for natural channels.

Figure 4-4 presents a number of typical photographs for natural channels; information for these photographs follows:

a. $n = 0.024$ (Columbia River at Vernita, Washington): The channel bottom consists of slime-covered cobbles and gravel, the steep left bank is composed of cemented cobbles and gravel, and the right bank consists of cobbles set in gravel.

b. $n = 0.030$ (Salt Creek at Roca, Nebraska): The bottom consists of sand and clay; the banks are smooth and free of vegetation.

c. $n = 0.032$ (Salt River below Stewart Mountain Dam, Arizona): The bottom and banks consist of smooth 0.15-m diameter cobbles, with few 0.45-m diameter boulders.

Table 4-2 Manning n for Different Channel Linings*

Channel	Type	n Depth Range, in mm		
		(0–150)	(150–600)	(>600)
Rigid	Concrete	0.015	0.013	0.013
	Grouted Riprap	0.040	0.030	0.028
	Stone Masonry	0.042	0.032	0.030
	Soil Cement	0.025	0.022	0.020
	Asphalt	0.018	0.016	0.016
Unlined	Bare Soil	0.023	0.020	0.020
	Rock Cut	0.045	0.035	0.025
Temporary	Woven Paper Net	0.016	0.015	0.015
	Jute Net	0.028	0.022	0.019
	Fiberglass Roving	0.028	0.021	0.019
	Straw with Net	0.065	0.033	0.025
	Curled Wood Mat	0.066	0.035	0.028
	Synthetic Mat	0.036	0.025	0.021
Gravel Riprap	25 mm D50	0.044	0.033	0.030
	50 mm D50	0.066	0.041	0.034
Rock Riprap	150 mm D50	0.104	0.069	0.035
	300 mm D50	–	0.078	0.040

* After Chen and Cotton [1988]

d. $n = 0.036$ (West Fork Bitterroot River near Conner, Montana): The bottom is gravel and boulders, $d_{50} = 1.72$m; left bank has overhanging bushes and the right bank has trees.

e. $n = 0.041$ (Middle Fork Flathead River near Essex, Montana): The bottom consists of boulders, $d_{50} = 1.4$ m; banks are composed of gravel and boulders and have trees and brushes.

f. $n = 0.049$ (Deep River at Ramseur, North Carolina): The bottom is mostly coarse sand and contains some gravel; the banks are fairly steep and have underbrush and trees.

g. $n = 0.050$ (Clear Creek near Golden, Colorado): The bottom and banks are composed of 0.7-m diameter angular boulders.

h. $n = 0.060$ (Rock Creek Canal near Darby, Montana): The bottom and banks consist of boulders $d_{50} = 2.1$ m.

i. $n = 0.070$ (Pond Creek near Loisville, Kentucky): The bottom is fine sand and silt; the banks are irregular with heavy growth of trees.

j. $n = 0.075$ (Rock Creek near Darby, Montana): The bottom consists of boulders, $d_{50} = 2.2$ m; the banks are composed of boulders and have brush and trees.

(a)   (b)

(c)   (d)

Fig. 4-4   Photographs for typical Manning n (After Barnes [1967])

Christensen [1984a] investigated the range of validity of the Manning equation assuming that the Nikuradse's equations [1932] for the friction factors of closed conduits are valid for the free-surface flows. By substituting the approximation

$$\frac{1}{\sqrt{f}} = 2.916(\frac{R}{k})^{1/6} \tag{4-23a}$$

for rough turbulent flows in circular conduits into Eq. 4-14 and noting that for closed conduits $R = D/4$, we obtain

$$V = 8.25\frac{\sqrt{g}}{k^{1/6}}S^{1/2}R^{2/3} \tag{4-23b}$$

A comparison of Eqs. 4-21 and 4-23b shows that they are identical if we replace $1/n$ of Eq. 4-21 by $8.25\sqrt{g}/k^{1/6}$ of Eq. 4-23b. This relationship between n and k_s is almost identical to the Strickler's formula reported by Forchheimer in 1930.

Equation 4-23b has the following advantages over Eq. 4-21: Manning n is difficult to estimate since it does not have any physical meaning. On the

Fig. 4-4   (Concluded)

other hand, k is physically based and is directly related to the size of surface roughness, which can be measured. In addition, since k is raised to the one-sixth power, an error in estimating its value has a considerably less effect on the computed value of V as compared to that introduced by a similar error in the estimation of n.

Manning coefficient, n, increases for very shallow depths where the height of lining roughness approaches the depth of flow. For lined channels, a constant n value is acceptable; however, to account for shallow flow depths, a higher n value should be considered [Chen and Cotton, 1988]. Henderson [1966] presented an equation for n that is a function of the riprap size only

$$n = C_m (3.28 d_{50})^{1/6} \tag{4-24a}$$

where d_{50} = mean gravel diameter, in m. For C_m for gravel-bed streams, Henderson [1966] recommended a value of 0.034, Hager [2001] suggested 0.039 and Maynord [1991] recommended 0.038. Based on the experimental data of Blodgett and McConaughy [1986], Chen and Cotton [1988] presented an equation for n as a function of hydraulic radius and tractive force

$$n = \frac{(R/0.3048)^{1/6}}{8.6 + 19.97 \log(R/d_{50})} \tag{4-24b}$$

where R = hydraulic radius, in m.

For vegetation-lined channels, a constant n may not be suitable due to significant variation in the amount of submergence of the vegetation with changes in flow and the resulting shear stress. Therefore, Chen and Cotton [1988] presented the following equation for n for grass-lined channels as a function of hydraulic radius and tractive force, based on the studies of Kouwen et al. [1969] and Kouwen et al. [1980]

$$n = \frac{(R/0.3048)^{1/6}}{C + 19.97 \log[(R/0.3048)]^{1.4} S_o^{0.4})} \tag{4-25}$$

where S_o is the channel bottom slope, and C is a dimensionless factor depending on the class of vegetation and R is in m. Soil Conservation Service (SCS) classified various types of vegetation into five categories [1954] depending on the maturity of grass, with Retardance Class A corresponding to the highest resistance and Class E to the lowest resistance. Table 4-3 [Akan, 2006] shows various Retardance Classes of common grass types with the corresponding C values taken from the design charts presented by Chen and Cotton [1988] utilizing the data of Kouwen et al. [1980].

Other Resistance Equations

In Europe, the following resistance equation has been widely used [Jaeger, 1961]:

$$V = k_s R^{2/3} S_f^{1/2} \qquad (4\text{-}26)$$

This is called the Strickler equation, and k_s is called the Strickler constant. In SI units, k_s may be computed from

Table 4-3   Empirical Coefficients for Resistance Equations*

Class Cover	Condition	C
A Weeping lovegrass	Excellent stand, tall (average 760 mm)	15.8
B Yellow Bluestem Ischaemum Kudzu Bermuda Grass Native Grass Mixture (little bluestem, bluestem, blue gamma, and other long and short Midwest grasses) Weeping lovegrass Lespedeza sericea Alfalfa Weeping lovegrass Kudzu Blue Gamma	Excellent stand, tall (average 910 mm) Very dense growth, uncut Good stand, tall (average 300 mm) Good stand, unmowed Good stand, tall (average 610 mm) Good stand, not woody, tall (average 480 mm) Good stand, uncut (average 280 mm) Good stand, unmowed (average 330 mm) Dense growth, uncut Good stand, uncut (average 280 mm)	23.0
C Crabgrass Bermuda grass Common Lespedeza Grass-Legume mixture–summer (orchard grass, redtop, Italian ryegrass, and common lespedeza) Centipedegrass Kentucky Bluegrass	Fair stand, uncut 250 to 1200 mm Good stand, mowed (average 150 mm) Good stand, uncut (average 280 mm) Good stand, uncut (150 to 200 mm) Very dense cover (average 150 mm) Good stand, headed (150 to 300 mm)	30.2
D Bermuda grass Common Lespedeza Buffalo grass Grass-legume mixture–fall, spring Lespedeza sericea	Good stand, cut to 60-mm height Excellent stand, uncut (average 110 mm) Good stand, uncut (80 to 150 mm) Good stand, uncut (100 to 130 mm) After cutting to 50-mm height. Very good stand before cutting.	34.6
E Bermuda grass Burned stubble	Good stand, cut to height 40-mm	37.7

* After Chen and Cotton [1988]

$$k_s = \frac{21.1}{k^{1/6}} \tag{4-27}$$

in which $k =$ the mean size of the wall roughness. Typical mean values for k for various materials, taken from Jaeger [1961], are listed in Table 4-4.

Table 4-4   Roughness sizes*

Material	k (mm)
Cast iron, new	0.5–1.0
Cast iron, somewhat rusty	1.0–1.5
Cement mortar, smoothed	0.3–0.8
Cement mortar, left rough	1.0–2.0
Rough wooden boards	1.0–2.5
Rough masonary (blockwork)	8.0–15.

* Compiled from data presented by Jaeger [1961]

A comparison of Eqs. 4-21 and 4-27 shows that the Manning and Strickler equations are similar and that $k_s = \frac{1}{n}$.

In Russia, the following equation for Chezy constant, C, has been widely used:

$$C = \frac{1}{n}R^a \tag{4-28}$$

in which $a = 1.3\sqrt{n}$ if $R > 1$ m and $a = 1.5\sqrt{n}$, if $R < 1$ m. This formula was proposed by Pavlovskii [Chow, 1959] in 1925.

4-4 Computation of Normal Depth

To analyze open channel flow, it is usually necessary to know the normal depth, y_n. A number of procedures for computing the normal depth in a given channel for a specified discharge are discussed in this section. We will consider only the Manning equation in our discussions since it is very widely used in the English-speaking countries. These discussions are valid for the Strickler equation as well if we replace n by $1/k_s$.

The Manning equation for uniform flow in terms of discharge may be written as

$$Q = VA = \frac{C_o}{n}AR^{2/3}S_o^{1/2} \tag{4-29}$$

in which $C_o = 1.49$ in Customary English units and $C_o = 1$ in SI units.

In this equation, A and R are function of the flow depth, y, and of the channel cross section, whereas n is a function of the flow surface and other factors discussed in the previous section. Thus, for a given channel section and specified bottom slope, only one discharge is possible for a given normal depth. However, if the value of this depth is known, then we can determine the corresponding discharge directly from Eq. 4-29. We may write this equation as

$$Q = KS_o^{1/2} \tag{4-30}$$

in which the conveyance factor, K, for the channel section is defined as

$$K = \frac{C_o}{n} AR^{2/3} \tag{4-31}$$

Note that K is a function of the normal depth, properties of the channel section and Manning n.

Equation 4-29 may be written as

$$AR^{2/3} = \frac{nQ}{C_o S_o^{1/2}} \tag{4-32}$$

in which the left-hand side is referred to as the *section factor*. Thus, for the specified values of n, Q, and S_o, we solve this equation to determine the normal depth in a given channel. This may be done by using the design charts presented by Chow [1959], by a trial-and-error procedure, or by using numerical methods for the solution of a nonlinear algebraic equation. These procedures are discussed in the following paragraphs.

Design Curves

These curves are presented in Fig. 4-5 for a trapezoidal and for a circular channel section. If we want to determine the normal depth for a specified discharge in a given channel section, then we know Q, n, and S_o. Therefore, we can compute the right-hand side of Eq. 4-32. Let us divide this computed value by $B_o^{8/3}$ if the channel cross section is trapezoidal and by $D_o^{8/3}$ if the channel cross section is circular. The resulting value is then equal to $AR^{2/3}/B_o^{8/3}$ for a trapezoidal section and equal to $AR^{2/3}/D_o^{8/3}$ for a circular cross section. Now, y_n/B_o or y_n/D_o corresponding to the value of $AR^{2/3}/B_o^{8/3}$ or $AR^{2/3}/D_o^{8/3}$ may be directly read from Fig. 4-5.

Trial-and-Error Procedure

Substitute expressions for the flow area A, hydraulic radius, R, and the values of n, Q, and S_o into Eq. 4-32 and then solve the resulting equation by a trial-and-error procedure.

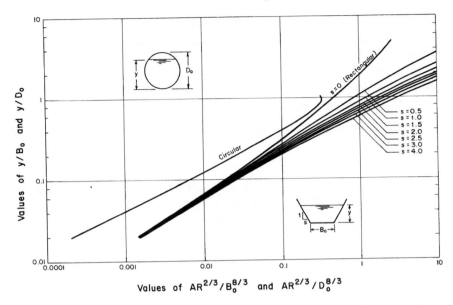

Fig. 4-5 **Curves for the computation of normal depth** (After Chow [1959])

Numerical Methods

Several methods, such as the bisection method, method of successive approximations, and the Newton method, are available [McCracken and Dorn, 1964, Chapra and Canale 2006] for solving Eq. 4-32. We discuss only the Newton method.

To determine y_n by this method, we write Eq. 4-32 as

$$F(y_n) = AR^{2/3} - \frac{nQ}{C_o S_o^{1/2}} = 0 \qquad (4\text{-}33)$$

For the Newton method, we need the first derivative of function F. An expression for this derivative may be obtained as follows:

$$\begin{aligned}
\frac{dF}{dy_n} &= \frac{d}{dy_n}\left(A^{5/3}P^{-2/3} - \frac{nQ}{C_o S_o^{1/2}}\right) \\
&= \frac{5}{3}P^{-2/3}A^{2/3}\frac{dA}{dy_n} - \frac{2}{3}A^{5/3}P^{-5/3}\frac{dP}{dy_n} \\
&= \frac{5}{3}BR^{2/3} - \frac{2}{3}R^{5/3}\frac{dP}{dy_n} \qquad (4\text{-}34)
\end{aligned}$$

since $dA/dy_n = B$. For a trapezoidal section having side slopes of s horizontal to 1 vertical, $dP/dy_n = 2\sqrt{1+s^2}$. Similar expressions for other channel sections may be obtained.

The following example will help in understanding these procedures for determining the normal depth. The computer programs available on the publisher website illustrate the application of the Newton and bisection methods.

Example 4-1

Compute the normal depth in a trapezoidal channel having a bottom-width of 10 m, side slopes of 2H to 1V and carrying a flow of 30 m^3/s. The slope of the channel bottom is 0.001 and $n = 0.013$.

Given:

$Q = 30$ m^3/s;
$n = 0.013$;
$B_o = 10$ m;
$s = 2$;
$S_o = 0.001$;
$C_o = 1.0$.

Determine:

$y_n = $?

Solution:

We use the above procedures one by one to determine y_n.

Design curves

By substituting the values of n, Q, and S_o into the right-hand side of Eq. 4-32, we obtain

$$\frac{nQ}{C_o S_o^{1/2}} = \frac{0.013 \times 30}{1. \times (0.001)^{1/2}}$$

$$= 12.33$$

Hence, it follows from Eq. 4-32 that

$$AR^{2/3} = 12.33$$

Now

$$\frac{AR^{2/3}}{B_o^{8/3}} = \frac{12.33}{(10)^{8/3}}$$

$$= 0.026$$

For $s = 2$ and $AR^{2/3}/B_o^{8/3} = 0.026$, we read from Fig. 4-5 that $y_n/B_o = 0.11$. Hence,

$$y_n = 1.1\,\text{m}.$$

Trial-and-Error Procedure

We earlier computed $AR^{2/3} = 12.33$ for the design curve procedure. By using the data for the channel, we obtain the following expressions for A and R:

$$A = \frac{1}{2}y_n(10 + 10 + 2sy_n)$$

$$= y_n(10 + 2y_n)$$

$$P = B + 2\sqrt{s^2 + 1}\,y_n$$

$$= 10 + 4.47y_n$$

$$R = \frac{y_n(10 + 2y_n)}{10 + 4.47y_n}$$

Now, substituting these expressions for A and R into $AR^{2/3} = 12.33$ and simplifying the resulting equation, we obtain

$$[y_n(20 + 2y_n)]^{5/3} - 12.33(10 + 4.47y_n)^{2/3} = 0$$

A trial-and-error solution of this equation yields

$$y_n = 1.09\,\text{m}$$

Numerical Method

To compute the normal depth in a trapezoidal channel, computer programs are developed using the Newton and the bisection methods. An initial estimate for the flow depth in the Newton method is used as 0.5 m; and YL = 0.5 m and YR = 10 m are used as initial estimates for the bisection method. The normal depth computed by both of these methods is 1.09 m.

4-5 Equivalent Manning Constant

In the previous discussion, we assumed that the flow surface at a channel cross section has the same roughness along the entire wetted perimeter. However, this is not always true. For example, if the channel bottom and sides are made from different materials, then the Manning n for the bottom and sides may have different values. To simplify the computations, it becomes necessary to determine a value of n, designated by n_e, that may be used for the entire section. This value of n_e is referred to as the equivalent n for the entire cross section.

Let us consider a channel section that may be subdivided into N subareas having wetted perimeter P_i and Manning constant, n_i, $(i = 1, 2, \cdots, N)$. By assuming that the mean flow velocity in each of the subareas is equal to the mean flow velocity [Horton, 1933; Einstein, 1934] in the entire section, the following equation may be derived:

$$n_e = \left(\frac{\sum P_i n_i^{3/2}}{\sum P_i} \right)^{2/3} \tag{4-35}$$

in which subscript i refers to values for the ith subarea. Similarly, the following expression for the equivalent Manning constant n_e may be derived by assuming that the total force resisting the flow is equal to the sum of forces resisting the flow in each subarea [Muhlhofer, 1933; Einstein and Banks, 1951]:

$$n_e = \frac{\left(\sum P_i n_i^2 \right)^{1/2}}{\left(\sum P_i \right)^{1/2}} \tag{4-36}$$

By utilizing the fact that the total discharge is equal to the sum of the discharge in each subarea, Lotter [1933] obtained the following equation for the equivalent Manning constant:

$$n_e = \frac{P R^{5/3}}{\left(\frac{\sum P_i R_i^{5/3}}{n_i} \right)} \tag{4-37}$$

Krishnamurthy and Christensen [1972] derived the following equation by assuming a logarithmic velocity distribution:

$$\ln n_e = \frac{\sum P_i y_i^{3/2} \ln n_i}{\sum P_i y_i^{3/2}} \tag{4-38}$$

in which y_i = flow depth.

By utilizing the data for 36 natural channel cross sections obtained by U.S. Gelogical Survey, Motayed and Krishnamurthy [1980] showed that the equivalent roughness computed by using Eq. 4-35 gives the least error of the above four equations for computing n_e.

4-6 Compound Channel Cross Section

A compound cross section may be defined as a section in which various subareas have different flow properties, e.g., surface roughness, etc. A natural stream having overbank flow during a flood (Fig. 4-6) is a typical example of a compound section. The roughness of the overbanks is usually higher than that of the main channel; and, therefore, the flow velocity in the main channel is higher than that in the overbank flow.

The analysis of flow in a compound section becomes complex if the flow in each subarea is considered separately. This requires the use of a two- or three-dimensional model or to apply a one-dimensional model separately to each subarea by considering the flow in each sub-area as parallel flow and allowing for the exchange of mass and momentum between the adjacent subareas.

In a straight channel, the water surface should be level over the entire cross section, since the pressure along any horizontal line must be constant although the flow velocity may vary from one subarea to the next. Due to different flow velocity, the level of the energy grade line is different in each subarea. Thus, there is no common level for the energy grade line for the entire section. To avoid this complexity, we derive in this section expressions for the energy coefficient, α, and for S_f in terms of the conveyance factor, K, of the subareas. With these expressions, the flow in a compound section may be computed without knowing the individual flows in each subarea.

Let us subdivide the compound section into N subareas. We want to derive an expression for the energy coefficient, α, such that the velocity head for the entire section $= \alpha V_m^2/2g$, in which V_m = mean flow velocity in the compound section. In Chapter 1, we derived the following expression for the velocity head coefficient,

$$\alpha = \frac{\sum_i^N V_i^3 A_i}{V_m^3 \sum_i^N A_i} \tag{4-39}$$

in which N = number of subareas. By substituting

$$V_m = \frac{\sum V_i A_i}{\sum A_i} \tag{4-40}$$

and $V_i = Q_i/A_i$ into Eq. 4-39, and simplifying the resulting equation, we obtain

$$\alpha = \frac{\sum (Q_i^3/A_i^2) . (\sum A_i)^2}{(\sum Q_i)^3} \tag{4-41}$$

Now, the flow in subarea i may be written as

$$Q_i = K_i S_{fi}^{\frac{1}{2}} \tag{4-42}$$

or

$$S_{fi}^{\frac{1}{2}} = \frac{Q_i}{K_i} \tag{4-43}$$

Let us assume that S_f has the same value for all subareas, i.e., $S_{fi} = S_f$, $(i = 1, 2, \cdots, N)$. Then, on the basis of Eq. 4-43, we may write the following equation for each subarea:

$$\frac{Q_1}{K_1} = \frac{Q_N}{K_N}$$

$$\frac{Q_2}{K_2} = \frac{Q_N}{K_N}$$

$$\cdots \qquad \cdots$$

$$\cdots \qquad \cdots$$

$$\cdots \qquad \cdots$$

$$\frac{Q_i}{K_i} = \frac{Q_N}{K_N}$$

$$\frac{Q_N}{K_N} = \frac{Q_N}{K_N} \tag{4-44}$$

It follows from this equation that

$$Q_1 = K_1 \frac{Q_N}{K_N}$$

$$Q_2 = K_2 \frac{Q_N}{K_N}$$

$$\cdots \qquad \cdots$$

$$\cdots \qquad \cdots$$

$$\cdots \qquad \cdots$$

$$Q_N = K_N \frac{Q_N}{K_N} \tag{4-45}$$

The addition of the preceding equations yields

$$Q = \sum Q_i = \frac{Q_N}{K_N} \sum K_i \tag{4-46}$$

By substituting this expression for $\sum Q_i$, and $Q_i = K_i(Q_N/K_N)$ into Eq. 4-41 and simplifying the resulting equation, we obtain

$$\alpha = \frac{\sum \frac{K_i^3}{A_i^2}(\sum A_i)^2}{(\sum K_i)^3} \tag{4-47}$$

The elimination of Q_N/K_N from Eqs. 4-44 and 4-46 and squaring both sides give

$$S_f = \left(\frac{\sum Q_i}{\sum K_i} \right)^2$$

$$= \frac{Q^2}{(\sum K_i)^2} \tag{4-48}$$

Now we have expressions for both α and S_f such that we do not have to know the flow in each sub-area, Q_i, $(i = 1, 2, \cdots, N)$, to compute α and S_f.

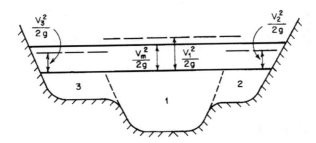

Fig. 4-6   Compound channel section

4-7 Summary

In this chapter, an expression relating the bottom shear stress to other flow variables was derived. Different empirical formulas for computing the friction losses were presented and procedures for computing the normal depth in a channel section for a specified discharge were outlined. Based on different simplifying assumptions, equations for computing the equivalent roughness coefficient were presented. The chapter concluded by deriving expressions for the energy coefficient and for the slope of the energy grade line in a compound channel section.

Problems

4-1 The flow depth in a long, 10-m wide, rectangular channel for a flow of 5 m^3/s is 4 m. Determine the flow depth if the flow is increased to 8 m^3/s.

4-2 Figure 4-7 shows the cross section of a long drainage channel with a free overfall at the downstream end. The channel is concrete-lined ($n = 0.013$) and has a bottom slope of 0.0003. For a discharge of 20 m^3/s, determine:

1. i The flow depth in the channel;
2. ii The flow depth just upstream of the fall.

4-3 In a long 10m-wide rectangular channel, flow depth for a discharge of 5 m^3/s is 4m. Determine the flow depth if the discharge is increased to 10 m^3/s.

4-4 A 5-m wide rectangular channel is carrying a flow of 5 m^3/s. If the Manning $n = 0.013$ and the bottom slope, $S_o = 0.001$, determine the normal depth.

4-5 Compute the normal depth in the channel of Example 4-1 for a discharge of 50 m^3/s.

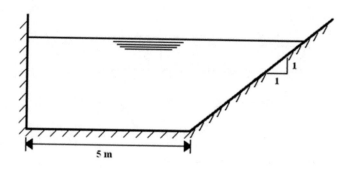

Fig. 4-7   Channel cross section

4-6 Develop a general-purpose computer program to compute the normal depth in a channel having a general cross section. Write the program such that A and R will be supplied by the user through a subroutine. Use this program to compute the normal depth in the channels of problems 4-4 and 4-7.

4-7 A channel with a cross section shown in Fig. 4-8 has a flow of 150 m³/s. The slope of the channel bottom is two per thousand, and the Manning n for the flow surfaces is 0.03. Compute the normal and critical depths in the channel.

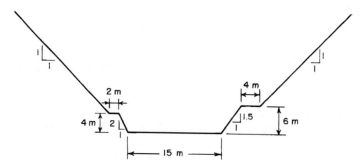

Fig. 4-8   Channel cross section for Prob. 4-7

4-8 A concrete-lined, trapezoidal, irrigation canal has a bottom width of 10 m, side slopes of 1H: 1V, and longitudinal bottom slope of 0.0005. If the canal is several kilometers long, determine the flow depth near the downstream end for a flow of 60 m³/s.

4-9 Analytically determine the depths at which the discharge and flow velocities are maximum in a circular conduit flowing partially full. Assume the flow is uniform and the Manning n does not vary with depth.

4-10 Prove that the most efficient cross sections for a given flow area are as follows:

1. Triangular section : vertex angle $= 90°$;
2. Trapezoidal section : half hexagon.

4-11 Prove that a semi-circle with its center at the middle of the water surface is the most efficient cross section.

4-12 For flow in a pipe flowing partially full, analytically prove that

$$\frac{Q_p}{Q_f} = \frac{\left(\theta - \frac{1}{2}\sin 2\theta\right)^{5/3}}{\pi\theta^{2/3}}$$

In this equation, the subscripts p and f refer to the pipe flowing partially full and full, respectively; $D_o =$ pipe diameter, and $y_n =$ normal depth. $S_o =$ the slope of the pipe bottom if the pipe is flowing partially full, and it is equal to the slope of the energy-grade line for full pipe flow. For a partially full pipe,

$$\theta = \cos^{-1}\left(1 - \frac{2y_n}{D_o}\right)$$

Experimental data for flow in partially full pipes may be approximated by the following equation [Christensen, 1984b]:

$$\frac{Q_p}{Q_f} = 0.46 - 0.5\cos\left(\frac{\pi y_n}{D_o}\right) + 0.04\cos\left(\frac{2\pi y_n}{D_o}\right)$$

For different values of y_n/D_o, compute Q_p/Q_f using the preceding expression derived analytically and from that based on the experimental results. Plot and compare these results.

Several authors state that the difference between these results is due to the variation of n with the flow depth. Do you agree with this explanation? If so, give your reasons.

Assuming that n varies with depth, compute and plot n_p/n_f with respect to y/D_o.

4-13 Compute the normal depth in an unlined tunnel having a standard horse-shoe section (Fig. 4-9), $d_o = 25$ m, carrying a flow of 800 m^3/s, and having a bottom slope of 0.0005. Assume n for the flow surface is 0.03.

4-14 Figure 4-10 shows the longitudinal profile of the Roman Aqueduct of Nimes [Hauck and Novak, 1987]. Assuming Manning n of 0.0125 and the channel width of 1.2 m for each segment of the aqueduct, compute the normal depth in different segments for flows of 210, 350, and 450 l/s.

4-15 The cross section of a drainage channel may be approximated as a trapezoidal section with the bottom width of 15 ft and side slope of 1.5H : 1V. If the channel drops 2.5 ft/mile, compute the flow depth for a flow of 150 ft^3/sec. Assume $n = 0.024$.

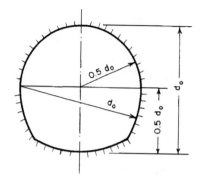

Fig. 4-9 Channel cross section for Prob. 4-13

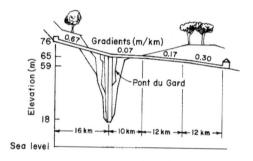

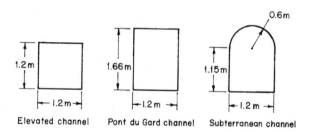

Fig. 4-10 **Roman Aqueduct of Nimes** (After Hauck and Novak [1987])

4-16 For the creek of Prob. 3-12, determine the flow depth if the bottom slope is 10 ft/mile and $n = 0.014$.

4-17 An 8-ft diameter concrete-lined sewer is laid at a bottom slope of 1 ft/mile. Find the flow depth for a flow of 30 ft^3/sec.

4-18 Determine the normal depth for the culvert of Problem 3-15 for a bottom slope of 2 ft/mile and $n = 0.014$.

4-19 The flow depth in a long trapezoidal channel (bottom width = 8 m, side slopes 1:1) for a flow of 28 m³/s is 3 m. The channel bottom slope is 0.0001. Determine the flow depth if the rate of discharge is doubled.

4-20 The flow depth at a section in a long rectangular channel changes from 4 ft to 5 ft. Determine the per cent change in the rate of discharge.

4-21 A boulder-lined drainage channel overflowed its banks during a spring runoff for flows exceeding 20 cfs. The channel is 15 ft wide, is rectangular in shape, and drops 10 ft/mile. If you are the design engineer, what will be your options to prevent flooding for this flow?

4-22 Is the flow subcritical or supercritical in a 4-m wide rectangular channel for a discharge of 9 m³/s. The bottom slope is 0.005 and $n = 0.014$.

4-23 Compute the critical and normal depths in a trapezoidal channel (bottom width = 20 ft, side slopes = 1.5H : 1V) for a flow of 220 ft³/sec. The bottom slope is 0.00032 and $n = 0.022$. Is the flow subcritical or supercritical?

4-24 The flow depth for a discharge of 15 m³/s in a long canal having a trapezoidal cross section (bottom width = 10 m; side slopes = 2H : 1V) is 2 m. If the discharge is increased to 20 m³/s, what will be the flow depth?

4-25 For a depth of 5 m in the symmetrical compound section shown in Fig. 4-11, determine:

1. Equivalent n_e;
2. Velocity-head coefficient, α;
3. Slope of the energy grade line, S_f.

Assume the flood plains and the main channel have the same bottom slope of 0.001, and Manning n for the main channel and for the floodplain are 0.021 and 0.039 respectively.

4-26 Compute the discharge in a 12-ft wide rock channel (n = 0.035) having a bottom slope of 0.001 and flow depth of 3 ft. What is the critical depth at this flow? Is the flow critical, subcritical, or supercritical?

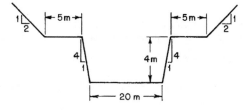

Fig. 4-11   Compound section

4-27 The crest of a 10-m wide long rectangular chute spillway is at El. 120 m. The water level upstream of the spillway is at El. 123 m. The bottom slope of the chute is 1 in 400 and Manning n is 0.013.

   i. Determine the discharge in the channel and the entrance flow depth;
   ii. Compute the flow depth at the downstream end of the chute.

4-28 A trapezoidal channel (bottom width = 5 m; side slopes = 2H : 1V) is carrying a flow of 10 m^3/s at a depth of 4 m. Determine the flow depth if the discharge is increased to 12 m^3/s.

4-29 Compute the critical and normal depths in an unlined horseshoe tunnel with a bottom slope of 0.001 and a concrete-lined invert. The design flow in m^3/s and the diameter D in m may be determined from the following expressions:

   Design discharge = 50 x;
   $D = 5x$ if $x \leq 5$; and
   $D = 2x$ if $x > 5$.

where x = (last digit of your social security number) +1 . Those who do not have social security number may use the last digit of the day of their date of birth.

4-30 The flow depth in a long trapezoidal channel (bottom width = 5 m; side slopes = 2H : 1V) for a flow of 5 m^3/s is 2 m. If the flow is doubled, what will be the flow depth?

4-31 The flow in a long trapezoidal channel (bottom width = 10 m; side slopes = 1.5H : 1V; bottom slope = 0.0002) is 160 m^3/s and the flow depth is 4 m. Determine the flow depth if the rate of discharge is reduced to 80 m^3/s.

4-32 Compute and plot the normal depth for the channel of Problem of 3-24 if the bottom slope is 0.000$||k||$; ($||k||$ means, replace the value of k (i.e., for k = 4, bottom slope is 0.0004) and Manning $n = 0.0||k||$. Start with the current year and compute until you are 25 years old.

References

Akan, A. O., 2006, *Open Channel Hydraulics,* Elsevier, Oxford, UK.
Anonymous, 1963, "Friction Factors in Open Channels," Task Force Report, *Jour. Hydraulics Division,* Amer. Soc. of Civ. Engs., vol. 89, no HY2, March, pp 97–143.
Barnes, H. H., 1967, *Roughness Characteristics of Natural Channels,"* U.S. Geological Survey, Water Supply Paper No. 1849, U. S. Government Press.
Berlamont, J. E., and Vanderstappen, N., 1981, "Unstable Turbulent Flow in Open Channels," *Jour. Hyd. Div.,* Amer. Soc. Civ. Engrs., vol. 107, no. 4, pp. 427–449.

Blodgett J. C. and McConaughy C. E., 1986, "Rock Riprap Design for Protection of Stream Channels near Highway Structures." U.S. Geological Survey, Prepared in Cooperation with the Federal Highway Administration, Sacramento, CA, Volume 2, Report No. 86–4128.

Chapra, S. C. and Canale, R. P., 2006, *Numerical Methods for Engineers*, 5th ed., McGraw-Hill, New York, NY.

Chen Y. H. and Cotton G. K., 1988, Design of Roadside Channels with Flexible Linings, Hydraulic Engineering Circular No. 15, Publication No. FHWA-IP-87-7, US Department of Transportation, Federal Highway Administration, McLean, VA.

Chow, V. T., 1959, *Open-Channel Hydraulics*, McGraw-Hill Book Co., New York, NY.

Christensen, B. A., 1984a, Discussion of "Flow Velocities in Pipelines," *Jour. Hyd. Engineering*, Amer. Soc. of Civ. Engrs., vol 110, no. 10, pp. 1510–1512.

Christensen, B. A., 1984b, "Analysis of Partially Filled Circular Storm Sewers," *Proc., Hydraulics Div. Conference*, Amer. Soc. of Civil Engineers, Aug., pp. 163–167.

Einstein, H. A., and Banks, R. B., 1951, "Fluid Resistance of Composite Roughness," *Trans.*, Amer. Geophysical Union, vol. 31, no. 4, pp. 603–610.

Einstein, H. A., 1934, "Der hydraulische oder Profil-Radius," *Schweizerische Bauzeitung*, Zurich, vol. 103, no. 8, Feb. 24, pp. 89–91.

Forchheimer, P., 1930, *Hydraulik*, Teubner Verlag, Berlin, Germany, p. 146.

Hager W. H., 2001, *Wastewater Hydraulics: Theory and Practice*. Springer-Verlag, New York, NY.

Hauck, G. F., and Novak, R. A., 1987, "Interaction of Flow and Incrustation in the Roman Aqueduct of Nimes," *Jour. Hydraulic Engineering*, Amer. Soc. Civil Engrs., vol 113, no. 2, pp. 141–157.

Henderson, F. M., 1966, *Open Channel Flow*, MacMillan Publishing Co., New York, NY.

Hollinrake, P. G., 1987, 1988, 1989, "The Structure of Flow in Open Channels," Vols. 1–3, Research Reports SR96, 153, 209, Hydraulic Research Ltd., Wallingford, UK.

Horton, R. A., 1933, "Separate Roughness Coefficients for Channel Bottom and Sides," *Engineering News Record*, vol. 111, no. 22, Nov. 30, pp. 652–653.

Jaeger, C., 1961, *Engineering Fluid Mechanics*, Blackie and Sons, London, UK.

Kouwen, N., Li, R. M., and Simons, D. B. 1980, "Velocity Measurements in a Channel Lined with Flexible Plastic Roughness Elements." Technical Report No. CER79-80-RML-DBS-11, Department of Civil Engineering, Colorado State University, Fort Collins, CO.

Kouwen N., Unny T.E., and Hill H. M. 1969, "Flow Retardance in Vegetated Channel." *Jour. Irrigation and Drainage Div.*, Amer. Soc. Civ. Engrs., vol. 95, no. 2, pp. 329–344.

Krishnamurthy, M., and Christensen, B. A., 1972, "Equivalent Roughness for Shallow Channels," *Jour. Hydraulics Div.*, Amer. Soc. Civil Engrs., vol. 98, no. 12, pp. 2257–2263.

Lotter, G. K., 1933, "Considerations on Hydraulic Design of Channels with Different Roughness of Walls," *Trans. All Union Scientific Research,* Institute of Hydraulic Engineering, vol. 9, Leningrad, Russia, pp. 238–241.

McCracken, D. D. and Dorn, W. S. 1964, *Numerical methods and FORTRAN programming,* John Wiley, New York, NY.

Maynord S. T., 1991, "Flow Resistance of Riprap," *Jour. Hydraulic Engineering,* Amer. Soc. Civil Engrs., vol. 117, no. 6, 687–695.

McWhorte J. C., Carpenter T. G., and Clark, R. N., 1968, "Erosion Control Criteria for Drainage Channels," Conducted for Mississippi State Highway Department in Cooperation with U.S. Department of Transportation, Federal Highway Administration, by the Agricultural and Biological Engineering Department, Agricultural Experiment Station, Mississippi State University, State College, MS.

Motayed, A. K., and Krishnamurthy, M., 1980, "Composite Roughness of Natural Channels," *Jour. Hydraulics Div.,* Amer. Soc. Civil Engrs., vol. 106, no. 6, pp. 1111–1116.

Muhlhofer, L., 1933, "Rauhigkeitsuntersuchungen in einem Stollen mit betonierter Sohle und unverkleideten Wanden," *Wasserkraft und Wasserwirtschaft,* vol. 28, no. 8, pp. 85–88.

Myers, W. R. C., 1978, "Momentum Transfer in a Compound Channel," *Jour. Hydraulic Research,* Inter. Assoc. Hydraulic Research, vol. 16, no. 2, pp. 139–150.

Nikuradse, J., 1932, "Gesetzmassigkeit der turbulenten Stromung in glatten Rohren," *Forschung Arb. Ing-Wesen,* vol. I, Heft 356, Berlin, Germany.

Rajaratnam, N., and Ahmadi, R. M., 1979, "Interaction Between Main Channel and Fllod-Plain Flows," *Jour. Hydraulics Div.,* Amer. Soc. Civil Engrs., vol. 105, no. 5, pp. 573–588.

Ramser, C. E., 1929, "Flow of Water in Drainage Channels," *Technical Bulletin No. 129,* U.S. Department of Agriculture, November.

Rouse, H., and Ince, S., 1963, *History of Hydraulics,* Dover Publications, New York, N.Y.

Scobey, F. C., 1939, "Flow of Water in Irrigation and Similar Canals," *Technical Bulletin No. 652,* U.S. Department of Agriculture, February.

Soil Conservation Service. (1954). *Handbook of Channel Design for Solid and Water Conservation,* SCS-TP-61. Stillwater, OK.

Sturm, T. W. and King, D. A., 1988, "Shape Effects on Flow Resistance in Horseshoe Conduits," *Jour. Hydraulic Engineering,* Amer. Soc. Civil Engrs., vol. 114, no. 11, pp. 1416–1429.

Sturm T. W., 2001, *Open Channel Hydraulics,* McGraw-Hill Book Co. New York, NY.

Thompson, G. T. and Roberson, J. A., 1976, "Theory of Flow Resistance for Vegetated Channels," *Trans.,* Amer. Soc. of Agric. Engrs. (Gen. Ed.), vol. 19, no. 2, pp. 288–293.

Vanoni, V. A., 1941, "Velocity Distribution in Open Channels," *Civil Engineering*, vol. 11, pp. 356–357.

GRADUALLY VARIED FLOW

Permeable Groynes and navigations masts and buoys along 2 km My Thuan reach of the Mekong River, with typical flow depth of 20 m (Courtesy, Eric J Lesleighter)

5-1 Introduction

In Chapter 4, we discussed uniform flow in which the flow depth remains constant with distance. Such flows occur only in long and prismatic channels (i.e., the channel cross section and bottom slope do not change with distance). In real-life projects, however, channel cross sections and bottom slopes are not constant with distance in natural channels and these are varied in constructed channels to suit the existing topographical conditions for economic reasons. In addition, hydraulic structures are provided for flow control. These changes in the channel geometry produce nonuniform flows while changing from one uniform-flow condition to another. As we discussed in Chapter 1, such flows are called *gradually varied flows* if the rate of variation of depth with respect to distance is small, and *rapidly varied flows* if the rate of variation is large. In other words, the flow depth changes gradually over a long distance in gradually varied flows and in a short distance in rapidly varied flows. Since the analysis of gradually varied flows is usually done for long channels, the friction losses due to boundary shear have to be included. These losses, however, may be neglected in the analysis of rapidly varied flows because the distances involved are short. In addition, the pressure distribution in gradually varied flow may be assumed hydrostatic because the streamlines are more or less straight and parallel. However, this is not the case in rapidly varied flows where significant acceleration normal to flow direction may be produced by sharp curvatures in the streamlines.

Steady, gradually varied flow is discussed in this chapter and the rapidly varied flow in Chapter 7. The gradually varied flow equations are first derived. The classification of various water surface profiles is then presented. This is followed by a presentation of procedures for qualitatively sketching the water-surface profiles and for determining the discharge from a reservoir. The water-surface profiles in compound channels are then discussed.

5-2 Governing Equation

The gradually varied flow equations in a prismatic channel having no lateral inflow or outflow are derived in this section by making the following simplifying *assumptions:*

1. The slope of the channel bottom is small.*
2. The channel is prismatic channel and there is no lateral inflow or outflow from the channel.
3. The pressure distribution is hydrostatic at all channel sections.

---

* For clarity of presentation, an exaggerated vertical scale is used in the illustrations of this chapter. Thus, the slope of the channel bottom, even though small, may appear to be large in these illustrations.

4. The head losses in gradually varied flow may be determined by using the equations for head losses in uniform flows.

These assumptions are usually valid for gradually varied flows. A channel with changing cross section or bottom slope may be divided into piecewise prismatic channels. The slope of the channel bottom may be assumed small if it is less than 5 percent. In such a case, $\sin\theta \simeq \tan\theta \simeq \theta$, in which θ = angle of the channel bottom with horizontal, and the flow depths measured vertically or normal to the bottom are approximately the same. The curvature of the streamlines in gradually varied flows is usually small and thus the assumption of hydrostatic pressure distribution is valid. The water-surface profiles measured during hydraulic model investigations and during field observations compare satisfactorily with those computed by using the head-loss equations for steady uniform flow.

By referring to Fig. 5-1, the total head at a channel section may be written as

$$H = z + y + \frac{\alpha V^2}{2g} \tag{5-1}$$

in which H = elevation of the energy grade line above the datum; z = elevation of the channel bottom; y = flow depth; V = mean flow velocity, and α = velocity-head coefficient. Let us consider distance, x, as positive in the down-

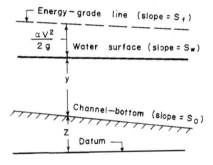

Fig. 5-1   Definition sketch

stream flow direction. By differentiating both sides of Eq. 5-1 with respect to x, and expressing V in terms of discharge, Q, we obtain

$$\frac{dH}{dx} = \frac{dz}{dx} + \frac{dy}{dx} + \frac{\alpha Q^2}{2g}\frac{d}{dx}\left(\frac{1}{A^2}\right) \tag{5-2}$$

Now, by definition

$$\frac{dH}{dx} = -S_f$$

$$\frac{dz}{dx} = -S_o \tag{5-3}$$

in which S_f = slope of the energy-grade line and S_o = slope of the channel bottom. There is a negative sign with S_f and S_o since both H and z decrease as x increases. Now,

$$
\begin{aligned}
\frac{d}{dx}\left(\frac{1}{A^2}\right) &= \frac{d}{dA}\left(\frac{1}{A^2}\right)\frac{dA}{dx} \\
&= \frac{d}{dA}\left(\frac{1}{A^2}\right)\frac{dA}{dy}\frac{dy}{dx} \\
&= -\frac{2B}{A^3}\frac{dy}{dx}
\end{aligned}
\tag{5-4}
$$

since $dA/dy = B$, as discussed in Chapter 2. Note that if the channel is not prismatic, then

$$
\frac{dA}{dx} = \frac{\partial A}{\partial x} + \frac{\partial A}{\partial y}\frac{dy}{dx}
$$

and Eqs. 5-4 and 5-5 are modified accordingly (see Problem 5-9).

By substituting Eqs. 5-3 and 5-4 into Eq. 5-2, and rearranging the resulting equation, we obtain

$$
\frac{dy}{dx} = \frac{S_o - S_f}{1 - (\alpha B Q^2)/(g A^3)}
\tag{5-5}
$$

This equation describes the rate of variation of y with x. By utilizing the expression for Froude number, $\mathbf{F}_r$, derived in Chapter 3, the second term in the denominator may be written as

$$
\begin{aligned}
\frac{\alpha B Q^2}{g A^3} &= \frac{(Q/A)^2}{(gA)/(\alpha B)} \\
&= \mathbf{F}_r^2
\end{aligned}
\tag{5-6}
$$

Hence, Eq. 5-5 becomes

$$
\frac{dy}{dx} = \frac{S_o - S_f}{1 - \mathbf{F}_r^2}
\tag{5-7}
$$

We will use this equation in the following sections to draw qualitative conclusions about the water-surface profiles.

5-3 Classification of Water-Surface Profiles

We use the following notation to designate different water surface profiles: A letter refers to the type of the channel bottom slope and a numeral to the relative position of the profile with respect to the critical-depth line (hereinafter called CDL) and the normal-depth line (hereinafter called NDL). The critical depth and the normal depth are y_c and y_n, respectively.

Channel-bottom slopes are classified into the following five categories: mild, steep, critical, horizontal (zero slope) and adverse (negative slope). The first letter of these names refers to the type, i.e., M for mild, S for steep, C for critical, H for horizontal and A for adverse slope.

The bottom slope of a channel is called as *mild* slope if the uniform flow is subcritical (i.e., $y_n > y_c$); for the specified discharge and Manning n; it is *critical* slope if the uniform flow is critical (i.e., $y_n = y_c$); and it is *steep slope* if the uniform flow is supercritical (i.e., $y_n < y_c$). It is apparent that the normal depth is infinite if the bottom slope is horizontal and it is nonexistent if the bottom slope is negative. To summarize, the channel bottom slope is called

- mild if $y_n > y_c$;
- steep if $y_n < y_c$; and
- critical if $y_n = y_c$.

Now, let us discuss how to designate the relative position of the surface profile. For the mild and steep slopes, the normal-depth and critical-depth lines divide the space above the channel bottom into three regions, as shown in Fig. 5-2. However, for the adverse, horizontal, and critical bottom slopes, there are only two regions since the normal depth does not exist, is infinite, or is the same as the critical depth, respectively. The region above both lines is designated as *Zone 1*; that between the upper and lower lines is designated as *Zone 2*, and the one between the lower line and the channel bottom is designated as *Zone 3*. Note that the upper line is the normal-depth line if the channel bottom slope is mild, and the upper line is the critical-depth line if the bottom slope is steep.

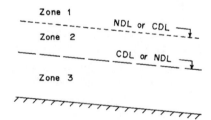

Fig. 5-2 Zones for classification of surface profiles

Thus, we have 13 different types of surface profiles: three for the mild slope, three for the steep slope, two for the critical slope (zone 2 does not exist since $y_n = y_c$ and we do not consider the critical-depth line as a surface profile); two for the horizontal slope (zone 1 does not exist since $y_n = \infty$), and two for the adverse slope (there is no zone 1, since y_n does not exist).

Figure 5-3 shows different zones and profiles for all five types of bottom slopes.

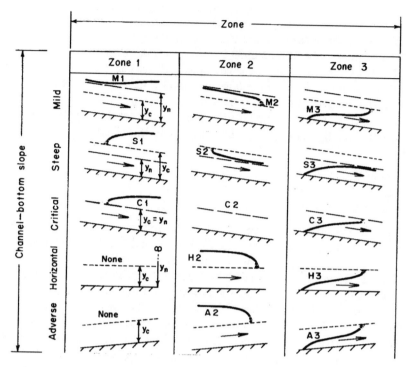

Fig. 5-3   **Water surface profiles**

5-4 General Remarks

By determining the sign of the numerator and the denominator of Eq. 5-7, we can make qualitative observations about various types of water-surface profiles. These observations allow us to sketch the profile without doing any detailed calculations. For example, they indicate whether the depth increases or decreases with distance, how the profiles end at the upstream and at the downstream limits, etc. First, let us make a few general remarks and then consider specific cases for illustration purposes.

The flow depth, y, increases with distance if dy/dx is positive and it decreases if dy/dx is negative. Thus, by determining the signs of the numerator and denominator of Eq. 5-7, we can say whether the flow depth for a particular profile increases or decreases with distance.

We discussed in Chapter 4 that the energy-grade line, water surface, and channel bottom are parallel to each other in uniform flow; i.e., $S_f = S_w = S_o$ when $y = y_n$. Therefore, it is clear from the Manning or Chezy equation that for specified discharge, Q,

$$S_f > S_o \text{ if } y < y_n. \tag{5-8}$$

and

$$S_f < S_o \text{ if } y > y_n \qquad (5\text{-}9)$$

By using these two inequalities, we determine the sign of the numerator of Eq. 5-7 and whether the flow is subcritical ($\mathbf{F}_r < 1$) or supercritical ($\mathbf{F}_r > 1$), we determine the sign of the denominator of Eq. 5-7.

Now, let us discuss how the surface profiles approach the normal and critical depths and the channel bottom.

As $y \to y_n$ (reads as y tends to y_n), $S_f \to S_o$. Therefore, it follows from Eq. 5-7 that $dy/dx \to 0$ provided $\mathbf{F}_r \neq 1$ (i.e., flow is not critical). In other words, the surface profile approaches the normal-depth line asymptotically.

As $y \to y_c$, $\mathbf{F}_r \to 1$ and the denominator of Eq. 5-7 tends to zero. Therefore, dy/dx tends to ∞ provided $S_f \neq S_o$. Thus, the water-surface profile approaches the critical-depth line vertically. Since a vertical water surface, is physically impossible, we may assume the water surface profile approaches the critical-depth line at a very steep slope. Therefore, the question arises as to why this conclusion about the vertical water surface derived theoretically is not realized in the real world. The reason for this discrepancy is that as soon as the water surface has a sharp curvature, the pressure distribution is not hydrostatic. Therefore, Eq. 5-7 is not valid, and any conclusions we draw from this equation become questionable. As we discussed in the previous chapters, a hydraulic jump is formed when the flow changes from supercritical to subcritical. In a hydraulic jump, the flow surface has a steep gradient since it passes through the critical depth line.

As $y \to \infty$, $V \to 0$, and consequently both $\mathbf{F}_r$ and S_f tend to zero. Hence, it follows from Eq. 5-7 that $dy/dx \to S_o$ for very large values of y. Since we are assuming that S_o is small, we may say that the water surface profile almost becomes horizontal as y becomes large.

Now, let us discuss what happens when the water surface approaches the channel bottom, i.e., $y \to 0$. From the Chezy equation, it follows that

$$S_f = \frac{Q^2}{C^2 A^2 R} \qquad (5\text{-}10)$$

in which $C = $ Chezy constant, and $R = $ hydraulic radius. For a very wide rectangular channel, $R \simeq y$. By substituting Eq. 5-10 into Eq. 5-5, replacing R by y, and simplifying the resulting equation, we obtain

$$\frac{dy}{dx} = \frac{gB(S_o C^2 B^2 y^3 - Q^2)}{C^2(gBy^3 - \alpha BQ^2)} \qquad (5\text{-}11)$$

It follows from this equation that as $y \to 0$,

$$\lim_{y \to 0} \frac{dy}{dx} = \frac{g}{\alpha C^2} \qquad (5\text{-}12)$$

Therefore, as $y \to 0$, the slope of the water surface profile is finite, has a positive value, and is a function of the Chezy constant, C, and the velocity-head coefficient, α.

However, if we use Manning equation instead of the Chezy equation, we find that $dy/dx \to \infty$ as $y \to 0$. This is left as an exercise for the reader to prove (see Problem 5-5).

To illustrate the application of the above general remarks, let us consider water surface profiles in a channel with mild slope. As we discussed above, $y_n > y_c$ if the slope is mild. Therefore, the flow depth, y, in the three zones is classified as follows:

- Zone 1: $y > y_n > y_c$;
- Zone 2: $y_n > y > y_c$; and
- Zone 3: $y_n > y_c > y$.

The qualitative characteristics of the water-surface profiles in each zone may be studied as follows.

Zone 1 (M1 Profile)

Since $y > y_n$ in Zone 1, $S_f < S_o$. Therefore, the numerator of Eq. 5-7 is positive. Similarly, $\mathbf{F}_r < 1$ since $y > y_c$. Therefore, the denominator of Eq. 5-7 is positive as well. Hence, it follows from Eq. 5-7 that

$$\frac{dy}{dx} = \frac{S_o - S_f}{1 - \mathbf{F}_r^2} = \frac{+}{+} = +$$

This means that y increases with distance x. As discussed previously, $y \to y_n$ asymptotically in the upstream direction and the water surface becomes almost horizontal as y becomes large in the downstream direction.

Zone 2 (M2 Profile)

In this case, $S_f > S_o$ since, $y < y_n$. Therefore, the numerator of Eq. 5-7 is negative. However, the denominator is positive, since $\mathbf{F}_r < 1$ because $y > y_c$. Hence, it follows from Eq. 5-7 that

$$\frac{dy}{dx} = \frac{S_o - S_f}{1 - \mathbf{F}_r^2} = \frac{-}{+} = -$$

Thus, y decreases as x increases. As discussed previously, $y \to y_n$ asymptotically; and $y \to y_c$ almost vertically.

Zone 3 (M3 Profile)

In Zone 3, $S_f > S_o$ since $y < y_n$. Therefore, the numerator of Eq. 5-7 is negative. The denominator is negative as well, since $\mathbf{F}_r > 1$ because $y < y_c$. Hence, it follows from Eq. 5-7 that

$$\frac{dy}{dx} = \frac{S_o - S_f}{1 - \mathbf{F}_r^2} = \frac{-}{-} = +$$

Thus, y increases as x increases.

As discussed previously, $y \to y_c$ almost vertically while the water surface profile approaches the channel bottom at a finite positive slope.

By using the preceding qualitative conclusions, the water-surface profiles in each region may be sketched as shown in Fig. 5-3. Note that the profiles are shown by dashed lines as they approach the critical-depth line and as they approach the channel bottom to indicate uncertainty in their shapes.

The qualitative characteristics of water-surface profiles for other types of channel bottom slopes may be studied in a similar manner. In general, the procedure is as follows: We first determine the signs of the numerator and of the denominator of Eq. 5-7, and hence determine the sign of dy/dx. Then, by utilizing the qualitative remarks made in the previous paragraphs, we sketch the water-surface profiles as they approach the normal- and critical-depth lines and the channel bottom.

The characteristics and shapes of various water-surface profiles and the situations in which they may occur in real life are presented in Fig. 5-4.

Note that H1 and A1 profiles do not exist since there is no Zone 1 in both cases. In addition, profile C2 actually represents uniform flow rather than gradually varied flow.

5-5 Sketching of Water-Surface Profiles

Any channel section at which there is a unique relationship between the flow depth and discharge is referred to as a *control*. The properties of surface profiles we discussed in the previous two sections are for a prismatic channel having a control section either at the upstream or at the downstream end. In real life, however, a channel system may have several control sections. In addition, a channel system having variable cross section or bottom slope may be divided into several prismatic channels. To qualitatively sketch the profiles in these cases, a number of guidelines are outlined in the following paragraphs and two examples are included for illustrative purposes.

Divide the channel system into prismatic channels and for the specified discharge, roughness coefficient, and channel cross section, compute the normal and critical depths in each channel. Now, using an exaggerated vertical scale, plot the channel bottom and the normal- and critical-depth lines. Then, on this diagram, mark the locations of controls – i.e., the locations where the water-surface profile passes through critical depth ($y = y_c$) and identify the channel reaches where the flow is expected to be uniform ($y = y_n$).

A downstream control governs if the flow is subcritical and an upstream control governs if the flow is supercritical. It is possible to have situations where part of the channel is governed by the upstream control and part of the channel is governed by the downstream control. In addition, a control at an intermediate location (e.g., a weir, sluice gate, spillway) may act as a control for both the upstream and downstream directions from the control location.

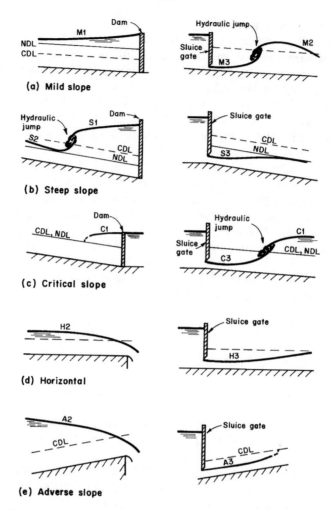

Fig. 5-4 Real-life cases of water-surface profiles (After Chow [1959])

At a channel entrance, the surface profile passes through the critical depth if the lake or reservoir level is higher than the critical-depth line and the channel bottom slope at the channel entrance is steep. To allow for the velocity head and losses at the channel entrance, the water surface at the upstream end of the channel may be slightly lower than the water level in the reservoir.

At a free overfall, the water surface passes through the critical-depth line approximately three to four times the critical depth upstream of the fall if the flow depth upstream of the fall is greater than the critical depth.

A hydraulic jump is formed whenever the flow changes from supercritical to subcritical flow. The exact location of the jump is determined by detailed calculations, as discussed in Chapter 8. However, an approximate location of the jump may be estimated by judgement while sketching the water-surface profiles.

The following two examples illustrate these procedures.

Example 5-1

Sketch the water-surface profile in the channels connecting the reservoirs, as shown in Fig. 5-5a. The bottom slope of channel 1 is steep and that of channel 2 is mild.

Solution:

Compute the critical and normal depths for each channel. Then plot the critical-depth line and the normal-depth line (marked as CDL and NDL, respectively in Fig. 5-5b).

The water depth at the channel entrance is equal to the critical depth, since the water level in the upstream reservoir is above the CDL of channel 1. Let us mark this water level at the channel entrance by a dot (5-5b). The water level at the downstream end of channel 2 is lower than the CDL. Therefore, the water surface passes through the CDL approximately three to four times the critical depth upstream of the entrance to the downstream reservoir. Let us again mark this water level at the downstream end by a dot, as shown in Fig. 5-5b.

In channel 1, the water surface at the entrance, after passing through the critical depth, tends to the normal depth. Thus, we have an S2 profile in channel 1. The flow decelerates downstream of the junction of channels 1 and 2 because of mild slope. Hence, the flow depth increases until a hydraulic jump is formed. The water surface follows the M2 profile downstream of the jump and the exact location of the jump is determined by detailed calculations as discussed in Chapter 8.

In the following example, a control gate is located at an intermediate location and depending upon the gate opening and the location of the other controls, several different profiles are possible.

Example 5-2

Sketch all possible water-surface profiles in the channel of Fig. 5-6. The channel is long and has a steep slope. Consider two different cases of gate opening.

Solution:

Several different water surface profiles are possible depending upon the location of the control sections as well as the gate opening. We may divide these profiles into the following two categories:

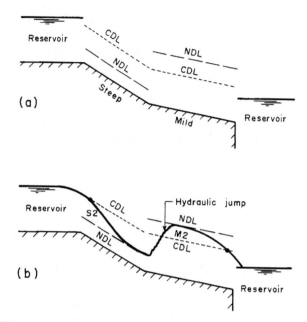

Fig. 5-5   Water surface profiles for Example 5-1

a. Control at the channel entrance;
b. Control at the gate.

Let us consider each of these cases one by one.

If the control is at the channel entrance, then the channel discharge is not controlled by the sluice gate. This is because we have supercritical flow in part of the channel length upstream of the sluice gate. Depending upon the gate opening, several different water-surface profiles are possible upstream of the gate, as shown in Fig. 5-6a and 5-6b. The flow depth approaches the NDL asymptotically, since the channel is long. Depending upon the gate opening, S1, S2, or S3 profiles are possible, as shown in Fig. 5-6.

The sluice gate controls the channel discharge only if the backwater from the gate extends to the channel entrance. In such a case, we have an S1 profile upstream of the gate in all cases (Fig. 5-6). However, downstream of the gate, the type of the surface profile depends on the gate opening, as shown in Fig. 5-7.

In this figure, a small drop in the water level at the channel entrance is shown to account for the entrance losses and the velocity head.

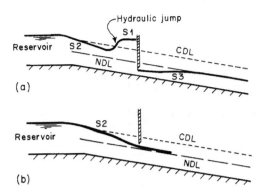

Fig. 5-6   Water surface profiles for control at channel entrance

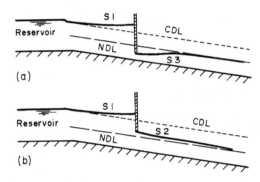

Fig. 5-7   Water surface profiles for control at gate

5-6 Discharge From a Reservoir

In the discussion for sketching the water surface profiles in the previous sections, we assumed that the channel discharge is known. However, this may not always be the case, as illustrated in the following example.

Let us consider a channel-reservoir system as shown in Fig. 5-8. The channel cross section, entrance loss coefficient, k, Manning n, and channel bottom slope, S_o, are specified. We want to determine the flow depth, y, and the discharge, Q, in the channel. The reservoir is large so that the flow velocity in the reservoir approaches zero. In addition, the reservoir water level is known and remains constant independent of the discharge in the channel.

Referring to Fig. 5-8, H_o, S_o, n, and the properties of the channel section are known and we want to determine y and Q.

For the specified flow variables and channel parameters, the channel bottom slope may be classified as

Steep;
Critical; or

Mild.

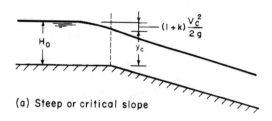

(a) Steep or critical slope

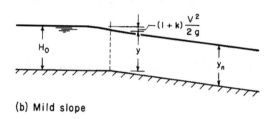

(b) Mild slope

Fig. 5-8   **Discharge from a reservoir**

The flow depth at the channel entrance is critical if the bottom slope is critical or steep and the reservoir water level is higher than the critical-depth line. However, normal depth occurs just downstream of the channel entrance if the bottom slope is mild.

To determine the type of bottom slope, we first determine the critical slope, S_c. If the flow depth at the channel entrance is critical and $\alpha = 1$, we can write

$$\frac{Q^2}{2gA_c^2} = \frac{D_c}{2} \tag{5-13}$$

and

$$H_o = y_c + (1+k)\frac{Q^2}{2gA_c^2} \tag{5-14}$$

in which k = entrance loss coefficient and both D_c = hydraulic depth and A_c = flow area corresponding to y_c. We solve these two equations for Q and y_c. If the slope of the channel bottom is equal to the critical slope, S_c, then the flow at this depth and discharge will be uniform. By utilizing this fact, we determine the value of S_c from the Manning equation

$$Q = \frac{1}{n}AR^{2/3}S_c^{\frac{1}{2}} \tag{5-15}$$

The channel bottom slope is critical if $S_o = S_c$; it is steep if $S_o > S_c$; and, it is mild if $S_o < S_c$.

The discharge and the flow depth we determined above are correct if the bottom slope is critical; while only the computed discharge is correct if the bottom slope is steep. The flow depth may be computed starting with the critical depth at the entrance. However, if the bottom slope is mild, then we solve the following two equations simultaneously to determine y and Q :

$$Q = \frac{1}{n}AR^{2/3}S_o^{\frac{1}{2}} \tag{5-16}$$

and

$$H_o = y + \frac{V^2}{2g} + k\frac{V^2}{2g}$$
$$= y + \frac{1+k}{2g}\left(\frac{Q}{A}\right)^2 \tag{5-17}$$

Eliminating Q from Eqs. 5-16 and 5-17, we obtain

$$H_o = y + \frac{1+k}{2gn^2}R^{\frac{4}{3}}S_o \tag{5-18}$$

Solution of this equation gives the flow depth in the channel. The discharge corresponding to this depth can now be determined from Eq. 5-16.

The following example illustrates this procedure.

Example 5-3

A 10-m wide, rectangular, concrete-lined channel (n = 0.013) has a bottom slope of 0.01 and a constant-level reservoir at the upstream end. The reservoir water level is 6.0 m above the channel bottom at entrance. Assuming the entrance losses and the approach velocity in the reservoir to be negligible, determine the channel discharge and qualitatively sketch the water surface profile.

Given:

$n = 0.013$;
$S_o = 0.01$;
$B = 10.0$ m;
$H_o = 6.0$ m;
Entrance losses are negligible.

Determine:

$Q = ?$
Water-surface profile.

Solution:

Let us assume the control is at the channel entrance, i.e., the bottom slope is steep or critical. Then,

$$y_c = \frac{2}{3}H_o$$
$$= \frac{2}{3} \times 6.$$
$$= 4.\,\mathrm{m.}$$

For critical flow, unit discharge,

$$q = \sqrt{gy_c^3}$$
$$= \sqrt{9.81(4.)^3}$$
$$= 25.06\,\mathrm{m^3/s/m.}$$

Hence,

$$Q = Bq$$
$$= 10 \times 25.06$$
$$= 250.6\,\mathrm{m^3/s}$$

Let us now determine the critical slope, S_c. This is the bottom slope for which we will have critical flow in the channel for $Q = 250.6\,\mathrm{m^3/s}$. Now, the Manning equation may be written as

$$Q = \frac{1}{n}AR^{\frac{2}{3}}S_c^{\frac{1}{2}}$$

or

$$S_c = \frac{n^2Q^2}{A^2R^{\frac{4}{3}}}$$
$$= \frac{(0.013)^2(250.6)^2}{(10 \times 4)^2[40/(10+8)]^{\frac{4}{3}}}$$
$$= 0.00229$$

Since, $S_c < S_o$, the slope of the channel bottom is steep and the channel discharge is 250.6 m^3/s.

To sketch the water-surface profile, we first determine the normal depth. by using any of the procedures we presented in Chapter 4. The flow area, A, and the hydraulic radius, R, corresponding to the normal depth satisfy the following equation

$$AR^{\frac{2}{3}} = \frac{nQ}{\sqrt{S_o}}$$

The substitution of the values of n, Q, and S_o and the expressions for A and R in terms of y_n into this equation gives

$$10y_n \left[\frac{10y_n}{10 + 2y_n} \right]^{\frac{2}{3}} = \frac{0.013 \times 250.6}{\sqrt{.01}} = 32.57$$

The solution of this equation by trial and error yields

$$y_n = 2.37\text{m}$$

The entrance flow depth will be critical and it will approach the normal depth asymptotically, as shown in Fig. 5-9.

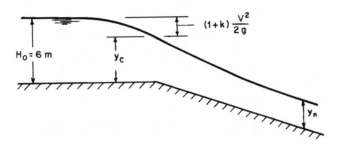

Fig. 5-9   **Water surface profile for Example 5-3**

5-7 Profiles in Compound Channels

The discussion of water-surface profiles in the previous sections is for channels having only simple cross sections. However, water-surface profiles in compound channels (channels having a compound cross section) need special treatment because there may be more than one critical depth, as we discussed in Chapter 3. Quintela [1982] briefly discussed the shapes of flow profiles in a compound channel for two different cases of steep and mild slopes. He illustrated the occurrence of rapidly-varied flow when the slope of the channel bottom changes from mild to steep. He tacitly assumed that the normal depth, y_n, is greater than the highest critical depth, y_{c_3}, if the slope is mild and that y_n is less than the lowest critical depth, y_{c_1}, if the slope is steep. However, another situation is possible when $y_{c_1} < y_n < y_{c_3}$, (Chaudhry and Bhallamudi [1988]). In this case, critical depth, y_{c_2} becomes very important (Fig. 5-10).

In this section, two examples are presented to show how multiple critical depths affect the water-surface profile in a compound channel. First, we consider a long channel with a free overfall at the downstream end and then a long channel with a reservoir at the upstream end.

Example 5-4

A long channel has a compound cross section (Fig. 5-11) and a free overfall at the downstream end. The channel discharge is 2.5 m³/s, and the Manning n for the main channel and for the floodplain are 0.013 and 0.0144, respectively. Discuss and sketch the water surface profiles for the following four channel bottom slopes, $S_0 = 0.0094, 0.0049, 0.0029$ and 0.001.

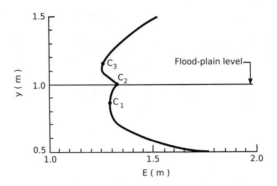

Fig. 5-10 Specific energy versus depth curves for compound channel

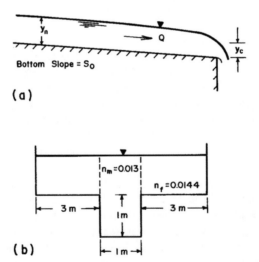

Fig. 5-11 Compound channel with a free overfall at outlet

Given:

$Q = 0.5$ m^3/s;
$n_m = 0.013$;
$n_f = 0.0144$;
Channel cross section, as shown in Fig. 5-11.

Determine:

Water-surface profiles for $S_o = 0.0094$, 0.0049, 0.0029, and 0.001.

Solution:

The specific energy and Froude number versus the flow depth for the channel cross section and for the specified discharges are plotted in Fig. 5-12. There are three critical depths: $y_{c_1} = 0.86$ m, $y_{c_2} = 1.002$ m and $y_{c_3} = 1.12$ m.

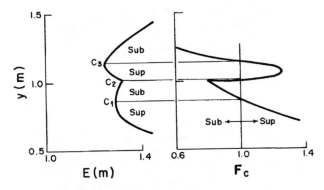

Fig. 5-12 E and F_c versus flow depth for channel of **Fig. 5-11**

Let us consider each of the channel bottom slopes one by one.

$S_o = 0.0094$

For this bottom slope, $y_n = 0.75$ m (determined by solving the Manning equation by trial and error). As shown in Fig. 5-12, $y_n < y_{c_1}$ and the Froude number is greater than 1. Therefore, the flow is supercritical, control is at the upstream end, and the free overfall at the downstream end does not affect the flow. Once the flow depth approaches the normal depth, it changes only slightly at the downstream end. The water-surface profile is shown in Fig. 5-13a.

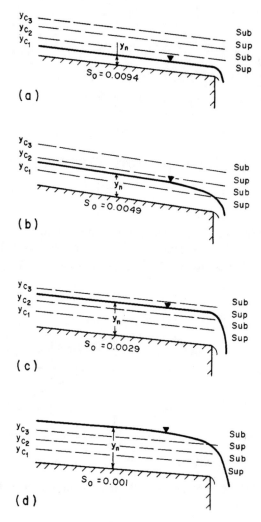

Fig. 5-13   **Water-surface profiles for channel of Fig. 5-11**

$S_o = 0.0049$

For this bottom slope $y_n = 0.97$ m. Thus, $y_{c_1} < y_n < y_{c_2}$ and $\mathbf{F}_{rc} < 1$. Therefore, the flow is subcritical (Fig. 5-12) and the control is at the downstream end. The water-surface profile may be computed starting at the downstream end with a depth equal to y_{c_1}. The water-surface profile is shown schematically in Fig. 5-13b.

$S_o = 0.0029$

For this slope, $y_n = 1.07$ m; i.e., $y_{c_2} < y_n < y_{c_3}$. Figure 5-12 shows that the Froude number for this normal depth is greater than 1. Therefore, the flow is supercritical; control is at the upstream end and the free overfall at the downstream end does not affect the flow in the channel. The water-surface profile is shown in Fig. 5-13c.

$S_o = 0.001$

For this slope, $y_n = 1.2$ m; i.e., $y_n > y_{c_3}$ and the flow is subcritical, as indicated by Fig. 5-12. The water-surface profile may be computed by starting at the downstream end with depth equal to y_{c_3}. The flow depth varies from normal depth, y_n at some upstream point to the critical depth, y_{c_3} near the downstream end. The flow varies rapidly from one critical depth, y_{c_3} to the other, y_{c_1} at the downstream end. The flow profile is shown in Fig. 5-13d.

Example 5-5

Sketch possible water-surface profiles in a long channel with a compound cross section and a reservoir at its upstream end, as shown in Fig. 5-14. The reservoir water surface is 1.2 m above the channel bottom at the channel entrance.

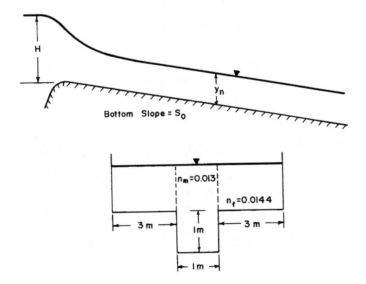

Fig. 5-14 Channel of Example 5-5

Given:

$H = 1.2$ m;
Channel cross section, as shown in Fig. 5-14.

Determine:

$Q = ?$
Water-surface profile.

Solution:

The method presented in the previous section to determine the channel discharge from a reservoir is followed. First, the channel bottom slope corresponding to the critical flow (critical slope) is calculated. If the actual channel-bottom slope is steeper than the critical slope, then the flow is supercritical. The flow is subcritical if the channel bottom slope is less than the critical slope. Since it is possible to have more than one critical slope for a compound channel, it is necessary to consider them in the analysis. The following discussion will illustrate this point.

Figure 5-15 shows the discharge and Froude number versus depth for the given reservoir level, $H = 1.2$ m. The discharge, Q, is calculated from the following energy equation

$$Q = \left[\frac{2(H - y)gA^2}{\alpha} \right]^{\frac{1}{2}}$$

(5-19)

in which y is the flow depth in the channel. The entrance losses and the velocity of approach are neglected in this equation.

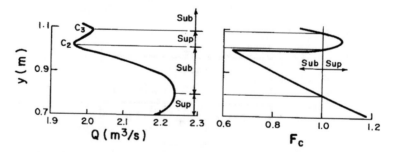

Fig. 5-15   Variation of Q and $\mathbf{F}_r$ with depth

Critical conditions occur when $\mathbf{F}_{rc} = 1$. It is clear from the discharge versus depth diagram (Fig. 5-15) that the discharge is not necessarily a local maximum at critical conditions. For example, it is actually minimum for y_{c_2}. As can be seen, there are three critical depths: $y_{c_1} = 0.8$ m, $y_{c_2} = 1.03$ m and $y_{c_3} = 1.10$ m. Corresponding to these critical depths, there are three critical discharges, $Q_{c_1} = 2.241$ m^3/s, $Q_{c_2} = 1.960$ m^3/s and $Q_{c_3} = 2.0$ m^3/s. Critical bottom slopes, S_c, corresponding to these critical discharges are determined from the Manning equation as $S_{c_1} = 0.0064$, $S_{c_2} = 0.0024$ and $S_{c_3} = 0.0015$, respectively.

Depending upon the channel bottom slope, S_o, the following four cases are possible.

$S_o > S_{c_1}$

This bottom slope is steeper than all three critical slopes and the flow in the channel is supercritical. Flow depth varies rapidly from the reservoir level to the critical depth, y_{c_3}, passes through the other two critical depths, y_{c_2} and y_{c_1} and approaches the normal depth as shown in Fig. 5-16a. The discharge in the channel is equal to the discharge corresponding to y_{c_1}, i.e., $Q = Q_{c_1} = 2.24$ m^3/s. The discharge is the same for all values of the bottom slope, $S_o > S_{c_1}$.

$S_{c_1} > S_o > S_{c_2}$

For this bottom slope, y_n is greater than y_{c_1} but less than y_{c_2}. Therefore, as indicated by Fig. 5-15, the Froude number is less than 1 and the flow is subcritical. The discharge in the channel depends on the channel bottom slope and may be determined by solving the energy equation simultaneously with the Manning equation for uniform flow. For $S_o = 0.0035$, the discharge is 2.08 m^3/s; the corresponding depth is 0.96 m and the flow profile is shown in Fig. 5-16b.

$S_{c_2} > S_o > S_{c_3}$

For this bottom slope, the normal depth, y_n is greater than y_{c_2} but less than y_{c_3}. Therefore, as indicated by Fig. 5-15, the Froude number is greater than 1 and the flow is supercritical. Flow depth passes through the critical depth y_{c_3} and then approaches the normal depth. The channel discharge is equal to the discharge corresponding to y_{c_3}, i.e., 2.0 m^3/s. The flow profile is shown in Fig. 5-16c.

$S_o < S_{c_3}$

In this case, the channel bottom slope is less than all critical slopes and the flow is subcritical. The channel discharge may be calculated as discussed previously for Case 2. The flow profile is shown in Fig. 5-16d.

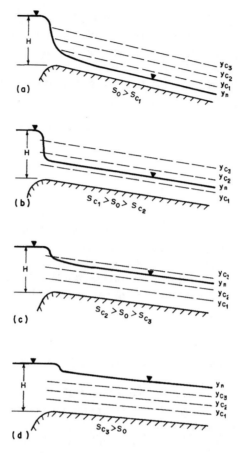

Fig. 5-16   Water surface profiles

5-8 Summary

In this chapter, an equation for the spatial variation of flow depth in gradually varied flow was derived. The classification of water-surface profiles was discussed and several general remarks were made on the properties of water-surface profiles. Procedures for sketching the water surface profiles in a channel were then outlined. A procedure for determining the discharge from a reservoir was presented. The properties of water surface profiles in a compound channel were discussed.

Problems

5-1 Prove that the gradually varied flow equation for a wide rectangular channel may be written as

$$\frac{dy}{dx} = S_o \frac{1 - (y_n/y)^{\frac{10}{3}}}{1 - (y_c/y)^3}$$

if Manning equation is used; and as

$$\frac{dy}{dx} = S_o \frac{1 - (y_n/y)^3}{1 - (y_c/y)^3}$$

if Chezy equation is used for the friction losses.

5-2 Derive the gradually varied flow equation for a prismatic channel having lateral flow of q per unit length. Assume that the lateral flow enters the channel perpendicular to the flow direction. Will this equation be valid if we had lateral outflow instead of lateral inflow?

5-3 Sketch the water-surface profiles in the channel system of Fig. 5-17. In this figure, NDL and CDL denote normal- and critical-depth lines, respectively.

5-4 If a sluice gate is used to control flow from a lake, should a gate be located near or at a long distance from the lake outlet. Why? If the gate is located at a long distance from the lake outlet, is there a situation in which the outflow from the lake does not depend upon the gate opening? Sketch all possible flow situations assuming the channel-bottom slope to be

  i. mild;
  ii. steep.

5-5 If the Manning equation is used to compute S_f, prove that the slope of the water surface in gradually varied flow, $dy/dx \to \infty$ as $y \to 0$.

5-6 A 5-m-wide rectangular concrete-lined canal takes off from a lake having a constant water level of 2 m above the channel bottom at the entrance. The channel is long, has a bottom slope of 0.004, and $n = 0.013$.

  i. If the head losses at the entrance are negligible, determine the discharge in the canal;
  ii. Compute the discharge if the bottom slope is changed to 0.001 and the entrance losses are $0.1V^2/(2g)$.

5-7 Lakes A and B are connected by a 10-m wide rectangular channel, as shown in Fig. 5-18. If n for the flow surfaces is 0.013, sketch the water-surface profile in the channel if the water level in Lake B is at

  i. El. 155.0;
  ii. El. 161.0.

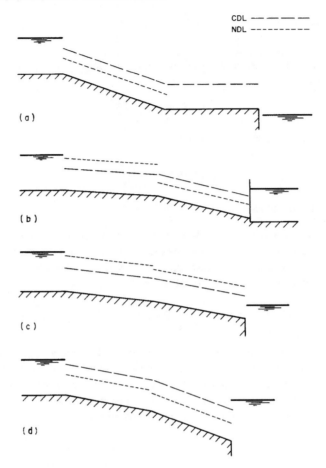

Fig. 5-17   Channel systems for Prob. 5-3

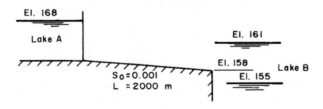

Fig. 5-18   Lake system for Prob. 5-7

5-8 A 15-m wide, 15-km long, concrete-line channel ($n = 0.013$) is planned for conveying water from reservoir X to reservoir Y. The water level in reservoir X is at El. 129.65 m and the level of the channel bottom at the entrance is at El. 121.4 m. Determine the channel discharge and sketch and label the type of water-surface profile for the following two cases.

1. The slope of the channel bottom is 0.001 and the water level in reservoir Y is at El. 109 m;
2. The slope of the channel bottom is 0.008 and the water level in reservoir Y is at El. 7 m.

Assume the entrance losses are negligible in both cases.

5-9 Prove that the following equation describes the gradually varied flow in a channel having variable cross section along its length

$$\frac{dy}{dx} = \frac{S_o - S_f + \frac{V^2}{gA}\frac{\partial A}{\partial x}}{1 - \frac{BV^2}{gA}}$$

5-10 For a wide rectangular channel, derive expressions for the channel bottom slope to be mild, steep, and critical.

5-11 A chute spillway is blasted through rock and is not lined. The bottom drops 1.5 ft in 20 feet. Determine the flow depth and the rate of discharge in the chute if the reservoir water level is 10 ft above the channel bottom at the entrance.

5-12 Name the water-surface profiles shown in Fig. 5-19.

5-13 Sketch the water surface profiles in the channels shown in Fig. 5-20.

5-14 The bottom slope of a long trapezoidal channel (bottom width = 15 ft, side slopes = 1:1) is suddenly changed from 0.0005 to 0.05. The flow in the channel is 800 ft^3/sec and the Manning n is 0.028. Compute the critical and normal flow depths in each channel reach and sketch the water surface profile.

5-15 Sketch the water surface profile for the channel of Prob. 5-14 if the channel downstream of the slope change is long and has a bottom slope of 0.0003. The bottom slope of the upstream channel is 0.05.

5-16 Sketch and label the types of water surface profiles in the channel of Fig. 5-21.

5-17 Sketch and label the type of water surface profiles in the channel shown in Fig. 5-22.

5-18 A mining company excavated a long 4-m wide, rectangular channel from a lake to their mining site. The water level in the lake is 3 m above the channel bottom at the entrance. If the bottom slope is 0.015 and Manning n is 0.025, determine the rate of discharge and the flow depth at the site.

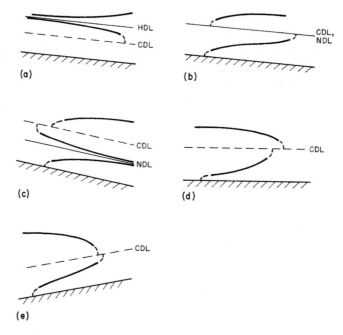

Fig. 5-19   Water-surface profiles

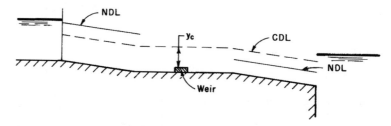

Fig. 5-20   Channels for Prob. 5-13

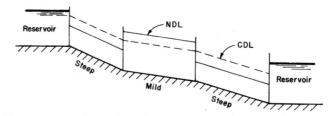

Fig. 5-21   Channel for Prob. 5-16

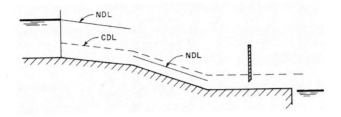

Fig. 5-22   Channel of Prob. 5-17

5-19 A concrete-lined channel with bottom width of 2 m and bottom slope of 0.001 is planned to take off from a large lake to a site located approximately 5 km from the lake. The side slopes are 1:1 up to 1.5 m depth and then the sides are vertical. The lake water level is 3 m above the channel bottom at the entrance. Determine the flow depth and the rate of discharge in the channel.

5-20 A long spillway chute is excavated in rock at a channel bottom slope of 1 in 100. The chute may be approximated as a 10-m wide rectangular cross section. If the water level in the upstream reservoir is 9 m above the bottom of the chute at its entrance, determine the discharge in the chute and the flow depth at the downstream end. The Manning n for the rock surfaces may be assumed as 0.025 and the losses at the entrance may be neglected.

5-21 Compute the rate of discharge and flow depth in a 8-m wide rectangular channel excavated through rock ($n = 0.035$) at a bottom slope of 0.02. The water level in the upstream lake is 7.5 m above the channel bottom at the entrance. Neglect head loss at the channel entrance.

References

Chaudhry, M. H., and Bhallamudi, S. M., 1988, "Critical Depth in Compound Channels", *Jour. Hydraulic Research,* Inter. Assoc. for Hydraulic Research, vol. 26, no. 4, pp. 377–395.

Chow, V. T., 1959, *Open-Channel Hydraulics*, McGraw-Hill Book Co., New York, NY.

Gill, H. A., 1976, "Exact solution of gradually varied flow," *Jour. Hyd. Div.,* Amer. Soc. Civil Engrs., vol. 102, no 9, pp.

Henderson, F. M., 1966, *Open-Channel Flow*, Macmillan Co., New York, NY.

Quintela, A.C., 1982, Discussion of Blalock , M.E. and Sturm , T.W., "Minimum Specific Energy in Compound Channel," *Jour. Hyd. Div.,* Amer. Soc. Civ. Engrs., vol. 108, no. 7, pp. 729–731 (see also Closure, vol. 109, 1983, pp. 483–486).

Sturm, T.W., Skolds, D.M., and Blalock, M.E., 1985, "Water Surface Profiles in Compound Channels," *Proc.,* Hydraulics and Hydrology in Computer Age, Amer. Soc. Civ. Engrs., Lake Buena Vista, FL, vol. 1, pp. 569–574.

COMPUTATION OF GRADUALLY VARIED FLOW

Glasscock Cutoff on the Mississippi River constructed in 1933 (Courtesy, US Army Corps of Engineers)

Supplementary Information The online version contains supplementary material available at (https://doi.org/10.1007/978-3-030-96447-4_6).

© Springer Nature Switzerland AG 2022
M. H. Chaudhry, *Open-Channel Flow*,
https://doi.org/10.1007/978-3-030-96447-4_6

6-1 Introduction

In the last chapter, we discussed how to qualitatively sketch water-surface profiles in channels having gradually varied flows. For engineering applications, however, it is necessary to compute the flow conditions in these flows. These computations, generally referred to as water-surface profile calculations, determine the water-surface elevations along the channel length for a specified discharge. The water-surface elevations are required for the planning, design, and operation of open channels to assess the effects of various engineering works and channel modifications. The addition of a dam, for example, raises water levels upstream of the dam and it is necessary to know the flow depths in the upstream area to determine the extent of flooding.

In addition, steady-state flow conditions are needed to specify proper initial conditions for the computation of unsteady flows.* Improper initial conditions introduce false transients into the simulation, which may lead to incorrect results. Unsteady-flow algorithms may be used directly to determine the initial conditions by continuing the computations until the flow conditions become steady. However, such a procedure is computationally inefficient and may not converge to the proper steady-state solution if the finite-difference scheme is not consistent as discussed in Chapter 14.

In this chapter, methods to compute gradually varied flows are presented. Preference is given to the methods suitable for a computer solution. Two traditional methods – commonly referred to as the direct and standard step methods – are first presented. The computations progress step by step from one section to the next in these methods. Then, numerical methods to integrate the governing differential equation are introduced. A procedure is then presented that computes the flow conditions at all specified locations of a channel system simultaneously instead of computing them from one section to the next.

6-2 General Remarks

The continuity, momentum, and energy equations relate various flow variables, such as the flow depth, discharge, flow velocity, at different sections of a channel and we solve these equations to determine the flow conditions. The channel cross section, Manning n, channel bottom slope, and the rate of discharge are usually known and we compute the flow depth and flow velocity at different channel sections in a specified channel length.

The rate of change of flow depth in gradually varied flows is usually small. Therefore, the assumption of hydrostatic pressure distribution is valid. In addition, we may introduce the velocity-head coefficient, α, to account for nonuniform velocity distribution and then use the mean flow velocity to compute the velocity head at a channel section. For a prismatic channel having

---

* These are discussed in Chapters 11 to 15.

no lateral inflows or outflows, the continuity equation between sections 1 and 2 (Fig. 6-1) may be written as

$$Q = V_1 A_1 = V_2 A_2 \qquad (6\text{-}1)$$

in which V = mean flow velocity; A = flow area; Q = rate of discharge; and the subscripts 1 and 2 refer to the variables for sections 1 and 2, respectively. Similarly, the energy equation between section 1 and 2 of a channel with small bottom slope may be written as

$$z_1 + y_1 + \alpha_1 \frac{V_1^2}{2g} = z_2 + y_2 + \alpha_2 \frac{V_2^2}{2g} + h_f \qquad (6\text{-}2)$$

in which z = elevation of the channel bottom above a specified datum; y = flow depth; and h_f = total head loss between sections 1 and 2. The head loss comprises the friction and form losses between these two sections.

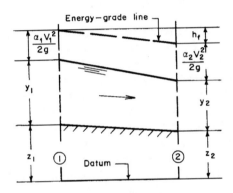

Fig. 6-1 Definition sketch

The following equation for gradually varied flow was derived in Chapter 5 by differentiating the energy equation

$$\frac{dy}{dx} = \frac{S_o - S_f}{1 - (\alpha B Q^2)/(g A^3)} \qquad (6\text{-}3)$$

in which x = distance along the channel (measured positive in the downstream direction); S_o = longitudinal slope of the channel bottom; S_f = slope of the energy grade line; B = top water-surface width, and g = acceleration due to gravity. If the momentum coefficient β is equal to unity, then this equation may also be derived by applying the Newton's second law of motion to a volume of water in a short channel length (see Prob. 6-1).

Equation 6-3 is a first-order ordinary differential equation in which x is the independent variable and y is the dependent variable. This equation describes the rate of variation of flow depth, y, with respect to distance, x. A close look at the terms of the right-hand side of this equation shows that this rate is a function of the properties of the channel section, flow depth and discharge. For a given channel section, the channel properties (e.g., top water-surface width, B, and the flow area, A) are functions of y only. The bottom slope, S_o, Manning n, and discharge, Q, are known. Therefore, for a specified discharge in a given channel, we may say that the right-hand side of Eq. 6-3 is a function of the flow depth, y which in turn is a function of distance, x. Let us designate this function as $f(x, y)$. Then, we may write Eq. 6-3 as

$$\frac{dy}{dx} = f(x, y) \tag{6-4}$$

in which

$$f(x, y) = \frac{S_o - S_f}{1 - (\alpha B Q^2)/(g A^3)} \tag{6-5}$$

This equation is integrated to determine the flow depth along a channel length. A closed-form solution of this equation is not available except for very simplified cases because $f(x, y)$ is a nonlinear function. Therefore, numerical methods are used for its integration. These methods yield the flow depth at discrete locations. Let us first discuss the basis of these methods.

Let the flow depth, y, be known at a given distance, x. Let us denote these known values by y_1 and x_1, respectively. To determine the water-surface profile, we may follow either of the following two procedures: Determine y_2 at specified location, x_2, or determine the location x_2 where specified flow depth y_2 will occur. Let us discuss each of them in more detail, starting with determining the value of y_2 at distance x_2.

By multiplying both sides of Eq. 6-4 by dx, and integrating, we obtain

$$\int_{y_1}^{y_2} dy = \int_{x_1}^{x_2} f(x, y) \, dx \tag{6-6}$$

By applying the limits of integration, this equation may be written as

$$y_2 = y_1 + \int_{x_1}^{x_2} f(x, y) \, dx \tag{6-7}$$

Computations progress in the downstream direction if dx is positive and they progress in the upstream direction if dx is negative. We can determine y_2 by numerically evaluating the integral term of this equation. Then, by successive application of this equation, we may compute the water-surface profile in the desired channel length.

To determine x_2 where the flow depth will be y_2, we may proceed as follows. Equation 6-3 may be written as

$$\frac{dx}{dy} = F(x, y) \tag{6-8}$$

in which

$$F(x, y) = \frac{1 - (\alpha B Q^2)/(g A^3)}{S_o - S_f} \tag{6-9}$$

By multiplying both sides of Eq. 6-8 by dy, integrating, and applying the limits of integration, we obtain

$$x_2 = x_1 + \int_{y_1}^{y_2} F(x, y)\, dy \tag{6-10}$$

The value of x_2 may be determined by numerically evaluating the integral term.

Instead of the differential equation, Eq. 6-3, we may use the energy equation, Eq. 6-2, between two adjacent sections for computing the water-surface profile. The main difficulty in the use of this equation is in determining the head losses, h_f between the two sections. We select an expression to approximate h_f, and then solve the resulting nonlinear algebraic equation to determine the flow depth at a specified location or to determine the location where a specified flow depth will occur. These procedures are discussed in the next two sections. Of course, similar difficulty arises if we use the governing differential equation, Eq. 6-3. The average value of S_f between the two sections may be used if the distance between the sections is short. This is usually a satisfactory approximation since short step lengths are used to numerically integrate Eq. 6-3.

Several procedures to compute the water-surface profiles have been developed [Bakhmeteff, 1932; Chow, 1959; Henderson, 1966; Eichert, 1970; Prasad, 1970; McBeans and Perkins, 1975; Chaudhry and Schulte, 1986; Schulte and Chaudhry, 1987]. Some earlier procedures used various varied-flow functions developed by integrating the differential equation (Eq. 6-3) describing the gradually varied flow. Several graphical and mathematical methods were developed for the integration of this equation or for solving the energy equation between the two adjacent sections [Bakhmeteff, 1932; Chow, 1959; Henderson, 1966]. Some of these methods have been used in various general-purpose computer programs for computing water-surface profiles [Soil Conservation Service, 1976; United States Army Corps of Engineers, 1982; Unites States Geological Survey, 1976]. Of these programs, HEC-RAS (the earlier versions of this program were called HEC-2) developed by the Army Corps of Engineers is most widely used.

We discussed in Chapters 3 and 5 that subcritical flow has a downstream control and the supercritical flow has an upstream control. To compute the water-surface profile, we start the computations at a location where the flow depth for the specified discharge is known. Consequently, we start the computations at a downstream control section if the flow is subcritical and proceed in the upstream direction. For supercritical flows, however, we start at an up-stream control section and compute the profile in the downstream direction. Unfortunately, this fact has been incorrectly attributed in many well-known publications to indicate that the computations become unstable or yield in-correct results if this convention is not followed. Other than the fact that the flow depth is known at a control section, there appears to be no reason why we should proceed in either the upstream or downstream direction. This is because all we are doing is either numerically solving a differential equation for the specified initial condition or solving a nonlinear algebraic equation. Whether we proceed in the positive or negative x direction should make little difference provided the computational step is properly selected in the computations based on the rate of change of flow depth with distance.

6-3 Direct-Step Method

In the previous section, we discussed how we may compute, from the known flow depth at a section, the location of an adjacent section where a specified depth will occur. Let us discuss this in more detail to develop a systematic procedure for the computations. Following Chow [1959], we call this procedure the *direct-step method*.

Referring to Fig. 6-2, let us say we know the flow depth at section 1 and we want to determine the location of section 2, where a specified flow depth, y_2 will occur in a given channel for a specified discharge, Q. In other words, the statement of our problem is as follows: The flow depth, y_1, at distance x_1 (i.e. section 1 in Fig. 6-2) is known; determine distance x_2 where a specified flow depth y_2 will occur. The properties of the channel section, S_o, Q, and n are known.

If $S_o =$ slope of the channel bottom, then referring to Fig. 6-2,

$$z_2 = z_1 - S_o(x_2 - x_1) \tag{6-11}$$

In addition, the specific energy

$$E_1 = y_1 + \frac{\alpha_1 V_1^2}{2g}$$

$$E_2 = y_2 + \frac{\alpha_2 V_2^2}{2g} \tag{6-12}$$

The slope of the energy grade line (for simplicity we will refer to it as the friction slope in the following discussion) in gradually varied flow may be computed with negligible error by using the corresponding formulas for friction slopes in uniform flow [Chow, 1959; Henderson, 1966]. However, since the flow depth, y, varies with distance, x, the friction slope S_f is a function of x as well. The following approximations have been used [United States Army Corps of Engineers, 1982] to select a representative value of S_f for the channel length between sections 1 and 2:

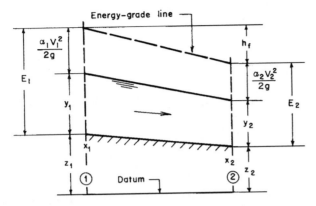

Fig. 6-2   Computation of distance for specified depth

Average friction slope

$$\bar{S}_f = \frac{1}{2}(S_{f_1} + S_{f_2})$$
(6-13a)

Geometric mean friction slope

$$\bar{S}_f = \sqrt{S_{f_1}S_{f_2}}$$
(6-13b)

Harmonic mean friction slope

$$\bar{S}_f = \frac{2S_{f_1}S_{f_2}}{S_{f_1} + S_{f_2}}$$
(6-13c)

By expanding the right-hand side of the above approximations in a Taylor series, we can prove (Prob. 6-15) that these three formulations for the approximation of the friction slope give identical results if the terms of the order $(\Delta S_f/S_{f1})^2$ and higher are neglected. In this expression, $\Delta S_f = S_{f2} - S_{f1}$. Laurenson [1986] showed that the average slope (Eq. 6-13a) gives the lowest maximum error although it is not always the smallest error. If the distance between sections 1 and 2 is short or the flow depths y_1 and y_2 are not significantly different, then Eq. 6-13a yields satisfactory results, in addition to being the simplest of the three approximations. Therefore, its use is recommended, and we will use it herein. Hence, an expression for h_f may be written as

$$h_f = \frac{1}{2}(S_{f1} + S_{f2})(x_2 - x_1)$$
(6-14)

Substitution of Eqs. 6-12 and 6-14 into Eq. 6-2 yields

$$z_1 + E_1 = z_2 + E_2 + \frac{1}{2}(S_{f1} + S_{f2})(x_2 - x_1) \tag{6-15}$$

By substituting the expression for z_2 from Eq. 6-11 into Eq. 6-15, and cancelling out z_1, we obtain

$$E_2 - E_1 = S_o(x_2 - x_1) - \frac{1}{2}(S_{f1} + S_{f2})(x_2 - x_1) \tag{6-16}$$

This equation may be written as

$$x_2 = x_1 + \frac{E_2 - E_1}{S_o - \frac{1}{2}(S_{f1} + S_{f2})} \tag{6-17}$$

Now, the location of section 2 is known. This is the starting value for the next step. Then, by successively increasing or decreasing the flow depth and determining where these depths will occur, the water-surface profile in the desired channel length may be computed. In Eq. 6-17, the direction of computations is automatically taken care of if proper sign for the numerator and for the denominator is used. Note that both the numerator and the denominator are very small and extreme care should be exercised in using the proper number of significant digits in the computations and rounding off the values.

There are two main *disadvantages* of this method. (1) The flow depth is not computed at the predetermined locations. Therefore, interpolations may become necessary if the flow depths are required at specified locations. Similarly, the cross-sectional information has to be estimated if such information is available only at the given locations. This may not yield accurate results in addition to requiring additional effort. (2) It is cumbersome to apply to nonprismatic channels.

The following example should help in understanding this computational procedure.

Example 6-1

A trapezoidal channel having a bottom slope of 0.001 is carrying a flow of 30 m^3/s. The bottom width is 10.0 m and the side slopes are 2H to 1V. A control structure is built at the downstream end which raises the water depth at the downstream end to 5.0 m. Compute the water surface profile. Manning n for the flow surfaces is 0.013 and $\alpha = 1$.

Given:

Bottom slope, $S_o = 0.001$;
Discharge, $Q = 30$ m^3/s;
Channel width, $B_o = 10.0$ m;
Manning $n = 0.013$;
Depth at the downstream end (i.e., at $x = 0$) = 5.0 m;
$\alpha = 1$.

Determine:

Water-surface profile in the channel.

Solution:

The normal depth, y_n, for this channel was computed in Example 4-1 as 1.16 m. The flow depth approaches the normal depth asymptotically at an infinite distance. Therefore, the computation of the surface profile may be stopped when the flow depth is within about five percent of the normal depth. We will continue the calculations in this example until $y = 1.05 y_n = 1.05 \times 1.16 = 1.21$, say 1.20 m.

We start the computations with a known depth of 5.0 m at the control structure and proceed in the upstream direction. Let us call the location at the control structure as $x = 0$. Since we are considering the distance in the downstream flow direction as positive, the values of x we determine from Eq. 6-17 are negative.

The calculations are done in a systematic manner, as shown in Table 6-1. The following explanatory remarks should be helpful to understand these calculations. In this discussion, the depth for the step under consideration is the current depth and the depth for the previous step as the previous depth.

Column 1, y

We first use large increments of change in y, i.e., 0.5 m and then decrease their size, i.e., 0.1 m, as the rate of variation of y with x becomes small.

Column 2, A

This is the flow area for the depth of column 1.

Column 3, R

Hydraulic radius, $R = A/P$, where P = wetted perimeter for the flow depth of column 1.

Column 4, V

Flow velocity, V is computed by dividing the specified rate of discharge, Q, by the flow area, A, of column 2.

Column 5, S_f

By using the specified value of Manning n, and the computed values of V of column 4 and R of column 3, this column is computed from the equation, $S_f = n^2 V^2 / (C_o^2 R^{1.33})$.

Column 6, $\bar{S}_f$

This is the average of S_f for the current depth and for the depth in previous step. This column is left blank for the first line since there is no previous depth when we start the computations. To indicate that this is an average slope, we list it between the lines corresponding to the current and the previous depths.

Column 7, $S_o - \bar{S}_f$

This is obtained by subtracting $\bar{S}_f$ of column 6 from the specified value of S_o.

Column 8, E

The specific energy, E, is computed for the selected value of y of column 1 and corresponding computed value of V of column 4, i.e., $E = y + \alpha V^2/(2g)$.

Column 9, $\Delta E = E_2 - E_1$

This column is obtained by subtracting E for the current depth from E for the previous depth. Again, since this column is the difference of E values corresponding to the current and the previous depths, we list its value between the lines for these depths.

Column 10, $\Delta x = x_2 - x_1$

The distance increment is computed from the equation, $\Delta x = (E_2 - E_1)/(S_o - \bar{S}_f)$, i.e., dividing column 9 by column 7.

Column 11, x_2

This is the distance where depth y will occur. It is obtained by algebraically adding Δx of column 10 to the x_2 value for the previous depth.

6-4 Standard Step Method

The procedure described in the previous section is not suitable if we want to determine the flow depth at specified locations or if the channel is non-prismatic (i.e., channel cross section and/or bottom slope vary with distance) and the channel cross sections are available only at some specified locations. In such cases, the procedure described in this section may be used. Following Chow [1959], we will call this method as the *standard step method* since this name has been widely used. A very popular computer program HEC-RAS

Table 6-1 Direct step method

$Q = 30 \ \text{m}^3/\text{s}; \ B_o = 10 \ \text{m}; \ s = 2; \ S_o = 0.001; \ n = 0.013; \ \alpha = 1.0; \ C_o = 1.0$

y	A	R	V	S_f	$\bar{S}_f$	$S_o - \bar{S}_f$	E	ΔE	Δx	x_2
(1)	(2)	(3)	(4)	(5)	(6)	(7)	(8)	(9)	(10)	(11)
5.00	100.0	3.09	0.30	0.000003			5.00459			0.0
					0.000004	0.000996		−0.49831	−500.5	
4.50	85.5	2.84	0.35	0.000005			4.50627			−500.5
					0.000007	0.000993		−0.49743	−500.8	
4.00	72.0	2.58	0.42	0.000008			4.00885			−1001.3
					0.000010	0.000990		−0.33743	−340.8	
3.66	63.4	2.40	0.47	0.000012			3.67142			−1342.1
					0.000014	0.000986		−0.32651	−331.3	
3.33	55.5	2.23	0.54	0.000017			3.34490			−1673.4
					0.000021	0.000979		−0.32499	−332.0	
3.00	48.0	2.05	0.63	0.000025			3.01991			−2005.4
					0.000030	0.000970		−0.24466	−252.3	
2.75	42.6	1.91	0.70	0.000035			2.77525			−2257.7
					0.000043	0.000957		−0.24263	−253.5	
2.50	37.5	1.77	0.80	0.000050			2.53262			−2511.2
					0.000063	0.000937		−0.23952	−255.5	
2.25	32.6	1.63	0.92	0.000075			2.29310			−2766.7
					0.000095	0.000905		−0.23459	−259.5	
2.00	28.0	1.48	1.07	0.000115			2.05851			−3025.9
					0.000142	0.000858		−0.18196	−212.1	
1.80	24.5	1.36	1.23	0.000169			1.87655			−3238.0
					0.000214	0.000786		−0.17371	−220.9	
1.60	21.1	1.23	1.42	0.000258			1.70284			−3459.0
					0.000337	0.000663		−0.15999	−241.4	
1.40	17.9	1.10	1.67	0.000416			1.54285			−3700.4
					0.000479	0.000521		−0.07188	−137.8	
1.30	16.4	1.04	1.83	0.000541			1.47097			−3838.2
					0.000629	0.000371		−0.06379	−171.9	
1.20	14.9	0.97	2.02	0.000717			1.40718			−4010.2

(originally called HEC-2), developed by the Hydrologic Engineering Center, United States Army Corps of Engineers [1982], is based on this method.

Referring to Fig. 6-3, the flow depth, y_1, for a specified discharge, Q, in a given channel at section 1 (distance x_1) is known; and we want to determine the flow depth at distance, x_2 (section 2). Let us assume that the values of the velocity-head coefficient, α, at section 1 and 2 are either known or we have determined their values as discussed in Sections 1-5 and 4-6. Since y_1 is known,

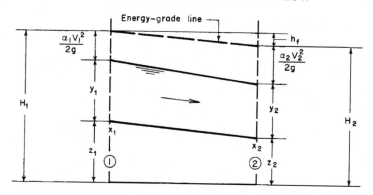

Fig. 6-3 **Computation of depth at specified location**

we can determine the flow velocity, V_1, at section 1 for the specified discharge, Q, from the continuity equation. Hence, the total head, H, at section 1

$$H_1 = z_1 + y_1 + \frac{\alpha_1 V_1^2}{2g} \qquad (6\text{-}18)$$

is known. According to the energy equation, the total head at section 2 is

$$H_2 = H_1 - h_f \qquad (6\text{-}19)$$

in which h_f = head losses (sum of the friction and form losses) between sections 1 and 2. By substituting the expression for h_f from Eq. 6-14 into Eq. 6-19, we obtain

$$H_2 = H_1 - \frac{1}{2}(S_{f1} + S_{f2})(x_2 - x_1) \qquad (6\text{-}20)$$

Substituting into Eq. 6-20 an expression for H_2 (similar to that for H_1 in Eq. 6-18) and transposing all terms to the left-hand side, we obtain

$$y_2 + \frac{\alpha_2 Q^2}{2g A_2^2} + \frac{1}{2}S_{f2}(x_2 - x_1) + z_2 - H_1 + \frac{1}{2}S_{f1}(x_2 - x_1) = 0 \qquad (6\text{-}21)$$

In this equation, A_2 and S_{f2} are functions of y_2 and all other quantities are either known or have already been calculated at section 1. Hence, y_2 may be determined by solving the following nonlinear algebraic equation

$$F(y_2) = y_2 + \frac{\alpha_2 Q^2}{2g A_2^2} + \frac{1}{2}S_{f2}(x_2 - x_1)$$

$$+ z_2 - H_1 + \frac{1}{2}S_{f1}(x_2 - x_1) = 0 \qquad (6\text{-}22)$$

Equation 6-22 may be solved for y_2 by a trial-and-error procedure, or by using the Newton-Raphson or bisection methods. Henderson [1966], Chaudhry and Schulte [1986] and Schulte and Chaudhry [1987] and French [1985] used the Newton-Raphson method to solve the energy equation. Subramanya [1986] used the Newton-Raphson method for solving for y_2 in wide rectangular channels; Paine [1992] used it for trapezoidal sections, and Rhodes [1993, 1995] for general sections. The use of the Newton-Raphson method is discussed here.

For this method, we need an expression for dF/dy_2. This expression may be obtained by differentiating Eq. 6-22 with respect to y_2, i.e.,

$$\frac{dF}{dy_2} = 1 - \frac{\alpha_2 Q^2}{g A_2^3} \frac{dA_2}{dy_2} + \frac{1}{2}(x_2 - x_1)\frac{d}{dy_2}\left(\frac{Q^2 n^2}{C_o^2 A_2^2 R_2^{\frac{4}{3}}}\right) \tag{6-23}$$

The last term of this equation may be evaluated as follows:

$$\begin{aligned}
\frac{d}{dy_2}\left(\frac{Q^2 n^2}{C_o^2 A_2^2 R_2^{\frac{4}{3}}}\right) &= \frac{-2Q^2 n^2}{C_o^2 A_2^3 R_2^{\frac{4}{3}}}\frac{dA_2}{dy_2} - \frac{4}{3}\frac{Q^2 n^2}{C_o^2 A_2^3 R_2^{\frac{4}{3}}}\frac{dR_2}{dy_2} \\
&= \frac{-2Q^2 n^2}{C_o^2 A_2^2 R_2^{\frac{4}{3}}}\frac{B_2}{A_2} - \frac{4}{3}\frac{Q^2 n^2}{C_o^2 A_2^2 R_2^{\frac{4}{3}}}\frac{1}{R_2}\frac{dR_2}{dy_2} \\
&= -2\left(S_{f2}\frac{B_2}{A_2} + \frac{2}{3}\frac{S_{f2}}{R_2}\frac{dR_2}{dy_2}\right) \tag{6-24}
\end{aligned}$$

Note that we have replaced dA_2/dy_2 by B_2 in this equation. By substituting Eq. 6-24 into Eq. 6-23, we obtain

$$\frac{dF}{dy_2} = 1 - \frac{\alpha_2 Q^2 B_2}{g A_2^3} - (x_2 - x_1)\left(S_{f2}\frac{B_2}{A_2} + \frac{2}{3}\frac{S_{f2}}{R_2}\frac{dR_2}{dy_2}\right) \tag{6-25}$$

The derivative dR_2/dy_2 of the last term in this equation may be evaluated as follows.

$$\begin{aligned}
\frac{dR_2}{dy_2} &= \frac{d}{dy_2}\left(\frac{A_2}{P_2}\right) \\
&= \frac{1}{P_2}\frac{dA_2}{dy_2} + A_2\frac{d}{dy_2}\left(\frac{1}{P_2}\right) \\
&= \frac{B_2}{P_2} - \frac{A_2}{P_2^2}\frac{dP_2}{dy_2} \tag{6-26}
\end{aligned}$$

For a rectangular channel, $dP_2/dy_2 = 2$ and for a trapezoidal channel, $dP_2/dy_2 = 2\sqrt{1 + s^2}$, in which $s =$ side slope of the channel (s horizontal to 1 vertical).

A step-by-step procedure for computing y_2 by using the Newton-Raphson method is as follows.

1. Calculate H_1 at section 1 from Eq. 6-18 for the known values of y_1 and z_1.

2. Estimate the flow depth at section 2. Let us designate this estimated flow depth and other quantities corresponding to this estimated depth by superscript *. At the beginning of the calculations, the rate of variation of y at x_1 may be determined from Eq. 6-3 by using $y = y_1$, i.e., $dy/dx = f(x_1, y_1)$. Then, the flow depth, y_2^*, may be computed from the equation $y_2^* = y_1 + f(x_1, y_1)(x_2 - x_1)$. During subsequent steps, however, y_2^* may be determined by extrapolating the change in the flow depth between the previous two sections computed during the preceding step.

3. By using the estimated value of flow depth, y_2^*, at section 2, compute B_2^*, A_2^*, R_2^*, and S_{f2}^*. The value of z_2 is either given in the available data or it may be computed from the known values of channel bottom slope and z_1.

4. Compute the value of $F(y_2^*)$ from Eq. 6-22 by using y_2^*, B_2^*, A_2^*, R_2^*, and S_{f2}^*.

5. Compute dF/dy_2 from Eq. 6-25 using y_2^* and the corresponding values of A_2^*, R_2^*, and S_{f2}^*, etc.

6. Then, a better estimate for y_2 can be computed from the equation

$$y_2 = y_2^* - \frac{F(y_2^*)}{[dF/dy_2]^*} \tag{6-27}$$

7. If $|y_2 - y_2^*| \leq \epsilon$, where ϵ is specified tolerance for the convergence of iterative solution (say, 0.001 m), then y_2^* is the flow depth, y_2, at section 2; otherwise, set $y_2^* = y_2$, and repeat the above steps 3 to 7 until a solution is obtained.

The following example should help in understanding this procedure.

Example 6-2

A trapezoidal channel having a bottom slope of 0.001 is carrying a flow of 30 m³/s. The bottom width is 10.0 m and the side slopes are 2H to 1V. At the downstream end, a control structure raises the water depth to 5.0 m. Determine the water-surface levels at 1, 2, and 4 km upstream of control structure. The Manning n for the flow surfaces is 0.013, $\alpha = 1.0$, and the elevation of the channel bottom at the downstream end is 0.0.

Given:

Bottom slope, $S_o = 0.001$;
Discharge, $Q = 30$ m³/s;
Channel width, $B_o = 10.0$ m;
Manning $n = 0.013$;
Depth at the downstream end (i.e., at $x = 0$) = 5.0 m;
$\alpha = 1.0$.

Determine:

Water-surface levels at 1, 2, and 4 km upstream of the control structure.

Solution:

Let us call the distance at the control structure as $x = 0$. Since the distance in the downstream flow direction is considered positive, the upstream distances where we want to determine the flow depths are negative.

Table 6-2 lists the calculations using a trial-and-error procedure.

The following explanatory comments should help to understand the computations of Table 6-2. We estimate the flow depth at 1 km upstream of the control structure and we then check whether this estimated flow depth satisfies Eq. 6-22 or not. If it does not satisfy this equation, then we discard the calculations corresponding to this estimated depth and estimate another value. This process is continued until a solution is obtained. Then, we consider each distance where we want to determine the water level (i.e., 2 and 4 km) one by one and repeat the above procedure.

Column 1, x. This is the specified location at which flow depth, y, is to be computed.

Column 2, y. This is the estimated flow depth.

Column 3, A. This is the flow area, A, for the flow depth of column 2.

Column 4, P. This is the wetted perimeter, P, for the flow depth of column 2.

Column 5, R. This is the hydraulic radius, R, corresponding to the flow depth of column 2 obtained by dividing column 3 by column 4.

Column 6, $R^{4/3}$. This column lists the value of R raised to the power 4/3.

Column 7, V. Flow velocity, $V = 30/A$, where A is listed in column 3. Velocity head corresponding to this depth is shown in column 8.

Column 8, $V^2/(2g)$. This column lists the value of velocity head, $V^2/(2g)$.

Column 9, z. This is the elevation of the channel bottom. It is computed from the known bottom elevation ($z_d = 0$) at the downstream end ($x_d = 0$) and the known channel bottom slope, S_o, of 0.001; i.e., $z = z_d - S_o(x - x_d)$.

Column 10, y. Total head of column 10 corresponds to the flow depth of column 2, velocity head of column 8, and the channel bottom level of column 9, i.e., $H = z + y + \alpha V^2/(2g)$.

Column 11, S_f. This is the slope of the energy grade line. It is computed by using the velocity of column 7, $R^{4/3}$ of column 6 and the known value of Manning n from the equation, $S_f = n^2 V^2/(C_o^2 R^{4/3})$.

Column 12, $\bar{S}_f$. This is the average of the S_f value for the flow depth at the current distance and that for the flow depth at the previous distance.

Column 13, Δx. This is the distance between the current location where we want to determine the flow depth and the location where we determined the flow depth during the previous step.

Column 14, h_f. The head losses in distance Δx are computed from the equation $h_f = \bar{S}_f \Delta x$, where $\bar{S}_f$ is given in column 12 and Δx is given in column 13.

Column 15, H. This is the elevation of the energy grade line computed by adding the head losses (h_f of column 14) to the elevation of the energy grade line (i.e., H of column 10 for the previous step) at the location where the flow depth was computed during the previous step.

Column 16. If the values of H in columns 10 and 15 are within an acceptable tolerance, then the estimated depth of column 2 is the flow depth at the location under consideration. We then proceed to compute the flow depth at the next selected location. However, if these values are not within the specified tolerance, then we discard the values corresponding to this flow depth and repeat the process with another value for the estimated depth.

6-5 Integration of Differential Equation

In Section 6-2, we discussed the computation of water-surface profile by integrating the differential equation, Eq. 6-3. We also mentioned that the integration has to be done numerically since $f(x, y)$ is a nonlinear function. In the following sections, we will present several numerical methods which may be used for this purpose. Some of these methods have been used in the past while others are being introduced for computing the water-surface profiles. We may classify these methods into the following two categories [McCracken and Dorn, 1964; Chapra and Canale, 1988]:

1. Single-step methods; and
2. Predictor-corrector methods.

Single-step methods are just like the step methods discussed in the previous two sections. The unknown depth at a section is expressed in terms of function, $f(x, y)$, at a neighboring point where the flow depth is either initially known or has been computed during the previous step. In a predictor-corrector method, a value of the unknown depth is first predicted by using the available information from the previous step. This predicted value is then refined by an iterative procedure during the corrector part until a solution is obtained with a specified accuracy. The details of both of these methods are presented in the following sections.

6-6 Single-step Methods

There are several single-step methods [McCracken and Dorn, 1964]. However, only the following four of these are presented here.

1. Euler method;

Table 6-2 Standard step method

$S_o = 0.001;\ Q = 30 \text{ m}^3/\text{s};\ B_o = 10.0 \text{ m};\ n = 0.013;\ y_o = 5.0 \text{ m};\ \alpha = 1;\ s = 2$

x (1)	y (2)	A (3)	P (4)	R (5)	$R^{1.33}$ (6)	V (7)	$V^2/(2g)$ (8)	z (9)	H (10)	S_f (11)	$\bar{S}_f$ (12)	Δx (13)	h_f (14)	H (15)	Difference (16)	Remarks (17)
0	5.000	100.0	32.36	3.090	4.484	0.300	0.0046	0.0	5.0046	0.00000379						
−1000	4.001	72.0	27.89	2.582	3.542	0.417	0.0088	1.0	5.0095	0.00000828	0.00000583	−1000	0.00583	5.0104	0.0009	
	4.002	72.04	27.90	2.583	3.543	0.416	0.0088	1.0	5.0104	0.00000827	0.00000583	−1000	0.00583	5.0104	0.0000	ok
−2000	3.003	48.10	23.43	2.052	2.607	0.624	0.0198	2.0	5.02333	0.00002525	0.00001676	−1000	0.01676	5.0272	0.0038	
	3.007	48.16	23.45	2.054	2.611	0.623	0.0198	2.0	5.0271	0.00002512	0.0000167	−1000	0.0167	5.0271	0.0000	ok
−3000	2.014	28.30	19.01	1.487	1.697	1.062	0.0574	3.0	5.0719	0.00011224	0.00006868	−1000	0.6868	5.0958	0.2390	
	2.038	28.68	19.11	1.501	1.718	1.046	0.0558	3.0	5.0935	0.00010763	0.00006638	−1000	0.00664	5.0935	0.0000	ok
−4000	1.078	13.10	14.82	0.884	0.848	2.290	0.2674	4.0	5.345	0.00104508	0.00057635	−1000	0.57635	5.6698	0.3249	
	1.232	15.40	15.51	0.990	0.986	1.954	0.1947	4.0	5.4263	0.00065429	0.00038096	−1000	0.38096	5.4744	0.0481	
	1.263	15.83	15.65	1.011	1.015	1.896	0.1832	4.0	5.4465	0.00059837	0.00035300	−1000	0.35300	5.4465	0.0000	ok

2. Modified Euler method;

3. Improved Euler method; and

4. Fourth-order Runge-Kutta method.

Let us now discuss how to use these methods to compute the water-surface profiles.

Referring to Fig. 6-4, let us say that we know the flow depth, y_i at distance x_i and that we want to determine the flow depth at distance x_{i+1}. Let $y = y(x)$ be the exact solution of the differential equation (Eq. 6-4). Then, the curve $y = y(x)$ represents the variation of y with respect to x.

Euler method

We compute the rate of variation of flow depth, y, with respect to distance, x, at distance x_i from Eq. 6-4, i.e.,

$$y_i' = \left.\frac{dy}{dx}\right|_i = f(x_i, y_i) \tag{6-28}$$

in which the subscript i refers to quantities at distance x_i, a prime, $\prime$, on y indicates a derivative of y with respect to x, and

$$f(x_i, y_i) = \frac{S_o - S_{fi}}{1 - Q^2 B_i/(gA_i^3)} \tag{6-29}$$

Since all the variables on the right-hand side of this equation are known, we can compute $f(x_i, y_i)$, which, on the basis of Eq. 6-28, is the rate of variation of y at point (x_i, y_i). By assuming that this rate of variation, y_i', is constant in the interval x_i to x_{i+1} (Line 1 in Fig. 6-4), we can determine the flow depth at x_{i+1} from the equation

$$y_{i+1} = y_i + y_i'\Delta x \tag{6-30}$$

in which $\Delta x = x_{i+1} - x_i$. The substitution of Eq. 6-28 into Eq. 6-30 yields

$$y_{i+1} = y_i + f(x_i, y_i)\Delta x \tag{6-31}$$

This is referred to as the Euler method. Now, y_{i+1} is known and we can determine y_{i+2} at distance x_{i+2} by repeating the same procedure.

Let us briefly discuss the accuracy of the Euler method. We may expand y_{i+1} in a Taylor series as

$$y_{i+1} = y_i + y_i'\Delta x + O\left(\Delta x\right)^2 \tag{6-32}$$

in which $O(\Delta x)^2$ means that the remaining terms are of the order of $(\Delta x)^2$ or smaller. A comparison of Eqs. 6-31 and 6-32 shows that we are including in our solution terms up to the first power of Δx. Therefore, this method is referred to as *first-order* accurate.

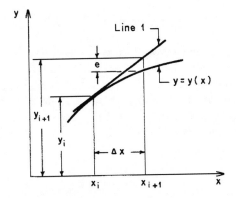

Fig. 6-4   **Geometrical representation of Euler method**

Equation 6-29 is the equation of a straight line, Line 1, shown in Fig. 6-4. Since, y' is function x and y, it may vary in the interval x_i to x_{i+1}, thereby introducing an error, e, by assuming it as constant during each step. Note that if the flow depth varies linearly, i.e., it is a straight line, then $e = 0$. Because of this error at each step, the numerically computed value may diverge from the correct solution. This method is usually unstable; i.e., a small error – round-off or truncation – is magnified as the value of x increases with repeated application of this procedure.

In the Euler method, we used the slope at only one point (x_i, y_i) to compute the value of y_{i+1}. By using the slope at more than one point, we may improve the accuracy of this method. Two such methods, improved Euler and modified Euler methods, are presented in the following paragraphs.

Improved Euler method

Let us call the flow depth at x_{i+1} obtained from the Euler method as y^*_{i+1}, i.e.,

$$y^*_{i+1} = y_i + y'_i \Delta x \tag{6-33}$$

By using this value y^*_{i+1}, we can compute the slope of the solution curve, $y = y(x)$, at $x = x_{i+1}$, i.e., $y'_{i+1} = f(x_{i+1}, y^*_{i+1})$ and to improve accuracy, let us use the average value of the slopes of the solution curve at x_i and at x_{i+1}. Then, we can determine the value of y_{i+1} from the equation

$$y_{i+1} = y_i + \frac{1}{2}(y'_i + y'_{i+1})\Delta x \tag{6-34}$$

This equation may be written as

$$y_{i+1} = y_i + \frac{1}{2}[f(x_i, y_i) + f(x_{i+1}, y^*_{i+1})]\Delta x \qquad (6\text{-}35)$$

This method is called the *improved* Euler method. By expanding Eq. 6-34 in Taylor series, we can show that this method is *second-order* accurate.

A geometrical representation of this method is shown in Fig. 6-5. In this figure, Line 1 is tangent at (x_i, y_i) and has a slope of y'_i whereas Line 2 is tangent at (x_{i+1}, y_{i+1}) and has a slope of y'_{i+1}. Line 3 is drawn through point (x_i, y_i) with an average slope of $\frac{1}{2}(y'_i + y'_{i+1})$.

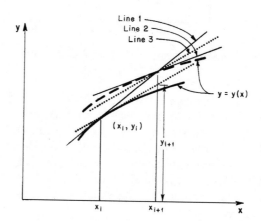

Fig. 6-5   Geometrical representation of the improved Euler method

Modified Euler Method

We may also improve the accuracy of the Euler method by using the slope of the curve $y = y(x)$ at $(x_{i+\frac{1}{2}}, y_{i+\frac{1}{2}})$, in which $x_{i+\frac{1}{2}} = \frac{1}{2}(x_i + x_{i+1})$ and $y_{i+\frac{1}{2}} = y_i + \frac{1}{2}y'_i\Delta x$. Let us call this slope as $y'_{i+\frac{1}{2}}$. Then,

$$y_{i+1} = y_i + y'_{i+\frac{1}{2}}\Delta x$$

or

$$y_{i+1} = y_i + f(x_{i+\frac{1}{2}}, y_{i+\frac{1}{2}})\Delta x \qquad (6\text{-}36)$$

This method is called the *modified* Euler method.

Figure 6-6 shows a geometrical representation of this method. In this figure, Line 1 is tangent at (x_i, y_i) and has a slope of y'_i, whereas Line 2 is tangent at $(x_{i+\frac{1}{2}}, y_{i+\frac{1}{2}})$ and has a slope of $y'_{i+\frac{1}{2}}$. Line 3 is drawn through point (x_i, y_i) with a slope of $y'_{i+\frac{1}{2}}$.

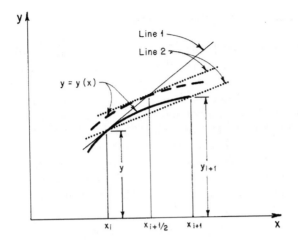

Fig. 6-6   Geometrical representation of the modified Euler method

Similar to the Improved Euler method, we can show by expanding the numerical solution in Taylor series, that the modified Euler method is second-order accurate.

Fourth-order Runge-Kutta Method

In the fourth-order Runge-Kutta method, a better representative slope of the solution curve $y = y(x)$ is determined from the following equations:

$$k_1 = f(x_i, y_i)$$
$$k_2 = f(x_i + \frac{1}{2}\Delta x, y_i + \frac{1}{2}k_1\Delta x)$$
$$k_3 = f(x_i + \frac{1}{2}\Delta x, y_i + \frac{1}{2}k_2\Delta x)$$
$$k_4 = f(x_i + \Delta x, y_i + k_3\Delta x) \tag{6-37}$$

Then

$$y_{i+1} = y_i + \frac{1}{6}(k_1 + 2k_2 + 2k_3 + k_4)\Delta x \tag{6-38}$$

As the name implies, this method is *fourth-order* accurate. Humpidge and Moss [1971] developed a general-purpose computer program based on this method to compute the water-surface profiles.

6-7 Predictor-Corrector Methods

In the numerical methods discussed in the previous section, we used the known information at point x_i and, to improve the accuracy, we used the value of the function $f(x, y)$ at more than one point, e.g., at x_i, x_{i+1} and $x_{i+\frac{1}{2}}$. In a predictor-corrector method, we do not compute the function at several points but rather predict the unknown flow depth first, correct this predicted value, and then recorrect this corrected value. This iterative procedure is continued until a solution of a desired accuracy is obtained.

Several predictor-corrector methods are reported in the literature. However, to conserve space, we present only one of these methods.

In the predictor part, let us use the Euler method to predict the value of y_{i+1}, i.e.,

$$y_{i+1}^{(0)} = y_i + f(x_i, y_i)\Delta x \qquad (6\text{-}39)$$

in which the superscript enclosed in the parenthesis indicates the number of the iteration (zero iteration is the initially estimated or predicted value). Then, we may correct it using the following equation

$$y_{i+1}^{(1)} = y_i + \frac{1}{2}[f(x_i, y_i) + f(x_{i+1}, y_{i+1}^{(0)})]\Delta x \qquad (6\text{-}40)$$

Now, we may recorrect $y_{i+1}^{(1)}$ again to obtain a better value

$$y_{i+1}^{(2)} = y_i + \frac{1}{2}[f(x_i, y_i) + f(x_{i+1}, y_{i+1}^{(1)})]\Delta x \qquad (6\text{-}41)$$

Thus, the jth iteration is

$$y_{i+1}^{(j)} = y_i + \frac{1}{2}[f(x_i, y_i) + f(x_{i+1}, y_{i+1}^{(j-1)})]\Delta x \qquad (6\text{-}42)$$

We continue this iterative procedure until $|y_{i+1}^{(j)} - y_{i+1}^{(j-1)}| \leq \epsilon$, where $\epsilon =$ specified tolerance. A similar method is used by Prasad [1970] to compute water-surface profiles, except that he compared the derivative, y_{i+1}', between two successive iterations instead of the depths.

6-8 Simultaneous Solution Procedure

The procedures presented in the previous sections are suitable for a single channel or for a series[†] channel system. However, to compute the gradually

---

[†] A channel system may be classified as series, parallel, branching, or network. These terms are borrowed from circuit theory in electrical engineering. In a series system, the outflow of one channel is inflow to the next and in a parallel system the channels are connected in a loop such that the flow divides at the point of separation and combines at the point of union, as shown in Fig. 6-7a.

varied flows in a channel system or in a channel network (Fig. 6-7) by these methods is difficult, if not impossible.

To illustrate this, let us consider the analysis of a simple parallel-channel system as shown in Fig. 6-7a. We first assume a discharge distribution, Q_1 and Q_2, in both channels so that the continuity equation is satisfied, i.e., $Q = Q_1 + Q_2$. Then, the water-surface profiles are computed in channel 1 for Q_1 and in channel 2 for Q_2 from the point of separation (point E) to the point of union (point F). The elevation of the energy grade line in the three channels at junction E must be the same for the computed water levels and corresponding flow velocities. This corresponds to identical water levels in all channels at the junction if the junction losses and the difference in the velocity heads in different channels at the junction are neglected. If the elevation of the energy-grade line is not the same, then other values of Q_1 and Q_2 are selected, and the entire procedure is repeated. It is clear that this is a time consuming process; for a complex network it is very difficult, if not impossible, to apply.

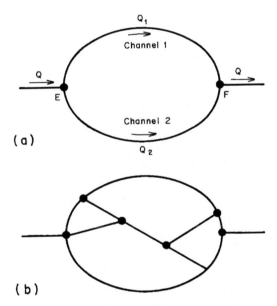

Fig. 6-7   **Parallel channel system and channel network**

We may compute gradually varied flows in single or series channels or in channel networks directly by using the simultaneous solution approach presented in this section. This approach utilizes the Newton-Raphson iterative procedure for the solution of a system of nonlinear equations [Epp and Fowler, 1970] and computes the flow conditions in the entire network simultaneously. Based on this method, Wylie [1972] presented an algorithm for channel systems. Chaudhry and Schulte [1986] used this method to analyze systems having two parallel channels; then, Schulte and Chaudhry [1987] extended this concept for application to channel networks. This solution procedure is presented in this section. Unlike the algorithm presented by Wylie, the governing equations are in terms of the commonly used variables, namely, flow depths and discharges. Therefore, this formulation is easier to understand and apply. In addition, a procedure is presented to number the nodes of a parallel system such that a banded matrix is obtained. This reduces the computational time and storage requirements and improves the accuracy of the computed results.

Some additional publications for simulating the gradually varied flow in channel networks are Choi and Molinas [1993], Kutija [1955], Nguyen and Kawano [1995], Sen and Garg [1998, 2002], and in tree-type branching channel system are Naidu et al. [1997], and in cyclic looped channel networks, Reddy and Bhallamudi [2004]

Governing Equations

Let us first present the notation we will use in the following discussion. We will use two subscripts to designate variables at different channel sections: The first subscript refers to the number of the channel, whereas the second subscript refers to the section number on that channel. For example, $y_{i,j}$ refers to the flow depth at section j of channel i. The only exception to this rule is the head loss term, $h_{f_{j,j+1}}$, which implies the losses between sections j and $j+1$.

Referring to the longitudinal profile of a channel shown in Fig. 6-8, the energy equation for the channel length (commonly termed a reach) between sections j and $j+1$ of channel i may be written as

$$z_{i,j} + y_{i,j} + \alpha_{i,j}\frac{Q_{i,j}^2}{2gA_{i,j}^2} = z_{i,j+1} + y_{i,j+1} + \alpha_{i,j+1}\frac{Q_{i,j+1}^2}{2gA_{i,j+1}^2} + h_{f_{j,j+1}}$$

$$(6\text{-}43)$$

As an approximation, the head losses between sections j and $j+1$ of channel i may be computed by using the average of the friction slopes at sections j and $j+1$. Replacing the flow velocity V by discharge, Q/flow area, A, Eq. 6-43 becomes

$$z_{i,j} + y_{i,j} + \alpha_{i,j}\frac{Q_{i,j}^2}{2gA_{i,j}^2} = z_{i,j+1} + y_{i,j+1} + \frac{Q_{i,j+1}^2}{2gA_{i,j+1}^2}$$
$$+ \frac{1}{2}\left(x_{i,j+1} - x_{i,j}\right)\left(\frac{Q_{i,j+1}^2 n_{i,j+1}^2}{C_o^2 A_{i,j+1}^2 R_{i,j+1}^{1.333}} + \frac{Q_{i,j}^2 n_{i,j}^2}{C_o^2 A_{i,j}^2 R_{i,j}^{1.333}}\right) \qquad (6\text{-}44)$$

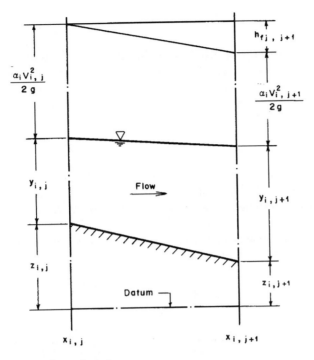

Fig. 6-8 Definition sketch

The second governing equation is the continuity equation

$$Q_{i,j} = Q_{i,j+1} \tag{6-45}$$

Equation 6-45 is valid if there is no lateral inflow or outflow between sections j and $j+1$ (Fig. 6-8). Although this equation may appear to be trivial at this stage, its inclusion in the system of governing equations becomes important while computing the water-surface profiles in branching systems or in channel networks. This will become apparent when we later discuss the analysis of branching systems and channel networks.

Single and Series Channels

In a series channel system, a number of channels are connected such that the outflow of one channel is the inflow into the next. Each channel may have different properties, e.g., cross-section, Manning n, bottom slope, etc. To facilitate an understanding of the computational procedure, we first consider a single channel and then a series channel system before studying the case of channel networks.

Figure 6-9 shows the longitudinal profile of channel i. The channel is subdivided into N_i reaches, where i refers to the channel number. If the first section is numbered 1, then the last section will be $N_i + 1$. For a single channel, the continuity equations need not be included in the system of equations, since the discharge at all sections is the same, i.e.,

$$Q_{i,1} = Q_{i,2} = \cdots = Q_{i,N_i+1} = Q_i \qquad (6\text{-}46)$$

The values of α and n are generally the same at different sections of a particular channel although they may be different for different channels. Therefore, we use only one subscript representing the channel number with these variables.

By writing the energy equation (Eq. 6-44) for each of the N_i reaches, we obtain the following system of equations

$$F_{i,1} = y_{i,2} - y_{i,1} + z_{i,2} - z_{i,1} + \frac{1}{2g}\left(\frac{\alpha_i Q_i^2}{A_{i,2}^2} - \frac{\alpha_i Q_i^2}{A_{i,1}^2}\right)$$
$$+ \frac{1}{2}\left(x_{i,2} - x_{i,1}\right)\left(\frac{Q_{i,2}^2 n_{i,2}^2}{C_o{}^2 A_{i,2}^2 R_{i,2}^{1.333}} + \frac{Q_{i,1}^2 n_{i,1}^2}{C_o{}^2 A_{i,1}^2 R_{i,1}^{1.333}}\right) = 0$$

$$F_{i,2} = y_{i,3} - y_{i,2} + z_{i,3} - z_{i,2} + \frac{1}{2g}\left(\frac{\alpha_i Q_i^2}{A_{i,3}^2} - \frac{\alpha_i Q_i^2}{A_{i,2}^2}\right)$$
$$+ \frac{1}{2}\left(x_{i,3} - x_{i,2}\right)\left(\frac{Q_{i,3}^2 n_{i,3}^2}{C_o{}^2 A_{i,3}^2 R_{i,3}^{1.333}} + \frac{Q_{i,2}^2 n_{i,2}^2}{C_o{}^2 A_{i,2}^2 R_{i,2}^{1.333}}\right) = 0$$

$$\cdots \qquad \cdots$$
$$\cdots \qquad \cdots$$
$$\cdots \qquad \cdots$$

$$F_{i,N_i} = y_{i,N_i+1} - y_{i,N_i} + z_{i,N_i+1} - z_{i,N_i}$$
$$+ \frac{1}{2g}\left(\frac{\alpha_i Q_i^2}{A_{i,N_i+1}^2} - \frac{\alpha_i Q_i^2}{A_{i,N_i}^2}\right)$$
$$+ \frac{1}{2}\left(x_{i,N_i+1} - x_{i,N_i}\right)\left(\frac{Q_{i,N_i+1}^2 n_{i,N_i+1}^2}{C_o{}^2 A_{i,N_i+1}^2 R_{i,N_i+1}^{1.333}} + \frac{Q_{i,N_i}^2 n_{i,N_i}^2}{C_o{}^2 A_{i,N_i}^2 R_{i,N_i}^{1.333}}\right) = 0$$

$$(6\text{-}47)$$

Since A and R are functions of the properties of the channel cross section and the flow depth, the preceding equations are only functions of the flow depth. However, we have N_i equations in $N_i + 1$ unknowns. Therefore, one more equation is needed to obtain a unique solution of the system of equations. This additional equation is provided by the *end condition*. For subcritical

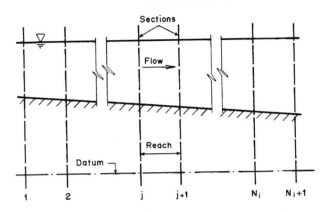

Fig. 6-9   Channel reaches

flows, the end condition is the specified flow depth, y_d, at the downstream end of the channel, i.e.,

$$F_{i,N_i+1} = y_{i,N_i+1} - y_d = 0 \qquad (6\text{-}48)$$

Similarly, the end condition for supercritical flow is a specified flow depth, y_u, at the upstream end of the channel, i.e.,

$$F_{i,1} = y_{i,1} - y_u = 0 \qquad (6\text{-}49)$$

For brevity, only the solution of a system having subcritical flow is discussed in the following paragraphs, i.e., Eqs. 6-47 and 6-48 describe the flow conditions in the channel system. These nonlinear equations may now be solved simultaneously using the Newton-Raphson method as follows. We are interested in determining corrections $\Delta y_{i,j}$ such that $y_{i,j}^{(1)} = y_{i,j}^{(0)} + \Delta y_{i,j}$ is a better estimate for the flow depth at section (i, j), where $y_{i,j}^{(0)}$ $(j = 1, 2, \ldots, N_i + 1)$ are the initial estimates for the flow depths (the superscript in the parenthesis indicates the number of the iteration).

By expanding Eqs. 6-47 and 6-48 in Taylor series and writing the system of equations in a matrix form, we obtain

$$
\begin{bmatrix}
\dfrac{\partial F_{i,1}}{\partial y_{i,1}} & \dfrac{\partial F_{i,1}}{\partial y_{i,2}} & \cdots & \dfrac{\partial F_{i,1}}{\partial y_{i,N_i+1}} \\[2ex]
\dfrac{\partial F_{i,2}}{\partial y_{i,1}} & \dfrac{\partial F_{i,2}}{\partial y_{i,2}} & \cdots & \dfrac{\partial F_{i,2}}{\partial y_{i,N_i+1}} \\[2ex]
\vdots & \vdots & \ddots & \vdots \\[2ex]
\dfrac{\partial F_{i,N_i+1}}{\partial y_{i,1}} & \dfrac{\partial F_{i,N_i+1}}{\partial y_{i,2}} & \cdots & \dfrac{\partial F_{i,N_i+1}}{\partial y_{i,N_i+1}}
\end{bmatrix}^{(0)}
\begin{pmatrix}
\Delta y_{i,1} \\[2ex]
\Delta y_{i,2} \\[2ex]
\vdots \\[2ex]
\Delta y_{i,N_i+1}
\end{pmatrix}
= -
\begin{pmatrix}
F_{i,1} \\[2ex]
F_{i,2} \\[2ex]
\vdots \\[2ex]
F_{i,N_i+1}
\end{pmatrix}^{(0)}
\tag{6-50}
$$

In this equation, the superscript 0 within the parenthesis indicates that the functions, $F_{i,j}$, and their partial derivatives are evaluated for the estimated flow depth, $y_{i,j}^{(0)}$.

The Jacobian matrix (matrix of partial derivatives) of the preceding system has an important characteristics. For each energy equation, all the partial derivatives are zero, except the partial derivative with respect to the flow depth at the section under consideration and with respect to the depth at the next section. For example, only the partial derivatives with respect to $y_{i,j}$ and $y_{i,j+1}$ of the energy equation for reach j between section j and $j+1$ are not zero. These non-zero partial derivatives are

$$
\frac{\partial F_{i,j}}{\partial y_{i,j}} = -1 + Q_i^2 \left(\frac{\alpha_i B_{i,j}}{g A_{i,j}^3} - \frac{2n_i^2 (x_{i,j+1} - x_{i,j})}{3C_o^2 A_{i,j}^2 R_{i,j}^{2.33}} \frac{dR_{i,j}}{dy_{i,j}} \right.
$$
$$
\left. - \frac{n_i^2 B_{i,j}(x_{i,j+1} - x_{i,j})}{C_o^2 A_{i,j}^3 R_{i,j}^{1.33}} \right)
\tag{6-51}
$$

$$
\frac{\partial F_{i,j}}{\partial y_{i,j+1}} = 1 - Q_i^2 \left(\frac{\alpha_i B_{i,j+1}}{g A_{i,j+1}^3} + \frac{2n_i^2 (x_{i,j+1} - x_{i,j})}{3C_o^2 A_{i,j+1}^2 R_{i,j+1}^{2.33}} \frac{dR_{i,j+1}}{dy_{i,j+1}} \right.
$$
$$
\left. + \frac{n_i^2 B_{i,j+1}(x_{i,j+1} - x_{i,j})}{C_o^2 A_{i,j+1}^3 R_{i,j+1}^{1.33}} \right)
\tag{6-52}
$$

where the value of dR/dy depends on the shape of the channel cross section, as discussed in Section 6-4.

It is clear from the above discussion that all nonzero partial derivatives lie on or near the principal diagonal of the Jacobian matrix. Therefore, the resulting matrix is banded, as shown in the following equation:

$$\begin{bmatrix} \dfrac{\partial F_{i,1}}{\partial y_{i,1}} & \dfrac{\partial F_{i,1}}{\partial y_{i,2}} & 0 & 0 & \cdots & 0 & 0 \\[2.2ex] 0 & \dfrac{\partial F_{i,2}}{\partial y_{i,2}} & \dfrac{\partial F_{i,2}}{\partial y_{i,3}} & 0 & \cdots & 0 & 0 \\[2.2ex] 0 & 0 & \dfrac{\partial F_{i,3}}{\partial y_{i,3}} & \dfrac{\partial F_{i,3}}{\partial y_{i,4}} & \cdots & 0 & 0 \\[2.2ex] \vdots & \vdots & \vdots & \vdots & \ddots & \vdots & \vdots \\[2.2ex] 0 & 0 & 0 & 0 & \cdots & \dfrac{\partial F_{i,N_i+1}}{\partial y_{i,N_i}} & \dfrac{\partial F_{i,N_i+1}}{\partial y_{i,N_i+1}} \end{bmatrix} \tag{6-53}$$

This particular Jacobian has two diagonal nonzero elements and it is referred to as a Jacobian of bandwidth two. The advantage of having such a banded matrix is that the computer memory required to store its elements and computational time required to invert it are significantly reduced. In addition, most computer systems have standard subroutines for the inversion of banded matrices.

The solution algorithm is as follows: The functions $F_{i,j}$ (Eq. 6-47) and the partial derivatives of the banded Jacobian (Eqs. 6-51 and 6-52) are computed for the estimated flow depths. Instead of the Jacobian of Eq. 6-50, the banded Jacobian of Eq. 6-53 is used, and the system is solved for the corrections, $\Delta y_{i,j}$, $(j = 1, 2, \ldots, N_i + 1)$. Then, better estimates of the flow depths are

$$y_{i,j}^{(1)} = y_{i,j}^{(0)} + \Delta y_{i,j} \tag{6-54}$$

If the absolute value of each of these corrections, $\Delta y_{i,j}$, is less than a specified tolerance, then the flow depths, $y_{i,j}^{(1)}$ computed from Eq. 6-54 are the desired solution. Otherwise, $y_{i,j}^{(0)}$ are set equal to $y_{i,j}^{(1)}$, and the previously described procedure is repeated until an acceptable solution is obtained. Good estimates of the initial flow depths are necessary for a rapid convergence of the iterations. The depth specified as the end condition may be specified as the initial estimate for the flow depths at different sections of the channel system.

Let us now discuss how to analyze a series system having M channels in series (Fig. 6-10a). First, we write the governing equations for all M channels and then solve them simultaneously by following the preceding procedure. We have $\sum_{i=1}^{M}(N_i + 1)$ sections on M channels. Therefore, we need as many equations to determine the depths at these sections. Since the discharge has the same value at all sections, we do not include it as an unknown. Therefore, we do not have to include the continuity equation in the governing equations. By writing the energy equation for all reaches of the system, we will have $\sum_{i=1}^{M} N_i$ equations. In addition, there are $M - 1$ channel junctions. For each of these junctions, we may write the energy equation as well. For example, this equation for the junction of channel i and $i + 1$ is

$$z_{i,N_i+1} + y_{i,N_i+1} + \frac{V_{i,N_i+1}^2}{2g} = z_{i+1,1} + y_{i+1,1} + (1+k)\frac{V_{i+1,1}^2}{2g} \tag{6-55}$$

in which k = coefficient of head losses at the junction. If the junction losses and the difference in the velocity heads at the junction are small, they may be neglected in this equation.

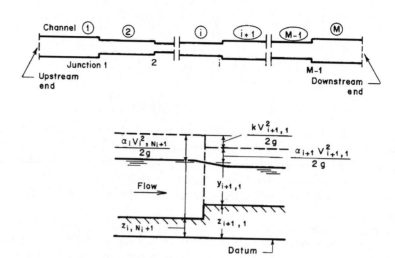

Fig. 6-10   Series channels

Thus, the energy equations for all reaches of M channels, the energy equations for the $M - 1$ channel junctions, and the end condition provide the necessary number of equations. These may be solved simultaneously to determine the depths at all sections of the system.

Channel Networks

Let us now discuss how to analyze channel networks (Fig. 6-7). We will consider only subcritical flow in the following discussion. For supercritical flow, additional constraints arise from the channel geometries at the branching nodes. The analysis of such a situation is beyond the scope of this section.

Let us consider the channel networks shown in Fig. 6-7. The flow in all channels is subcritical. In addition to the flow depths, the discharges in the individual channels are not known as well. Therefore, the continuity equation (Eq. 6-45) for each channel reach is also included to obtain the necessary number of equations.

Let us write the energy equation (Eqs. 6-47) and the continuity equation (Eq. 6-45) for N_i reaches of channel i. The resulting system of equations is

$$F_{i,1} = y_{i,2} - y_{i,1} + z_{i,2} - z_{i,1} + \frac{\alpha_i}{2g}\left(\frac{Q_{i,2}^2}{A_{i,2}^2} - \frac{Q_{i,1}^2}{A_{i,1}^2}\right)$$

$$+ \frac{1}{2}\left(x_{i,2} - x_{i,1}\right)\left(\frac{Q_{i,2}^2 n_{i,2}^2}{C_o{}^2 A_{i,2}^2 R_{i,2}^{1.333}} + \frac{Q_{i,1}^2 n_{i,1}^2}{C_o{}^2 A_{i,1}^2 R_{i,1}^{1.333}}\right) = 0$$

$$F_{i,2} = Q_{i,2} - Q_{i,1} = 0$$

$$F_{i,3} = y_{i,3} - y_{i,2} + z_{i,3} - z_{i,2} + \frac{\alpha_i}{2g}\left(\frac{Q_{i,3}^2}{A_{i,3}^2} - \frac{Q_{i,2}^2}{A_{i,2}^2}\right)$$

$$+ \frac{1}{2}\left(x_{i,3} - x_{i,2}\right)\left(\frac{Q_{i,3}^2 n_{i,3}^2}{C_o{}^2 A_{i,3}^2 R_{i,3}^{1.333}} + \frac{Q_{i,2}^2 n_{i,2}^2}{C_o{}^2 A_{i,2}^2 R_{i,2}^{1.333}}\right) = 0$$

$$F_{i,4} = Q_{i,3} - Q_{i,2} = 0$$

$$\cdots \quad \cdots \quad \cdots$$
$$\cdots \quad \cdots \quad \cdots$$
$$\cdots \quad \cdots \quad \cdots$$

$$F_{i,2N_i-1} = y_{i,N_i+1} - y_{i,N_i} + z_{i,N_i+1} - z_{i,N_i}$$

$$+ \frac{\alpha_i}{2g}\left(\frac{Q_{i,N_i+1}^2}{A_{i,N_i+1}^2} - \frac{Q_{i,N_i}^2}{A_{i,N_i}^2}\right)$$

$$+ \frac{1}{2}\left(x_{i,N_i+1} - x_{i,N_i}\right)\left(\frac{Q_{i,N_i+1}^2 n_{i,N_i+1}^2}{C_o{}^2 A_{i,N_i+1}^2 R_{i,N_i+1}^{1.333}} + \frac{Q_{i,N_i}^2 n_{i,N_i}^2}{C_o{}^2 A_{i,N_i}^2 R_{i,N_i}^{1.333}}\right) = 0$$

$$F_{i,2N_i} = Q_{i,N_i+1} - Q_{i,N_i} = 0 \tag{6-56}$$

For illustration purposes, let us consider the simplest type of network having two parallel channels, as shown in Fig. 6-11. Writing the energy and the continuity equations for the remaining three channels, $i+1$, $i+2$, and $i+3$, of this system in the same manner as Eqs. 6-56 gives a total of $2(N_i + N_{i+1} + N_{i+2} + N_{i+3})$ equations (here the subscripts refer to the channel number). Since the flow depth and the rate of discharge are the two unknowns for each section, we have $2(N_i + N_{i+1} + N_{i+2} + N_{i+3} + 4)$ unknowns. Therefore, we need eight additional equations for a unique solution. Two of these equations are given by the *end conditions*. These end conditions for subcritical flows are the specified flow depth, y_d, and the specified discharge, Q_d, at the downstream end of channel $i + 3$, i.e.,

$$F_{i+3,2N_{i+3}+1} = y_{i+3,N_{i+3}+1} - y_d = 0 \tag{6-57}$$

$$F_{i+3,2N_{i+3}+2} = Q_{i+3,N_{i+3}+1} - Q_d = 0 \tag{6-58}$$

In addition, three equations are provided by the energy equations at the junction of channels i, $i+1$, and $i+2$. Similarly, three equations are available at the junction of channels $i+1$, $i+2$, and $i+3$ (see Fig. 6-12).

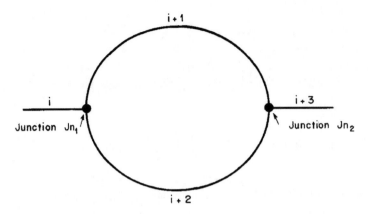

Fig. 6-11   Parallel-channel system

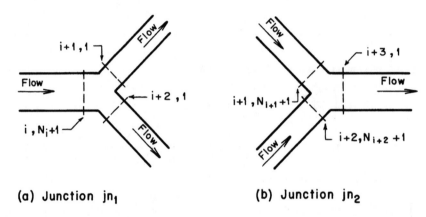

(a) Junction jn₁   **(b) Junction jn₂**

Fig. 6-12   Channel junctions

Assuming no lateral inflow, the continuity equation and the two energy equations at junction, jn_1, of channels i, $i + 1$, and $i + 2$ (Fig. 6-12a) may be written as

$$F_{jn_1,1} = Q_{i,N_i+1} - Q_{i+1,1} - Q_{i+2,1} = 0 \qquad (6\text{-}59)$$

$$F_{jn_1,2} = z_{i,N_i+1} + y_{i,N_i+1} + \frac{Q_{i,N_i+1}^2}{2gA_{i,N_i+1}^2}$$

$$- z_{i+1,1} - y_{i+1,1} - (1+k)\frac{Q_{i+1,1}^2}{2gA_{i+1,1}^2} = 0 \qquad (6\text{-}60)$$

$$F_{jn_1,3} = z_{i,N_i+1} + y_{i,N_i+1} + \frac{Q_{i,N_i+1}^2}{2gA_{N_i+1}^2}$$

$$- z_{i+2,1} - y_{i+2,1} - (1+k)\frac{Q_{i+2,1}^2}{2gA_{i+2,1}^2} = 0 \qquad (6\text{-}61)$$

where the subscript jn_1 identifies the upstream junction. The junction losses and the differences in the velocity heads at the junction may be neglected if they are small.

Similarly, the following three equations are available at the downstream junction (Fig. 6-12b), in which we use subscript jn_2 to denote values for the junction.

$$F_{jn_2,1} = Q_{i+3,1} - Q_{i+1,N_{i+1}+1} - Q_{i+2,N_{i+2}+1} = 0 \qquad (6\text{-}62)$$

$$F_{jn_2,2} = z_{i+1,N_{i+1}+1} + y_{i+1,N_{i+1}+1} + \frac{Q_{i+1,N_{i+1}+1}^2}{2gA_{i+1,N_{i+1}+1}^2}$$

$$- z_{i+3,1} - y_{i+3,1} - (1+k)\frac{Q_{i+3,1}^2}{2gA_{i+3,1}^2} = 0 \qquad (6\text{-}63)$$

$$F_{jn_2,3} = z_{i+2,N_{i+2}+1} + y_{i+2,N_{i+2}+1} + \frac{Q_{i+2,N_{i+2}+1}^2}{2gA_{i+2,N_{i+2}+1}^2}$$

$$- z_{i+3,1} - y_{i+3,1} - (1+k)\frac{Q_{i+3,1}^2}{2gA_{i+3,1}^2} = 0 \qquad (6\text{-}64)$$

For complex networks (Fig. 6-7b), Eqs. 6-59 to 6-61 or Eqs. 6-62 to 6-64 are included for the branching junction of three channels; and Eq. 6-55 is included for a series junction of two channels.

Now, the system of equations is solved simultaneously as follows. We first estimate $y_{l,j}^{(0)}$ as well as $Q_{l,j}^{(0)}$, ($l = i, i+1, \cdots, i+3$ and $j = 1, 2, \cdots, N_i + 1$). Reasonable initial estimated values for the flow depths may be obtained by setting all of them equal to the downstream depth specified as the end condition. It is desirable to satisfy continuity at each node by choosing appropriate discharge estimates as well as the respective flow directions. However, a correct solution is obtained even if the initial estimates do not satisfy the continuity condition, although convergence in this case is slower. To account for the reverse flow, i.e.., flow direction opposite to the assumed one, the energy equation may be written as

$$F_{i,k} = y_{i,j+1} - y_{i,j} + z_{i,j+1} - z_{i,j}$$

$$+ \frac{\alpha_i}{2g} \left(\frac{Q_{i,j+1}|Q_{i,j+1}|}{A_{i,j+1}^2} - \frac{Q_{i,j}|Q_{i,j}|}{A_{i,j}^2} \right)$$

$$+ \frac{1}{2}(x_{i,j+1} - x_{i,j}) \left(\frac{n_{i,j+1}^2 Q_{i,j+1}|Q_{i,j+1}|}{C_o^2 A_{i,j+1}^2 R_{i,j+1}^{1.33}} + \frac{n_{i,j}^2 Q_{i,j}|Q_{i,j}|}{C_o^2 A_{i,j}^2 R_{i,j}^{1.33}} \right) = 0$$

$$(6\text{-}65)$$

in which $Q_{i,j}^2$ is replaced by $Q_{i,j}|Q_{i,j}|$ to yield correct sign for the head-loss term.

By expanding in Taylor series, a matrix system similar to Eq. 6-50 is obtained. For each energy equation, there are now four nonzero partial derivatives, namely, the partial derivatives with respect to the flow depth and with respect to the discharge at the section under consideration as well as the partial derivatives with respect to the corresponding variables for the adjacent section. Thus, for an energy equation $F_{i,k}$ between sections j and $j+1$ of channel i, the following nonzero partial derivatives are obtained:

$$\frac{\partial F_{i,k}}{\partial y_{i,j}} = -1 + Q_{i,j}^2 \left(\frac{\alpha_i B_{i,j}}{g A_{i,j}^3} - \frac{2n_i^2(x_{i,j+1} - x_{i,j})}{3C_o^2 A_{i,j}^2 R_{i,j}^{2.33}} \frac{dR_{i,j}}{dy_{i,j}} \right.$$

$$\left. - \frac{n_i^2 B_{i,j}(x_{i,j+1} - x_{i,j})}{C_o^2 A_{i,j}^3 R_{i,j}^{1.33}} \right)$$

$$\frac{\partial F_{i,k}}{\partial Q_{i,j}} = 2Q_{i,j} \left(-\frac{\alpha_i}{2g A_{i,j}^2} + \frac{n_i^2(x_{i,j+1} - x_{i,j})}{2C_o^2 A_{i,j}^2 R_{i,j}^{1.33}} \right)$$

$$\frac{\partial F_{i,k}}{\partial y_{i,j+1}} = 1 - Q_{i,j+1}^2 \left(\frac{\alpha_i B_{i,j+1}}{g A_{i,j+1}^3} + \frac{2n_i^2(x_{i,j+1} - x_{i,j})}{3C_o^2 A_{i,j+1}^2 R_{i,j+1}^{2.33}} \frac{dR_{i,j+1}}{dy_{i,j+1}} \right.$$

$$\left. + \frac{n_i^2 B_{i,j+1}(x_{i,j+1} - x_{i,j})}{C_o^2 A_{i,j+1}^3 R_{i,j+1}^{1.33}} \right)$$

$$\frac{\partial F_{i,k}}{\partial Q_{i,j+1}} = 2Q_{i,j+1} \left(-\frac{\alpha_i}{2g A_{i,j+1}^2} + \frac{n_i^2(x_{i,j+1} - x_{i,j})}{2C_o^2 A_{i,j+1}^2 R_{i,j+1}^{1.33}} \right) \qquad (6\text{-}66)$$

in which subscript k refers to the equation number, and its value is not identical to that of j. Similarly, the nonzero partial derivatives for any continuity equation $F_{i,k+1}$ are those with respect to the discharges of the adjacent sections, i.e.,

$$\frac{\partial F_{i,k+1}}{\partial Q_{i,j}} = -1$$

$$\frac{\partial F_{i,k+1}}{\partial Q_{i,j+1}} = 1 \qquad (6\text{-}67)$$

Similar equations may be written for the remaining three channels. The partial derivatives at the junctions are (values in the parenthesis apply if the losses and the difference in the velocity heads at the junction are neglected):

$$\frac{\partial F_{jn}}{\partial y_{i,j}} = -1 + \frac{Q_{i,j}^2 B_{i,j}}{gA_{i,j}^3} \qquad (\text{or} = -1)$$

$$\frac{\partial F_{jn}}{\partial y_{i,j+1}} = 1 - \frac{(1+k)Q_{i+1,1}^2 B_{i+1,1}}{gA_{i+1,1}^3} \qquad (\text{or} = 1)$$

$$\frac{\partial F_{jn}}{\partial Q_{i,j}} = -\frac{Q_{i,j}}{gA_{i,j}^2} \qquad (\text{or} = 0)$$

$$\frac{\partial F_{jn}}{\partial Q_{i+1,1}} = \frac{(1+k)Q_{i+1,1}}{gA_{i+1,1}^2} \qquad (\text{or} = 0) \tag{6-68}$$

Note that the above equations are written by assuming the flow direction from section (i, j) to section $(i + 1, 1)$.

If the equations for a channel network are arbitrarily arranged, then all the nonzero elements of the Jacobian matrix may not necessarily lie on or near the principal diagonal as was the case for a series system. This results in increased storage requirements, increased computer time, and, most probably, reduced accuracy. For the parallel-channel network shown in Fig. 6-11, these limitations may be avoided by arranging the governing equations as discussed in the following paragraphs.

For each reach of channel i, the energy and continuity equations are written consecutively from section 1 to section $N_i + 1$. Then, Eqs. 6-59 through 6-61, i.e., $F_{jn_1,k}$, $(k = 1, 2, 3)$ are written for the upstream junction. The energy equation between sections 1 and 2 of channel $i + 1$ is written, followed by the energy equation between sections 1 and 2 of channel $i+2$. Then, the continuity equation between sections 1 and 2 of channel $i + 1$ is written, followed by the continuity equation between sections 1 and 2 of channel $i + 2$. These four equations are repeated in the same alternating sequence for the remaining reaches of the parallel channels, $i + 1$ and $i + 2$. In order to have such a numbering system, it is necessary that the number of sections on each of the parallel channels be the same. Next, Eqs. 6-62 through 6-64 are written for the downstream junction, jn_2. Finally, the energy and continuity equations for channel $i + 3$ are written similar to those for channel i. If the governing equations are arranged in this manner, then the resulting Jacobian is banded with a bandwidth of seven. The remainder of the solution algorithm is identical to that described previously. For a parallel-channel system with M parallel channels, this arrangement of equations results in a Jacobian of bandwidth $3M + 1$.

For complex channel networks, however, no generalized procedure is available for arranging the governing equations to produce a Jacobian matrix of minimum bandwidth. This is because the system is asymmetric. However, by manually numbering each channel and each node from the upstream end to the downstream end in a semi-systematic manner, a banded matrix of small bandwidth may be obtained. It may not, however, have the minimum band width. A number of procedures are available to reduce the bandwidth of non-symmetric matrices [Berztiss, 1971; Deo, 1974]. However, they are generally very complex and the effort required to apply them exceeds the additional time and storage required for the solution of a non-minimized Jacobian. Therefore, the use of these methods does not appear to be cost effective, and they are not presented herein.

Example 6-3

Figure 6-13 shows a channel network. The channel data are listed in Table 6-3. All channels have subcritical flow and trapezoidal cross sections having side slopes of 1.5 horizontal to 1 vertical. The flow depth at the downstream end (Node F) is 5 m for a flow of 250 m^3/s. The form loss coefficient, k, for all junctions is 0.0 and the velocity-head coefficient, α, for all channels is 1. Compute the flows and depths in different channels of the system.

Solution

The channels and nodes are numbered as shown in Fig. 6-13. The iterative procedure is started by assuming all flow depths equal to 5.0 m. The discharges are given random values without satisfying continuity at the branching junctions. The assumed flow directions in different channels are as shown by the arrows of Fig. 6-13. A tolerance of 0.0005 for both $\Delta y_{i,j}$ and $\Delta Q_{i,j}$ is specified for the convergence of the solution procedure. The iterations converge after only six iterations. The computed discharges and flow depths are listed in Table 6-4.

Table 6-3   Channel Data

Channel	Length (m)	Width (m)	Reach (m)	n	So
1	200.0	30.0	50.0	0.013	0.0005
2	200.0	40.0	50.0	0.013	0.0005
3	200.0	20.0	50.0	0.012	0.0005
4	100.0	20.0	25.0	0.014	0.0005
5	100.0	20.0	25.0	0.013	0.0005
6	100.0	25.0	25.0	0.013	0.0005
7	100.0	30.0	25.0	0.014	0.0005
8	300.0	50.0	75.0	0.014	0.0005

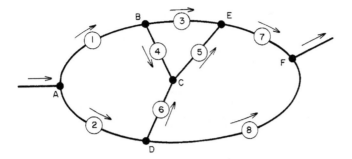

Fig. 6-13   **Network of Example 6-3**

These values were verified by computing the flow profiles by using the fourth-order Runge-Kutta method. Instead of using graphical or other manual methods to determine the discharge in various channels, the values calculated by the procedure of this section were used directly. The water levels computed by the Runge-Kutta method were identical (within the specified tolerance) to those listed in Table 6-4. The computational time for each individual channel was approximately the same for both methods. It is not possible to estimate the number of trial-and-error iterations required to determine the discharge in the Runge-Kutta method. Assuming m such iterations, the computational time for the Runge-Kutta method is at least m times that required by the procedure presented in this section.

Practical Applications

Parallel-channel systems occur frequently in nature, mainly in the form of flow around islands. However, channel networks are not as common as parallel systems, although they occur in a braided river system, e.g., in a delta. A common design problem is to provide cutoff channels in a meandering stream for flood control. The allowable flow rate and water levels in the old stream dictate the design of the new channel. By using the algorithm presented in this section, different designs may be modeled efficiently.

The algorithm of this section may also be used to determine other channel parameters, such as the roughness coefficients. For example, if the flow depths in a channel system are known for a specified discharge, then the governing equations may be solved simultaneously to directly determine Manning n instead of using a trial-and-error procedure.

Table 6-4   Network of Example 6-3

Section	Channel 1 $Q = 95.748 \text{ m}^2/\text{s}$		Channel 2 $Q = 154.252 \text{ m}^2/\text{s}$		Channel 3 $Q = 55.003 \text{ m}^2/\text{s}$		Channel 4 $Q = 40.655 \text{ m}^2/\text{s}$	
	Depth (m)	Distance (m)	Depth (m)	Distance (m)	Depth (m)	Distance (m)	Depth (m)	Distance (m)
1	4.754	0.0	4.754	0.0	4.853	0.0	4.853	0.0
2	4.779	50.0	4.779	50.0	4.877	50.0	4.865	25.0
3	4.803	100.0	4.803	100.0	4.902	100.0	4.677	50.0
4	4.828	150.0	4.828	150.0	4.927	150.0	4.800	75.0
5	4.853	200.0	4.852	200.0	4.951	200.0	4.902	100.0

Section	Channel 5 $Q = 52.669 \text{ m}^2/\text{s}$		Channel 6 $Q = 12.014 \text{ m}^2/\text{s}$		Channel 7 $Q = 107.762 \text{ m}^2/\text{s}$		Channel 8 $Q = 142.238 \text{ m}^2/\text{s}$	
	Depth (m)	Distance (m)	Depth (m)	Distance (m)	Depth (m)	Distance (m)	Depth (m)	Distance (m)
1	4.902	0.0	4.852	0.0	4.051	0.0	4.352	0.0
2	4.914	25.0	4.864	25.0	4.986	25.0	4.860	75.0
3	4.927	50.0	4.877	50.0	4.979	50.0	4.926	150.0
4	4.939	75.0	4.889	75.0	4.988	75.0	4.963	225.0
5	4.952	100.0	4.902	100.0	5.000	100.0	5.000	300.0

6-9 Summary

A number of procedures to compute steady-state, gradually varied flows were presented in this chapter. The first two methods – direct and standard step – solve the energy equation between two consecutive channel sections. Then, a number of numerical methods were presented to integrate the governing differential equation. These methods do not allow a direct solution of parallel-channel systems or complex channel networks. Therefore, trial-and-error procedures have to be used. For the analysis of these channel systems, a simultaneous solution algorithm based on the Newton-Raphson method was presented.

Problems

6-1 Derive Eq. 6-3 by using the principle of conservation of momentum. (Hint: Apply Newton's second law of motion to a short channel length, Δx).

6-2 A 10-m wide, rectangular, concrete-lined canal has a bottom slope of 0.01 and a constant-level lake at the upstream end. The water level in the lake is 6.0 m above the bottom of the canal at the entrance. If the entrance losses are negligible, determine

   i. The flow depth 800 m downstream of the canal entrance; and

   ii. The distance from the lake where the flow depth is 2.5 m.

6-3 A trapezoidal channel having a bottom slope of 0.001 is carrying a flow of 75 m³/s. The channel bottom is 50 m wide, $n = 0.025$, and the channel side slopes are 1.5 horizontal to 1 vertical. If a control structure is built at the downstream end that raises the water depth at the downstream end to 12 m, determine the amount by which the channel banks must be raised along its length. Assume the channel had uniform flow prior to the construction of the control structure.

6-4 A 5 km long lined canal has a free overfall at the lower end and a constant-level reservoir at the upper end. If the critical depth at the fall is 4 m, determine the minimum water level in the lake assuming $n = 0.013$ and the head losses at the entrance $= 0.2V^2/(2g)$. The canal bottom width is 8.0 m, side slopes are 1.5H : 1V, and the channel bottom slope is 0.0001.

6-5 Figure 6-14 shows the cross section of a natural stream. The channel bottom slope is 0.0002, $n = 0.035$, and the discharge is 80 m³/s. If the flow depth at a bridge crossing is 8.0 m, determine the flow depth 3.0 km upstream of the bridge.

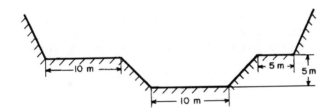

Fig. 6-14   Channel cross section for Prob. 6-5

6-6 Develop computer programs to compute the water-surface profile in a trapezoidal channel having a free overfall at the downstream end. To compute the profile, use the following methods:

   i. Direct step method;

   ii. Standard Step method;

   iii. Euler method;

iv. Modified Euler method; and

v. Fourth-order Runge-Kutta method.

6-7 By using the computer programs of Prob. 6-6, compute, plot, and compare the water-surface profile in a trapezoidal channel having the following data:

Bottom width = 20.0 m;

Side slopes = 2 horizontal to 1 vertical;

Manning $n = 0.013$;

Discharge = 30 m^3/s;

Channel bottom slope = 0.00015; and

A free-overfall at the downstream end.

Select appropriate values for the flow depths in method (i) and appropriate distance locations in methods (ii) to (v) of Prob. 6-6.

6-8 Investigate the sensitivity of the computed water level at a distance of 5 km upstream from the outfall by using different increments for the flow depth and different distance locations in Prob. 6-4.

6-9 Write a computer program to compute the water-surface profile in a trapezoidal channel having steep bottom slope. The water-level in a constant-level lake located at the upper end is 1.0 m above the normal depth in the channel. Neglect the entrance losses.

6-10 Use the computer program of Prob. 6-9 to determine the flow depth 0.5 km downstream from the lake outlet.

6-11 Verify the validity of the computer programs developed in Prob. 6-6 by comparing the computed results with those obtained from the following equation for a very wide channel derived by Bresse [Bresse, 1980; Rouse, 1950]:

$$x = \frac{y}{S_o} - y_n \left[\frac{1}{S_o} - \frac{C^2}{g} \right] \phi$$

In this equation, the Bresse function is

$$\phi = \frac{1}{6} \ln \left[\frac{w^2 + w + 1}{(w-1)^2} \right] - \frac{1}{\sqrt{3}} \tan^{-1} \left[\frac{\sqrt{3}}{2w+1} \right] + C_1$$

$w = y/y_n$; C_1 = a constant of integration; y = flow depth at distance x; y_n = normal depth; S_o = bottom slope; and C = Chezy constant.

[Hint: Solve and compare the results for flow in a 500-m wide, rectangular channel with $S_o = 0.0008$ and $C = 100$ for a flow of $Q = 500$ m^3/s. Assume the critical depth occurs 10 m upstream of the free overfall].

6-12 By using the Euler, modified Euler, and fourth-order Runge-Kutta methods, compute the flow depth 1 km upstream of the fall of Prob. 6-11 by using $\Delta x = 25, 50, 100$, and 200 m. By assuming the results of the Bresse equation to be exact, compute and plot the error versus Δx.

6-13 A meandering river (Fig. 6-15) has a bottom slope of 0.1 m/km. The stations as marked are distances, in km, along the river from point A. To reduce flooding, it is planned to provide cutoffs as shown. The river bottom at point F is at El. 500 and the water level at F for a flow of 500 m³/s is at El. 505. The Manning n for the river channel and for the cutoff are 0.050 and 0.020 respectively. The river channel is 500 m wide and the cutoffs are 150 m wide.

Compute the water level at point A if

   i. There are no cutoffs;
  ii. Only cutoff BD is provided;
 iii. Both cutoffs BD and CE are provided.

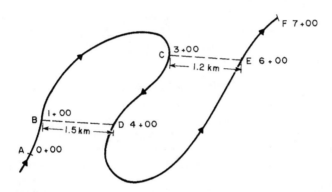

Fig. 6-15   **Meandering river and cutoffs**

6-14 Figure 6-16 shows the tailrace system of a G.M. Shrum Hydroelectric Power Plant. If the water level at the downstream weir is at El. 504.00 for a flow of 1688 m³/s, determine the water levels in each manifold.

6-15 Prove that all three formulations for the approximation of the friction slope (Eqs. 6-13a to 6-13c) give identical results if the terms of the order $(\Delta S_f/S_{f1})^2$ and higher are neglected. In this expression, $\Delta S_f = S_{f2} - S_{f1}$.

6-16 In order to reduce the height of the Pont du Gard for the Roman Aqueduct of Nimes (Fig. 4-10), the bottom slope of the upstream part was increased [Hauck and Novak, 1987]. This resulted in reducing the available bottom slope downstream of the crossing. There has been heavy incrustation of the aqueduct due to low velocities and due to some other factors. If you were the designer rehabilitating the aqueduct, list the modifications you would propose to maximize the flow capacity with minimum structural modification

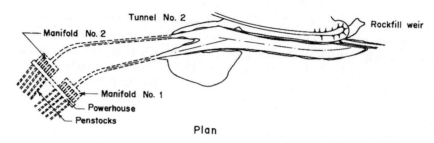

Plan

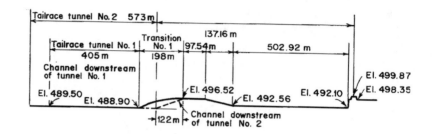

Center line profile

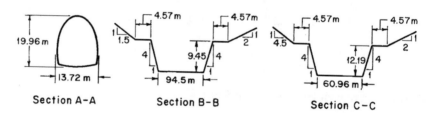

Section A-A   Section B-B   Section C-C

Fig. 6-16   **Tailrace system of a G.M. Shrum Hydropower Plant**

to the channel cross sections (you may raise or lower the channel). Analyze and compare the effect of these modifications on the flow capacity, costs, and available freeboard along the length. Which modification do you prefer and why?

By assuming the incrustation progressed as shown in Fig. 6-17, compute and plot the water levels in the aqueduct for flows of 210 and 450 l/s.

6-17 Plot the water surface profile in the outlet of Problem 3-18.

6-18 Debris accumulation at a bridge raised the water level to 12 ft. The trapezoidal flood channel is 20 ft wide at the bottom, has side slopes of 2H :

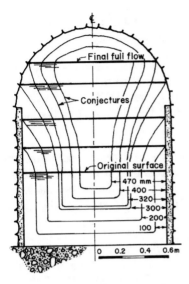

Fig. 6-17   **Possible scenarios for progressive incrustation** (After Hauck and Novak, 1987)

1V, and the channel bottom slope is 0.0003. How far will the effect of clogging extend for a flow of 800 ft³/sec?

6-19 A 10-ft square box culvert is 150 ft long and is laid at a slope of 0.01. The flow depth upstream of the entrance is 15 ft. The accumulation of debris at a channel crossing 0.5 miles downstream of the culvert raises the water level by 5 ft at the crossing. The channel is trapezoidal in shape, 10 ft wide at the bottom and has side slopes of 1.5H : 1V. If the channel bottom level drops 1.2 ft from the culvert exit to the crossing, compute and plot the water surface profiles in the channel and inside the culvert. Assume the channel had uniform flow prior to the accumulation of debris.

6-20 A 10-m wide, rectangular, concrete-lined channel ($n = 0.013$) has a bottom slope of 0.01. There is a constant-level lake at the upstream end with the lake water surface 5 m above the channel bottom. If the flow depth at the channel entrance is critical, determine the locations where the flow depth is 3.9, 3.7, 3.5, 3.3, and 3.0 m.

6-21 A rectangular canal is 10 m wide and carries a flow of 50 m³/s. The bottom and sides of the canal are concrete-lined, the longitudinal bottom slope is 0.0006 and the canal ends in a free outfall. What is the depth of flow 2 km upstream of the fall? Assume the concrete lining has deteriorated somewhat due to weathering.

If the flow depth is critical at a distance of $4y_c$ upstream of the fall, compute the water surface profile in the canal.

6-22 A trapezoidal channel with bottom width of 10 m and side slopes of 1.5H : 1V is carrying a flow of 80 m^3/s. The channel bottom slope is 0.002 and $n = 0.015$. A dam is planned that will raise the flow depth to 10 m. Compute the flow depth in the channel 250, 500, and 750 m upstream of the dam.

6-23 The normal depth in a 10-m wide rectangular channel having a bottom slope of 0.001 is 2 m. The Manning n for the flow surfaces may be assumed to be 0.020. The construction of a bridge raises the water level upstream of the bridge by 1 m. Determine the distance from the bridge where the flow depth in 2.5 m.

6-24 Compute the discharge in a rock channel ($n = 0.035$) having a bottom slope of 0.001 and flow depth of 3 ft. What is the critical depth at this flow? Is the flow critical, subcritical, or supercritical?

6-25 The normal depth in a 10 m wide rectangular channel with a bottom slope of 0.001 is 2 m. The Manning n for the flow surfaces is 0.020. The channel is constricted to build a bridge which raises the water level on the upstream side of the bridge by 2 m. Determine the distance from the bridge where the flow depth is 3.5 m.

6-26 Compute the water surface profile in the trapezoidal channel of Problem 4-32 with the depth at the downstream end raised above the normal depth by $f+5$. m, where f is the last digit in the year of your date of birth (i.e., $f = 2$ in 2002.) Compute the depth until the depth is $1.1y_n$, where y_n is the normal depth.

Graduate students are required to use any two methods of their choice for computing the water surface profile; undergraduates use only one method.

References

Bakhmeteff, B. A., 1932, *Hydraulics of Open Channels,* McGraw-Hill Book Co., New York, NY.

Berztiss, A.T., 1971, *Data Structures–Theory and Practice,* Academic Press, Inc., New York, NY.

Boudine, E.J., 1861, De l'axe hydraulique des cours d'eau contenus dans un lit prismatique et des dispositifs realisant, en pratique, ses formes diverses (The flow profiles of water in a prismatic channel and actual dispositions in various forms), *Annales des travaux publiques de Belgique,* Brussels, vol. 20, 397–555.

Bresse, J.A.C., 1860, *Cours de Mecanique Appliquee, Hydraulique,* 2e. partie, Mallet-Bachelier, Paris, France.

Chapra, S.C., and Canale, R.P., 1988, *Numerical Methods for Engineers,* second edition, McGraw-Hill Book Co., New York, NY.

Chaudhry, M. H., and Schulte, A., 1986, "Computation of Steady-State, Gradually Varied Flows in Parallel Channels," *Canadian Jour. of Civil Engineering,* vol. 13, no. 1, pp. 39–45.

Choi, G.W., and Molinas, A. (1993). "Simultaneous Solution Algorithm for Channel Network Modeling." *Water Resources Research,* vol. 29, no. 2, pp. 321–328.

Chow, V. T., 1959, *Open-Channel Hydraulics,* McGraw-Hill Book Co., New York, NY.

Davis, D.W., and Burnham, M.W., 1987, "Accuracy of Computed Water Surface Profiles," *Proc., National Hydraulic Engineering Conference,* Amer. Soc. Civil Engrs., pp. 818–823.

Deo, N., 1974, *Graph Theory with Applications to Engineering and Computer Science,* Prentice-Hall, Inc., Englewood Cliffs, NJ.

Eichert, Bill S., 1970, "Survey of Programs for Water-Surface Profiles," *Jour. Hyd. Div.,* Amer. Soc. Civ. Engrs., no. 2, pp. 547–563.

Epp, R., and Fowler, A.G., 1970, "Efficient Code for Steady- State Flows in Networks," *Jour. Hyd. Div.,* Amer. Soc. Civ. Engrs., vol. 96, no. HY1, Jan., pp. 43–56.

Federal Emergency Management Agency, 1995, "Guidelines and Specifications for Study Contractors," January.

French, R. H., 1985, *Open-Channel Hydraulics,* McGraw-Hill Book Co., New York, NY.

Hauck, G.F., and Novak, R.A., 1987, "Interaction of Flow and Incrustation in the Roman Aqueduct of Nimes," *Jour. Hydraulic Engineering,* Amer. Soc. Civil Engrs., vol 113, no 2, pp. 141–157.

Henderson, F. M., 1966, *Open-Channel Flow,* Macmillan Publishing Co., New York, NY.

Humpidge, H. B., and Moss, W. D., 1971, "The Development of a Comprehensive Computer Program for the Calculation of Flow Profiles in Open Channels," *Proc. Inst. Civ. Engrs.,* vol. 50, Sept., pp. 49–65.

Kumar, A., 1978, "Integral Solutions of the Gradually Varied Equations for Rectangular and Triangular Channels," *Proc. Inst. Civ. Engrs.,* vol. 65, pt. 2, Sept., pp. 509–515.

Kumar, A., 1979, "Gradually Varied Surface Profiles in Horizontal and Adversely Sloping Channels," *Proc. Inst. Civ. Engrs.,* vol. 67, pt. 2, Jun., pp. 435–452.

Kutija, V. (1995). "A Generalized Method for the Solution of Flows in Networks," *Jou. Hyd. Research,* vol. 33, no. 4, pp. 535–555.

Laurenson, E.M., 1986, "Friction Slope Averaging in Backwater Calculations," *Jour. Hydraulic Engineering,* Amer. Soc. Civil Engrs., vol. 112, no. 12, pp. 1151–1163.

McBeans, E., and Perkins, F., 1975, "Numerical Errors in Water Profile Computation," *Jour. Hyd. Div.,* Amer. Soc. Civ. Engrs., vol. 101, no. 11, pp. 1389–1403.

McCracken, D. D. and Dorn, W. S., 1964, *Numerical Methods and FORTRAN Programming,* John Wiley and Sons, Inc., New york, NY.

Molinas, A., and Yang, C.T., 1985, "Generalized Water Surface Profile Computations," *Jour. Hydraulic Engineering,* Amer. Soc. Civil Engrs., vol. 111, no. 3, pp. 381–397.

Naidu, B. J.., Bhallamudi, S. M., and Narasimhan, S. (1997). "GVF Computation in Tree-Type Channel Networks," *Jour. Hyd. Engineering,* Amer. Soc. Civ. Engrs., vol. 123, no. 8, pp. 700–708.

Nguyen, Q. K., and Kawano, H., 1995,. "Simultaneous Solution for Flood Routing in Channel Networks," *Jour. Hyd. Engineering,* Amer. Soc. Civ. Engrs., vol. 121, no. 10, pp. 744–750.

Paine, J. N., 1992, "Open-Channel Flow Algorithm in Newton-Raphson Form," *Jour. Irrigation and Drainage Engineering,* Amer. Soc. Civ. Engrs., vol. 118, no. 2, pp. 306–319.

Prasad, R., 1970, "Numerical Method of Computing Flow Profiles," *Jour. Hyd. Div.,* Amer. Soc. Civ. Engrs., vol. 96, no. 1, pp. 75–86.

Reddy, H. P., and Bhallamudi, S. M., 2004, "Gradually Varied Flow Computation in Cyclic Looped Channel Networks," *Jour. Irrigation and Drainage Engineering,* Amer. Soc. Civ. Engrs., vol.130, no. 5, pp. 424–431.

Rhodes, D. G., 1993, Discussion of "Open Channel Flow Algorithm in Newton-Raphson Form," by John N. Paine, *Jour. Irrigation and Drainage Engineering,* Amer. Soc. Civ. Engrs., vol. 119, no. 5, pp. 914–922.

Rhodes, D.G., 1995, "Newton-Raphson Solution for Gradually Varied Flow," *Jour. Hyd. Research,* International Assoc. for Hydraulic Research, vol. 33, no. 2, pp. 213–218.

Rouse, H., (ed.) 1950, *Engineering Hydraulics,* John Wiley & Sons, New York, NY.

Schulte, A. M., and Chaudhry, M.H., 1987, "Gradually Varied Flows in Open Channel Networks," *Jour. of Hydraulic Research* , Inter. Assoc. for Hydraulic Research, vol. 25, no. 3, pp. 357–371.

Sen, D. J., and Garg, N.K., 1998, "Efficient Solution Technique for Dendritic Channel Networks using FEM," *Jour. Hyd. Engineering,* Amer. Soc. Civ. Engrs., vol. 124, no. 8, pp. 831–839.

Sen, D. J., and Garg, N.K., 2002, "Efficient Algorithm for Gradually Varied Flows in Channel Networks," *Jour. Irrigation and Drainage Engineering,* Amer. Soc. Civ. Engrs., vol. 128, no. 6, pp. 351–357.

Soil Conservation Service, 1976, *WSP-2 Computer Program*, Technical Release No. 61.

Subramanya, K., 1986, *Flow in Open Channels,* Tata McGraw-Hill, New Delhi, India, pp. 149–155.

United States Army Corps of Engineers, 1982, *HEC-2, Water Surface Profiles, User's Manual,* Hydrologic Engineering Center, Davis, CA.

Unites States Geological Survey, 1976, "Computer Applications for Step-Backwater and Floodway Analysis," *Open file Report* 76–499.

Wylie, E.B., 1972, "Water Surface Profiles in Divided Channels," *Jour. of Hyd. Research*, Inter. Assoc. for Hydraulic Research, vol. 10, no. 3, pp. 325–341.

7

RAPIDLY VARIED FLOW

Arial view of Oroville lake and flow through damaged spillway on April 21, 2017 (Courtesy, California Department of Water Resources)

© Springer Nature Switzerland AG 2022
M. H. Chaudhry, *Open-Channel Flow*,
https://doi.org/10.1007/978-3-030-96447-4_7

7-1 Introduction

The streamlines in the uniform and gradually varied flows we considered in the previous chapters are either parallel or may be assumed as parallel. Therefore, the accelerations in these flows is negligible and the pressure distribution may be assumed as hydrostatic. The analyses in which the pressure distribution is hydrostatic is referred to as the *shallow-water theory*. However, as we discussed in Chapters 1 and 5, many times the streamlines have sharp curvatures, thereby making the assumption of hydrostatic pressure distribution invalid. In addition, the flow surface may become discontinuous if the flow depth changes rapidly such that the surface profile breaks, e. g., in a hydraulic jump.

Due to nonhydrostatic pressure distribution, rapidly varied flow cannot be analyzed by using the same approach as that for parallel flow. In the past, these flows have been mainly investigated experimentally; and empirical relationships and other information in the form of charts and diagrams have been developed. Each particular phenomenon has been studied more or less in isolation, and a considerable amount of information is available for typical design applications.

The rapidly varied flows have been analyzed based on the Boussinesq and Fawer [1937] assumptions. In the Boussinesq assumption, the vertical flow velocity is assumed to vary linearly from zero at the channel bottom to the maximum at the free surface. In the Fawer assumption, this variation is assumed to be exponential.

The rapidly varied flow usually occurs in a short distance. Therefore, the losses due to shear at the channel boundaries are small and may be neglected in a typical analysis. However, because of sudden changes in the channel geometry, the flow may separate, and eddies and swirls may form. These phenomena complicate the flow pattern, and it becomes difficult to generalize the velocity distribution at a cross section. Even in those cases where it may be possible to approximate the velocity distribution, the energy coefficient, α, and the momentum coefficient, β, cannot be easily estimated and usually have a value significantly higher than unity. Because of separation zones, and formation of rollers and eddies, it becomes difficult to define the flow boundaries and to determine the average flow variables for a cross section.

In this and in the following chapter, we discuss rapidly varied flows. The material presented in this chapter follows the traditional approach. First, we discuss how the application of common conservation laws for the analysis of rapidly varied flows requires special consideration. Then, empirical information on the transitions, hydraulic jump, spillways, and energy dissipators is included for design purposes. Chapter 8 deals mainly with the numerical computation of these flows.

7-2 Application of Conservation Laws

As we discussed in the last section, the assumption of hydrostatic pressure distribution may not be valid in rapidly varied flows because the streamlines are not parallel, due to variation in the cross-sectional shape and size, due to change in the flow direction, or for other reasons. Therefore, special care should be exercised while applying the conservation laws of mass, momentum, and energy for the analysis of rapidly varied flows. To illustrate this, we discuss a number of typical situations in the following paragraphs [Ippen, 1950].

Let us consider the flow conditions at three sections in a constant-width channel with a sudden step in its bottom (Fig. 7-1). Sections 1 and 3 are located at some distance upstream and downstream of the step but section 2 is located just downstream of the step. Let us assume the flow is uniform at section 1 and it becomes uniform again at section 3. The flow separates at the step due to sudden change in the bottom profile, and the velocity distribution at section 2 is as shown in this figure (Fig. 7-1). At sections 1 and 3, the pressure distribution may be assumed hydrostatic and the velocity in most of the cross section is approximately equal to the mean velocity.

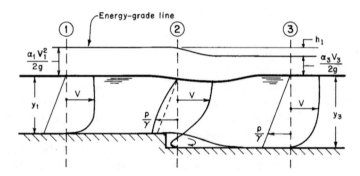

Fig. 7-1   Definition sketch for abrupt drop

Let us now consider the application of the three conservation laws one by one to the flow at sections 1, 2, and 3.

The volumetric rate of flow through a cross section may be written as

$$Q = \int_A \mathbf{v} \cdot d\mathbf{A} \tag{7-1}$$

in which Q = volumetric rate of flow and $\mathbf{v}$ = flow velocity at any point in the cross section. Note that the dot product automatically takes care of whether vector $\mathbf{v}$ and $\mathbf{A}$ are orthogonal or not. If the flow velocity v is normal to the flow area dA, then we may write this equation as

$$Q = \int_A v dA \tag{7-2}$$

Assuming the mass density to be constant and the flow to be steady, the volumetric flow rate at section 1 is equal to the flow rate at section 3, i. e., $Q_1 = Q_3$, or $V_1 A_1 = V_3 A_3$ in which V = mean flow velocity at the section and the subscripts 1 and 3 refer to the variables for cross sections 1 and 3, respectively. This is commonly known as the *continuity equation*. However, note that the flow velocity at section 2 is negative in part of the channel depth. In this case, we cannot express the volumetric flow rate in terms of the mean flow velocity. Thus, it is necessary to know the velocity distribution to compute the rate of discharge at section 2.

For computing the rate of momentum flux at any section, we may write

$$\text{Rate of momentum flux in the } x\text{-direction} = \rho \int_A v_x (v dA) \qquad (7\text{-}3)$$

in which v_x = component of the flow velocity in the x direction. To evaluate this integral, the distribution of flow velocity and its component in the x direction should be known. Similar equations may be written for the momentum flux in the y and z directions.

To account for the nonuniform velocity distribution, we may write

$$\text{Rate of momentum flux} = \rho Q (\beta V_x) \qquad (7\text{-}4)$$

in which β = momentum coefficient as defined in Chapter 1 and V_x = mean flow velocity in the x direction. Equation 7-4 may be used for sections 1 and 3, but it is unsuitable for section 2, for which Eq. 7-3 must be used.

Note that the pressure distribution may not be hydrostatic in the region of curvilinear flow because of local accelerations. Therefore, the pressure distribution should be known to determine the force acting on a section. For example, the pressure force acting at sections 1 and 3 may be determined by assuming hydrostatic pressure distribution at these sections. However, this is not the case at section 2, where the pressure distribution should be known. Such a distribution may not be known without measurements on the proto-type or without tests on a hydraulic model. For this reason, the application of the momentum equation in the regions having rapidly varied flow is not simple and straightforward.

For the energy in curvilinear flow, we cannot write the following expression

$$H = z + y + \alpha \frac{V^2}{2g} \qquad (7\text{-}5)$$

for the total head, since the pressure distribution is not hydrostatic. In such a case, the energy flux at any section may be written in the power form as

$$P = \rho g \int_A (z + \frac{p}{\gamma} + \frac{v^2}{2g}) v dA \qquad (7\text{-}6)$$

To evaluate this integral, the velocity and pressure distribution at the section should be known.

In summary, it is not usually possible to utilize the concept of mean quantities at a cross section in rapidly varied flows; instead, the distributions of velocity and pressure are needed to properly use the conservation laws properly. Such a distribution may not be available for universal applications. Two widely used velocity distributions at a cross section are the Boussinesq and Fawer [1937] assumptions. To simplify the analysis, however, we usually select sections away from the regions of rapidly varied flow and then apply the conservation laws to these sections.

7-3 Channel Transitions

A channel transition is a local change in the channel characteristics (usually area, shape and/or direction) that results in changing the flow from one state to another. Typical examples of channel transitions are contractions, expansions, and bends. We will call a reduction in the cross-sectional area in the direction of flow a channel *contraction* and an increase in the area a channel *expansion*. A transition in which there is a unique relationship between the flow depth and the rate of discharge is referred to as a *control*. As discussed in Chapter 5, critical flow occurs at a control, which may be natural or artificial. For example, structures such as spillways, weirs etc. are artificial controls while free falls are natural controls.

A transition is usually provided to change the channel alignment and/or cross section. The design requirements of a transition may be to minimize head losses, to dissipate energy, or to reduce flow velocities to prevent scour and erosion. Because of the unique relationship between the flow depth and the rate of discharge, a control may be used to measure the rates of discharge in a channel.

In this section, a number of general remarks on transitions are made. We first consider subcritical flow in contractions and expansions. The generation of shock waves in supercritical flows at changes in the channel geometries is then discussed.

General Remarks

The design and construction of transitions usually impose conflicting requirements. For example, for the ease and economy of construction, a transition should be simple and may thus have discontinuous boundaries. However, if the design objective is to minimize head losses in the transition, then its cross-sectional area, size, shape, and configuration should change gradually, and the boundaries should be streamlined. Such a design prevents eddy formation and flow separation, and reduces the possibility of cavitation. Vittal and Chiranjeevi [1983] presented a procedure for the design of transitions having subcritical flow; and Sturm [1985] for the design of a contraction having supercritical flow.

Because of large changes in the flow boundaries in a short distance, acceleration plays a dominant role in the flow through transitions as compared to the shear resistance at the channel boundaries. Therefore, the validity of the assumption of one-dimensional flow becomes questionable. In addition, the effects of curvilinear streamlines and the possibility of flow separation and cavitation has to be considered.

In flows having significant vertical acceleration, the flow velocity and pressure varies in the direction of flow as well as in the vertical direction. In such a situation, we may have to consider the flow as two- or three- dimensional. The energy losses in a transition are usually small and may be neglected. The flow may thus be assumed to be irrotational and a flow net may be drawn to analyze flow condition in a transition. Such analyses help in identifying the regions of sudden pressure changes. A sudden pressure drop usually indicates the possibility of cavitation; whereas a sudden pressure rise shows the possibility of flow separation, instability, and vibrations.

In rapid transitions, the velocity distribution may not necessarily be uniform at a channel cross section. There may even be negative flow velocities in part of the channel depth. In such situations, it may be difficult to compute the total head at the section even though the flow depth at the section is known.

The boundary-layer theory may be utilized to predict flow separation. In such analyses, it is necessary to take into consideration the losses due to shear at the boundaries. If the pressure gradient at the boundary is negative – i.e., the flow is accelerating – then the momentum of the boundary layer is augmented, since the acceleration is greater than the boundary shear. For a positive pressure gradient, however, the momentum is further reduced. Therefore, a negative pressure gradient indicates a stable boundary layer and a positive pressure gradient usually leads to flow separation.

Now, let us discuss flow through transitions. In many cases, sharp water surface disturbances in the form of cross waves (like shock waves in gas dynamics) are present in supercritical flows. These require special consideration in the analysis. As a result, we will discuss subcritical and supercritical flows separately in the following paragraphs.

Subcritical Flow

We will first discuss channel expansions and then channel contractions.

Expansions

A channel expansion may be due to an increase in the channel width, a drop in the channel bottom, or a combination of these two, as shown in Fig. 7-2. An expansion having abrupt geometrical changes is called a *sudden expansion* and it is called a *gradual expansion* if the changes occur over a finite distance.

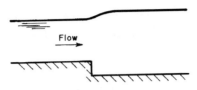

(a) Longitudinal section

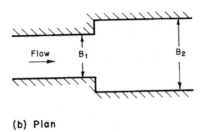

(b) Plan

Fig. 7-2   Channel expansion

Channel expansions are utilized in many hydraulic structures, such as flumes, outlets, siphons, and aqueducts. The design of a transition involves the selection of its shape so that flow does not separate and the form losses are minimized. To determine the optimum shape, experimental studies were conducted by several investigators [Smith and Yu, 1966; Mazumdar, 1967; Nashta and Garde, 1988]. The use of a number of devices to control separation has been investigated by Smith and Yu [1966], Rao and Seetharamiah [1969], Skogerboe et al. [1971], and Mazumdar [1967]. Several experimental investigations [Abbott and Kline, 1962; Graber, 1982] show that the flow downstream of an expansion becomes unsymmetrical for $B_2/B_1 \leq 1.5$ where B_1 = channel width upstream of the transition and B_2 = channel width downstream of the transition. It is found that the long and short eddy regions of the confined jet usually remain stable, although they may be interchanged by temporarily inserting a vane in the flow [Nashta and Garde, 1988].

Experimental data show that the shape of the separating streamline downstream of an expansion does not depend on the Reynolds number; this shape may be determined from the following equation

$$\frac{B_x - B_1}{B_2 - B_1} = \frac{x}{L}[1 - (1 - \frac{x}{L})^m] \tag{7-7}$$

in which B_x = twice the distance of the separating streamline from the transition centerline at distance x from the transition; L = distance of the point where the streamline meets the boundary; and m = an exponent which varies between 0.6 and 0.66. An expansion shaped according to this equation gives a smaller separation zone and thus reduces head losses.

Let us consider the flow through a sudden expansion (Fig. 7-3), in which the channel width increases from B_1 to B_2 and the channel sections 1, 2, and 3 are located as shown.

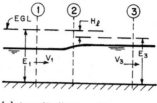

(a) Longitudinal section

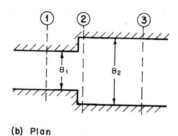

(b) Plan

Fig. 7-3   Definition sketch

Let us make the following assumptions: (1) $E_2 = E_1$, (2) $F_{s1} = F_{s2}$, (3) $y_1 = y_2$, (4) the width of the jet at section 2 is equal to B_1 and (5) the Froude number, $\mathbf{F}_{r1}$ is small so that $\mathbf{F}_{r1}^4$ and higher powers may be neglected. Based on these assumptions, Henderson [1966] showed that

$$E_1 - E_3 = \frac{V_1^2}{2g}\left[\left(1 - \frac{B_1}{B_2}\right)^2 + 2\mathbf{F}_{r1}^2(B_2 - B_1)\frac{B_1^3}{B_2^4}\right] \qquad (7\text{-}8)$$

The second term inside the bracket is small if $\mathbf{F}_{r1} < 0.5$ or if $B_2/B_1 > 1.5$. In most practical situations, the former condition is satisfied. If $B_2/B_1 < 1.5$, then the overall energy loss in the transition is small, and as a result this term becomes insignificant. Hence, the head loss, H_l, in a sudden expansion may be written as

$$H_l = \frac{(V_1 - V_3)^2}{2g} \qquad (7\text{-}9)$$

Experimental results of Formica [1955] show that Eq. 7-9 overestimates the head losses by about 10 per cent.

The head losses may be reduced by providing a gradual expansion. Normally, a taper of 1 in 4 is recommended. The head loss in such a transition is

$$H_l = 0.3\frac{(V_1 - V_3)^2}{2g} \tag{7-10}$$

More gradual tapering of walls than a taper of 1 in 4 does not significantly reduce the head losses in subcritical flow, although construction costs may be substantially increased.

Contractions

The channel contraction may comprise of a reduction in the channel width, raising the channel bottom, or a combination of the two (Fig. 7-4). An abrupt change in the cross section is called a *sudden* contraction, whereas if the change occurs over a distance, it is called a *gradual* contraction.

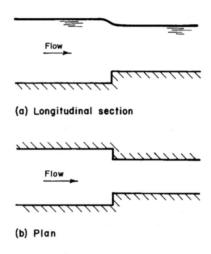

(a) Longitudinal section

(b) Plan

Fig. 7-4   Channel contractions

The head losses in channel contraction are less than those in the channel expansions. According to tests conducted by Formica [1955], the head loss through a sudden or square-edged contraction is

$$H_l = 0.23\frac{V_3^2}{2g} \tag{7-11a}$$

The head loss through a contraction when the edges are rounded is

$$H_l = 0.11\frac{V_3^2}{2g} \tag{7-11b}$$

in which V_3 is the velocity at a section in the narrower section where the flow has almost become uniform. Yarnell's [1934] test results indicate higher loss coefficients: 0.35 for square and 0.18 for rounded-edged contractions instead of those given in Eqs. 7-11a and 7-11b.

A contraction may choke the flow, as discussed in Chapter 2 if the channel width is reduced too much, since the energy may not be sufficient to pass the required amount of discharge per unit width.

7-4 Supercritical Flow

In supercritical flow through transitions, complications may arise due to the formation of shock waves on the flow surfaces. These waves are generated at a change in the flow boundary, as discussed in the following paragraphs. This treatment closely follows Henderson [1966].

To illustrate the generation and formation of shock waves, let us consider the movement of an observer through a stationary fluid. Let us assume that the observer is traveling at velocity V and creates a disturbance (say by detonating an explosive charge) while arriving at different locations, marked as 1, 2, 3, $\cdots$ in Fig. 7-5. Let us denote the celerity (i.e., velocity with respect to the medium in which the disturbance is traveling) of the disturbance by c. Then, depending upon the relative magnitude of the velocity of the observer V to that of celerity, c, three different situations are possible, as shown in Fig. 7-5. These are discussed in the following paragraphs.

The disturbance created by the observer at each location travels outwards, as shown in Fig. 7-5. If $V < c$, the observer lags behind the front of the disturbance and the front of the disturbances generated at different locations forms a circular pattern (Fig. 7-5a). For the case of $V = c$, the disturbance moves outward from the point of generation. Because the celerity of the disturbance and the velocity of the observer are equal, the waves generated at different locations combine and form a wave front (Fig. 7-5b). When $V > c$, the observer moves ahead of the front of the disturbance and when it reaches at location 4, an envelop tangent to the fronts of the disturbance may be drawn. This forms an oblique wave front as shown in Fig. 7-5c.

Now, let us see whether there is a relationship between the wave celerity, c, the velocity of the observer, V, and the angle, β, between the wave fronts. Referring to Fig. 7-5c, let the observer travel from location A_1 to A_4 in time Δt. During this time, the disturbance travels from A_1 to D_1. In terms of velocities and the distances traveled during time, Δt, we may write

$$
\begin{aligned}
\sin \beta &= \frac{A_1 D_1}{A_1 A_4} \\
&= \frac{c \Delta t}{V \Delta t} \\
&= \frac{c}{V}
\end{aligned}
\tag{7-12}
$$

Note that in this case the location of the wave front varies from one time to the next. If the fluid were in motion and the observer is stationary at

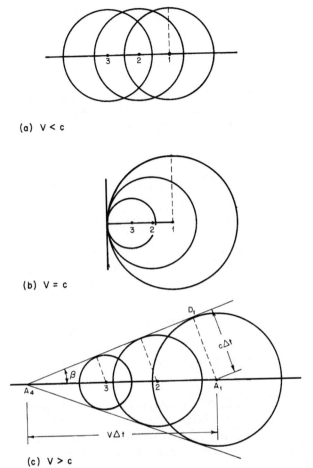

(a) V < c

(b) V = c

(c) V > c

Fig. 7-5   Propagation of a disturbance

one location, then a standing oblique wave front is formed. The cause of the disturbance may be a change in the alignment of the side wall, an irregularity on the wall surface, etc.

   If the disturbance may be assumed to be a long wave of small amplitude, then as we proved in Chapter 3, $c = \sqrt{gy}$, where $y =$ flow depth. Based on this relationship, we may write Eq. 7-12 as

$$\sin \beta = \frac{c}{V} = \frac{1}{\mathbf{F}_r} \tag{7-13}$$

in which $\mathbf{F}_r =$ Froude number. However, if the disturbance is either large in magnitude or cannot be assumed to be a long wave, then Eq. 7-13 cannot be used. In the following section, we consider a large disturbance produced by a wall deflecting inwards to the flow.

Oblique Hydraulic Jump

Let the alignment of the vertical side wall of a channel change inwards into flow by angle $\Delta\theta$, as shown in Fig. 7-6. This produces an oblique, positive wave front. Let the height of this standing wave be Δy and let it make an angle β with the wall.

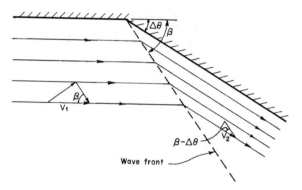

Fig. 7-6   Oblique hydraulic jump

Referring to Fig. 7-6, there is a component of velocity parallel to the wave front. Since there is no force acting parallel to and along the wave front, the tangential velocities on either side of the front should be same even though the flow depths differ by height Δy. Hence, we may write for the tangential components

$$V_1 \cos\beta = V_2 \cos(\beta - \Delta\theta) \tag{7-14}$$

The components of flow velocity normal to the wave front are $V_1 \sin\beta$ and $V_2 \sin(\beta - \Delta\theta)$. Then, it follows from the continuity equation that

$$y_1 V_1 \sin\beta = y_2 V_2 \sin(\beta - \Delta\theta) \tag{7-15}$$

We may apply the momentum equation as we did for the hydraulic jump in Chapter 2, except that in this case we replace V_1 by the normal component, $V_1 \sin\beta$. Thus, Eq. 2-58 for the oblique jump may be written as

$$\frac{V_1^2 \sin^2\beta}{gy_1} = \frac{1}{2}\frac{y_2}{y_1}\left(\frac{y_2}{y_1} + 1\right) \tag{7-16}$$

It follows from this equation that

$$\sin\beta = \frac{1}{\mathbf{F}_{r1}}\sqrt{\frac{1}{2}\frac{y_2}{y_1}\left(\frac{y_2}{y_1} + 1\right)} \tag{7-17}$$

Note that for a small-amplitude wave, this equation becomes $\sin \beta = 1/\mathbf{F}_{r1}$, which is the same as Eq. 7-13.

Division of Eq. 7-15 by Eq. 7-14 and simplification of the resulting equation yields

$$\frac{y_2}{y_1} = \frac{\tan \beta}{\tan(\beta - \Delta\theta)} \tag{7-18}$$

By substituting $y_2 = y_1 + \Delta y$ and doing some algebraic manipulations, Eq. 7-18 may be reduced to

$$\frac{\Delta y}{y} = \frac{\sec^2 \beta \tan \Delta\theta}{\tan \beta - \tan \Delta\theta} \tag{7-19}$$

For small values of $\Delta\theta$, $\tan \Delta\theta \simeq \Delta\theta$ and $\tan \Delta\theta$ is very small as compared to $\tan \beta$. Then, in the limit as $\Delta\theta \to 0$, the preceding equation becomes

$$\frac{dy}{d\theta} = \frac{2y}{\sin 2\beta} \tag{7-20}$$

By combining Eqs. 7-13 and 7-20, we obtain

$$\frac{dy}{d\theta} = \frac{V^2}{g} \tan \beta \tag{7-21}$$

This equation defines the variation of flow depth along a curved wall. Thus, for a finite change in the wall angle, $\Delta\theta$, there is a change in flow depth, Δy. This change in the flow depth may be assumed to be carried along an oblique front in the form of a shock wave even though the change in the flow depth is continuous.

The loss of energy in such a continuous change of flow depth is small and may be neglected. Thus, the specific energy, E, is constant. From the expression for E it follows that $V = \sqrt{2g(E - y)}$. Substitution of this expression for V into Eq. 7-21 and elimination of β from the resulting equation and Eq. 7-13 yield

$$\frac{dy}{d\theta} = \frac{2(E - y)\sqrt{y}}{\sqrt{2E - 3y}} \tag{7-22}$$

Integration of this equation and substituting for E in terms of y and $\mathbf{F}_r$ give

$$\theta = \sqrt{3} \tan^{-1} \frac{\sqrt{3}}{\sqrt{\mathbf{F}_r^2 - 1}} - \tan^{-1} \frac{1}{\sqrt{\mathbf{F}_r^2 - 1}} + \theta_o \tag{7-23}$$

where θ_o is the integration constant obtained by substituting $\theta = 0$ for the initial depth $y = y_1$. This equation may be used to calculate the change in depth caused by the change in the direction of wall.

In the next few sections, we present empirical information on weirs, hydraulic jump, spillways, and energy dissipators.

7-5 Weirs

In the laboratory and in the field, weirs have been used for measuring discharge in open channels for over two hundred years. Weirs may be classified as sharp-crested and as broad-crested. In the former, a thin vertical plate is fixed to the channel bottom and sides whereas a broad-crested weir comprises of a sudden rise in the channel bottom for some distance. A brief description of each follows.

Sharp-Crested Weirs

A sharp-crested weir usually comprises of a thin plate mounted perpendicular to the flow direction. The top of the plate has a beveled, sharp edge which makes the nappe spring clear from the plate. These weirs may be rectangular or triangular in shape. The latter are used mainly for measuring small rates of discharge. There are no contractions in the lateral direction if the rectangular weir occupies the full width of the channel. This is called a *suppressed* weir. The basic theoretical development on weirs are based on the assumption that the pressure above and below the nappe is atmospheric. Therefore, it is necessary to vent the underside of the nappe so that the lower side of the nappe is at atmospheric pressure. If venting is not done, then the pressure under the jet may not be atmospheric, and assuming it as such in the discharge computations may produce erroneous results.

Let us consider the flow over a rectangular weir, as shown in Fig. 7-7, neglect the viscous losses and assume the pressure at all points in the nappe is atmospheric. Then, the flow velocity at point a is $\sqrt{2gh}$, where $h =$ the vertical distance below the energy-grade line, discharge per unit width may then be determined from

$$q = \int_{h_o}^{H_o+h_o} \sqrt{2gh}\,dh = \frac{2}{3}\sqrt{2g}\left[(H_o + h_o)^{3/2} - h_o^{3/2}\right] \qquad (7\text{-}24)$$

in which $h_o = V_o^2/(2g)$. Theoretically speaking, it would be more appropriate to include the energy coefficient, α, in the velocity head to account for the nonuniform velocity distribution. However, experimental results show that α varies between 1.00 and 1.08 [Ranga Raju, 1981] for flows approaching a weir and we may thus safely assume it to be equal to 1.

To account for contraction and other effects, we may introduce discharge coefficient, C_d. This equation may then be written as

$$q = \frac{2}{3}C_d\sqrt{2g}H_o^{3/2} \qquad (7\text{-}25)$$

Based on Rehbock's experimental results [Henderson, 1966], C_d may be approximated as

$$C_d = 0.611 + 0.08\frac{H_o}{P} \qquad (7\text{-}26)$$

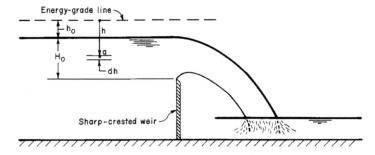

Fig. 7-7   Sharp-crested weir

in which P = height of the weir above the channel bottom (Fig. 7-7). Measurements by Rouse [1950] indicate that this formula is valid for $H_o/P < 5$. It is approximate upto $H_o/P = 10$ when C_d is about 1.135 [Henderson, 1966]. For $H_o/P > 15$, the weir becomes a sill and the discharge may be computed from the critical flow equation assuming $y_c = H_o$.

By proceeding similarly for a *triangular weir,* it may be shown that [Henderson, 1966]

$$Q = \frac{8}{15} C_c \tan \frac{\alpha}{2} \sqrt{2g} H_o^{5/2} \qquad (7\text{-}27)$$

in which C_c = contraction coefficient and α = weir angle. The contraction coefficient for the most commonly used 90°-weir is approximately equal to 0.585. By substituting this value into Eq. 7-27 and simplifying, we obtain

$$Q = 2.5 H_o^{5/2} \qquad (7\text{-}28)$$

An extensive amount of information on weirs is available in the literature. For details, see Bos [1976], Ackers et al. [1978], and Kabos [1984].

Broad-Crested Weirs

We showed in Chapter 3 the flow may become critical if the channel bottom is raised by a specified amount. If the raised part of the channel is of sufficient length in the direction of flow, then the flow may be critical and the streamlines may be parallel to the weir. This may be utilized for flow measurement as follows.

Let us assume the flow on the weir is critical and the losses between the weir and the location where the upstream flow depth is measured are negligible. Then we may write the energy equation between these two sections (Fig. 7-8) as

$$H + \frac{V^2}{2g} = \frac{3}{2} y_c \qquad (7\text{-}29)$$

in which H = upstream flow depth above the crest. By assuming the velocity of approach, V, to be negligible, this equation for the discharge per unit width, q, may be written as

$$q = \frac{2}{3}H\sqrt{\frac{2}{3}gH} \qquad (7\text{-}30)$$

However, a general equation for the discharge may be written as

$$Q = CB\sqrt{g}H^{3/2} \qquad (7\text{-}31)$$

in which $B =$ channel width and C is a coefficient introduced to take into consideration the effects of various simplifying assumptions. If W is the height of the weir above the channel bottom, then $V = Q/[B(H + W)]$. Hence, it follows from Eqs. 7-29 and 7-31 that [Ranga Raju, 1981]

$$\frac{H}{H+W} = \frac{\sqrt{3}(C^{3/2} - \frac{2}{3})^{1/2}}{C} \qquad (7\text{-}32)$$

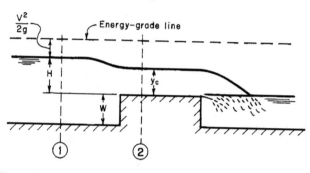

Fig. 7-8 **Broad-crested weir**

It is unlikely in real-life that the ideal situation – where the flow is critical as well as parallel to the crest – ever occurs [Henderson, 1966]. Instead there are two other possibilities: If the weir length in the direction of flow is short, then the flow depth on the crest varies with distance, the flow may be curvilinear, and it may even separate. Thus, errors may be introduced in the computation of discharge by assuming it to be critical parallel flow. Similarly, if the weir is too long, then the viscous effects are not negligible and a correction becomes necessary.

A weir may be assumed to be long if $L/H > 3$, where $L =$ length of the weir in the direction of flow. In such a case, the value of H is reduced by $\delta^\star$ to account for the viscous effects, where $\delta^\star$ is the maximum displacement thickness of the boundary layer. For details, see Hall [1962].

The downstream submergence may affect the rate of discharge. A weir may be assumed to discharge freely if the tailwater level is lower than $0.8H$ above the weir crest [Henderson, 1966].

7-6 Hydraulic Jump

As we discussed in Chapters 2 and 5, a hydraulic jump is formed whenever flow changes from supercritical to subcritical flow. In this transition from the supercritical to subcritical flow, water surface rises abruptly, surface rollers are formed, intense mixing occurs, air is entrained, and usually a large amount of energy is dissipated. By utilizing these characteristics, a hydraulic jump may be used to dissipate energy, to mix chemicals, or to act as an aeration device.

In this section, we discuss different characteristics of hydraulic jump and present some empirical information. The computation of flows having a jump by using modern finite difference techniques is outlined in the next chapter.

A jump in a horizontal, rectangular channel is referred to as a *classical jump*. Several experimental investigations have been conducted to determine the characteristics of such a jump and an extensive amount of literature is available: Rouse et al. [1958], Rajaratnam [1967], Rajarantnam and Murahari [1974], McCorquodale and Khalifa [1983], Madsen and Svendsen [1983], Hughes and Flack [1984], McCorquodale [1986], Hager [1988, 1989, 1992], Gharangik and Chaudhry [1991], Younus and Chaudhry [1994], Gunal and Narayanan [1996]Khan and Steffler [1996]and Ead and Rajaratnam [2002]. Also see review articles by Rajaratnam [1967]; McCorquodale [1986]; and monograph by Hager [1992].

Ratio of Sequent Depths

Equation 2-61 relates the flow depths on the upstream and downstream sides of a classical jump in terms of the Froude number at the entrance to the jump, $\mathbf{F}_{r1}$. We will call $\mathbf{F}_{r1}$ in the following discussion as the approach Froude number. From this equation, it can be proved that for $\mathbf{F}_{r1} > 2$,

$$y_r = \sqrt{2}\mathbf{F}_{r1} - \frac{1}{2} \tag{7-33}$$

in which the ratio of the sequent depths, $y_r = y_2/y_1$. This is a linear relationship between the ratio y_r and the approaching Froude number.

Note that we neglected the shear stress at the channel boundaries in the derivation of Eq. 2-61. Several experimental investigations show that this equation is valid in spite of this assumption, even for cases where the flow enters the jump at a steep angle. However, by utilizing extensive experimental data, Rajaratnam [1965] showed that y_r computed from an equation similar to Eq. 2-61 compares better with the experimental results if the boundary shear stress is included than that determined from an equation in which these losses are neglected. The viscous effects become important as the Reynolds number $\mathbf{R}_{e1} = V_1 y_1/\nu$ becomes small or as both $\mathbf{F}_{r1}$ and y_1/B (B = channel width) become too large [Hager and Bremen, 1989]. Such a condition is possible on the scale models. As a rough guide, the equations derived by neglecting the viscous losses may be used with confidence for $\mathbf{F}_{r1} < 12$ if $\mathbf{R}_{e1} > 10^5$.

Length of Jump

The length of a jump is needed to select the apron length and the height of the side walls of a stilling basin. To determine the length of a jump during laboratory investigations, it is difficult to mark the beginning and the end of a jump because of highly turbulent flow surface, formation of roller and eddies and air entrainment. In addition, the surface disturbances are of random nature, and the time-averaged quantities may not always give consistent results. The length of the roller may be taken to the point where the flow velocity at the top reverses and the jet continues.

For practical applications, experimental data have been summarized in a non-dimensional form relating the approach Froude number, $\mathbf{F}_{r1}$, and L/y_1 or L/y_2, where $L =$ length of the jump. Although satisfactory correlation has been observed for L/y_1, considerable amount of disagreement exists between the data reported by different researchers for L/y_2. Figure 7-9 shows the curve recommended by the U.S. Bureau of Reclamation.

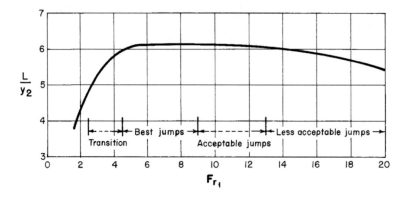

Fig. 7-9 Length of hydraulic jump (After Peterka [1958])

The following equation [Hager, 1992a] for the length of the roller, L_r, gives good results if $y_1/B < 0.1$

$$\frac{L_r}{y_1} = -1.2 + 160 \tanh \frac{\mathbf{F}_{r1}}{20} \tag{7-34}$$

In addition, by using the criterion that the turbulence has diminished at the end of the jump, Hager [1992a] developed the following equation for the length of the jump

$$\frac{L}{y_1} = 220 \tanh \frac{\mathbf{F}_{r1} - 1}{22} \tag{7-35}$$

or simply, $L = 6y_2$ for $4 < \mathbf{F}_{r1} < 12$.

Jump Profile

The information on the jump profile is needed to determine the weight of water in a dissipator to counteract the uplift force if the basin floor is laid on a permeable foundation. Also, the height of the side walls may be varied for economic reasons if the water profile is known.

Figure 7-10 shows the jump profiles for different approach Froude numbers [Bakhmeteff and Matzke, 1936]. For design purposes, the vertical pressure on the basin floor may be assumed to be the same as that corresponding to the hydrostatic pressure for the profile depth. This has been confirmed by several experimental investigations.

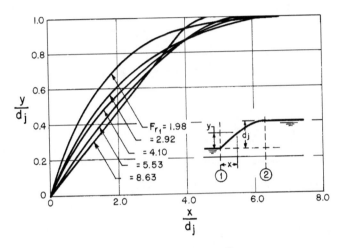

Fig. 7-10   Jump profile (After Peterka [1958])

Based on extensive laboratory experiments, Hager [1991] developed the following empirical relationship for the flow depth, y, at distance, x, from the beginning of the jump

$$Y = \tanh(1.5X) \qquad (7\text{-}36)$$

in which $X = x/L_r$; $Y = (y - y_1)/(y_2 - y_1)$ and L_r = length of the roller.

Jump types

Hydraulic jump occurs in four distinct forms [Peterka, 1958] depending upon the approach Froude number, $\mathbf{F}_{r1}$, as shown in Figure 7-11. Each of these forms has a distinct flow pattern, formation of rollers and eddies, etc. The energy dissipation in the jump depends upon the flow pattern and the strength of the rollers. The range of Froude number listed in the following paragraphs for various types of jumps is not precise, and there is some overlap from one type to the other.

(a) Pre-jump , very low energy loss (F_{r1} = 1.7 to 2.5)

(b) Transition , rough water surface (F_{r1} = 2.5 to 4.5)

(c) Good jump , least affected by tail water (F_{r1} = 4.5 to 9.0)

(d) Effective but rough (F_{r1} > 9.0)

Fig. 7-11   Jump types (After Peterka [1958])

Weak jump ($1 < F_{r1} < 2.5$)

For $1 < F_{r1} < 1.7$, y_1 and y_2 are approximately equal to each other and only a slight ruffle is formed on the surface. This undulation results in very little energy dissipation. However, as F_{r1} approaches 1.7, a number of small rollers are formed on the water surface, although the downstream water surface remains smooth. The energy loss is low in this jump.

Oscillating jump ($2.5 < F_{r1} < 4.5$)

The jet at the entrance to the jump oscillates from the bottom to the top at an irregular period. Turbulence may be near the channel bottom at one instant and at the water surface the next. These oscillations result in the formation of irregular waves, which may persist for a long distance downstream of the jump. They may cause considerable damage to the channel banks. Therefore, this range of F_{r1} should be avoided while designing an energy dissipator.

Steady jump ($4.5 < \mathbf{F}_{r1} < 9$)

For this range, the jump forms steadily at the same location, and the position of the jump is least sensitive to the downstream flow conditions. The jump is well balanced and the energy dissipation is considerable.

Strong jump ($\mathbf{F}_{r1} > 9$)

In this case, the difference between the conjugate depths is large. At irregular intervals, slugs of water roll down the front of the jump face into high-velocity jet and generate additional waves. The jump action is very rough and the dissipation rate is high.

Energy loss

The difference between the total head upstream and downstream of the jump is the energy loss in the jump. For a horizontal channel bottom, this is the same as the difference in the specific energy upstream and downstream of the jump. We may derive an expression for this head loss, h_l, as follows:

$$h_l = E_1 - E_2 = (y_1 - y_2) + \frac{V_1^2}{2g} - \frac{V_2^2}{2g} \tag{7-37}$$

in which the subscripts 1 and 2 refer to the quantities upstream and downstream of the jump. By substituting $q = V_1 y_1 = V_2 y_2$ and doing a number of algebraic manipulations, Eq. 7-37 may be written as

$$h_l = \frac{(y_2 - y_1)^3}{4y_1 y_2} \tag{7-38}$$

This is a theoretical expression for the head losses in a classical jump. Figure 7-12, based on experimental results, shows the energy dissipation in a jump for different values of the approach Froude number.

Jump Location

As we discussed in the previous paragraphs, a hydraulic jump is formed at a location where the flow depths upstream and downstream of the jump satisfy the equation for the sequent depth ratio (Eq. 2-61). We will illustrate the location of jump formation by considering the flow downstream of a sluice gate. Let the flow depth at the sluice outlet be y_1 and the sequent depth corresponding to this depth be y_2. There are several different possibilities for the formation of jump, depending upon the tailwater depth, y_d.

The jump is formed on the apron if the downstream depth, y_d, is equal to the depth y_2 required by Eq. 2-61 (Fig. 7-13a). If y_d is less than y_2, then the jump moves downstream to a point where the upstream depth y_1' is the

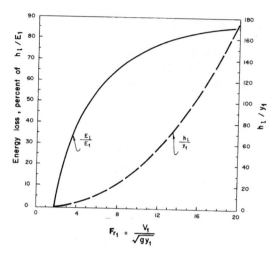

Fig. 7-12 **Energy dissipation in jump** (After Peterka [1958])

sequent depth to y_d, (Fig. 7-13b). In this figure, we have used a broken line to show the sequent depth y_2 required for the depth y_1 at the sluice outlet. However, if the tailwater depth is higher than the required amount, then the jump is pushed back, as shown in Fig. 7-13c. This is called *submerged,* or *drowned, jump.*

Tailwater level plays a significant role in the formation of jump at a particular location. In most practical situations, the tailwater level depends upon the channel discharge, Q. A curve between Q and the tailwater level is referred to as the tailwater rating curve. Similarly, we may prepare a curve between y_2 and Q, which we will refer to as the jump curve. Depending upon these two curves, five different flow situations are possible [Leliavsky, 1955]. These are shown in Fig. 7-14, in which a full line is used for the tailwater curve and a broken line is used for the jump curve. These five cases are discussed in the following paragraphs.

In case (a), the tailwater rating curve and the jump curve coincide for all rates of discharge. The requirements for the sequent depth are always satisfied and the jump forms at the same location. This is an ideal situation which rarely occurs in nature.

The jump curve is always above the tailwater curve for case (b). The downstream depth is less than the required sequent depth and the jump moves further downstream. To ensure jump formation on the apron, a sill may be used.

In case (c) the tailwater curve is always above the jump curve. Thus the downstream depth is more than that required by the sequent depth. The jump moves upstream and may drown. The jump may be controlled at the desired

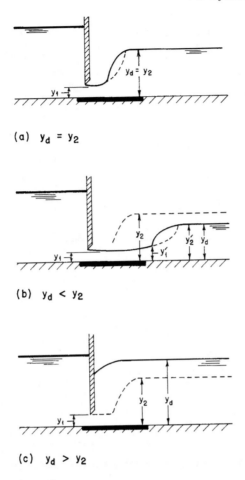

(a) $y_d = y_2$

(b) $y_d < y_2$

(c) $y_d > y_2$

Fig. 7-13   Jump location

location by providing a drop in the channel bottom or by letting the jump form on a sloping apron.

The tailwater curve is below the jump curve at low discharges and above it for higher discharges in case (d). The stilling basin may be designed so that the jump is formed in the basin at low rates of discharges and the jump moves on to a sloping apron at higher discharges.

Case (e) is opposite to case (d) in the sense that the tailwater curve is above the jump curve at low discharges and below the jump curve at high discharges. A stilling pool may be designed in this case to form the jump at high discharges.

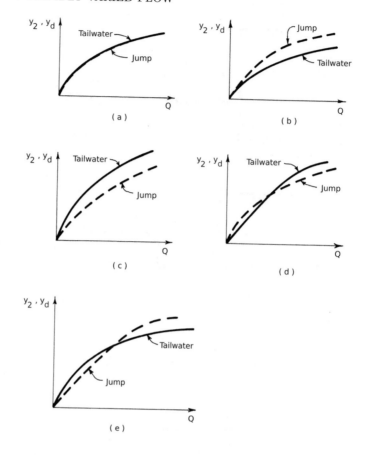

Fig. 7-14   Effect of tailwater level on jump formation

Control of Jump

The location of a jump may be controlled by providing a number of appurtenances, such as baffle blocks, sill, drop or rise in the channel bottom. The sill may be sharp- or broad-crested weir. Typically, the flow in the vicinity of these appurtenances is rapidly varied and the velocity distribution is not uniform. As a result, it becomes difficult to apply the momentum equation to analyze accurately the formation of jump. Therefore, laboratory experiments are done to develop empirical relationships for universal applications and model studies are conducted for specific projects.

In the following paragraphs, we discuss the control of jump by means of a sharp-crested weir and by an abrupt rise.

Sharp-crested weir

Figure 7-15 shows the relationship between different variables for jump control by means of a sharp-crested weir. This diagram, developed by Forster and Skrinde [1950] from results of extensive laboratory tests, may be used to determine the effectiveness of the weir for jump formation provided the weir is not submerged. For the known value of the approach Froude number, $\mathbf{F}_{r1}$, the distance X between the toe of the jump and the weir may be determined from this figure. For different X/y_2 ratios, the values may be interpolated between the curves.

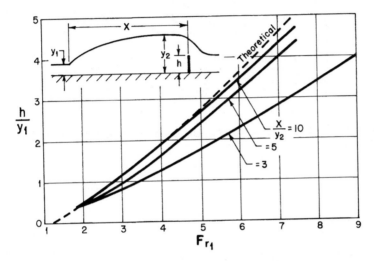

Fig. 7-15   Control of hydraulic jump by sharp-crested weir (After Forster and Skrinde 1950)

Abrupt rise

Based on laboratory experiments and theoretical analysis, Forster and Skrinde [1950] prepared Fig. 7-16 for the control of jump by means of an abrupt rise. The jump was formed at a distance, $x = 5(h + y_3)$ upstream of the rise. This diagram may be used to predict the performance of an abrupt rise if the values of V_1, y_1, y_2, y_3, and h are known. The definition of these variables should be clear from Fig. 7-16.

An abrupt rise increases the drowning effect if a point lies above the line $y_2 = y_3$. The region between $y_2 = y_3$ and $y_3 = y_c$ lines is further divided by the h/y_1 curves. A point lying on these curves between these two lines represents the condition when the jump is formed at $x = 5(h + y_3)$. A point above the h/y_1 curve shows a condition in which the jump is forced upstream and may be drowned. The condition where the rise is too low and the jump is forced downstream and may eventually be washed out is represented by points below the h/y_1 curves.

Points below the $y_3 = y_c$ line represent supercritical flow downstream of the rise. In this condition, the rise acts as a weir and Fig. 7-15 may be used for the analysis.

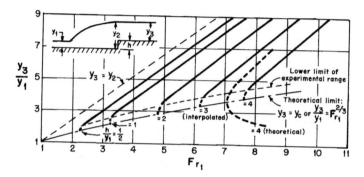

Fig. 7-16   Control of hydraulic jump by abrupt rise (After Forster and Skrinde [1950])

7-7 Spillways

A spillway is used to release surplus or flood water or for other controlled releases, such as for irrigation, navigation, or environmental considerations [Zipparro and Hansen 1993; Hager, 1992a; Hager and Vischer 1998]. Spillways may be classified into different categories using different criterion for such classification. For example, based on *function*, a spillway may be classified as service, auxiliary, or emergency; and based on the structural components, it may be called an overflow, chute, or tunnel spillway. Utilizing the type of inlet, a spillway may be classified as orifice, siphon, side channel, or morning glory.

The overflow spillway is one of the common types of spillway. Information on this spillway is presented here. For details on other types of spillways, see Chow [1959], Peterka [1958], and Zipparro and Hansen [1993].

Overflow Spillway

An overflow spillway is used on concrete-gravity, arch and buttress dams where part of the dam length may be used for spillway. Because of the shape, it is also called an ogee spillway.

An overflow spillway has three main parts: the crest, the sloping face, and the energy dissipator at the toe. The first two are discussed in this section, and the third is discussed in Section 7-8.

Crest of Overflow Spillway

The pressures on the crest are atmospheric if the crest shape is the same as the underside of the nappe of a jet over a sharp-crested weir. These pressures may be above atmospheric (positive) or sub-atmospheric (negative) depending upon the shape of the crest relative to the underside of the nappe over a sharp-crested weir (Fig. 7-17). The shape of the crest is based on the design head,

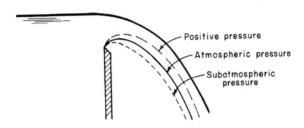

Fig. 7-17   Pressure on spillway crest

H_d, which is selected for a given site such that the minimum pressure at the crest is higher than -6 m in order to prevent cavitation. Figure 7-18 may be used to select H_d so that this condition for the minimum pressure is satisfied. In this figure, $H =$ maximum head on the crest = maximum upstream level $-$ crest level. Usually, the design head is selected such that $1.3 < H/H_d < 1.5$. In this range, acceptable levels of sub-atmospheric pressures are produced on the spillway face. This results in an increase in the discharge capacity of the spillway and at the same time does not result in cavitation on the spillway. The crest shape may then be determined from the empirical relationships presented in Fig. 7-19. These relationships were developed by the U.S. Army Corps of Engineers based on extensive laboratory investigations.

Rating Curve

A curve between the upstream reservoir level and the spillway discharge is called the *rating curve*. Discharge through an overflow spillway under a given total head on the spillway crest may be written as [Roberson et al., 1998]

$$Q = CL_e\sqrt{2g}H_e^{1.5} \tag{7-39}$$

in which $L_e =$ effective spillway length; $C =$ coefficient of discharge; and $H_e =$ total head on the crest $= H + V_o^2/(2g)$. Normally the velocity of approach,

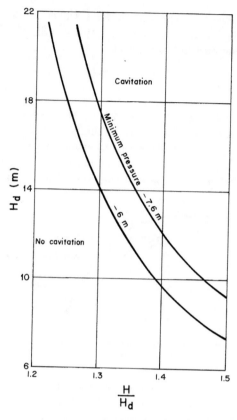

Fig. 7-18   Design head for overflow spillway

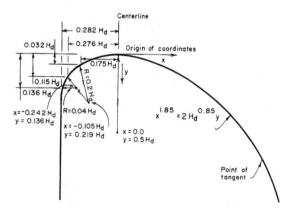

Fig. 7-19   Crest shape

V_o, for an overflow spillway is small. Therefore, the velocity head is negligible, and H_e may be taken as the difference between the upstream level and the crest level, i.e., $H_e = H$.

Through extensive laboratory tests, the U. S. Bureau of Reclamation [1973] has compiled information on the coefficient of discharge. The value of this coefficient, C_o, for the design head, H_d, is presented in Fig. 7-20(a) and its value C for other value of H is given in Fig. 7-20(b). To account for contractions at the piers and the abutments, the effective length, L_e, may be computed from the following equation:

$$L_e = L_n - 2(Nk_p + k_a)H_e \qquad (7\text{-}40)$$

in which L_n = net spillway length between the piers; k_p = pier coefficient; and k_a = abutment contraction coefficient. The pier coefficient may be determined from Fig. 7-21. For abutments having large radius of curvature, k_a is almost equal to zero and may be neglected. The values for k_a for unsymmetrical approach conditions may be determined from U. S. Army Corps of Engineers [1977].

Water-Surface Profile

The water-surface profile is required to set the height of the side walls and to select the elevation of the trunnion axis for the radial gates or for any other structure in the vicinity of the water surface. For preliminary design purposes, Fig. 7-22 presents data taken from the U. S. Army Corps of Engineers [1977]. Hydraulic model studies are normally conducted for large spillways.

Downstream Face

The downstream face of a spillway usually has a very steep angle, such as 0.8H : 1V to 0.6H : 1V. Because of this steepness, some difficulties arise in the analysis of flow conditions utilizing the procedures discussed in the earlier chapters. The development of boundary layer starts at the crest and the point where it intersects the flow surface is called the *transition point*. Downstream of the transition point, the turbulence is fully developed, large amount of air is entrained and bulking of flow occurs. This results in increasing the flow depth than that if no air were entrained. The U.S. Army Corps of Engineers recommend increasing the flow depth by 20 per cent to account for bulking [U.S. Army Corps of Engineers, 1965].

Several investigations have been conducted on the air entrainment or insufflation in high-velocity flows. A brief description is included on the topic in Chapter 10; for details, the reader should refer to a monograph prepared by Falvey [1980].

To determine the flow velocity at the toe of a spillway, no precise calculation procedures are presently available. Based on experience, computations,

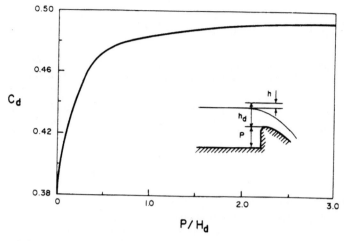

(a) Discharge coefficient at design head

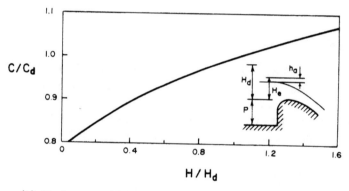

(b) Discharge coefficient at other heads

Fig. 7-20   Discharge coefficient (After U. S. Bureau of Reclamation [1973])

and experimental results, Bradley and Peterka [1957] presented a graph for this purpose. Another viable procedure is to assume a value for the head losses on the spillway face (say 5 to 10 per cent of the difference between the upstream reservoir level and the tailwater level) and then compute the theoretical velocity at the toe from the energy equation. This procedure is simpler and may be used for typical engineering applications.

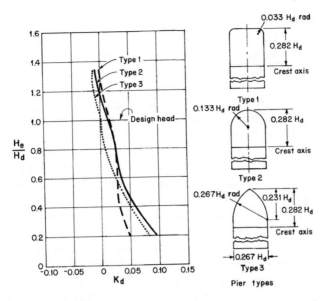

Fig. 7-21   Pier coefficient (After U.S. Army Corps of Engrs. [1977])

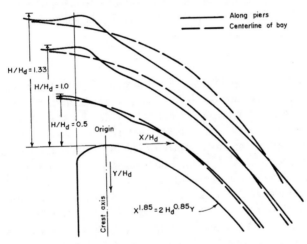

Fig. 7-22   Water-surface profile (After U.S. Army Corps of Engrs. [1977])

7-8 Energy Dissipators

The flow velocity at the toe of a high-head spillway is usually high and may cause serious scour and erosion of the downstream channel if proper precautions are not taken. For this purpose, energy dissipators are provided to dissipate sufficient amount of energy before water enters the downstream channel. In order to have an idea about the amount of energy dissipation, let us compute its value at the toe of the Grand Coulee dam, located on the Columbia River in the United States of America. The design discharge for the spillway is 28,320 m^3/s, and the upstream and the tailwater levels for this flow are 393.8 m and 308.23 m, respectively. Assuming no losses on the spillway face, the amount of energy at the toe = $\rho g Q H$, where H is the difference between the headwater and tailwater levels. Substitution of the values of different variables into this expression yields an energy of 23 GW. This should give the reader an idea about the amount of energy involved and clearly shows that even excellent rock may be eroded if proper measures are not taken.

Three types of energy dissipators [Hager 1992; Hager and Vischer, 1995] have been commonly used: stilling basins, flip buckets, and roller buckets [Mason, 1982]. Each dissipator has certain advantages and disadvantages and may be selected for a particular project depending upon the site characteristics. A brief description of these dissipators is given in the following paragraphs.

Stilling Basins

The hydraulic jump is used for energy dissipation in a stilling basin. Typically, this basin may be used for heads less than 50 m. At higher heads, cavitation becomes a problem. A concrete apron is provided for the length of the jump and the invert level of the apron is set such that the downstream water level provides the necessary sequent depth for the flow depth and the Froude number at the entrance to the jump. Long apron lengths and low apron levels are needed for such a stilling basin. Low apron levels require large amount of excavation and concrete. For economic reasons as well as to make the stilling basin operate efficiently over a wide range of flows, other devices and accessories may be provided to stabilize the jump, to reduce the length of the jump, and to permit the apron at a higher elevation. These devices include chute blocks, baffle blocks, and end sills (see Fig. 7-23).

The chute blocks serrate the flow entering the basin and lift up part of the jet. This produces more eddies increasing energy dissipation, the jump length is decreased, and the tendency of the jump to sweep out of the basin is reduced. The baffle blocks stabilize the jump and dissipate energy due to impact. The sill mainly stabilizes the jump and inhibits the tendency of the jump to sweep out.

Based on laboratory tests and field experience, several standardized stilling basins have been developed. Notable among these are: St. Anthony Falls stilling basin; eight stilling basins developed by the U.S. Bureau of Reclamation

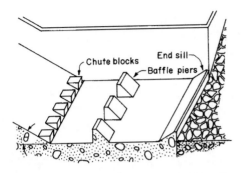

Fig. 7-23   Stilling basin accessories (After Peterka [1958])

(each suitable for a certain range of head), and a basin recommended by the U.S. Army Corps of Engineers [Murphy, 1974]. Only details of the last basin are given here; for information on the other basins, the reader should consult Chow [1959] and the monograph prepared by the U.S. Bureau of Reclamation [Peterka, 1958].

Figure 7-24 shows a stilling basin [Murphy, 1974] suitable for high-head installations. To design this basin, first the design and standard project flood are selected. The design flood may be less than the probable maximum flood, and the standard project flood may be less than the design flood. By assuming the head losses on the spillway face to be a certain percentage of the total head (say 5-10 per cent), the values of V_1 and y_1 are computed for the design flood. Then, the values of Froude number, $\mathbf{F}_{r1} = V_1/\sqrt{gy_1}$, and the sequent depth y_2 are determined. Similarly, the sequent depth, y_2' is computed for the standard project flood. The dimensions of the basin and of the appurtenances are then determined from the following relationships:

$$L_1 = 1.5y_2 \quad \text{for} \quad \mathbf{F}_{r1} \leq 4.6;$$

$$= [1.5 + \frac{1}{11}(\mathbf{F}_{r1} - 4.6)]y_2 \quad \text{for} \quad \mathbf{F}_{r1} > 4.6;$$

$$L_2 = 2.5h$$

$$h = \frac{1}{6}y_2 \quad \text{for} \quad \mathbf{F}_{r1} \leq 4.6;$$

$$= [1. + 0.13(\mathbf{F}_{r1} - 4.6)]y_1 \quad \text{for} \quad \mathbf{F}_{r1} > 4.6;$$

$$h_s = \frac{1}{2}h;$$

$$d_2 \geq 0.85y_2;$$

$$\simeq y_2'$$

$$L \geq L_1 + y_2$$

$$\geq 4y_2.$$

Baffle block rows 1 and 2 are staggered and the baffle block width is less than or equal to h. The spacing between the blocks is to be at least equal to the baffle block width.

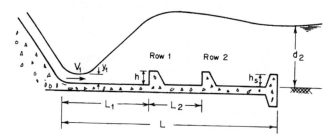

Fig. 7-24   U.S. Army Corps of Engineers stilling basin

Flip Buckets

The flip bucket energy dissipator is suitable for sites where the tailwater depth is low (which would require a large amount of excavation if a hydraulic jump dissipator were used) and the rock in the downstream area is good and resistant to erosion. The flip bucket, also called ski-jump dissipator, throws the jet at a sufficient distance away from the spillway where a large scour hole may be produced. Initially, the jet impact causes the channel bottom to scour and erode. The scour hole is then enlarged by a ball-mill motion of the eroded rock pieces in the scour hole. A plunge pool may be excavated prior to the first spill for controlled erosion and to keep the plunge pool in a desired location.

A small amount of the energy of the jet is dissipated by the internal turbulence and the shearing action of the surrounding air as it travels in the air. However, most of the energy of the jet is dissipated in the plunge pool.

During the operation of a flip bucket, a large amount of spray is produced, which may be undesirable for roads, bridges, and electrical equipment, such as transmission lines or grid stations. In addition, large water-level fluctuations may be produced in the tailrace area by the plunging jet and the associated return currents and eddies. These water level oscillations near the draft tube exits may impair the operation of Francis turbines, since these oscillations may result in load swings.

The horizontal throw distance, x_b, from the bucket lip to a point where the jet impinges the river bottom (Fig. 7-25) may be computed from the following equation

$$\frac{x_b}{h_o} = \sin 2\alpha + 2\cos\alpha\sqrt{\sin^2\alpha + \frac{y_b}{h_o}} \qquad (7\text{-}41)$$

in which y_b = the height of the bucket lip above the river bottom, α = bucket angle with the horizontal axis, and h_o = velocity head at the bucket lip.

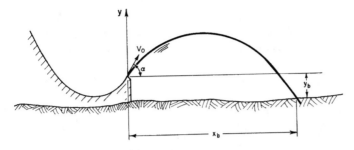

Fig. 7-25   **Definition sketch of the flip bucket**

If the bucket lip is at the river bottom level, then $y_b = 0$, and Eq. 7-41 simplifies as

$$\frac{x_b}{h_o} = 2\sin 2\alpha \qquad (7\text{-}42)$$

The throw distance is normally less than the distance computed from this equation since the air resistance and other losses are neglected in the derivation of this equation. For design purposes, this distance may be reduced by 20 per cent.

It is clear from Eq. 7-42 that the throw distance is maximum for $\alpha = 45°$. However, experience has shown that a lip angle of $20°$–$30°$ should be preferred since a bucket having this angle produces scour holes of shallower depth due to low angle of impact than that produced by a $45°$ bucket. The bucket radius is usually $10-20$ m, and the height of the bucket lip is set approximately 10 per cent of the bucket radius above the bucket invert.

Scour depth

A plunge pool is said to be *stable* if the scour has reached equilibrium conditions. No procedures are presently available to accurately determine the maximum depth of the stable plunge pool downstream of a flip bucket. However, based on laboratory studies and field information obtained on several projects, a number of empirical formulas have been developed for predicting the maximum depth of scour of a stable plunge pool. Mason and Arumugam [1985] list and compare these formulas. They also presented the following new formula which gives better comparison with both hydraulic-model and prototype results:

$$D_m = K \frac{q^a H^b h^{0.15}}{g^{0.3} d^{0.1}} \qquad (7\text{-}43)$$

in which D_m = maximum scour depth measured from the tailwater surface; q = flow intensity per unit width; h = tailwater depth above unscoured bottom level; H = head difference between the headwater and the tailwater levels; d_m = mean bed material size; and in SI Units, $a = 0.60 - 300/H$; $b = 0.15 -$

$H/200$; and $K = 6.42 - 3.1h^{0.1}$. Mason and Arumugam [1985] used a constant value for d_m of 0.25 m for prototypes. However, for rocky bottoms, d_m, may be taken as the size of blocks into which the rock may be assumed to fracture.

Roller Buckets

A roller bucket may be used for energy dissipation if the downstream depth is significantly greater than that required for the formation of a hydraulic jump. In this dissipator, the dissipation is caused mainly by two rollers: counterclockwise roller near the water surface above the bucket and a roller on the channel bottom downstream of the bucket. The movement of these rollers along with the intermixing of the incoming flows results in the dissipation of energy.

The Grand Coulee spillway on the Columbia River in the U.S.A. was the first spillway to have a roller bucket energy dissipator. Since then two types of roller buckets – solid and slotted – have been developed through hydraulic model studies and used successfully on several projects. In a solid bucket, the ground roller may bring material towards the bucket and deposit it in the bucket during periods of unsymmetrical operation. In a slotted bucket, part of the flow passes through the slots, spreads laterally, and is distributed over a greater area. Therefore, the flow concentration is less than that in a solid bucket. In addition, any material that might get into the bucket is washed out. The sweepout in a slotted bucket occurs at a slightly higher tailwater level than that in a solid bucket. Experience with both buckets indicates that a slotted bucket is preferred over the solid bucket.

The tailwater depth is the difference between the tailwater and the bucket invert level. The *minimum tailwater limit* is the tailwater depth just safely above the depth required for sweepout. At sweepout, a jet at the bottom scours the channel bottom and an unstable condition develops in the bucket, thereby causing excessive scour and erosion. Because of this, a bucket may not be designed for both submerged and flip action. A *safe lower limit*, T_{min}, has been established by adding a small factor of safety to the sweepout tailwater depth (Fig. 7-26a) determined from an extensive series of tests.

Similarly, an *upper tailwater limit* (Fig. 7-26b) was determined by raising the tailwater level until the flow dived from the apron lip.

Design procedure

Figure 7-27 shows the typical parameters for a roller bucket. The steps for the design of a bucket for a specified discharge, Q, and the spillway width, B, are as follows:

1. By assuming a value for the head losses on the spillway (say, 5–10 per cent), compute the flow velocity at the toe and hence determine $y_1 = Q/(BV_1)$, and the Froude number, $\mathbf{F}_r$.

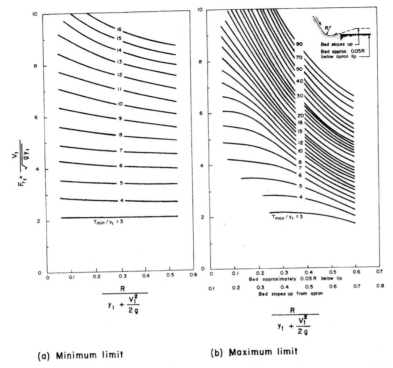

(a) Minimum limit   (b) Maximum limit

Fig. 7-26   Minimum and maximum tailwater limits (After Peterka [1958])

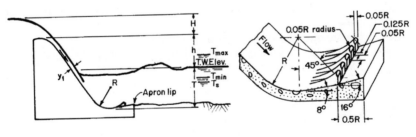

Fig. 7-27   Definition sketch (After Peterka [1958])

2. For $\mathbf{F}_r$ computed in step 1, determine the factor $R/(y_1 + V_1^2/2g)$ from Fig. 7-28 and hence the minimum bucket radius, R, and select a value for R.

3. For the computed values of $\mathbf{F}_r$ and $R/(y_1 + V_1^2/2g)$, determine T_{min}/y_1 from Fig. 7-26a, and hence compute the minimum tailwater limit, T_{min}. Similarly, determine the maximum tailwater limit, T_{max}, from Fig. 7-26b.

4. Set the bucket invert elevation so that the tailwater is between T_{min} and T_{max}. If possible, keep the apron lip and the bucket invert above the river

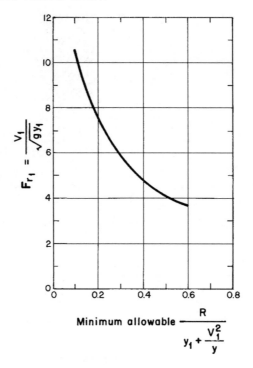

Fig. 7-28   Minimum bucket radius (After Peterka [1958])

bottom. For best performance, set the bucket invert so that the tailwater depth $\simeq T_{min}$.

5. Check the sweepout condition from Fig. 7-29.
6. Determine the tooth size, spacing, and other dimensions from Fig. 7-27.

7-9 Summary

In this chapter, we discussed the differences between the analysis of rapidly and gradually varied flows. It was shown that it is necessary to know the pressure and velocity distributions in rapidly varied flows to apply the conservation laws of mass, momentum, and energy. The generation of shock waves in supercritical flows was discussed. Empirical relationships derived from the laboratory and field investigations on transitions, weirs, and hydraulic jump were presented. Salient features and the design of spillways and energy dissipators were discussed.

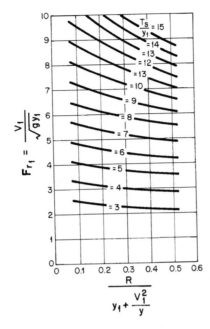

Fig. 7-29   Tailwater sweepout depth (After Peterka [1958])

Problems

7-1 If the flow depth at the entrance to a constant-width, rectangular transition with bottom slope S_o, is equal to the critical depth, y_c, then prove that the flow depth y in the transition is given by the equation

$$S_o x = y + \frac{1}{2}\frac{y_c^3}{y^2} - \frac{3}{2}y_c$$

7-2 For a horizontal, frictionless channel transition shown in Fig. 7-30, prove that

$$\theta x = \frac{Q}{y\sqrt{2g(H_o - y)}} - \frac{Q}{\sqrt{g(2E/3)^3}}$$

in which y = flow depth at distance x, Q = rate of discharge, and H_o = total head. Assume the flow is critical at the entrance of the transition.

7-3 Prove that for large values of $\mathbf{F}_{r1}$, the ratio of the sequent depths ($y_r = y_2/y_1$) and $\mathbf{F}_{r1}$ are linearly related. By using this relationship, show that

$$\frac{\Delta H}{H_1} = \left(1 - \frac{\sqrt{2}}{\mathbf{F}_{r1}}\right)^2$$

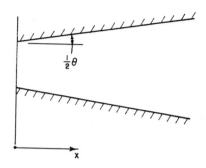

Fig. 7-30   Channel transition of Prob. 7-2

where $\Delta H = H_1 - H_2$, H = total head, and the subscripts 1 and 2 refer to the quantities upstream and downstream of the jump.

7-4 By applying the momentum principle, show that the head loss in a sudden expansion (Fig. 7-3) is $(V_1 - V_3)^2/(2g)$.

7-5 Prove that for a hydraulic jump, the ratio of the sequent depths, $y_r = y_2/y_1$, for $\mathbf{F}_{r1} > 2$ may be approximated as

$$y_r = \sqrt{2}\mathbf{F}_{r1} - \frac{1}{2}$$

7-6 Show that the head loss, h_l, in a classical hydraulic jump is

$$h_l = \frac{(y_2 - y_1)^3}{4y_1 y_2}$$

7-7 Determine the throw distance for a 20° flip bucket located at the end of a chute spillway. The bucket lip is 60 ft below the reservoir level. Assume the energy losses are 10 per cent of the total available head.

7-8 Design a stilling basin for the spillway of Problem 7-7. Assume a suitable tailrace rating curve.

7-9 Design a flip bucket, a roller bucket, and a flip bucket for an overflow spillway (six 50-ft wide bays with 10-ft wide piers) for a flow of 360000 ft^3/sec. The difference between the upstream and downstream water levels at this discharge is 140 ft.

Which of these energy dissipators do you prefer for each of the following sites?

  i. Located in cold climate with a bridge downstream of the spillway;
 ii. Excellent rock conditions;
iii. Large tailrace submergence.

7-10 Compute and plot the rating curve for the spillway of Problem 7-9.

7-11 Determine the rating curve for an overflow spillway having four 50-ft wide bays. The piers are of type 2, the design head for the crest is 32 ft, the maximum reservoir level is at El. 948 and the crest is at El. 900. The river bottom upstream of the spillway is at El. 836.

7-12 An expression for the discharge, Q, through an overflow spillway may be written as

$$Q = CL_e H^{1.5}$$

In the customary English units (i.e., Q in ft^3/sec and L_e and H in ft), $C = 3.8$ for a given spillway at a design head of 10 m. Will the value of C be the same in SI units (i.e., Q in m^3/s, and L_e and H in m), If not, what is the corresponding value in the SI units. Determine the discharge for $L_e = 100$ m and $H = 10$ m.

7-13 Suppose you are the design engineer for the energy dissipators of three large projects. Each project has the following typical characteristics. Which type of energy dissipator would you select for each project? List, in point form, the other characteristics the project should have so that your selected energy dissipator will operate properly.

   i. Shallow water depth and good rock conditions in the tailrace area;
   ii. High tailwater depth, rock quality is average;
  iii. The variation of tailwater depth with respect to the spillway discharge is small and the river bottom is highly erodible.

7-14 For the reservoir levels at El. 150 and 160 ft, determine the discharge through an overflow five-bay spillway having type 2 piers. Assume the abutment contraction coefficient to be zero. The crest is at El 120, river bottom at El 20, and the crest design head is 25 ft. Each bay is 60 ft wide.

7-15 Compute the spillway discharge for the upstream water levels at El. 300 and 310 m. The spillway length is 20 m, there are no piers and no abutment contractions. The crest is at El 280 and is shaped for a design head of 18 m and the river bottom is at El. 244.

7-16 Determine the discharge through an overflow spillway having no piers for the upstream reservoir level at El. 620 and 648 ft. The abutments are shaped to prevent contractions. The design head is selected such that it is 0.75 of the maximum head. If the maximum permissible reservoir level for the design flow of 60,000 ft^3/sec is El. 652, select the appropriate spillway length and the crest level.

7-17 Determine the location of the hydraulic jump in the channel of Prob. 5-15.

7-18 The longitudinal section of a spillway is shown in Fig. 7-31. Its width remains constant at 100 ft, there are no piers and the crest is shaped for a 24 ft design head. Compute:

i. The spillway discharge for the maximum reservoir level of El. 165;

ii. The water surface profile in the chute;

iii. Required downstream water level for the formation of hydraulic jump in the stilling basin;

iv. The energy losses in the jump.

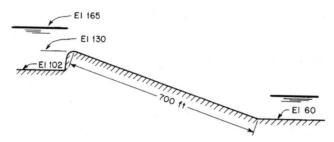

Fig. 7-31   Spillway of Prob. 7-18

7-19 If the downstream water level for the design discharge of 50,000 ft³/s is at El. 72 ft, design an energy dissipator for the spillway of Prob. 7-18.

7-20 The water level in the reservoir upstream of a 50-m wide spillway is at El. 200 m and water level in the downstream channel for a design flow of 2700 m³/s is at El. 50 m. If the stilling basin width is the same as that of the spillway, determine the floor level of the basin so that the jump is formed in the basin. Assume the losses in the spillway are negligible and no basin appurtenances such as baffle or chute blocks are to be provided to stabilize the jump.

What would be the basin level if you could use an end sill or one-row of baffle blocks and an end sill?

References

Abbott, D. E., and Kline, S. J., 1962, "Experimental investigation of subsonic turbulent flow over single and double backward facing step," *Jour. Basic Engineering,* Amer. Soc. Mech. Engrs., vol 84, pp. 317–325.

Ackers, P., White, W. R., Perkins, J. A., and Harrison, A. J. M., 1978, *Weirs and Flumes for Flow Measurement,* Wiley and Sons, New York, N.Y.

Advani, R. M., 1962, "A New Method for Hydraulic Jump in Circular Channels," *Water Power,* vol. 16, no. 9, pp. 349–350.

Ali, K. H. M., and Ridgeway, A., 1977, "Hydraulic Jump in Trapezoidal and Triangular Channels," *Proc.,* Institution of Civil Engineers, vol. 63, pp. 203–214, 761–767.

Anderson, V. M., 1978, "Undular Hydraulic Jump," Jour., *Hydraulics Div.,* Amer Soc. Civ. Engrs, vol. 104, no. 8, pp. 1185–1188 (Discussion, 1979, HY9, 1208–1211; 1980, HY7, pp. 1252–1254).

Bakhmeteff , B. A., and Matzke , A. E., 1936, "The Hydraulic Jump in Terms of Dynamic Similarity," *Trans.* Amer. Soc. Civ. Engrs., vol 100, pp. 630–680.

Bakhmeteff, B. A., and Matzke, A. E., 1938, "The Hydraulic Jump in Sloped Channels," *Trans.,* Amer. Soc. Mech. Engrs., HYD-60-1, pp. 111–118.

Bos, M. G., 1976, *Discharge Measurement Structures,* Report 4, Laboratorium voor Hydraulica en Afvoerhydrologie, Landbauwhogeschool, Wageningen, the Netherland.

Bradley, J. N., and Peterka, A. J. , 1957, "Hydraulic Design of Stilling Basins: Short Stilling Basins for Canal Structures, Small Outlet Works, and Small Spillways (Basin III)," *Journal of the Hydraulics Division,* 83(5), 1403–1.

Brooks, P., 2001, "Experimental Study of Erosional Cyclic Steps," *Ph. D. thesis,* University of Minnesota.

Cassidy, J. J., 1970, "Designing Spillway Crest for High Head Operations," *Jour. Hyd. Div.,* Amer. Soc. Civ. Engrs. vol 96, Hy3, pp. 745–753.

Chee, S. P., and Padiyar, P. V., 1969, "Erosion at the Base of Flip Buckets," *Engineering Journal,* Inst. of Canada, Nov.

Chow, V. T., 1959, *Open-Channel Hydraulics,* McGraw-Hill, New York, NY.

Ead, S. A., and Rajaratnam, N., 2002, "Hydraulic Jumps on Corrugated Beds," *Jour. Hyd. Engineering,* Amer. Soc. Civ. Engrs., vol. 128, no. 7, pp. 656–663.

Elevatorski, E. A., 1958, "Trajectory Bucket-type Energy Dissipators," *Jour. Power Div.,* Amer. Soc. Civil Engrs., Feb., pp. 1553–1 to 17.

Falvey, H. T., 1980, "Air-water Flow in Hydraulic Structures," *Engineering Monograph no. 41,* U.S. Bureau of Reclamation, Denver, Colorado.

Fawer, C., 1937, "Etude de quelques écoulements permanents á filets courbes," Ph.D. thesis, University of Lausanne.

Formica, G., 1955, "Esperienze preliminari sulle perdite di carico nei canali dovute a cambiamenti di sezione," (Preliminary Tests on Head Losses in Channels due to Cross-sectional Changes), *L'Energia Elletrica,* Milan, vol. 32, no. 7, July, p. 554.

Forster, J.W., and Skrinde, R.A., 1950, "Control of the Hydraulic Jump by Sills," *Trans.,* Amer. Soc. Civil Engrs., pp. 973–1022.

Garg, S. P., and Sharma, H. R., 1971, "Efficiency of Hydraulic Jump," *Jour., Hydraulics Div.,* Amer. Soc. Civ. Engrs., Vol. 97, HY3, pp. 409–420 (Discussions: Vol. 97, HY9, 1570–1573; HY10,1790–1795, HY11, 1923; HY12, 2107–2110; Vol. 98, HY1, 278–284; Vol. 99, HY3, 527–529.)

Graber, S. D., 1982, "Asymmetric flow in symmetric expansion," *Jour. Hydraulic Engineering,* Amer. Soc. Civil Engrs., vol 108, no 10, pp. 1082–1101.

Gharangik, A. M., and Chaudhry, M. H., 1991, "Numerical Simulation of Hydraulic Jump," *Jour. Hydraulic Engineering,* Amer. Soc. Civ. Engrs., vol. 117, no. 9, pp. 1195–1121.

Gunal, M., Narayanan, R.,1996, "Hydraulic Jump in Sloping Channels," *Jour. Hydraulic Engineering,* Amer. Soc. Civ. Engrs., vol. 122, no. 8, pp. 436–442.

Hager, W. H., 1986, "Discharge Measurement Structures," *Communication 1*, Chaire de Constructions Hydrauliques, Ecole Polytechnique Federale de Lausanne, Lausanne, Switzerland.

Hager, W. H., 1988, "B-jump in Sloping Channel," *Jour Hydraulic Research*, Inter. Assoc. Hyd. Research, vol. 26, no. 5, pp. 539–558.

Hager, W. H., 1991, "Experiments on Standard Spillway Flows," *Proc.*, Institution of Civil Engrs., vol 91, Sept., pp. 399–416.

Hager, W. H., 1992, *Energy Dissipaters and Hydraulic Jump*, Kluwer Academic Publishers, London, UK.

Hager, W. H., 1992a, *Spillways, Shock Waves, and Air Entrainment*, International Commission on Large Dams, Paris, France.

Hager, W. H., and Sinniger, R., 1985, "Flow Characteristics in a Stilling Basin With an Abrupt Bottom Rise," *Jour. Hydraulic Research*, Inter. Assoc. Hyd. Research, Vol. 23, no. 2, pp. 101–113 (Discussion: Vol. 24, No. 3, 207–215.)

Hager, W. H., and Bremen, R., 1989, "Classical Hydraulic Jump: Sequent Depths Ratios," *Jour Hyd. Research*, Inter. Assoc. Hyd. Research, Vol. 27, No. 5, pp. 565–585.

Hager, W. H. and D. L. Vischer, 1995, *Energy Dissipators*, Balkema, Rotterdam.

Hager, W. H. and D. L. Vischer, 1998, *Dam Hydraulics*, John Wiley & Sons, Chickester, UK.

Hall, G. W. 1962, "Discharge Characteristics of Broad-Crested Weirs Using Boundary Layer Theory," *Proc.* Institution of Civil Engrs. (London), vol. 22, June, 177.

Henderson, F. M., 1966, *Open Channel Flow*, MacMillian, New York, NY.

Hughes, W. C., and Flack, J. E., 1984, "Hydraulic Jump Properties over a Rough Bed," *Jour. Hydraulic Engineering*, Amer. Soc. Civ. Engrs., vol. 110, no. 12, pp. 1755–1771.

Ippen, A. T., 1950, "Channel Transitions and Controls," in H. Rouse (ed.), *Engineering Hydraulics*, Chap. 8, John Wiley & Sons, New York, N.Y.

Kabos, H., 1984, *Symposium on Scale Effects in Modeling Hydraulic Structures*, Chapter 2, International Assoc. for Hydraulic Research.

Kao, D.T. and Dean, P.S., 1976, "Spatially Varied Flow in Channel Transitions," *Rivers 76*, Amer. Soc. Civ. Engrs., pp. 1551–1571.

Khan, A. A., and Steffler, M., 1996, "Physical Based Hydraulic Jump Model for Depth-Averaged Computations," *Jour. Hydraulic Engineering*, Amer. Soc. Civ. Engrs., vol. 122, no. 10, pp. 540–548.

Kindsvater, C. E., 1944, "The Hydraulic Jump in Sloping Channels," *Trans.* Amer. Soc. Civ. Engrs., vol. 109, pp. 1107–1154.

Leliavsky, S., 1955, *Irrigation and Hydraulic Design*, vol. 1, Chapman and Hall, London.

Lenau, C. W., and Cassidy, J. J., 1969, "Flow through Spillway Flip Buckets ," *Jour. Hyd. Div.*, Amer. Soc. Civil Engrs., HY2, March, pp. 633–648.

Madsen, P. A., and Svendsen, I. A., 1983, "Turbulent Bores and Hydraulic Jumps," *Jour. Fluid Mechanics*, vol. 129, no. 4, pp. 1–25.

Mason, P. J., 1982, "The Choice of Hydraulic Energy Dissipator for Dam Outlet Works Based on a Survey of Prototype Usage," *Proc.*, Institution of Civil Engineers, vol 72, Part 1, pp. 209–219 (see also discussion, vol 74, 1983, pp. 123–126).

Mason, P. J., and Arumugam, K., 1985, "Free Jet Scour Below Dams and Flip Buckets," *Jour. Hydraulic Engineering,* Amer. Soc. of Civil Engrs., vol 111,no. 2, pp. 220–235.

Martin, R., 1975, "Scouring of Rocky River Beds by Free-Jet Spillways," *Water Power and Dam Construction*, April.

Mazumdar, S.K., 1967, "Optimum length of transition in open channel expansive subcritical flow," *Jour.* Institution of Engrs. (India), vol 48, no 3, pp. 463–476.

McCorquodale, J.A., 1986, "Hydraulic Jump and Internal Flows," Chapt. 6 in *Encyclopedia of Fluid Mechanics 1*, N. P. Cheremisinoff (ed.), Gulf Publishing Co., Houston, U.S.A.

McCorquodale, J. A., and Khalifa, A., 1983, "Internal Flow in Hydraulic Jumps," *Jour. Hydraulic Engineering*, vol. 109, no. 5, pp. 684–701.

Mehrotra, S. C., 1976, "Length of Hydraulic Jump," *Jour. Hyd. Div.*, Amer. Soc. Civ. Engrs., vol. 102, HY7, pp. 1027–1033.

Mehta, P.R., 1979, "Flow characteristics in two-dimensional expansion," *Jour. Hyd. Div.*, Amer Soc Civ Engrs., vol 105, no 5, pp. 501–516.

Mazumder, S.K., and Ahuja, K.C., 1978, "Optimum Length of Contracting Transition in Open Channel Subcritical Flow," *Jour. Inst. Eng (India)*, Civ. Eng. Div., vol. 58, Part CI 5, March, pp. 218–223.

Murphy, T.E., 1974, "Spillway Stilling Basin, Hydraulic Jump Type," *Memorandum*, Waterways Experiment Station, 25 June, 11 pp.

Nashta, C.F., and Garde, R.J., 1988, "Subcritical flow in rigid- bed open channel expansions," *Jour. Hydraulic Research,* Inter. Assoc. Hydraulic Research, vol 26, no. 1, pp. 49–65 (see also discussion vol. 27, no. 4, pp. 556–558).

Ohtsu, I., 1976, "Free Hydraulic Jump and Submerged Hydraulic Jump in Trapezoidal and Rectangular Channels," *Trans.*, Japanese Soc. Civ. Engineering, Vol. 8, pp. 122–125.

Peterka, A. J., 1958, *Hydraulic Design of Stilling Basins and Energy Dissipators,* U.S. Bureau of Reclamation, Denver, Col (7th printing in 1983.)

Rajaratnam, N., 1965, "The Hydraulic Jump as a Well Jet," *Jour. Hydraulics Division*, Amer Soc. Civil Engrs., vol. 91, no. 5, pp. 107–132.

Rajaratnam, N., 1967, "Hydraulic Jump," *Advances in hydrosciences*, V. T. Chow (ed.), vol. 4, Academic Press, pp. 197–280.

Rajaratnam, N., and Subramanya, K., 1968, "Profile of Hydraulic Jump," *Jour., Hyd. Div.*, Amer. Soc. Civ. Engrs., vol. 94, HY3, pp. 663–673.

Rajaratnam, N., Subramanya, K., and Muralidhar, D., 1968, "Flow Profile over Sharp-crested Weirs," *Jour. Hyd. Div.*, Amer Soc Civ. Engrs., vol. 94, HY3, pp. 843–847.

Rajarantnam, N., and Murahari, V., 1974, "Flow Characteristics of Sloping Channel Jumps," *Jour. Hyd. Div.*, Amer. Soc. Civ. Engrs., vol. 100, no. 6, pp. 731–740.

Rama-Murthy, A.S., Basak, S., and RamaR., 1970, "Open channel expansions fitted with local hump," *Jour. Hyd. Div.*, Amer Soc Civ Engrs., vol 96, no 5, pp. 1105–1113.

Ranga Raju, K. G., 1981, *Flow Through Open Channels,* Tata McGraw Hill, New Dehli, India.

Rao, N. S. L., 1975, *Theory of Weirs,* Advances in Hydroscience, V. T. Chow (ed.), vol. 10, Academic Press, New York, N. Y.

Rao, B.V. and Seetharamiah, K., 1969, "Separation control devices in diverging channels," *Proc. 13th Congress,* Inter Assoc Hyd Research, Kyoto, vol 1, pp. 113–121.

Reese, A.J., and Maynard, S.T., 1987, "Design of Spillway Crests," *Jour. Hydraulic Engineering,* Amer. Soc. Civil Engrs., vol 113, no 4, pp. 476–490.

Roberson, J.A.,, Cassidy, J.J., and Chaudhry, M.H., 1998, *Hydraulic engineering,* 2nd ed., John Wiley & Sons, New York, NY.

Rouse, H., Bhoota, B. V., and Hsu, E. Y., 1951, "Design of Channel Expansions," *Trans. Amer. Soc. Civil Engrs.,* vol. 116, p. 347

Rouse, H., Siao, T. T., and Nagratnam, S., 1958, "Turbulent Characteristics of the Hydraulic Jump," *Jour. Hydraulic Division,* Amer. Soc. Civil Engrs., vol. 84, no. 1, pp. 1–29.

Skogerboe, G.V., Austin, L.H., and Bennel, R.S., 1971, "Energy loss analysis for open channel expansion," *Jour. Hyd. Div.,* Amer. Soc. Civ. Engrs., vol 97, no 10, pp 1719–1736.

Smith, C.D., and Yu, J.N., 1966, "Use of baffles in open channel transitions," *Jour. Hydraulics Div.,* Amer Soc Civ Engrs., vol 92, no 2, pp 1–16.

Sturm, T. W., 1985, "Simplified Design of Contractions in Supercritical Flow," *Jour. Hyd. Engineering,* Amer. Soc. Civ. Engrs., vol 111, pp. 871–875 (Discussion: vol 113, HY4, pp. 539–543).

U.S. Army Corps of Engineers, 1965, "Hydraulic Design of Spillways," *Engineering Manual* EM1110-2-1603.

U. S. Army Corps of Engineers, 1977, *Hydraulic Design Criteria,* Waterways Experiment Station, Vicksburg, Miss.

U. S. Bureau of Reclamation, 1973, *Design of Small Dams,* $2^n d$ ed., Denver, CO.

Vittal, N., and Chiranjeevi, V.V., 1983, "Open Channel Transitions: Rational Method of Design," *Jour. Hydraulic Engineering,* Amer. Soc. Civil Engrs., vol. 109, no. 1, pp. 99–115.

Vittal, N., 1978, "Direct Solution to Problems of Open Channel Transitions," *Jour. Hyd. Div.,* Amer. Soc. Civ. Engrs., vol. 104, no. 11, pp. 1485–1494.

White, W. R., 1977, "Thin Plate Weirs," *Proc.* Institution of Civil Engineers, vol. 63, no. 2, pp. 255–269.

Yarnell, D. L., 1934, "Bridge Piers as Channel obstructions," *Technical Bulletin no. 422,* U.S. Department of Agriculture, Nov.

Younus, M. and Chaudhry, M. H., 1994, "A Depth Averaged k-e Model for the Computation of Free Surface Flow," *Jour. Hydraulic Research,* vol. 32, no. 3, pp. 415–444.

Zipparro, V. and Hansen, H., 1993, *Davis' Handbook of Applied Hydraulics,"* 4th ed., McGraw-Hill Book Co., New York, NY.

8

COMPUTATION OF RAPIDLY VARIED FLOW

Diamond surface waves for a flow of 1,540 m³/s in the 55-m wide chute of the spillway of Bennet Dam, BC, Canada (Courtesy, B. C. Hydro and Power Authority, BC, Canada)

© Springer Nature Switzerland AG 2022
M. H. Chaudhry, *Open-Channel Flow*,
https://doi.org/10.1007/978-3-030-96447-4_8

8-1 Introduction

Typical examples of natural and man-made open channels having rapidly varied flows are mountainous streams, rivers during periods of high floods, spillway chutes, conveyance channels, sewer systems, and outlet works. Unlike the case of gradually varied flows, a number of difficulties, such as the formation of roll waves, air entrainment, and cavitation, are encountered in the analysis of these flows. In addition, instabilities may develop if the Froude number exceeds a critical value, giving rise to roll waves or slug flow. Standing wave and large surface disturbances, commonly referred to as *shocks* or *standing waves,* are important aspects of rapidly varied flows and need to be considered in the analysis and design.

To compute supercritical flow in channel expansions, including the effects of bottom slope and friction, Liggett and Vasudev [1965] numerically integrated the steady, two-dimensional, shallow-water equations. However, these and many other procedures suitable for gradually varied flows cannot be used to compute flows with shocks or standing hydraulic jumps. By using shock-tracking techniques, Pandolfi [1975] analyzed flow around a blunted obstacle in a supercritical stream. Demuren [1979] computed the sub- and super-critical steady flows by using methods developed by Patankar and Spalding and compared the computed and experimental results. Although the agreement between computed and experimental results is fair, the ability of the numerical scheme to handle discontinuities is not clearly demonstrated. The method of characteristics was used for the analysis of two-dimensional supercritical flows by Bagge and Herbich [1967] , Herbich and Walsh [1972], Villegas [1976], and Dakshinamoorthy [1977]. Ellis and Pender [1982] used an implicit method of characteristics to compute high-velocity flows in channels of arbitrary alignment and slope. Like other characteristic-based procedures, this method cannot compute oblique jumps and it requires many interpolations which may seriously affect the accuracy of the solution. Jimenez and Chaudhry [1988], Bhallamudi and Chaudhry [1992], and Gharangik and Chaudhry [1991] utilized shock-capturing finite-difference methods to analyze rapidly varied flows. Tseng et al. [2001] used high-resolution shock capturing schemes for simulating one-dimensional, rapidly varied flow incorporating the method of characteristics for the unsteady boundary conditions. This chapter is based mainly on the papers published by the author with Jimenez, Gharangik, and Bhallamudi.

In this chapter,* we present finite-difference methods for the computation of rapidly varied flows. These are shock-capturing methods and do not require any special treatment if a shock develops in the solution. Three different formulations are discussed. The St. Venant equation, also referred to as the shallow-water equations, are assumed to describe these flows in the first two formulations and Boussinesq terms are included in the third to account

---

* The material presented in this chapter will be easier to follow if the reader first becomes familiar with the unsteady flow equations and the finite-difference methods of Chapters 11–15.

for nonhydrostatic pressure distribution. The validity of these computational procedures is verified by comparing the computed results with the analytical solutions and with the experimental measurements.

8-2 Governing Equations

The flow conditions are function of time in unsteady flows. If this function is a constant, then steady flow may be considered as a special case of unsteady flow. We may, therefore, solve the unsteady flow equations to analyze steady flows. This may be done by computing the flow conditions in the channel system for a sufficient length of time until steady-state conditions are obtained. This procedure offers a number of advantages for the solution of the governing equations; we discuss this in more detail later in this section.

The St. Venant equations (see Chapter 15 for their derivation and the simplifying assumptions upon which they are based) describing the two-dimensional unsteady flows may be written in a vector form as

$$\frac{\partial \mathbf{U}}{\partial \mathbf{t}} + \frac{\partial \mathbf{E}}{\partial \mathbf{x}} + \frac{\partial \mathbf{F}}{\partial \mathbf{y}} + \mathbf{S} = 0 \tag{8-1}$$

in which

$$\mathbf{U} = \begin{Bmatrix} h \\ uh \\ vh \end{Bmatrix}; \quad \mathbf{E} = \begin{Bmatrix} uh \\ u^2h + \frac{1}{2}gh^2 \\ uvh \end{Bmatrix};$$

$$\mathbf{F} = \begin{Bmatrix} vh \\ uvh \\ v^2h + \frac{1}{2}gh^2 \end{Bmatrix}; \quad \mathbf{S} = \begin{Bmatrix} 0 \\ -gh(S_{ox} - S_{fx}) \\ -gh(S_{oy} - S_{fy}) \end{Bmatrix} \tag{8-2}$$

in which t = time; u = depth-averaged flow velocity in the x direction; v = depth-averaged flow velocity in the y direction; h = water depth measured vertically; g = acceleration due to gravity; $S_{o(x,y)} = \sin \alpha_{(x,y)}$ = channel bottom slope in the (x, y) directions; $\alpha_{(x,y)}$ = angles between the bottom of the channel and the (x, y) directions; $S_{f(x,y)}$ = friction slopes in the (x, y) directions and (x, y) coordinate system is as shown in Fig. 8-1. The friction slope S_f is calculated from the following steady state formulas

$$S_{f_x} = \frac{n^2 u \sqrt{u^2 + v^2}}{C_o^2 h^{1.33}}; \quad S_{f_y} = \frac{n^2 v \sqrt{u^2 + v^2}}{C_o^2 h^{1.33}} \tag{8-3}$$

in which n = Manning coefficient and C_o = correction factor for units (C_o = 1 in SI units and C_o = 1.49 in the English units).

Of all the simplifying assumptions made to derive the St. Venant equations, the hydrostatic pressure distribution is probably the weakest one for the present application. The pressure distribution is hydrostatic at all points except in the vicinity of a shock, such as surge wave, hydraulic jump, etc. Although some details are lost in the vicinity of the shock if these equations are used for the analysis of rapidly varied flows, the overall results are adequate for engineering purposes (Cunge [1975]). Liggett and Vasudev [1965] showed by using dimensional arguments that this assumption is valid as long as the "shallowness" parameter h_o/l_o is small, where h_o and l_o are water depth and characteristic length. Following the theory of Engelund and Munch-Petersen [1953], Jimenez and Chaudhry [1988] showed that the shallow-water theory may reasonably represent smooth, steady, supercritical flow if the Froude number is not close to 1 and the depth-to-width ratio is of the order of 0.1 or less.

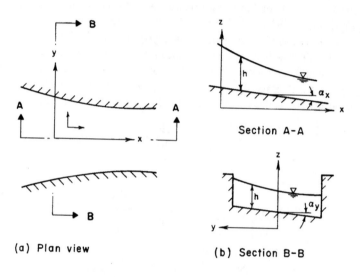

(a) Plan view   (b) Section B-B

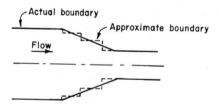

(c) Approximation of boundary

Fig. 8-1   Notation

There are three independent variables in Eq. 8-1, namely x, y, and t. As we discussed in Chapter 1, the flow is steady if the variation of flow variables with respect to time is zero. Thus, we may deduce equations describing steady flow from Eq. 8-1 by dropping the derivative term, $\partial \mathbf{U}/\partial t = 0$. In other words, the equation describing steady, two-dimensional flow in open channels is

$$\frac{\partial \mathbf{E}}{\partial \mathbf{x}} + \frac{\partial \mathbf{F}}{\partial \mathbf{y}} + \mathbf{S} = 0 \qquad (8\text{-}4)$$

Characteristic directions

According to the theory of characteristics, the characteristic directions, λ_i, of Eq. 8-4 are given by the eigenvalues of the matrix of coefficients of the nondivergent form of these equations [Jimenez and Chaudhry, 1988], i.e. ,

$$\lambda_1 = \frac{v}{u} \qquad (8\text{-}5)$$

$$\lambda_{2,3} = \frac{uv \pm gh\sqrt{\mathbf{F}_r^2 - 1}}{u^2 - gh} \qquad (8\text{-}6)$$

in which $\lambda_i = (dy/dx)_i$ are the slopes of the characteristics lines and $\mathbf{F}_r$ is the local Froude number, given by

$$\mathbf{F}_r = \frac{\sqrt{u^2 + v^2}}{\sqrt{gh}} = \frac{V}{\sqrt{gh}} \qquad (8\text{-}7)$$

in which V is the magnitude of the velocity vector. It follows from Eqs. 8-6 that Eq. 8-4 is

- *Hyperbolic* if $\mathbf{F}_r > 1$;
- *Parabolic* if $\mathbf{F}_r = 1$; and
- *Elliptic* if $\mathbf{F}_r < 1$.

Note that Eq. 8-5 defines the direction of the streamlines. If θ is the angle between the velocity vector and the x axis, then

$$u = V \cos \theta$$
$$v = V \sin \theta \qquad (8\text{-}8)$$

For an infinitely wide channel, the angular position of a small stationary wave [Engelund and Munch-Petersen, 1953]

$$\mu = \sin^{-1} \frac{1}{\mathbf{F}_r} \qquad (8\text{-}9)$$

By substituting Eqs. 8-8 and 8-9 into Eqs. 8-5 and 8-6, re-arranging, and simplifying, we obtain the following expressions for the characteristic directions:

$$\left(\frac{dy}{dx}\right)_1 = \tan \theta$$
$$\left(\frac{dy}{dx}\right)_{2,3} = \tan(\theta \pm \mu) \qquad (8\text{-}10)$$

These equations permit a clear graphical interpretation. Figure 8-2 shows a streamline passing through point P, which makes an angle θ with the x axis. It is clear from Eqs. 8-10 that, in addition to the streamline, two more characteristics pass through P: one at angle μ above the streamline C^+, and the other at angle μ below the streamline C^-. For the physical meaning of the Mach lines, it may be shown from more elementary considerations (see Henderson [1966], p. 239) that they define the locus of weak disturbances originating at point P. In other words, they bound the zone of influence of P.

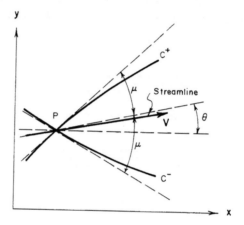

Fig. 8-2   Characteristic directions

Coordinate Transformations

Most real-life channels have irregular or curvilinear channel geometries. The inclusion of boundaries becomes a problem while analyzing these channels by the finite-difference methods. The grid points usually do not coincide with the boundaries, thus requiring interpolation procedures which have proven to deteriorate the solution [Roache, 1972]. To avoid this problem, the coordinates may be transformed such that the coordinate axes coincide with the boundaries. For example, the simple coordinate transformation of Fig. 8-3 yields good results in many cases.

Let us consider a symmetrical transition as shown in Fig. 8-3a. Due to symmetry, only one half of the channel may be analyzed. The following transformation of independent variables x and y converts the physical domain into a rectangular computational domain in coordinates ξ and η (Fig. 8-3b):

$$\xi = x$$

$$\eta = \frac{y}{f(x)} \tag{8-11}$$

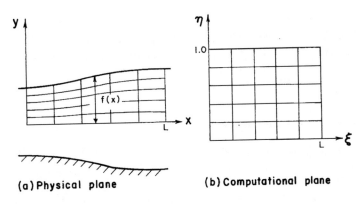

(a) Physical plane   (b) Computational plane

Fig. 8-3   Coordinate transformation

in which $f(x)$ is the distance between the symmetry line and the left boundary at distance x (Fig. 8-3a). This transformation allows a uniformly spaced grid in the computational plane such that the boundaries now coincide with the lines $\eta = 0$ and $\eta = 1$. To do this, the governing equations are first transformed in terms of ξ and η by applying the chain rule of partial differentiation. Then, the resulting equations may be cast into conservation form by means of algebraic manipulations outlined by Anderson et al. [1984]. This process gives the following divergent-free equation in the transformed coordinates:

$$\frac{\partial \mathbf{U}}{\partial t} + \frac{\partial \mathbf{G}}{\partial \xi} + \frac{\partial \mathbf{H}}{\partial \eta} + \widehat{\mathbf{S}} = 0 \tag{8-12}$$

in which

$$\mathbf{G} = f(\xi)\mathbf{E}$$
$$\mathbf{H} = \mathbf{G} - \eta\, f'(\xi)\mathbf{E}$$
$$\widehat{\mathbf{S}} = f(\xi)\mathbf{S} \tag{8-13}$$

and $f'(\xi) = df/d\xi$. Note that Eq. 8-12 is analogous to the original equation, Eq. 8-1. The former is in the x and y coordinates whereas the latter is in the ξ and η.

Polar coordinates combined with transformations of the type given by Eq. 8-11 may be used to obtain a rectangular computational domain for the channels with general geometries.

8-3 Computation of Supercritical Flow

As we discussed in the last section, Eq. 8-4 is hyperbolic if $\mathbf{F}_r > 1$. Thus, a channel having supercritical flow throughout its length may be analyzed by using the steady form of the shallow-water equations. In the x-y coordinates, we obtained Eq. 8-4 from Eq. 8-1. A similar equation may be obtained from Eq. 8-12 in the ξ-η coordinates. These equations are hyperbolic as long as the flow is supercritical. This offers special advantages for their numerical solution in the sense that we solve the equations directly to obtain the flow conditions and not in time until steady conditions are reached.

Finite-difference methods

The x-y plane (ξ-η plane for the transformed coordinates) is divided into a computational grid to solve Eq. 8-4 by the finite-difference methods. We will use the following notation to identify variables at different grid points. The superscript k and subscript j indicate nodes in the x and y directions, respectively. As discussed previously, when the local Froude number is greater than 1, the system of equations describing steady flows (Eqs. 8-4) are hyperbolic [Abbott, 1979]. Therefore, a marching procedure may be used to integrate them. The solution is obtained by starting at the upstream end of the channel and advancing the computations first to $x_0 + \Delta x$, then, to $x_0 + 2\Delta x$, and so on. In this case, the x direction is called the *marching direction*. This direction may be any one as long as the system is hyperbolic with respect to that particular marching direction [Kutler, 1975]. What this means is that the disturbances originating in the flow field should not travel opposite to the marching direction. According to Eqs. 8-9, Eq. 8-7 is hyperbolic with respect to the x-direction if

$$u^2 - gh > 0 \qquad (8\text{-}14)$$

Thus, the coordinates are selected such that the marching direction is aligned with the predominant flow direction. Otherwise, the requirement given by Eq. 8-14 may not be fulfilled, even if the flow is supercritical.

It is desirable to use shock-capturing or through methods, since a complex oblique jump pattern may develop in many situations involving supercritical flow. The channels or structures having rapidly varied flow are usually short, and probably less than few hundreds marching steps are sufficient to compute the water surface profiles in them. Therefore, explicit methods suitable for hyperbolic systems may be utilized. Two explicit schemes – Lax and MacCormack schemes – were used by Jimenez and Chaudhry [1988]. The Lax scheme is first-order accurate, and the MacCormack scheme is second-order accurate. They are probably the simplest of the available explicit, dissipative numerical schemes and have been widely used in fluid flow applications [Anderson et al., 1984]. We discuss only the application of the MacCormack scheme in this chapter; readers interested in the application of the Lax scheme should see Jimenez and Chaudhry [1988].

MacCormack Scheme

The MacCormack scheme is a two-step predictor-corrector scheme [Anderson et al., 1984]. Considering x as the marching direction, application of the scheme to Eq. 8-4 yields the following equations:

Predictor

$$\mathbf{E}_j^* = \mathbf{E}_j^k - \frac{\Delta x}{\Delta y}(\mathbf{F}_{j+1}^k - \mathbf{F}_j^k) - \Delta x \mathbf{S}_j^k \qquad (8\text{-}15)$$

Corrector

$$\mathbf{E}_j^{**} = \mathbf{E}_j^k - \frac{\Delta x}{\Delta y}(\mathbf{F}_j^* - \mathbf{F}_{j-1}^*) - \Delta x \mathbf{S}_j^* \qquad (8\text{-}16)$$

$$\mathbf{E}_j^{k+1} = \frac{1}{2}(\mathbf{E}_j^* + \mathbf{E}_j^{**}) \qquad (8\text{-}17)$$

An asterisk (*) indicates the values at the end of the predictor part and (**) refers to the values at the end of the corrector part. Another variation of the method is to use backward differences for the y derivative in the predictor part and forward differences in the corrector part. In some applications, alternation of both variations every other integration step is recommended [Roache, 1972]. Kutler [1975] showed that the shock resolution is best in problems involving discontinuities when the difference in the predictor part is in the direction of the propagation of discontinuity. The validity of this conclusion has been verified for the present application by Jimenez and Chaudhry [1988].

We consider the ξ coordinate as the marching direction if the equations are in the transformed coordinates and we use the MacCormack scheme to integrate the steady form of Eq. 8-12 (i.e., equation obtained by substituting $\partial \mathbf{U}/\partial t = 0$ into Eq. 8-12). The stability requirement of the equations in the transformed coordinates is modified accordingly.

Stability

The *Courant-Friedrichs-Lewy condition* (CFL for short) has to be satisfied for the above scheme to be stable. The CFL condition for Eq. 8-1 may be written as [Anderson, et al., 1984]

$$C_n = |\lambda_{max}|\frac{\Delta x}{\Delta y} \leq 1 \qquad (8\text{-}18)$$

in which $|\lambda_{max}|$ is the maximum absolute value of the characteristic slopes, $|\lambda_i|$, and C_n is referred to as the *Courant number*. It follows from Eq. 8-9 that

$$|\lambda_{max}| = \frac{|uv| + gh\sqrt{\mathbf{F}_r^2 - 1}}{u^2 - gh} \tag{8-19}$$

The truncation error in the MacCormack scheme is the smallest when the largest possible value of the Courant number, compatible with the above stability condition, is used [Anderson, et al., 1984]. However, note that CFL is derived by neglecting the head-loss term and by using a linearized form of the governing equations.

Experience has shown that this stability condition is adequate for the analyses of systems having low head losses, although additional stability criteria may have to be satisfied for systems having large losses. Since the head losses are usually very small in a typical channel having rapidly varied flows, the previously mentioned CFL stability condition should be sufficient for the present application. The step size should be chosen so that Eq. 8-18 is satisfied at all points in the y direction since $|\lambda_{max}|$ is a local function of h, u, and v. If this condition is not satisfied, then the step size should be reduced and the calculations repeated after each integration step to avoid instability.

Boundary Conditions

The above finite-difference equations are used at the interior grid points. To start the computations, we specify the initial conditions and we include the boundary conditions to simulate the boundaries of the channel and the inflow and outflow conditions at the ends of the channel.

Proper inclusion of the boundaries is very important for a successful application of any numerical technique, especially for hyperbolic systems, in which an error introduced at the boundaries is propagated and reflected throughout the grid. These errors may cause instability in many cases [Anderson et al., 1984].

For the initial conditions, we specify all three variables (h, u, v) at all grid points. It is sufficient to analyze one-half of a symmetrical system by means of a symmetrical boundary at the symmetry plane. In addition, we have to specify the boundary conditions for the channel boundaries.

For a *solid boundary* we enforce the condition that there is no mass flow through it. This may be done by the following equation, referred to as the *slip condition*

$$\frac{v}{u} = \tan\theta \tag{8-20}$$

in which θ is the angle between the wall and the x-axis. A *symmetry boundary* is similar to a solid boundary in that the normal velocity with respect to the symmetry plane should be zero. In addition, it is required that the normal gradients of all variables with respect to the symmetry plane vanish.

Several wall boundary techniques enforce in one way or another the basic requirement given by Eq. 8-20. The problem arises in applying this equation at the grid points along the wall. The values of all the variables are required for this purpose, and Eq. 8-20 does not provide all the needed information. Thus, these values are computed using information from the interior points plus the boundary condition. Abbett [1971] developed a technique that has proven to be successful in many supersonic flow computations. This procedure was adapted for the analysis of supercritical flows by Jimenez and Chaudhry [1988] and is discussed here. For simplicity, let us assume that the wall under consideration is aligned with the x-axis. Thus, the boundary condition is given by $v = 0$.

The basic idea in the Abbett procedure [Abbett, 1971; Kutler, 1975] is to apply the numerical scheme up to the wall using one-sided differences as a first step. Then, to enforce the surface tangency requirement, a simple wave is superimposed on the solution to make the flow parallel to the wall. The details of the method are presented with reference to the MacCormack scheme.

Let us assume that the solution is being advanced from grid k to grid $k + 1$ (Fig. 8-4a). Since the first step of the MacCormack scheme as given by Eq. 8-15 involves a forward difference, it can be applied at the wall to yield the predicted value $\mathbf{E}^\star_{j=1}$. It is followed by the corrector step (Eq. 8-16); however, the backward difference in the y-direction is replaced by a forward difference to yield the first corrected value $\widetilde{\mathbf{E}}_1 = \frac{1}{2}(\mathbf{E}^\star_1 + \mathbf{E}^{\star\star}_1)$. We will use a tilde $\widetilde{\mathbf{E}}$ on various variables to indicate values corresponding to $\widetilde{\mathbf{E}}$. From this value of $\widetilde{\mathbf{E}}_1$, we determine $\widetilde{h}$, $\widetilde{u}$, and $\widetilde{v}$ at grid $k + 1$. Generally, the resulting velocity vector, $\widetilde{V}$, will not be tangent to the wall. Let this angle between $\widetilde{V}$ and the wall be $\Delta\theta$. Then,

$$\Delta\theta = \tan^{-1} \frac{\widetilde{v}}{\widetilde{u}} \qquad (8\text{-}21)$$

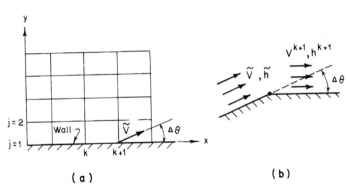

Fig. 8-4 **Abbett procedure for solid walls**

If $\Delta\theta$ is positive, an expansion wave is required to rotate $\widetilde{V}$ so that it becomes tangent to the wall; and a contraction wave is necessary for the same purpose if $\Delta\theta$ is negative. Figure 8-4 shows a situation in which $\Delta\theta > 0$. This is equivalent to conditions produced by a wall turning away from the flow. The situation for $\Delta\theta < 0$ corresponds to a wall turning into flow.

A comparison of three available procedures for computing $\Delta\theta$ gave similar results for $|\Delta\theta| < 5°$ and $2 < \mathbf{F}_r < 8$ [Jimenez and Chaudhry, 1988]. Among these procedures, the following does not require an iterative solution like the other two and is presented here.

Knapp [1951] found from experiments on curved channels that the following equation gives good results:

$$\frac{h^{k+1}}{\widetilde{h}} = \widetilde{\mathbf{F}}_r^2 \sin^2(\widetilde{\mu} - \frac{\Delta\theta}{2}) \tag{8-22}$$

in which $\widetilde{\mu} = \sin^{-1}(1/\widetilde{\mathbf{F}}_r)$. This expression is obtained by assuming a constant magnitude of velocity, $\widetilde{V} = V^{k+1}$, through the cross-wave.

The Abbett procedure may be applied to a curved wall, in which the angle $\Delta\theta$ includes the deviation of the wall because of its curvature.

Verification

The results for several cases computed by using the MacCormack scheme were compared with the analytical solutions and with the experimental results [Jimenez and Chaudhry, 1987]. Only two comparisons are presented here. The first is with the analytical solution and the second with the experimental measurements.

Oblique Hydraulic Jump

As we discussed in the last chapter, an oblique hydraulic jump or standing wave is produced when a vertical boundary is deflected inward into the flow, e.g., as in a channel contraction. This causes an abrupt depth increase, which is propagated from the point of deflection in the wall to the interior of the flow field at angle β with respect to the flow direction. Equation 7-17 is the analytical solution for the problem if the bottom friction and slope of the channel bottom are neglected.

Figure 8-5a shows the grid system used in the numerical computations. The x-axis is aligned with the wall downstream of the deflection point. The variables at $x = 0$ are specified as $h = h_o; u = V_o\cos\theta$; and $v = -V_o\sin\theta$, where h_o and V_o are the approach flow depth and velocity, and θ is the angle of wall deflection. Along the lower boundary ($y = L_y$), a zero-order extrapolation from the immediate interior point is used, i.e., $\mathbf{E}_{N_y}^{k+1} = \mathbf{E}_{N_y-1}^{k+1}$. This is a good approximation, since the flow field remains undisturbed upstream of the wave front.

For the example shown here, $\mathbf{F}_{r0} = V_o/\sqrt{gh_o} = 4$ and $\theta = 12°$. The analytical solution for this is $h_1/h_o = 1.987$; $\beta = 25.505°$; and $V_1/V_o = 0.9282$, in which h_1 and V_1 are the downstream flow depth and velocity, respectively.

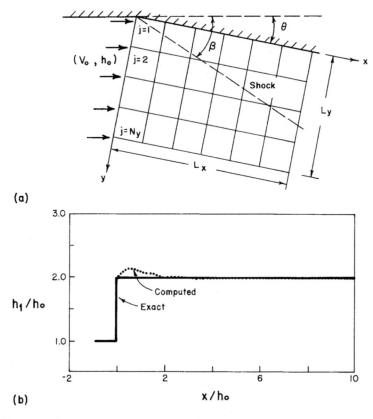

(a)

(b)

Fig. 8-5 Oblique jump

Figure 8-5(b) shows the water surface profiles along the wall computed by using $C_n = 0.98$. Since the Abbet procedure incorporates Eq. 7-17, the abrupt initial jump is computed exactly; it is then followed by a small overshoot which is rapidly corrected. The same trend was observed for different Froude numbers and deflection angles.

A three-dimensional plot of the computed water surface for $C_n = 0.98$ is shown in Figure 8-6. The analytical solution is indicated by a dashed line. The figure illustrates the shock-capturing capabilities of the scheme. The strength of the shock and its direction are accurately predicted. The former is obtained exactly up to the third significant digit; the error in the latter is not easy to determine because of the spreading of the shock over 2 or 3 grid points. As mentioned earlier, the best resolution of the shock is obtained when the finite-difference approximation in the predictor part is in the direction of propagation of the discontinuity [Kutler, 1975]. For example, for the results shown in Fig. 8-6, the overshoot on the back side of the jump is approximately 4 per cent of $h_1 - h_o$, as compared to the opposite alternative (using backward finite-

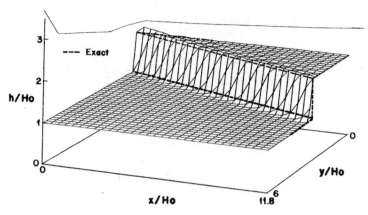

Fig. 8-6   Computed water surface profile

differences in the predictor part for this example), for which the overshoot is about 25 per cent.

Circular-Arc Contraction

The flow in a contraction composed of circular arcs (Fig. 8-7) is analyzed. The results compared here are for an initial Froude number of 4 and an initial water depth of 0.030 m. The flow at the entrance of the contraction is uniform. This was the condition in the experiment for the assumed Manning n of 0.012 and for a bottom slope of $S_o = 0.072$. In the computations, a constant depth and uniform velocity distribution were assumed at the upstream section and 21 grid points in the y direction (for half-channel width) and a Courant number of 0.98 were used.

Figure 8-7(b) compares the computed water surface profile at the wall with the experimental results reported by Ippen et al. [1951, Fig. 38] and Fig. 8-7(c) shows a three-dimensional plot of the computed water surface profile. The walls of the channel are not shown in this plot. The comparison of water depths in the length of the contraction, including the first peak, is good. Downstream of the transition, however, the disagreement between the experimental and computed results becomes large. This example shows that although a solution of the shallow-water equations simulates the general features of the flow, the prediction of the maximum water levels is unsatisfactory. This is because the disturbances as well as the depth to width ratio ($h/b \approx 0.2$ for the downstream channel) are large.

These comparisons show that this scheme gives satisfactory results if the assumption of hydrostatic pressure distribution is valid.

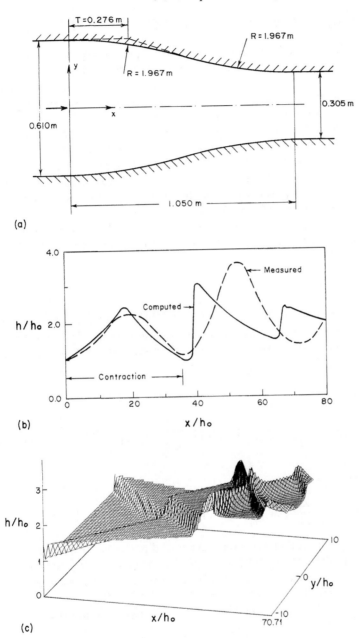

Fig. 8-7 Circular-arc contraction

8-4 Computation of Sub- and Supercritical Flows

In this section, two-dimensional, depth-averaged, *unsteady* flow equations in a transformed coordinate system (Eq. 8-12) are solved numerically to analyze flows in channel expansions and contractions. An unsteady flow model is used to obtain steady flow solutions by treating the time variable as an iteration parameter and letting the solution converge to the steady state. Unlike the steady model of the last section, which can be used only for supercritical flows, the unsteady model is capable of simulating both sub- and supercritical flows. The approximation of the side wall boundaries of a physical system as shown by dotted lines in Fig. 8-1(c) may introduce large errors. It is better in such cases to convert the physical domain into a rectangular computational domain using simple transformations discussed in Section 8-2.

We first describe the details of the MacCormack scheme and procedures for including the initial and boundary conditions and then present results for its verification.

Numerical Solution

The MacCormack scheme [MacCormack, 1969] is used to integrate numerically the transformed form of the governing equations (Eq. 8-5). Referring to the finite-difference grid shown in Fig. 8-8, these finite-difference approximations are as follows.

Predictor

$$\mathbf{U}_{i,j}^{\star} = \mathbf{U}_{i,j}^{k} - \frac{\Delta t}{\Delta \xi} \left(\mathbf{G}_{i+1,j}^{k} - \mathbf{G}_{i,j}^{k} \right) - \frac{\Delta t}{\Delta \eta} \left(\mathbf{H}_{i,j+1}^{k} - \mathbf{H}_{i,j}^{k} \right) - \mathbf{S}_{i,j}^{k} \Delta t \quad (8\text{-}23)$$

Corrector

$$\mathbf{U}_{i,j}^{\star\star} = \mathbf{U}_{i,j}^{k} - \frac{\Delta t}{\Delta \xi} \left(\mathbf{G}_{i,j}^{\star} - \mathbf{G}_{i-1,j}^{\star} \right) - \frac{\Delta t}{\Delta \eta} \left(\mathbf{H}_{i,j}^{\star} - \mathbf{H}_{i,j-1}^{\star} \right) - \mathbf{S}_{i,j}^{\star} \Delta t \quad (8\text{-}24)$$

in which the subscripts i and j refer to the grid points in the ξ and η directions, respectively. The superscript k refers to the variable at the known time level, $*$ refers to the variables computed at the end of the predictor part and $**$ refers to the variables at the end of the corrector part.

Now, $\mathbf{U}$ at the unknown time level $k+1$ is determined from

$$\mathbf{U}_{i,j}^{k+1} = \frac{1}{2} \left(\mathbf{U}_{i,j}^{\star} + \mathbf{U}_{i,j}^{\star\star} \right) \quad (8\text{-}25)$$

As we discussed in Section 8-3, there are two other formulations of the MacCormack scheme in addition to that given here. These are to use backward finite differences in the predictor part and forward finite differences in the corrector part or to alternate the direction of differencing from one time step to the next. All of these three alternatives gave similar results for the steady state solutions studied by Bhallamudi and Chaudhry [1992].

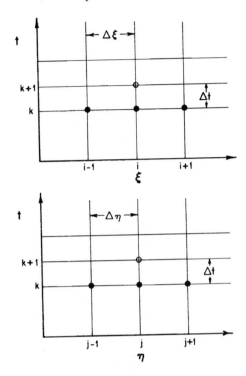

Fig. 8-8 Computational grid

Initial and Boundary Conditions

To start the unsteady state computations, the values of u, v and h at time $t = 0$ are specified at all the grid points. In the present application, specification of their approximate values is sufficient, since these are needed only to start the computations that are continued until the solution converges to a steady state.

Typical boundaries for a channel may be included in the analysis as follows.

Inflow and outflow boundaries

The specification of inflow and outflow boundary conditions and at the upstream and downstream ends depends on whether the flow is subcritical or supercritical [Stoker, 1957; and Verboom et al., 1982].   For two-dimensional supercritical flow, three boundary conditions have to be specified at the inflow boundary and none at the outflow boundary. For two-dimensional subcritical flow, however, two conditions are specified at the inflow boundary and one at the outflow boundary. The details of these open boundary specifications and the approximations are discussed later (in the verification section) with reference to the particular problem solved.

Symmetry boundary

A reflection procedure is used at a symmetry boundary [Roache, 1972]. In this procedure, the nonconservative flow variables u and h at the imaginary reflection points shown in Fig. 8-9(a) are specified as even functions with respect to the symmetry line. However, the normal velocity is specified as an odd function so that the average normal velocity at the boundary is zero. Note that the reflection procedure is exact for a symmetry line.

Solid side wall boundary

A slip condition is used as the boundary condition for a side wall. Therefore, the resultant velocity at a solid wall is tangent to it. The reflection procedure used herein for the solid side wall is approximate and is not exact as was the case for a symmetry boundary procedure.

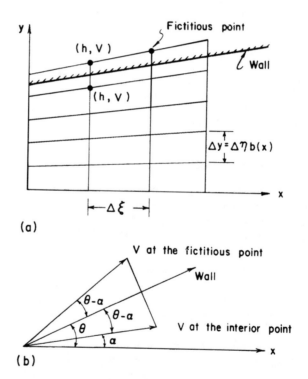

Fig. 8-9 Imaginary reflection point

Referring to Fig. 8-9(b), the flow depth and the magnitude of resultant velocity at an imaginary reflection point are specified equal to their values at the corresponding interior grid point. The direction of the flow velocity, however, is determined such that the normal velocity at the wall is zero. If θ is the angle between the wall and the x-axis and α is the angle between the resultant velocity at the interior point and the x axis (Fig. 8-9), then the velocity components u_o and v_o at the reflection point are given by

$$u_o = V \cos(2\theta - \alpha) \tag{8-26}$$

$$v_o = V \sin(2\theta - \alpha) \tag{8-27}$$

where V = resultant velocity at the interior point. Equations 8-26 and 8-27 are derived for a channel expansion; similar equations may be written for a channel contraction.

Stability

The MacCormack scheme is stable if the Courant-Friedrichs-Lewy(CFL) condition is satisfied. This condition for two-dimensional flow in the transformed coordinates may be written as [Roache, 1972]

$$C_n = \frac{(V + \sqrt{gh})\triangle t}{b(x)\triangle \xi \triangle \eta} \sqrt{\triangle \xi^2 + [b(x)\triangle \eta]^2} \leq 1 \tag{8-28}$$

where V = resultant velocity at the grid point. The numerical scheme is stable only if the above condition is satisfied at every grid point. Since the preceding condition is heuristic and is derived from a linearized form of the governing equations for one-dimensional flows, some numerical experimentation should be done before selecting an actual upward limit of the value of C_n.

Artificial Viscosity

The dispersive errors in the MacCormack scheme produce high-frequency oscillations near the steep gradients. To dampen these oscillations, [Anderson et al., 1984] a procedure developed by Jameson et al. [1981] is used. This procedure smooths large gradients and leaves the smooth areas relatively undisturbed.

   In this procedure, we first compute the following parameters using the computed values of h at $k + 1$ time level.

$$\nu_{\xi_{i,j}} = \frac{|h_{i+1,j} - 2h_{i,j} + h_{i-1,j}|}{|h_{i+1,j}| + |2h_{i,j}| + |h_{i-1,j}|}$$

$$\nu_{\eta_{i,j}} = \frac{|h_{i,j+1} - 2h_{i,j} + h_{i,j-1}|}{|h_{i,j+1}| + |2h_{i,j}| + |h_{i,j-1}|} \tag{8-29a}$$

At the points where $h_{i,j-1}$ does not exist, we use

$$\nu_{\eta_{i,j}} = \frac{|h_{i,j+1} - h_{i,j}|}{|h_{i,j+1}| + |h_{i,j}|}$$

and where $h_{i,j+1}$ does not exist, we use

$$\nu_{\eta_{i,j}} = \frac{|h_{i,j-1} - h_{i,j}|}{|h_{i,j-1}| + |h_{i,j}|} \tag{8-29b}$$

Then, we determine from the following equations

$$\epsilon_{\xi_{i-\frac{1}{2},j}} = \kappa \max \left(\nu_{\xi_{i-1,j}}, \nu_{\xi_{i,j}} \right)$$

$$\epsilon_{\eta_{i,j-\frac{1}{2}}} = \kappa \max \left(\nu_{\eta_{i,j-1}}, \nu_{\eta_{i,j}} \right) \tag{8-30}$$

where κ is a dissipation constant. The final values of the variable f at the new time step are computed from the following equation:

$$f_{i,j}^{k+1} = f_{i,j}^{k+1} + \left[\epsilon_{\xi_{i+\frac{1}{2},j}} \left(f_{i+1,j}^{k+1} - f_{i,j}^{k+1} \right) - \epsilon_{\xi_{i-\frac{1}{2},j}} \left(f_{i,j}^{k+1} - f_{i-1,j}^{k+1} \right) \right]$$
$$+ \left[\epsilon_{\eta_{i,j+\frac{1}{2}}} \left(f_{i,j+1}^{k+1} - f_{i,j}^{k+1} \right) - \epsilon_{\eta_{i,j-\frac{1}{2}}} \left(f_{i,j}^{k+1} - f_{i,j-1}^{k+1} \right) \right] \tag{8-31}$$

in which f refers to u, v, and h. Equation 8-31 should be viewed as a FOR-TRAN replacement statement. The dissipation constant, κ, is used to regulate the amount of dissipation

The above procedure is equivalent to adding second-order dissipative terms to the original governing equations. The actual numerical eddy viscosity coefficient in the ξ-direction is of the order of $\kappa \nu_\xi \Delta \xi^2 / \Delta t$. This indicates that the influence of κ on the results depends upon the gradients in the flow depth as well as on the grid size. As can be seen, its influence in the smooth regions is minimal since ν tends to be zero in such a case. A numerical grid is chosen such that the model gives convergent results and a finer grid does not improve the results significantly. The value of κ is selected such that it is as small as possible and at the same time smooths the high-frequency oscillations. A value of 0.3 for κ is recommended for an initial trial.

Verification

The computed results are compared with the laboratory test data for two cases; for other comparisons, see Bhallamudi and Chaudhry [1992].

Supercritical flow in symmetrical contraction

In the laboratory tests reported by Ippen, et al. [1951] on supercritical flow in the symmetrical, straight-wall contraction shown in Fig. 8-10, the upstream depth, h_o, was 0.0305 m and the upstream Froude number, $\mathbf{F}_{ro}$, was equal to 4.0. The computations were done using a grid $\triangle \xi$ = 0.0483 m and $\triangle \eta$ = 0.0476. The dissipation coefficient, κ, was 0.8, the Courant number was equal to 0.80, and the friction and bottom slopes were assumed to be zero.

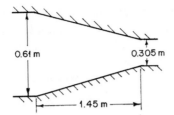

(a) Contraction

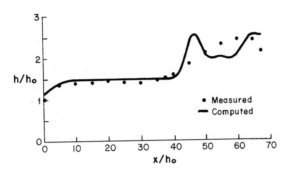

(b) Water level along wall

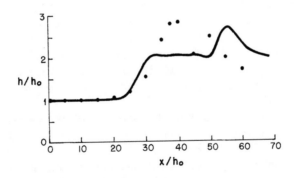

(c) Water level along centerline

Fig. 8-10   Supercritical flow in a contraction

A depth of 0.0305 m, streamwise velocity of 2.188 m/s, and zero transverse velocity were specified at every grid point as the initial conditions. Starting with these initial values, the flow conditions were computed up to 3 s when the flow became steady. At the upstream boundary, h, u, and v were specified as 0.0305 m, 2.188 m/s and zero, respectively. No condition was specified at the downstream boundary. The variables at the downstream end were, however, extrapolated from the interior points.

As shown in Fig. 8-10, the agreement between the computed water-surface profiles is good along the walls; and at the centerline, where the flows are smooth. However, this is not the case for the centerline water surface profile in the vicinity of strong shocks. Although the computed maximum height of the shock is about the same as that in the experiment, the computed location differs significantly. Thus, the computed results may be used confidently for selecting the wall height; however, they are not accurate in the middle of the channel, which is more of an academic interest. The differences between the computed and measured results at the centerline of the transition may be due to the assumption of hydrostatic pressure distribution not being valid near steep gradients and the exclusion of the effects of air entrainment.

Hydraulic jump in a gradual expansion

The simulation of both sub- and supercritical flows is demonstrated by simulating a hydraulic jump in a gradual expansion. Figure 8-11 shows the general dimensions of the channel and the transition. The specified flow conditions in the channel were as follows: discharge = 0.007 m^3/s; depth at the upstream end = 0.06 m; depth at the downstream end = 0.07 m; channel bottom slope = 0.00017, and Manning $n = 0.015$. For these conditions, the flow is supercritical at the inflow section (Froude number = 1.52) and subcritical at the outflow section (Froude number = 0.268).

In the computations, the flow depth $h = 0.07$ m, streamwise velocity $u = 0.222$ m/s and transverse velocity $v = 0$ m/s were specified as the initial conditions at all grid points. Since the flow is supercritical at the upstream end, three conditions have to be specified for the upstream boundary condition. For this purpose, the following values were used: $u = 1.167$ m/s, $v = 0.0$ m/s and $h = 0.06$ m. The flow is subcritical at the downstream end. Therefore, only one condition $u = 0.222$ m/s was imposed as the downstream boundary, and the flow depth was determined from the positive characteristic equation. The length of the channel downstream of the transition was long enough so that the transverse velocity at the downstream end could be assumed as zero. A grid with $\triangle\xi = 0.15$ m and $\triangle\eta = 0.0476$ was used. Computations were performed with Courant number = 0.90 and $\kappa = 0.003$. Because of the presence of bottom slope and friction, the numerical oscillations were not significant. Therefore, a small amount of artificial viscosity was sufficient to obtain satisfactory results. Computations were done up to $t = 7$ seconds when the solution converged to a steady state. The computed water surface profile along the channel centerline

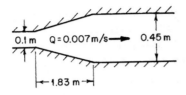

(a) Expansion

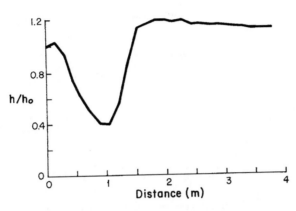

(b) Variation of water depth with distance

Fig. 8-11 **Hydraulic jump in gradual expansion**

is plotted in Fig. 8-11. Since the flow is supercritical at the inflow section and subcritical at the outflow section, a hydraulic jump is formed, as shown. This illustrates the ability of the model to handle mixed supercritical and subcritical flows.

8-5 Simulation of Hydraulic Jump

To determine the jump location in a channel, Chow [1959] computed the water surface profiles for supercritical flow starting from the upstream end and the subcritical flow starting from the downstream end. The jump is formed at a location where the specific forces on both sides of the jump are equal. Mc-Corquodale and Khalifa [1983] used the strip-integral method to compute the jump length, water surface profile, and pressures at the bottom. To solve the St. Venant equation numerically, Abbott et al. [1969] used a finite-difference method and Katopodes [1984] used the finite-element method. In these simulations, the computations were continued until a steady state was reached. The location of the hydraulic jump is automatically computed as part of the solution. The water surface in rapidly varied flows, however, has steep gradients and the assumption of hydrostatic pressure distribution may not be valid [Basco, 1983]. If we include the additional terms in the gradually varied flow equations to allow for nonhydrostatic pressure distribution, the resulting equations are referred to as the *Boussinesq equations.*

In this section, these equations are solved to compute the formation of hydraulic jump in a rectangular channel. The inclusion of initial and boundary conditions is discussed, and the importance of the Boussinesq terms is investigated.

Governing Equations

The Boussinesq equations[†] for one-dimensional flow in vector form may be written as

$$\frac{\partial \mathbf{U}}{\partial t} + \frac{\partial \mathbf{E}}{\partial x} = \mathbf{S} \tag{8-32}$$

in which

$$\mathbf{U} = \begin{Bmatrix} h \\ uh \end{Bmatrix}; \quad \mathbf{E} = \begin{Bmatrix} uh \\ u^2 h + \frac{1}{2} gh^2 - E \end{Bmatrix}; \quad \mathbf{S} = \begin{Bmatrix} 0 \\ gh(S_o - S_f) \end{Bmatrix};$$

and

$$E = \frac{1}{3} h^3 \left[\frac{\partial^2 u}{\partial x \partial t} + u \frac{\partial^2 u}{\partial x^2} - \left(\frac{\partial u}{\partial x} \right)^2 \right] \tag{8-33}$$

In this equation, E is called the Boussinesq term. It is introduced by the second-order term of pressure distribution along the water depth. It is clear that Eqs. 8-32 are reduced to the St. Venant equations if the Boussinesq term, E, is omitted from these equations.

Numerical Solution

The first and second-order numerical schemes yield satisfactory results for the solution of St. Venant equations. However, the Boussinesq equations describing rapidly varied flow have third-order terms; and considerable effort must be expended to reduce the truncation errors while approximating these terms by finite differences [Abbott, 1979]. Therefore, it is necessary to employ third- or higher-order accurate methods to solve these equations numerically. For this reason, the two-four scheme developed by Gottlieb and Turkel [1976] is used herein to solve these equations at the interior computational nodes.

The following finite-difference approximations are used in the two-four scheme.

Predictor

$$\mathbf{U}_i^* = \mathbf{U}_i^k + \frac{1}{6} \frac{\Delta t}{\Delta x} \left[\mathbf{E}_{i+2}^k - 8\mathbf{E}_{i+1}^k + 7\mathbf{E}_i^k \right] + \Delta t \mathbf{S}_i^k \tag{8-34}$$

[†] For the derivation of these equations and the simplifying assumptions on which they are based, see Chapter 12.

Corrector

$$\mathbf{U}_i^{**} = \frac{1}{2}\left(\mathbf{U}_i^* + \mathbf{U}_i^k\right) + \frac{1}{12}\frac{\Delta t}{\Delta x}\left[-7\mathbf{E}_i^* + 8\mathbf{E}_{i-1}^* - \mathbf{E}_{i-2}^*\right] + \frac{1}{2}\Delta t\mathbf{S}_i^* \quad (8\text{-}35)$$

The term $\partial^2 u/\partial x^2$ is approximated by using a three-point central finite-difference approximation in both the predictor and corrector parts. To approximate the term $(\partial u/\partial x)^2$, a forward finite-difference approximation in the predictor part and a backward finite-difference approximation in the corrector part are used.

To dampen the high-frequency oscillations near the steep gradients, a procedure [Jameson et al., 1981] is outlined in the following paragraph.

A parameter ν_i is first computed using the computed flow depths at $k+1$ time level

$$\nu_i = \frac{|h_{i+1} - 2h_i + h_{i-1}|}{|h_{i+1}| + 2|h_i| + |h_{i-1}|}$$

$$\nu_{i+\frac{1}{2}} = \kappa \max\left(\xi_{i+1}, \xi_i\right) \quad (8\text{-}36)$$

in which κ is used to regulate the amount of dissipation. The computed variables u and h are then modified as

$$f_i^{k+1} = f_i^{k+1} + \xi_{i+\frac{1}{2}}\left(f_{i+1}^{k+1} - f_i^{k+1}\right) - \xi_{i-\frac{1}{2}}(f_i^{k+1} - f_{i-1}^{k+1}) \quad (8\text{-}37)$$

in which f refers to both u and h; and this equation should be viewed as a FORTRAN statement.

Initial and Boundary Conditions

For the initial conditions, the flow at time $t = 0$ is assumed to be supercritical in the entire channel. By starting with the specified flow depth and velocity at section 1 (Fig. 8-12), the initial steady-state flow depth and flow velocity at all computational nodes are determined by numerically integrating the equation describing the gradually varied flow (Eq. 5-5). Since the computations are continued until steady conditions are reached, it is sufficient to specify only the approximate values of the initial flow depths and velocities.

The flow conditions at the boundaries are computed as follows. At the *upstream boundary*, the flow depth, h, and velocity, u, are specified equal to their initial values and they remain unchanged during the computations. At the downstream boundary, a constant flow depth is specified and the flow velocity is calculated from the characteristic form of Eq. 8-32 using a forward finite-difference approximation, i.e., the velocity, u_{i+1}^{k+1} at the unknown time level $k+1$ is determined from the following equation [Chaudhry, 2014]

$$u_{i+1}^{k+1} = u_i^k - \left(\frac{g}{c}\right)_i^k\left(h_{i+1}^{k+1} - h_i^k\right) + g\Delta t\left(S_o - S_f\right)_i \quad (8\text{-}38)$$

in which $c = \sqrt{gh}$ = celerity of a small gravity wave.

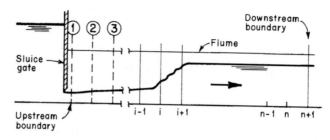

Fig. 8-12   Definition sketch

Stability Conditions

The two-four scheme is stable if the following CFL condition is satisfied at each grid point:

$$\Delta t = C_n \frac{\Delta x}{|u| + \sqrt{gh}} \tag{8-39}$$

In this equation, C_n is the desired Courant number which must be less than or equal to 2/3 for the two-four scheme [Gottlieb and Turkel, 1976].

Computational Procedure

The channel is divided into a number of equal-length reaches. Because the approximation of a second-order partial derivative requires values at the two neighboring nodes, it is not possible to calculate the variables at the computational nodes near the boundaries. Therefore, the flow equations at these nodes are first solved by neglecting the Boussinesq terms and by using the second-order MacCormack scheme for their solution. This should not significantly effect the overall accuracy of the solution in the region of interest, since the boundary nodes are located away from the jump location.

In the computations, the size of time step was restricted by the Courant stability condition and the spatial grid size. The Courant number was set equal to 0.65, since best results are obtained when it is approximately equal to 2/3. To smooth high-frequency oscillations near the jump, the dissipation coefficient, κ, in the Jameson's formula (Eq. 8-37) was determined by a trial-and-error procedure. Trials values ranging from 0.01 to 0.05 indicated that a value of 0.03 provided the best results.

Several runs with different values of the upstream flow depth, velocity, Froude number, and downstream flow depth were made. The Manning n for the flume was determined by trial and error so that the computed water-surface profile matched with the measured water levels in the flume during the initial steady supercritical flow. The n values varied from 0.008 to 0.011 depending upon the flow depth, since the bottom of the flume is made up of metal and the sides are made up of glass. The initial steady state depth and velocity at every computational node were first computed by assuming the flow to be supercritical throughout the flume. Then, the unsteady computations were started by increasing the downstream depth to the value measured during the experiment. The computations were continued until they converged to the final steady state for the specified end conditions.

Results

The size of the spatial grid, Δx, was varied from 0.15 m to 0.6 m. Simulations were also done by the second-order MacCormack scheme, neglecting the Boussinesq term. No definite trend could be established that would indicate that reducing the value of Δx in the second-order method gave results tending towards those obtained by the fourth-order method. Thus, the computation of various nonlinear terms of the governing equations appears to play a more important role than the truncation errors introduced by the numerical scheme.

Figure 8-13 shows the water surface profiles at different times following an increase in the downstream depth at time $t = 0$. The jump travels from the downstream end towards the upstream end and then moves back and forth until it stabilizes in one location.

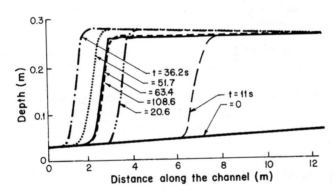

Fig. 8-13 **Water surface profile at different times for $\mathbf{F}_r = 7$**

When the numerical solution converged to a steady state, the Boussinesq term is found to be small relative to the other spatial derivative terms in the vicinity of the hydraulic jump, and it is almost negligible in the regions away from the jump. The Boussinesq term at locations away from the jump virtually becomes zero, although the values of the other terms remain approximately the same. This is to be expected, since the flow surface in regions away from the jump is more or less smooth. The pressure distribution in such flows is hydrostatic and thus the Boussinesq term is negligible.

The computed results are compared with the measured results in Fig. 8-14. To conserve space, only the comparisons for $\mathbf{F}_r = 2.3$ and 7 are included herein; for similar comparisons for other values of $\mathbf{F}_r$, see Gharangik and

Chaudhry [1991]. The comparison of the computed and measured results generally shows that the fourth-order accurate numerical models with or without Boussinesq terms give approximately the same results for all Froude numbers tested.

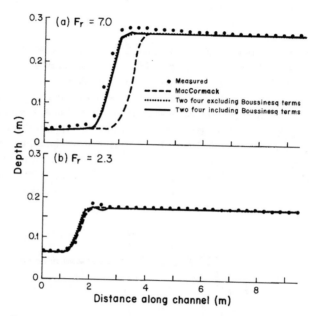

Fig. 8-14   Comparison of computed and measured jump profiles

8-6 Summary

In this chapter, we discussed the numerical modeling of rapidly varied flows. Three different formulations were presented. In the first, a steady form of the shallow water equations was numerically integrated. This is valid only for supercritical flows. In the second formulation, the unsteady gradually varied equations were solved with time as the iteration parameter. Since the pressure distribution in these two formulations is assumed hydrostatic, the computed results agree satisfactorily with the analytical solutions and with the experimental measurements in the regions where this assumption is valid. In the third formulation, the Boussinesq equations were solved by an explicit scheme which is second-order accurate in time and fourth-order in space. The simulation of the formation of hydraulic jump in a rectangular channel was used for illustration purposes and for comparing the computed results with those measured in a laboratory flume.

Problems

8-1 Develop a computer program to compute supercritical flows in a contraction by using the Lax and MacCormack schemes. Compare the computed results for a channel with the width changing from 4 m to 3 m in a distance of 5 m. The flow depth at the upstream end of the contraction is 2 m and the Froude number is 4.

8-2 By using the computer programs of Problem 9-1, compute and compare the maximum height of the shock wave for different lengths of the contraction.

8-3 Write a computer program to compute the flow profile in a rectangular channel with supercritical flow in the upper part and subcritical flow in the lower part.

References

Abbett, M., 1971, "Boundary Conditions in Computational Procedures for Inviscid, Supersonic Steady Flow Field Calculations," Aerotherm Report 71-41.

Abbott, M. B., 1979, *Computational Hydraulics; Elements of the Theory of Free Surface Flow*, Pitman Publishing Limited, London.

Abbott, M. B., Marshall, G., and Rodenhuis, G. S., 1969, "Amplitude-Dissipative and Phase-Dissipative Scheme for Hydraulic Jump Simulation," *Proc. 13th Congress,* Inter. Assoc. Hyd. Research, Tokyo, vol. 1, Aug., pp. 313-329.

Anderson, D. A., Tannehill, J. D., and Pletcher, R. H., 1984, *Computational Fluid Mechanics and Heat Transfer*, McGraw-Hill, New York, NY.

Bagge, G. and Herbich, J. B., 1967, "Transitions in Supercritical Open-Channel Flow," *Jour. Hydr. Div.,* Amer. Soc. Civ. Engrs., vol. 93, no. 5, pp. 23-41.

Basco, D. R., 1983, "Introduction to Rapidly-Varied Unsteady, Free-Surface Flow Computation," *Water Resources Investion Report,* U.S. Geological Survey, Report No. 83-4284.

Bhallamudi, S. M., and Chaudhry, M. H., 1992, "Computation of Flows in Open-Channel Transitions," *Jour. Hydraulic Research,* Inter. Assoc. Hyd. Research, vol. 30, no. 1, pp. 77-93.

Chaudhry, M. H., 2014, *Applied Hydraulic Transients*, 3rd ed., Springer New York Heidelberg Dordrecht London.

Chow, V. T., 1959, *Open Channel Hydraulics*, McGraw-Hill Book Co., New York, NY.

Cunge, J., 1975, "Rapidly Varying Flow in Power and Pumping Canals," in *Unsteady Flow in Open Channels,* (Eds. Mahmood, K. and Yevjevich, V.), Water Resources Publications, pp. 539-586.

Dakshinamoorthy, S., 1977, "High Velocity Flow through Expansions," *Proc. 17th Congress,* Inter. Assoc. Hyd. Research, Baden-Baden, vol.2, pp. 373-381.

Demuren, A. O., 1979, "Prediction of Steady Surface-Layer Flows," Ph.D. dissertation, University of London.

Ellis, J. and Pender G., 1982, "Chute Spillway Design Calculations," *Proc. Inst. Civ. Engrs.,* Part 2, vol. 73, June, pp. 299-312.

Engelund, F. and Munch-Petersen, J., 1953, "Steady Flow in Contracted and Expanded Rectangular Channels," *La Houille Blanche,* vol. 8, no. 4, Aug-Sept, pp. 464-474.

Fennema, R. J. and Chaudhry, M. H., 1986, "Explicit Numerical Schemes for Unsteady Free–Surface Flows with Shocks," *Water Resources Research,* vol. 22, no. 13, pp. 1923-1930.

Fennema, R. J. and Chaudhry, M. H., 1990, "Numerical Solution of Two-Dimensional Transient Free-Surface Flows," *Jour. of Hydr. Eng.,* Amer. Soc. Civ. Engr., Vol. 116, Aug., pp. 1013-1034.

Garcia, R. and Kahawita, R. A., 1 986, "Numerical Solution of the St. Venant Equations with the MacCormacK Finite-Difference Scheme," *Int. Jour. Numer. Meth. in Fluids,* vol. 6, pp. 259-274.

Gharangik, A. and Chaudhry, M. H., 1991, "Numerical Simulation of Hydraulic Jump," *Jour. Hydraulic Engineering,* Amer. Soc. Civ. Engrs., vol 117, no. 9, pp. 1195-1211.

Gottlieb, D. and Turkel, E., 1976, "Dissipative Two-Four Methods for Time-Dependent Problems," *Mathematics of Computation,* Vol. 30, No. 136, Oct., pp. 703-723.

Henderson, F. M., 1966, *Open Channel Flow,* MacMillan, New York, NY.

Herbich, J. B. and Walsh, P., 1972, "Supercritical Flow in Rectangular Expansions," *Jour. Hydr. Div.,* Amer. Soc. Civ. Engrs., vol. 98, no. 9, Sept., pp. 1691-1700.

Ippen, A. T., 1951, *et al., Proceedings of a Symposium on High-Velocity Flow in Open Channels, Trans.* Amer. Soc. Civ. Engrs., vol. 116, pp. 265-400.

Jameson, A., Schmidt, W., and Turkel, E., 1981, "Numerical Solutions of the Euler equations by Finite Volume Methods Using Runge-Kutta Time-Stepping Schemes," *AIAA 14th Fluid And Plasma Dynamics Conference,* Palo Alto, California, AIAA-81-1259.

Jimenez, O. F. and Chaudhry, M. H., 1988, "Computation of Supercritical Free-Surface Flows," *Jour. of Hydr. Eng.,* Amer. Soc. Civ. Engr., vol. 114, no. 4, Apr., pp. 377-395.

Katopodes, N. D., 1984, "A Dissipative Galerkin Scheme for Open-Channel Flow," *Jour. Hyd. Engineering,* Amer. Soc. Civ. Engrs., vol. 110, no. 4, April, pp. 450-466.

Knapp, R. T., 1951, "Design of Channel Curves for Supercritical Flow," *Symposium on High-Velocity Flow in Open Channels,* Trans. Amer. Soc. Civ. Engrs., vol. 116, pp. 296-325.

Kutler, P., 1975, "Computation of Three-Dimensional, Inviscid Supersonic Flows," in *Progress in Numerical Fluid Dynamics*, Lecture Notes in Physics No. 41, Springer-Verlag, pp. 287-374.

Liggett, J. A. and Vasudev, S. U., 1965, "Slope and Friction Effects in Two Dimensional, High Speed Flow," *Proc. 11th Int. Congress*, Inter. Assoc. Hyd. Research, Leningrad, vol. 1, paper 1.25.

MacCormacK, R. W., 1969, "The Effect of Viscosity in Hypervelocity Impact Cratering," *Amer. Inst. Aero. Astro.*, Paper 69-354, Cincinnati, Ohio.

McCorquodale, J. A. and Khalifa, A., 1983, "Internal Flow in Hydraulic Jumps," *Jour. Hyd. Engineering*, Amer. Soc. Civ. Engrs., vol. 109, no. 5, May, pp. 684-701.

McCowan , A. D., 1987, "The Range of Application of Boussinesq Type Numerical Short Wave Models," *Proc. 22^{n}d Congress*, Inter. Assoc. Hyd. Research, pp. 378-384.

Pandolfi, M., 1975, "Numerical Experiments on Free Surface Water Motion with Bores," *Proc. 4th Int. Conf. on Numerical Methods in Fluid Dynamics*, Lecture Notes in Physics No. 35, Springer-Verlag, pp. 304-312.

Roache, P. J., 1972, *Computational Fluid Dynamics*, Hermosa Publishers, Albuquerque, NM.

Stoker, J. J., 1957, *Water Waves*, Interscience Publishers, New York, NY.

Tseng, M. H., Hsu, C. A, and Chu, C. R., 2001, "Channel Routing in Open-Channel Flows with Surges," *Jour. Hyd. Engineering*, Amer. Soc. Civ. Engrs., vol. 127, no. 2, pp. 115-122.

Villegas, F., 1976, "Design of the Punchiná Spillway," *Water Power & Dam Construction*, Nov. 1976, pp. 32-34.

Verboom, G. K., Stelling , G. S. and Officier, M. J., 1982, "Boundary Conditions for the shallow water Equations," *Engineering Applications of Computational Hydraulics*, Vol. 1, (Abbott, M. B. and Cunge, J. A., eds.), Pitman, Boston.

9

CHANNEL DESIGN

Peace Canyon Project spillway (design discharge is 12, 280 m^3/s), powerhouse (four 175 MW Francis turbines rated at 39.6 m and 514 m^3/s) and the upstream dyke constructed from spoil material and designed using model studies to prevent vortices at the power intakes (Courtesy, B. C. Hydro and Power Authority, BC, Canada)

© Springer Nature Switzerland AG 2022
M. H. Chaudhry, *Open-Channel Flow*,
https://doi.org/10.1007/978-3-030-96447-4_9

9-1 Introduction

The design of a channel involves the selection of channel alignment, shape, size, and bottom slope and whether the channel should be lined to reduce seepage and/or to prevent the erosion of channel sides and bottom. Since a lined channel offers less resistance to flow than an unlined channel, the channel size required to convey a specified flow rate at a selected slope is smaller for a lined channel than that if no lining were provided. Therefore, in some cases, a lined channel may be more economical than an unlined channel.

Procedures are not presently available for selecting optimum channel parameters directly. Each site has unique features that require special considerations. Typically, the design of a channel is done by trial and error. Channel parameters are selected and an analysis is done to verify that the operational requirements are met with these parameters. A number of alternatives are considered, and their costs are compared. Then, the most economical alternative that gives satisfactory performance is selected. In this process, it is necessary to include the maintenance costs while comparing different alternatives. Similarly, the costs of energy required if pumping is involved and, for power canals, the amount of revenues produced by hydropower generation must be included in the overall economic analysis.

The channel design may be divided into two categories, depending upon whether the channel boundary is erodible or non-erodible. For erodible channels, flow velocities are kept low so that the channel bottom and sides are not eroded. The minimum flow velocity in flows carrying a large amount of sediment should be such that the material being transported is not deposited in the channel.

In this chapter, we first consider the design of rigid-boundary channels and then the design of erodible channels.

9-2 Rigid-Boundary Channels

In the design of a rigid-boundary channel, the channel cross section and size are selected such that the required discharge is carried through the channel for the available head with a suitable amount of freeboard. The *freeboard* is defined as the vertical distance between the design water surface and the top of the channel banks. Freeboard is provided to allow for unaccounted factors in design, uncertainty in the selection of different parameters, disturbances on the water surface, etc.

The *channel alignment* is selected so that the channel length is as short as possible and at the same time meets other site restrictions and requirements, such as accessibility, right of way, and balancing of cut and fill amounts. The bottom slope is usually dictated by the site topography whereas the selection of channel shape and dimensions take into consideration the amount of flow to be carried, the ease and economy of construction and the hydraulic efficiency of the cross section. A triangular channel is used for small rates of discharge, and a trapezoidal cross section is generally used for large flows. For structural reasons, channels excavated through mountains or built underground usually have a circular or horseshoe shape. Normally, the Froude number is kept low (approximately up to 0.3) so that the flow surface does not become rough, especially downstream of obstructions and bends. Similarly, the flow velocity is selected such that the lining is not eroded and any sediment carried in the flow is not deposited.

Normally, these channels are designed based on the assumption of uniform flow, although in some situations gradually varied flow calculations may be needed to assess the suitability of selected channel size for extreme events.

The maximum permissible velocity is not usually a consideration in the design of rigid boundary channels if the flow does not carry large amounts of sediments. However, if the sediment load is large, then flow velocities should not be too high to avoid erosion of the channel. The minimum flow velocity should be such that sediment is not deposited, aquatic growth is inhibited, and sulfide formation does not occur. The lower limit for the minimum velocity depends upon the particle size and the specific gravity of sediments carried in the flow. The channel size does not have significant effect on the lower limit. Generally, the minimum velocity in a channel is about 0.6 to 0.9 m/s. Flow velocities of 12 m/s have been found to be acceptable in concrete channels if the water is not carrying large concentrations of sediment. The channel invert may be eroded at much lower velocities than this value if the flow carries sand or other gritty material.

The channel side slopes depend upon the type of soil in which the channel is constructed. Nearly vertical channel sides may be used in rocks and stiff clays, whereas side slopes of 3 horizontal to 1 vertical may be needed in sandy soils. For lined channels, U. S. Bureau of Reclamation recommends a value of 1.5 horizontal to 1 vertical.

To allow for waves and water surface disturbances, a suitable amount of freeboard should be provided. It is not possible to specify a general formula for determining the freeboard under general conditions. As a rough estimate, the following formula, suggested by the U. S. Bureau of Reclamation, may be used:

$$F_b = \sqrt{ky} \qquad (9\text{-}1)$$

in which F_b = freeboard in m, y = flow depth, in m, and k = coefficient varying from 0.8 for a flow capacity of about 0.5 m^3/s to 1.4 for a flow capacity exceeding 85 m^3/s. Table 9-1 lists recommended freeboards for canals based on recommendations of the Central Board of Irrigation and Power, India [1968; Ranga Raju, 1983]. These values are somewhat less than those given by Eq. 9-1.

Steps to design a rigid-boundary channel are as follows:

Table 9-1   Suggested Freeboard*

Discharge (m^3/s)	< 0.75	0.75 to 1.5	1.5 to 85	> 85
Freeboard (m)	0.45	0.60	0.75	0.90

* After Ranga Raju [1983]

1. Select a value of roughness coefficient n for the flow surface and select bottom slope S_o based on topography and other considerations listed in the previous paragraphs.
2. Compute section factor from $AR^{2/3} = nQ/(C_o S_o)$, in which $A =$ flow area, $R =$ hydraulic radius, $Q =$ design discharge, and $C_o = 1$ for SI units and $C_o = 1.49$ in U. S. customary units.
3. Determine the channel dimensions and the flow depth for which $AR^{2/3}$ is equal to the value determined in step 2. For example, for a trapezoidal section, select a value for the side slope s and compute several different ratios of bottom width B_o and flow depth y for which $AR^{2/3}$ is equal to that determined in step 2. Select a ratio B_o/y that gives a cross section near to the best hydraulic section (see Section 9-3).
4. Check that the minimum velocity is not less than that required to carry the sediment to prevent silting.
5. Add a suitable amount of freeboard.

The following example illustrates this procedure.

Example 9-1

Design a trapezoidal channel to carry a discharge of 10 m^3/s. The channel will be excavated through rock by blasting. The topography in the area is such that a bottom slope of 1 in 4000 will be suitable.

Given:

$Q = 10$ m^3/s;
Flow surface is blasted rock;
$S_o = 0.00025$.

Solution:

For the blasted rock surface, $n = 0.030$ and the side slopes may be almost vertical. Let us select a value for the side slope s as 1 horizontal to 4 vertical. The substitution of these values into the Manning equation yields

$$AR^{\frac{2}{3}} = \frac{nQ}{C_o S_o^{0.5}}$$
$$= \frac{0.030 \times 10}{(0.00025)^{\frac{1}{2}}}$$
$$= 18.97$$

Since the channel section is almost rectangular, let us select $B_o = 2y$. Then, $A = (B_o + \frac{1}{4}y)y = 2.25y^2; P = B_o + \frac{1}{2}\sqrt{17}y = 4.06y; R = (2.25y^2)/(4.06y) = 0.55y$. Hence,

$$AR^{\frac{2}{3}} = (2.25y^2)(0.55y)^{\frac{2}{3}}$$
$$= 1.518y^{2.67}$$
$$= 18.97$$

Solving this equation for y, we get $y = 2.57$ m. Then, $B_o = 2 \times 2.57 = 5.14$ m. For ease of construction, let us use $B_o = 5$ m. Then, the corresponding value of y for which $AR^{\frac{2}{3}} = 18.97$ is determined by trial and error as 2.64 m. Based on Eq. 9-1, the freeboard $= \sqrt{0.8 \times 2.64} = 1.45$ m. As compared to this value, a freeboard of 0.75 m selected from Table 9-1 appears to be more appropriate. Therefore, total depth $= 2.64 + 0.75 = 3.39 \simeq 3.4$ m.

The flow area for a flow depth of 2.64 m is 14.94 m^2. Therefore, the flow velocity $= 10/14.94 = 0.67$ m/s. This is close to the minimum allowable flow velocity; thus, a bottom width of 5 m and a cross section depth of 3.4 m is satisfactory.

9-3 Most Efficient Hydraulic Section

A cross section that gives maximum section factor, $AR^{2/3}$, for a specified flow area, A, is called the *most efficient hydraulic section* or *best hydraulic section*. Since Q is proportional to $AR^{2/3}$ for a given channel (i.e., n and S_o are specified) and $R = A/P$, we can say that the most efficient hydraulic section is the one that yields the minimum wetted perimeter, P for a given A.

Theoretically speaking, the most efficient hydraulic section yields the most economical channel. However, it must be kept in mind that the above formulation is oversimplified. For example, we did not take into consideration the possibility of scour and erosion which may impose restrictions on the maximum flow velocity. And, for channel excavation we have to take into account the amount of overburden, viability of changing the bottom slope to suit the existing topographical conditions for minimizing the amount of excavation, ease of access, transportation of the excavated material to the disposal site, the viability of the matching of the cut and fill volumes, etc. In addition, for a lined channel, the cost of lining as compared to the per unit cost of excavation has to be taken into consideration for an overall economical design.

The proportions for common cross sections so that they are the most efficient are derived in the following paragraphs.

Rectangular Section

For a rectangular channel, $A = By$ and $P = B + 2y$. For the best hydraulic section, we want to determine the ratio of B and y such that P is minimum for constant A. Now, P can be written in terms of A and y as

$$P = \frac{A}{y} + 2y \tag{9-2}$$

Differentiating this expression for P with respect to y and then equating the resulting expression to zero, we obtain

$$\frac{dP}{dy} = \frac{-A}{y^2} + 2 = 0 \tag{9-3}$$

or

$$\frac{A}{y^2} = 2 \tag{9-4}$$

But $A = By$. Therefore,

$$\frac{By}{y^2} = 2 \tag{9-5}$$

or

$$y = \frac{1}{2}B \tag{9-6}$$

Thus, a rectangular cross section is the most efficient when the flow depth is one-half the channel width.

Triangular Section

Let us consider a symmetrical triangular section having side slope s horizontal to 1 vertical. Then,

$$A = sy^2$$
$$P = 2(\sqrt{1 + s^2})y \tag{9-7}$$

Substituting for y in terms of s and A, we obtain

$$P = 2\sqrt{1 + s^2}(\frac{A}{s})^{1/2} \tag{9-8}$$

Taking the square of both sides, this equation becomes

$$P^2 = 4(s + \frac{1}{s})A \tag{9-9}$$

As we discussed previously, P should be minimum for a given A for the most efficient hydraulic section. For this condition, $dP/ds = 0$. By differentiating Eq. 9-9, we obtain

$$2P\frac{dP}{ds} = 4(1 - \frac{1}{s^2})A = 0 \qquad (9\text{-}10)$$

Hence, it follows from Eq. 9-10 that $s = 1$. Thus, a triangular section with the sides inclined at $45°$ is the most efficient triangular section.

Trapezoidal section

For a trapezoidal section (Fig. 9-1)

$$P = B_o + 2\sqrt{1 + s^2}\,y$$
$$A = (B_o + sy)y \qquad (9\text{-}11)$$

The elimination of B_o from the above two equations and the simplification of the resulting equation yield

$$P = \frac{A}{y} + y(2\sqrt{1 + s^2} - s) \qquad (9\text{-}12)$$

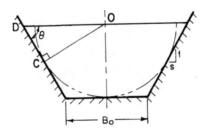

Fig. 9-1   Trapezoidal section

If both A and y are constants and s is variable, then the condition for the most efficient section is $\partial P/\partial s = 0$. Hence, differentiating Eq. 9-12 with respect to s, equating the resulting equation to zero and simplifying, we obtain

$$s = \frac{1}{\sqrt{3}}$$

or

$$\theta = 60° \qquad (9\text{-}13)$$

Now, let us consider A and s to be constants and y to be variable. Then, the condition for the most efficient section is $dP/dy = 0$. Differentiating Eq. 9-12 with respect to y, equating the resulting equation to zero, and simplifying, we obtain

$$B_o = 2(\sqrt{s^2 + 1} - s)y \tag{9-14}$$

Based on this equation, the top water-surface width is

$$B = B_o + 2sy = 2\sqrt{s^2 + 1}y \tag{9-15}$$

Thus, the top water-surface width is twice the length of the sloping side. In other words, these derivations show that the most efficient section is one-half of a hexagon.

Referring to triangle OCD of Fig. 9-1, and substituting expression for B from Eq. 9-15, we obtain

$$\begin{aligned}
OC &= OD \sin \theta \\
&= \frac{1}{2} B \sin \theta \\
&= y
\end{aligned} \tag{9-16}$$

Thus a circle with radius y and with center at O is tangential to the channel bottom and sides.

9-4 Erodible Channels

If the channel bottom or sides are erodible, then the design requires that the channel size and bottom slope are selected so that channel is not eroded. Two methods have been used for the design of these channels: the permissible velocity method and the tractive force method. Both methods are discussed in this section.

Permissible Velocity Method

In the permissible velocity method, the channel size is selected such that the mean flow velocity for the design discharge under uniform flow conditions is less than the permissible flow velocity. The *permissible velocity* is defined as the mean velocity at or below which the channel bottom and sides are not eroded. This velocity depends primarily upon the type of soil and the size of particles even though it has been recognized that it should depend upon the flow depth as well as whether the channel is straight or not. This is because, for the same value of mean velocity, the flow velocity at the channel bottom is higher for low depths than that at large depths. Similarly, a curved alignment induces secondary currents. These produce higher flow velocities near the channel sides, which may cause erosion.

A trapezoidal channel section is usually used for erodible channels. To design these channels, first an appropriate value for the side slope is selected so that the sides are stable under all conditions. Table 9-2, compiled from data given by Fortier and Scobey [1926], lists recommended slopes for different materials.

Table 9-2   Suggested Side Slopes*

Material	Side slope
Rock	Nearly vertical
Stiff clay	$\frac{1}{2}$ to 1:1
Firm soil	1:1
Loose sandy soil	2:1
Sandy loam	3:1

* After Fortier and Scobey [1926]

The maximum permissible velocities for different materials are presented in Table 9-3. The values listed in this table are for a straight channel having a flow depth of about 1 m. As a rough estimate, Lane [1955] suggested reducing these values by 5 per cent for slightly sinuous channels, 13 per cent for moderately sinuous channels, and 22 per cent for very sinuous channels. For other flow depths, these velocities may be multiplied by a correction factor, k, to determine the permissible flow velocity [Mehrotra, 1983]. For very wide channels, $k = y^{1/6}$.

The steps for the design of a channel using permissible velocity are as follows:

1. For the specified material, select value of Manning n (from Table 4-1), side slope s (from Table 9-2), and the permissible velocity, V (from Table 9-3);
2. Determine the required hydraulic radius, R, from Manning equation, and the required flow area, A, from the continuity equation, $A = Q/V$;
3. Compute the wetted perimeter, $P = A/R$;
4. Determine the channel bottom width, B_o, and the flow depth, y, for which the flow area A is equal to that computed in step 2 and the wetted perimeter, P, is equal to that computed in step 3;
5. Add a suitable value for the freeboard using Table 9-1.

The following example illustrates this procedure.

Example 9-2

Design a channel to carry a flow of 6.91 m^3/s. The channel will be excavated through stiff clay at a channel bottom slope of 0.00318.

Table 9-3   **Recommended Permissible Velocities**[*]

Material	V (m/s)
Fine sand	0.6
Coarse sand	1.2
Earth	
Sandy silt	0.6
Silt clay	1.1
Clay	1.8
Grass-lined earth (slopes < 5 per cent)	
Bermuda grass	
Sandy silt	1.8
Silt clay	2.4
Kentucky Blue grass	
Sandy silt	1.5
Silt clay	2.1
Poor rock (usually sedimentary)	
Soft sandstone	2.4
Soft shale	1.1
Good rock (usually igneous or hard metamorphic)	6.1

[*] After U. S. Army Corps of Engineers [1970]

Given:

$Q = 6.91$ m^3/s;
$S_o = 0.00318$;
Channel material is clay.

Determine:

$B_o = ?$
Flow depth = ?

Solution:

For stiff clay, $n = 0.025$, suggested side slope, $s = 1:1$ (from Table 9-2), and the permissible flow velocity (from Table 9-3) is 1.8 m/s. Hence,

$$A = 6.91/1.8 = 3.83 \text{ m}^2$$

Substituting values for V, n, and S_o into Manning equation, and solving for R, we get R = 0.713 m. Hence,

$$P = \frac{3.83}{0.713} = 5.37 \text{m}$$

Substitution into expressions for P and A from Eq. 9-11 and equating them to the values computed above, we obtain

$$B_o + 2.83y = 5.37$$
$$(B_o + y)y = 3.83$$

Elimination of B_o from these two equations yields

$$1.83y^2 - 5.37y + 3.83 = 0$$

Solution of this equation gives $y = 1.22$ m. The freeboard from Eq. 9-1 may be calculated as $\sqrt{0.8 \times 1.22} = 0.99$ m, whereas the suggested value in Table 9-1 is 0.75 m. Let us select a freeboard of 0.75 m. Then, the depth of the section $= 1.22 + 0.0.75 = 1.97$ m. Select a depth of 2.0 m and bottom width of 1.9 m.

Tractive Force Method

As compared to the permissible velocity, the scour and erosion process may be viewed in a more rational fashion by considering the forces acting on a particle lying on the channel bottom or on the channel sides. The channel is eroded if the resultant of forces tending to move the particle is greater than the resultant of forces resisting the motion; otherwise, it is stable. This concept, referred to as the tractive force approach, was introduced by du Boys in 1879 and re-stated by Lane in 1955 [Chow, 1959].

The force exerted by the water flowing on the channel bottom and sides is called *tractive force* or *drag force*. This is the force due to shear stress. In uniform flow in a straight channel, this force is equal to the component of weight of water acting in the direction of flow.

Let us consider a channel with bottom slope S_o. The weight of water in a reach of length L of this channel is γAL, in which $A =$ flow area. Now, the component of the weight of water in the downstream direction is γALS_o, in which $\gamma =$ specific weight of water. In uniform flow this component of the weight of water is equal to the tractive force that acts over the wetted perimeter, P. Then, the average unit tractive force or shear stress, $\tau_o = \gamma ALS_o/(PL) = \gamma RS_o$, in which $R =$ hydraulic radius. In very wide channels, $R \simeq y$. Hence, $\tau_o = \gamma yS_o$.

The distribution of unit tractive force or shear stress over the channel perimeter is not uniform. Although many attempts have been made to determine this distribution, they have not been conclusive. As an approximation for a trapezoidal channel [Lane, 1955], τ_o at the channel bottom may be assumed to be equal to γyS_o, and at the channel sides to be equal to $0.76\gamma yS_o$.

The shear stress at which the channel material just moves from a stationary condition is called *critical stress*, τ_c. The critical stress is a function of the material size and the sediment concentration. In addition, the critical stress at the channel sides is less than that at a level surface because the component of the weight along the side slope tends to roll the material down the slope, thereby causing instability.

Let us consider a particle lying on the channel side, as shown in Fig. 9-2. Let the side slope be θ, a = effective area and W_s = submerged weight of the particle, ϕ = angle of repose of the particle, and τ_s = shear stress on the channel sides. Two forces tending to move the particle are the tractive force, $a\tau_s$, due to flowing water and the component of the weight of particle along the side slope, $W_s \sin \theta$. The resultant of these two forces is

$$R = \sqrt{W_s^2 \sin^2 \theta + a^2 \tau_s^2} \qquad (9\text{-}17)$$

The normal force, $W_s \cos \theta \tan \phi$, resists the particle motion. In this expression, ϕ = angle of repose of the bank material. At the point of impending motion, the resultant of the forces causing motion is equal to the resultant of forces resisting the motion. Thus, for the impending motion,

$$W_s \cos \theta \tan \phi = \sqrt{W_s^2 \sin^2 \theta + a^2 \tau_s^2} \qquad (9\text{-}18)$$

It follows from this equation that

$$\tau_s = \frac{W_s}{a} \cos \theta \tan \phi \sqrt{1 - \frac{\tan^2 \theta}{\tan^2 \phi}} \qquad (9\text{-}19)$$

For the impending motion of a particle on a level surface

$$W_s \tan \phi = a\tau_l \qquad (9\text{-}20)$$

in which τ_l = shear stress at impending motion of a particle on a level surface. Then

$$\tau_l = \frac{W_s}{a} \tan \phi \qquad (9\text{-}21)$$

It follows from Eqs. 9-19 and 9-21 that

$$K = \frac{\tau_s}{\tau_l} = \cos \theta \sqrt{1 - \frac{\tan^2 \theta}{\tan^2 \phi}} \qquad (9\text{-}22)$$

which may be simplified as

$$K = \sqrt{1 - \frac{\sin^2 \theta}{\sin^2 \phi}} \qquad (9\text{-}23)$$

This is the reduction factor for the critical stress on the channel sides.

The effect of the angle of repose should be considered only for coarse, noncohesive materials. For cohesive and fine noncohesive materials, the gravity component causing the particle to roll down the side slope is much smaller

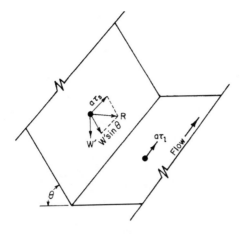

Fig. 9-2   Forces acting on a particle

than the cohesive forces and may thus be neglected. Figure 9-3 shows the curves prepared by the U. S. Bureau of Reclamation for the angle of repose for noncohesive material larger than 5 mm in diameter. The diameter in this figure is the diameter of a particle such that 25 per cent of material by weight is larger than this diameter.

The critical shear stress for non-cohesive material is shown in Fig. 9-4 and that for cohesive material in Fig. 9-5. These values are for straight channels. Lane recommended reducing these values by 10 per cent for slightly sinuous channels, 25 per cent for moderately sinuous channels, and 40 per cent for very sinuous channels.

The procedure for designing a channel by the tractive force approach involves the selection of a cross section such that the unit tractive force acting on the channel sides is equal to the permissible shear stress for the channel material. Then, we check that the unit tractive force on the channel bottom is less than the permissible stress.

The design steps are as follows:

1. For the channel material, select a side slope from Table 9-2, the angle of repose from Fig. 9-3 and the critical shear stress from Fig. 9-4 for noncohesive materials and from Fig. 9-5 for cohesive materials. Determine the permissible shear stress by taking into consideration whether the channel is straight or not.
2. For the noncohesive material, compute the reduction factor, K, from Eq. 9-23 and then determine the permissible shear stress for the sides by multiplying by K the permissible stress determined in step 1.
3. Equate the permissible stress for the sides determined in step 2 to $0.76\gamma y S_o$ and determine y from the resulting equation.

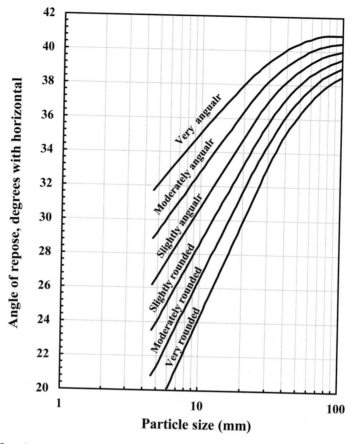

Fig. 9-3   Angles of repose for non-cohesive material (After U.S. Bureau of Reclamation)

4. For y determined in step 3 and for the selected values of the Manning n and the side slope, s, compute the bottom width, B_o, from Manning equation for the design discharge.
5. Now, check that the shear stress on the bottom, $\gamma y S_o$, is less than the permissible shear stress of step 1.

The following example illustrates this procedure.

Example 9-3

Design a straight trapezoidal channel for a design discharge of 10 m³/s. The bottom slope is 0.00025 and the channel is excavated through fine gravel having particle size of 8 mm. Assume the particles are moderately rounded and the water carries fine sediment at low concentrations.

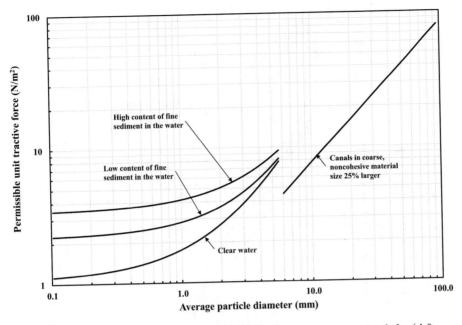

Fig. 9-4 **Permissible shear stress for noncohesive materials** (After U.S. Bureau of Reclamation)

Given:

$Q = 10 \text{ m}^3/s$;
$S_o = 0.00025$;
Material: Fine gravel, moderately rounded; and
Particle size $= 8$ mm.

Determine:

$B_o = ?$
Flow depth $= ?$

Solution:

For fine gravel, $n = 0.024$ and $s = 3\text{H} : 1\text{V}$. Therefore, $\theta = \tan^{-1} \frac{1}{3} = 18.4°$
From Fig. 9-3, $\phi = 24.°$ Hence,

$$K = \sqrt{1 - \frac{\sin^2 \theta}{\sin^2 \phi}} = \sqrt{1 - \frac{0.1}{0.16}} = 0.63$$

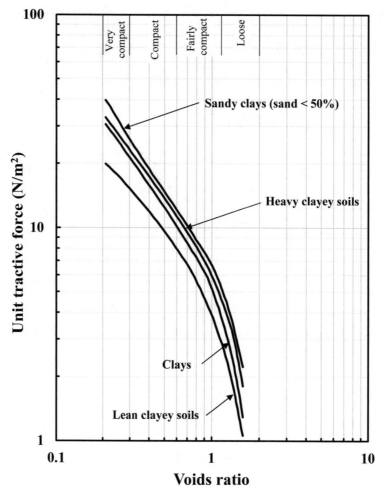

Fig. 9-5   Permissible shear stress for cohesive materials (After U.S. Bureau of Reclamation)

From Fig. 9-4, the critical shear stress = 0.15 lbs/ft^2 = 7.18 N/m^2. Since the channel is straight, we do not have to make a correction for the alignment. The permissible shear stress for the channel side is 7.18× 0.63 = 4.52 N/m^2.

Now, the unit tractive force on the side = $0.76\gamma y S_o = 0.76 \times 999 \times 9.81y \times 0.00025 = 1.862y$. By equating the unit tractive force to the permissible stress, we obtain

$$1.862y = 4.52$$

or

$$y = 2.43\text{m}$$

The channel bottom width, B_o, needed to carry 10 m^3/s may be determined from the Manning equation

$$\frac{1}{n}(B_o + sy)y \left(\frac{(B_o + sy)y}{B_o + 2\sqrt{1 + s^2}y}\right)^{\frac{2}{3}} \sqrt{S_o} = Q$$

By substituting $n = 0.024$, $s = 3$, $y = 2.43$, $S_o = 0.00025$, and $Q = 10$ m^3/s, and solving for B_o, we obtain

$$B_o = 8.24\text{m}$$

For a selected freeboard of 0.75 m, the depth of section $= 2.43 + 0.75 = 3.2$ m. For ease of construction, select a bottom width, $B_o = 8.25$ m.

9-5 Alluvial Channels

An alluvial channel is defined as a channel in which the flow transports sediment having the same characteristics as that of the material in the channel bottom. Such a channel is said to be *stable* if the sediment inflow into a channel reach is equal to the sediment outflow. Thus, the channel cross section and the bottom slope do not change due to erosion or deposition.

Two approaches have been used for the design of stable alluvial channels: (1) tractive force method; and (2) regime theory. The tractive force approach is more rational, since it utilizes the laws governing sediment transport and resistance to flow. The regime theory is purely empirical in nature and was developed based on observations on a number of irrigation canals in the Indo-Pakistan subcontinent. Since the sediment concentration in these canals is usually less than 500 ppm by weight, the regime theory should be assumed to be applicable to channels carrying similar concentration of sediment load.

The tractive force approach was discussed in the previous section. A brief outline of the regime theory follows; readers interested in the details, should refer to Blench [1957], Simons and Albertson [1963], and Brandon [1987].

Regime Theory

Lacey [1930] defined a *regime* channel as a channel carrying a constant discharge under uniform flow in an unlimited incoherent alluvium having the same characteristics as that transported without changing the bottom slope, shape, or size of the cross section over a period of time.

Two types of equations have been extensively used for design in India and Pakistan. These are the Kennedy and the Lacey equations. The main limitation of the Kennedy equations is that they do not specify a stable width, thereby making an infinite number of depth to width ratios possible. However, experience shows that stability is possible only if the width does not vary over a wide range: Sides are scoured in a very narrow channel, whereas deposition occurs in a very wide channel. To take this factor into account, Lindley introduced a relationship between the nonsilting, nonscouring velocity and the bottom width. Lacey developed the following equations based on the analysis of a large amount of data collected on several irrigation canals in the Indian subcontinent:

$$P = 4.75\sqrt{Q}$$
$$f_s = 1.76d^{1/2}$$
$$R = 0.47\left(\frac{Q}{f_s}\right)^{1/3}$$
$$S = 3 \times 10^{-4}f_s^{5/3}Q^{1/6} \tag{9-24}$$

in which P = wetted perimeter, in m; R = hydraulic radius, in m; Q = flow, in m^3/s; d = diameter of sediment, in mm; and f_s = silt factor which takes into consideration the effect of sediment size on the channel dimensions. The particle size and silt factors for various materials are listed in Table 9-4.

Table 9-4   Particle size and silt factors for various materials[*]

Material	Size (mm)	Silt factor
Small boulders, cobbles, shingles	64–256	6.12 to 9.75
Coarse gravel	8–64	4.68
Fine gravel	4–8	2.0
Coarse sand	0.5–2.0	1.44–1.56
Medium sand	0.25–0.5	1.31
Fine sand	0.06–0.25	1.1–1.3
Silt (colloidal)		1.0
Fine silt (colloidal)		0.4–0.9

[*] After Gupta [1989]

Combining these equations, the following resistance equation is obtained:

$$V = 10.8R^{2/3}S^{1/3} \tag{9-25}$$

Example 9-4

By using the regime approach, determine the cross section of an alluvial channel for a design flow of 8 m^3/s. The sediment carried by water is 0.4 mm sand.

Given:

$Q = 8 \text{ m}^3/\text{s};$
$d = 0.4 \text{ mm}.$

Determine:

$B_o = ?$
$y = ?$

Solution:

$$P = 4.75\sqrt{8} = 13.44 \text{ m}$$

$$f_s = 1.76(0.4)^{\frac{1}{2}} = 1.11$$

$$R = 0.47\left(\frac{8}{1.11}\right)^{0.333} = 0.907 \text{ m}$$

Let the side slopes be 1H to 2 V. Then,

$$A = (B_o + 0.5y)y = PR = 13.44 \times 0.907 = 12.18 \text{ m}$$
$$P = B_o + 2\sqrt{1 + (0.5)^2}y$$
$$= B_o + 2.24y = 13.44 \text{ m}$$

Substituting this expression for B_o into equation for A, we obtain

$$1.74y^2 - 13.44y + 12.18 = 0$$

Solution of this equation yields $y = 1.05$ m. Let us use a freeboard of 0.6 m. Then, the depth of the section is $1.05 + 0.6 = 1.65$ m and the width, $B_o = 13.44 - 2.24 \times 1.05 = 11.1$ m.

Now,

$$S = 3 \times 10^{-4} \times (1.11)^{1.67}(8)^{0.167}$$
$$= 5.05 \times 10^{-4}$$

9-6 Summary

In this chapter, the design of channels with rigid boundaries which do not erode or scour was discussed. Then, equations were derived for the most efficient hydraulic section that conveys the maximum discharge for a specified cross sectional area for different cross sectional shapes. The following two design procedures were presented for the design of erodible channels: permissible velocity and tractive force. Regime theory was briefly introduced for the alluvial channels.

Problems

9-1 Design a power canal to carry a flow of 50 m^3/s. The canal will be excavated through competent rock by blasting and will have a bottom slope of 0.0002.

9-2 Design an irrigation canal to irrigate 100 km^2 of farmland. The water demand is 0.1 m^3/s/km^2 of land. The topography is flat and a bottom slope of 1 m per 2 km will be appropriate. The soil through which the channel is to be excavated is stiff clay.

9-3 A channel has to be designed for drainage to carry flow from 200 km^2. If the flow/km^2 is 0.5 m^3/s/km^2, determine the channel size using (a) the permissible velocity method, (2) the tractive force method, (3) the regime theory. The material size is 2 mm and $S_o = 0.00002$.

9-4 Design a storm sewer for a new housing subdivision. The area of the subdivision is 4 km^2. The storm runoff may be taken as 0.15 m^3/s/km^2. The topography is such that a channel bottom slope of 1 m in 2 km will be economical.

9-5 Design an irrigation canal for a design discharge of 1100 ft^3/s. The general slope in the area is 2 ft/mile and the soil is clay.

9-6 Design a flood control channel to carry a flow of 500 ft^3/sec. A bottom slope of 0.003 will balance the cut and fill volumes. Assume the bottom and sides are paved with kiln-dried bricks.

9-7 A 2-km long horse-shoe tunnel drilled through sound rock is to be used for river diversion for the construction of a dam. The bottom level at the tunnel inlet is at El. 100 and at the exit at El. 98.5. For a design flow of 100 m^3/s, design the tunnel so that it will be free flow. The downstream water level at design discharge is at El. 102. Plot the water surface profile in the tunnel.

For a flow of 150 m^3/s the downstream level is at El. 105. Determine the water surface profile in the tunnel for this flow.

9-8 Design a grass-lined channel to carry a flow of 100 cfs. A 3 per cent bottom slope is suitable for the terrain.

9-9 Design a canal to carry 30 m^3/s. The canal will be excavated through rock and the local topography allows a bottom slope of 0.0003.

References

Amer. Soc. Civil Engrs., 1979, *Design and Construction of Sanitary and Storm Sewers.*

Blench, T., 1957, *Regime Behaviour of Canals and Rivers,* Butterworth, London.

Brandon, T. W. (ed.), 1987, *River Engineering Part 1, Design Principles,* vol.7, Institution of Water and Environmental Management, London.

Central Board of Irrigation and Power, 1968, *Current Practices in Canal Design in India,* New Delhi, India, June.

Chow, V. T., 1959, *Open Channel Hydraulics,* McGraw-Hill Book Co., New York, NY.

Chang, H. H., 1990, "Hydraulic Design of Erodible Bed Channels," *Jour. Hydraulic Engineering,* Amer. Soc. Civil Engrs., vol. 116, no. 1, pp. 87-101.

Fortier, S. and Scobey, F. C., 1926, "Permissible Canal Velocities," *Trans.* Amer. Soc. Civil Engrs., vol. 89, pp. 940-956.

Gupta, R. S., 1989, *Hydrology and Hydraulic Systems,* Prentice Hall, Englewood Cliffs, NJ.

Lane, E. W., 1955, "Stable Channel Design," *Trans.* Amer. Soc. of Civil Engrs.

Lacey, G., 1930, "Stable Channels in Alluvium," Paper 4736, *Proc.,* Institute of Civil Engrs., vol. 229, London.

Mehrotra, S. C., 1983, "Permissible Velocity Correction Factors," *Jour. Hydraulic Engineering,* Amer. Soc. Civil Engrs., vol. 109, no. 2, Feb., pp. 305-308.

Ranga Raju, K.G., 1983, *Flow Through Open Channels,* Tata McGraw Hill, New Delhi, India.

Simons, D. B., and Albertson, M. L., 1963, "Uniform Water Conveyance Channels in Alluvial Material," *Trans.,* Amer Soc. Civil Engrs, vol. 128, no. 1, pp. 65–107.

U. S. Army Corps of Engineers, 1970, *Hydraulic Design of Flood Control Channels,* Report EM 1110-2-1601.

10

STEADY FLOW SPECIAL TOPICS

Panama canal, 82-km long, connects the Atlantic Ocean with the Pacific Oceans. Canal locks at each end lift ships 26 m above the sea level. Its construction was completed in 1914 and it was expanded in 2016 to allow larger ships (Courtesy, Panama Canal Authority)

© Springer Nature Switzerland AG 2022
M. H. Chaudhry, *Open-Channel Flow*,
https://doi.org/10.1007/978-3-030-96447-4_10

10-1 Introduction

In this chapter, we briefly discuss the flows in a channel connecting two reservoirs, air entrainment, flow through culverts, and flow measurement. Most of this discussion involves application of the material presented in the previous chapters.

10-2 Flow in a Channel Connecting Two Reservoirs

Several different flow situations are possible, depending upon the system configuration and parameters. For example, the slope of the channel bottom may be mild or steep; the channel length may be short or long (a channel is considered short if the gradually varied flow profile extends to the upstream reservoir). The discussion in this section mainly deals with short channels, unless stated otherwise. We first consider a channel having mild bottom slope and then a channel with steep slope. Both qualitative discussion and procedures for flow computation are presented. In these discussions, we neglect the entrance and exit losses and the velocity head at the channel entrance and exit.

A. Mild bottom slope

The following three cases are possible:

1. Upstream reservoir level constant, downstream reservoir level variable;
2. Downstream reservoir level constant, upstream reservoir level variable;
3. Constant discharge, both upstream and downstream reservoir levels variable.

We discuss each of these cases one by one.

1. Upstream reservoir level constant, downstream reservoir level variable

Figure 10-1 shows the channel system for this case. We are interested in plotting a curve between the channel discharge and the downstream reservoir level. This curve is referred to as the *delivery curve* for the channel.

There is no channel flow and the water surface in the channel is level (Line *ab* in Fig. 10-1) if the downstream reservoir level is at the same elevation as the upstream reservoir level. For the downstream reservoir levels above point *b*, water flows in the upstream direction. We refer to the flow from the downstream reservoir towards the upstream reservoir as negative.

The channel discharge increases as we lower the water level in the downstream reservoir below point *b*. When the downstream level is at point *c*, the

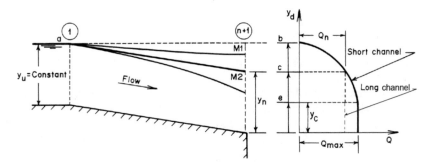

Fig. 10-1   Upstream reservoir level constant, downstream reservoir level variable

depth at the downstream end is the same as that at the upstream end, the flow surface in the channel is parallel to the channel bottom, and the flow in the channel is uniform. Let us call the discharge when the flow is uniform Q_n. As we lower the downstream water level further, two types of flow situations are possible, depending upon the channel length: If the channel is long (i.e., the water-surface profile from the downstream reservoir does not extend to the upstream reservoir), then the channel discharge does not increase above Q_n. However, if the channel is short, then the discharge increases as the downstream reservoir level is lowered below point c, as shown in Fig. 10-1. The channel discharge keeps on increasing until we reach point e when the downstream flow depth corresponds to the critical depth. If we lower the water level further, it results in a free overfall and does not contribute to an increase in the channel discharge.

The preceding discussion has been mainly qualitative. Let us now discuss how to compute the channel discharge and the flow depth along the channel length.

As we discussed above, the channel discharge is zero if the water surfaces in the upstream and the downstream reservoirs are at the same level (i.e., at point b in Fig. 10-1).

The flow is uniform if the flow depths at the channel entrance and at the channel outlet are equal (i.e., at point c in Fig. 10-1). In this case, the flow depth at all channel sections is equal to the normal depth, y_n. The rate of discharge may be directly computed for this flow depth from the Manning equation or any other similar equation.

To determine the channel discharge and the water levels in the channel if the specified downstream reservoir level is between points c and b or below point c, we may use any of the following two procedures:

1. Trial-and-error approach;
2. Simultaneous solution approach.

For the trial-and-error approach, we assume a value for the channel discharge and determine the value of y_c corresponding to this discharge. Then, we compute the water-surface profile in the channel, starting with a trial flow depth y_d at the downstream end. If the flow depth at the upstream end computed by this procedure is equal to the specified value y_u, then the assumed rate of discharge and the computed water levels are correct. Otherwise, we select another value for the discharge and repeat this procedure.

In the simultaneous-solution procedure discussed in Chapter 6, we divide the channel into n reaches and call the section at the upstream end 1 and at the downstream end $n+1$. The lengths of these reaches may not be equal. For the specified values of y_u and y_d (for plotting the delivery curve, we select several values of y_d and repeat this procedure for each depth one by one) and the channel parameters, we want to solve the resulting system of equations to determine the flow depths at $n+1$ sections and the rate of discharge, Q. Thus, there are $n+2$ unknowns and we need $n+2$ equations. Two of these equations are provided by the upstream- and the downstream-end conditions, i.e., if the entrance and exit losses and the velocity head are neglected, then

$$y_1 = y_u \tag{10-1}$$
$$y_{n+1} = y_d \tag{10-2}$$

The remaining n equations are obtained by writing the energy equation between two consecutive channel sections, i.e.,

$$z_i + y_i + \frac{\alpha_i Q^2}{2gA_i^2} = z_{i+1} + y_{i+1} + \frac{\alpha_{i+1} Q^2}{2gA_{i+1}^2} + h_{f_i}$$

$$(i = 1, 2, \cdots, n)$$

in which z = elevation of the channel bottom above a specified datum; h_{f_i} = head losses between sections i and $i+1$, and the subscripts i, $i+1$, etc. refer to quantities for i and $i+1$ sections.

For water levels in the downstream reservoir at or below point e, the flow depth at the downstream end of the channel is equal to y_c. Therefore, we replace Eq. 10-2 by the following equation for the critical depth at a general cross section:

$$D_{n+1} = \frac{Q^2}{gA_{n+1}^2} \tag{10-3}$$

in which D_{n+1} = hydraulic depth at section $n+1$. Other than specifying the downstream flow depth by Eq. 10-4 instead of Eq. 10-2, we proceed similarly as discussed in the previous paragraphs.

2. Downstream reservoir level constant, upstream reservoir level variable

In this case, there is no flow in the channel if the water surface in the upstream reservoir is at the same level as that in the downstream reservoir (point a in Fig. 10-2). The channel flow is negative for the upstream reservoir level below this point.

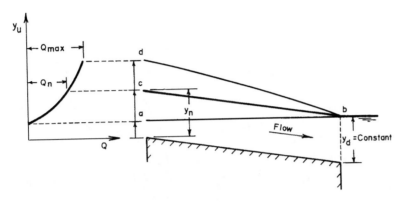

Fig. 10-2   Downstream reservoir level constant, upstream reservoir level variable

The rate of discharge increases as we raise the upstream reservoir level above point a. At point c, the flow depth at the upstream end is the same as that at the downstream end. In such a situation, the flow is uniform and the flow depth at all channel sections is equal to the normal depth. With further increase in the upstream reservoir level, the channel discharge keeps on increasing until point d when the downstream flow depth, y_d, corresponds to the critical depth for the channel discharge. Any further increase in the upstream water level results in raising the water surface profile in the entire channel.

To compute the rate of discharge and the water surface profile in the channel, we may use the simultaneous solution approach or compute them directly as follows. For the direct computations, we first determine Q_{max} assuming y_d as the critical depth. Then, for different values of discharge and of the downstream flow depth, we compute the water surface profiles starting at the downstream end with the selected value of y_d. This gives the upstream depth for the selected values of discharge and y_d. For the simultaneous-solution approach, two equations are provided by the upstream and the downstream end conditions, i.e., $y_1 = y_u$ and $y_{n+1} = y_d$ and the remaining n equations are obtained by writing the energy equation between two consecutive channel sections. The resulting system of equations is solved for the rate of discharge, Q, and the flow depths at $n+1$ cross sections.

3. Variable upstream and downstream reservoir levels, constant rate of discharge

In this case, y_u and y_d are both variable and the rate of discharge is specified.

Referring to Fig. 10-3, a line drawn at 45° and with an intercept of S_oL on the y_d axis indicates that the upstream and downstream reservoirs have the same water levels. Therefore, all curves for different discharges approach this line asymptotically. A line drawn through the origin at 45° represents the uniform flow conditions since $y_u = y_d$.

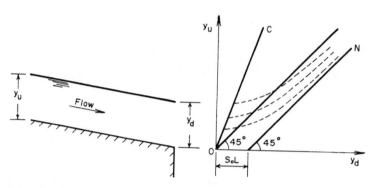

Fig. 10-3 **Both reservoir levels variable, constant discharge**

The curve OC represents the critical flow conditions. This is for the case when $y_d = y_c$ for the specified discharge and y_1 is equal to the given value of y_u. If the channel is long, then the Q-contours extend to the left of the C-curve and they are parallel to the y_2 axis.

To plot the curves shown in Fig. 10-3, we select a value for the rate of discharge, Q. Then, starting with flow depth y_d, we compute the water surface profile in the channel and determine the flow depth y_u. For this rate of discharge, we also compute the point on the C-curve by first determining the critical flow depth for the specified rate of discharge and then determining the corresponding upstream flow depth. A point corresponding to the uniform flow will lie on the 45° line. By repeating this procedure for two or three values of downstream flow depths, we have sufficient number of points to draw the curve.

B. Steep bottom slope

The control in a channel with steep bottom slope is located at the upstream end. For the flow depth in the downstream reservoir lower than the flow depth at the downstream end of the channel, the water levels in the channel do not depend upon the tailwater levels (Fig. 10-4). However, as the tailwater level is raised above the flow depth at the downstream end, it starts to affect the water levels in the downstream portions of the channel, and a hydraulic jump may be formed. Upstream of the hydraulic jump, however, the water levels are not affected by the tailwater levels. As we raise the tailwater level further, the jump moves upstream until it reaches the upstream end. In such a case, the inlet becomes submerged and the channel discharge becomes dependent upon the tailwater level.

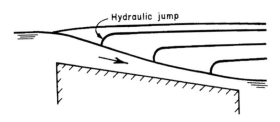

Fig. 10-4   Channel with steep bottom slope

To compute the water-surface profile in such a channel, we first compute the rate of discharge using the critical flow conditions at the channel entrance. Then, starting with the critical depth at the upstream end, we compute the water-surface profile in the channel proceeding in the downstream direction. If the tailwater levels are high, then a hydraulic jump may be formed in the channel. In such a case, we compute the water surface profile in the lower part of the channel by starting at the downstream end with the specified flow depth and proceeding in the upstream direction. The location of the jump is determined by matching the specific forces upstream and downstream of the jump. Another procedure available for computing the water surface profile and the location of the jump is to solve the unsteady flow equations. For this purpose, we solve the continuity and momentum equations subject to the end conditions and continue the computations until the solution converges to the steady-flow conditions. This was discussed in detail in Chapter 8.

10-3 Air Entrainment in High-Velocity Flow

As the boundary layer from the channel bottom intersects the water surface [Falvey, 1980; Wood, 1991], air is entrained in high-velocity flow [Lane, 1939] although no air is entrained in slow-moving water even when the boundary layer intersects the surface. Thus, some degree of turbulence must be exceeded for the air entrainment to begin. We refer to air entrainment as a process in which air enters the body of water [Falvey, 1980; Wood, 1991].

The appearance of "white water" as such does not necessarily imply air entrainment, since it may be due to reflections coming from different angles. This conclusion has been confirmed by high-speed photography.

A designer needs to know the volume of entrained air to select the height of side walls. It is also helpful in assessing the cavitation potential since air near the channel bottom reduces the possibility of cavitation. In addition, the friction losses are reduced by the presence of air next to the channel bottom and sides, as discussed later in this section.

The turbulent open-channel flow involving air entrainment may be divided into the following four vertical zones [Killen and Anderson, 1969], as shown in Fig. 10-5:

1. Upper zone;
2. Mixing zone;
3. Underlying zone; and
4. Air-free zone.

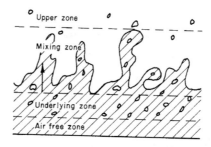

Fig. 10-5   Vertical zones of aerated flow (After Killen and Anderson [1969])

The upper zone comprises a small mass of flying water particles ejected from the mixing zone. This zone is not important for engineering applications. The mixing zone has surface waves of random amplitude and frequency. To prevent overtopping of the side walls, it is necessary to take the height of these waves into consideration. The knowledge of the characteristic of this zone is important, since all air entrained into flow or released from it passes through this zone. The surface waves do not penetrate the underlying zone, and the air concentration at any depth in this zone is determined by the number and the size of air bubbles. A number of correlations for the air-concentration distribution have been developed by using the turbulent-boundary layer theory, although it must be pointed out that they are not totally reliable. The air-free zone exists only in that part of the channel where aeration is still developing. At the interface, the air concentration as well as the rate of change in concentration with depth is small.

The development of self-aerating flows in a wide channel may be divided into four regions [Wood, 1991], as shown in Fig. 10-6:

1. Non-aerated flow region;

2. Partially aerated flow region;
3. Fully aerated flow region; and
4. Uniform fully developed aerated flow region.

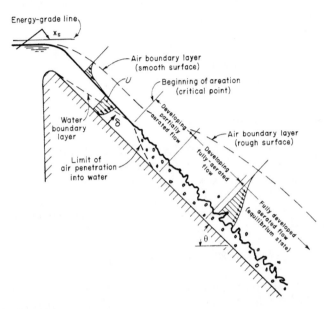

Fig. 10-6 Development of aerated flow regimes (After Falvey [1980])

In the nonaerated flow region, the turbulent boundary layer does not reach the flow surface and no air is entrained. In the partially and fully aerated flow regions, the air-concentration profiles vary with distance. However, the air does not reach the channel bottom in the partially aerated region, which is the case in the fully aerated region. In the uniform flow region, the flow conditions reach an equilibrium state and are constant with distance.

Nonaerated flow region

As discussed earlier, the air entrainment starts at a point where the boundary layer from the bottom intersects the top water surface. This point is referred to as the *point of inception*. Several authors have presented empirical relationships for the location of this point based on experimental data. Keller and Rastogi [1975] computed the developing boundary layer by solving the time-averaged, two-dimensional Navier Stokes equations. On the basis of these studies and other results for the values of different constants, Wood [1991] presented the following equation

$$\frac{\delta}{x_s} = 0.021 \left(\frac{x_s}{h_s}\right)^{0.11} \left(\frac{k_s}{x_s}\right)^{0.10} \tag{10-4}$$

in which δ = boundary layer thickness (defined as the perpendicular distance from the channel bottom to a depth where the velocity is 99 per cent of the free streamline velocity); k_s = equivalent sand roughness; x_s = distance along the slope along which boundary layer grows (Fig. 10-6); and h_s = static head at the inception point ($\sin\theta = h_s/x_s$). This equation shows the relative importance of various parameters upon which the growth of boundary layer depends and is applicable for slopes of $5°$ to $70°$.

Uniform aerated region

In this region, the mean flow properties, such as the flow depth and the depth-averaged air concentration, $\bar{C}$, are function of the discharge/unit width, q; channel bottom slope, S_o; bottom roughness; and the fluid properties.

By using the data obtained by Straub and Anderson [1958] on an artificially roughened 0.46-m-wide flume, Hager [1991] developed the following empirical expressions for this region.

$$\text{Average air concentration, } \bar{C} = 0.75(\sin\theta)^{0.75} \tag{10-5}$$

This cross-sectional average was obtained by integrating the function $C(y)$ from the channel bottom to a depth, y_r, where the air concentration is 90 per cent. For the uniform aerated flow depth in a wide rectangular channel

$$y_{99} = y_w + 1.35 y_w \left[\frac{\sin^3\theta y_w}{n^2 g^3}\right]^{1/4} \tag{10-6}$$

in which y_{99} = flow depth corresponding to 99 per cent air concentration; y_w = depth corresponding to pure water; n = Manning constant; and g = acceleration due to gravity.

Friction factor

The friction factor in a uniform aerated flow, f_e, is constant for the mean air concentration less than 30 per cent [Wood, 1991]. As the air concentration increases, the value of f_e decreases rapidly. If f is the friction factor for flow with no entrained air, then f_e may be determined from the following equation [Hager, 1992]

$$f_e = \frac{f_w}{1 + 10\bar{C}^4} \tag{10-7}$$

Velocity distribution

For the mean air concentration up to 50 per cent, the velocity distribution may be approximated as

$$\frac{V}{V_r} = \left(\frac{y}{y_r}\right)^{0.16} \tag{10-8}$$

in which y_r and V_r are the flow depth and flow velocity where the air concentration is 90 per cent.

The flow velocity V_m of an air-water mixture may be related to the velocity of air and water phases as

$$V_m \rho_m = \rho_w V_w(1 - C) + \rho_a V_a C \tag{10-9}$$

in which the subscripts $m, a,$ and w refer to the quantities for the air-water mixture, air, and water respectively, and

$$\rho_m = \rho_w(1 - C) + \rho_a C \tag{10-10}$$

Flow calculations

For a given channel bottom slope, the mean air concentration may be computed from Eq. 10-5. The friction factor for the aerated flow, f_e, may be determined from Eq. 10-7 if the friction factor for the flow of water, f_w, is available. The bulk flow depth, y_e, may then be computed from

$$y_e = \left(\frac{f_e q^2}{8 g S_o}\right)^{1/3} \tag{10-11}$$

The average water velocity, $V_w = q/y_e$ and the flow depth at 90 per cent air concentration, $y_r = y_e/(1 - \bar{C})$. The data obtained at Aviemore dam shows that $V_r = 1.2 V_w$. Now, the velocity distribution may be obtained from Eq. 10-8.

10-4 Flow Through Culverts

A short passage way for flow under a highway, railroad, or other embankment is referred to as a culvert. The culvert may be circular, rectangular, arch, or elliptical in shape. A rectangular culvert is referred to as a box culvert.

Although a culvert is a simple hydraulic structure, the computation of flow conditions through it may be somewhat complex. This is because several different flow conditions are possible and these conditions depend upon several parameters. The culvert may flow full, or partially full throughout its length, or in part of the length. The control may be at the upstream end (called *inlet control*) or it may be at the downstream end (called *outlet control*). Depending upon the head and tailwater levels, the control may shift from the inlet to the outlet and vice versa as these water levels change.

The flow through culvert has been classified into several types for analysis purposes [Chow, 1959; Henderson, 1966; Normann et al., 1985]. We may analyze these flows by utilizing the concepts we presented in the previous chapters. The discussion in the following paragraphs should be helpful for such analyses.

A curve relating the headwater level to the rate of discharge through the culvert is referred to as the *rating curve* or the *performance curve*. To determine the flow capacity, the curves for both inlet and outlet control may be plotted and the curve which gives lower discharge may be selected in those situations where it is not certain whether the control is at the inlet or at the outlet.

A culvert does not flow full even if the entrance is submerged if the head H at inlet is less than $1.5D$, where D is the height of the culvert at the entrance and H = the upstream water level − the culvert invert level. Similarly, a culvert having a square-edged entrance may not flow full even if the headwater is higher than the top of the culvert because of the flow contraction at the top.

Inlet control

For the case of inlet control, the flow through the culvert mainly depends upon the inlet conditions, e.g., area, shape and configuration at the inlet. The flow in the culvert is supercritical and thus it is independent of the conditions in the culvert or in the tailwater area.

Figure 10-7 shows different possible flow conditions for the inlet control. The flow depth at the entrance is equal to the critical depth in case the entrance is not submerged. In such a case, the rate of discharge may be computed from the weir equation. For a submerged entrance, the flow springs clear from the top of the culvert if $H < 1.5D$. This limit may be even higher for a square-edged entrance. Then, the flow passes through the critical depth and the flow depth tends towards the normal depth. The flow in such a situation may be computed by using the orifice equation. A hydraulic jump may form inside the culvert depending upon the tailwater level. The downstream part of the culvert may be primed if the outlet is submerged.

The rate of discharge through a box culvert may be computed from the following equations [Henderson, 1966].

Unsubmerged entrance ($H < 1.2D$)

$$Q = \frac{2}{3}CBH\sqrt{\frac{2}{3}gH} \qquad (10\text{-}12)$$

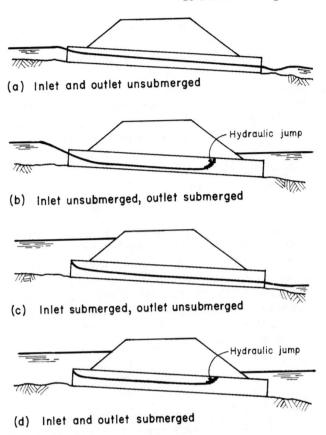

(a) Inlet and outlet unsubmerged

(b) Inlet unsubmerged, outlet submerged

(c) Inlet submerged, outlet unsubmerged

(d) Inlet and outlet submerged

Fig. 10-7   **Flow conditions for inlet control** (After Normann et al. [1985])

in which B = culvert width and the coefficient C accounts for the contraction on the sides. For square-edged sides, $C = 0.9$; and for slightly rounded sides, $C = 1$.

Submerged entrance ($H > 1.2D$)

In this case, the discharge may be computed from the orifice equation

$$Q = CBD\sqrt{2g(H - CD)} \tag{10-13}$$

in which C accounts for the contractions at the sides and the top. For a square-edge entrance, $C = 0.6$; for rounded edges, $C = 0.8$.

Outlet control

For the case of outlet control, the culvert either flows full or partially full. In the latter case, the flow is subcritical. The flow capacity depends upon the culvert area, shape, length, bottom slope, head losses in the culvert, and the headwater and tailwater levels.

Different flow conditions for the outlet control are shown in Fig. 10-8. The flow depth at the exit is critical if the tailwater level is at or below the critical depth. To compute the water surface profile in a free flow culvert, we start with the critical depth or with the tailwater level if it is higher than the critical depth, Then, we follow the procedure outlined in Section 10-1 for the case of constant downstream reservoir level. For a pressurized culvert, we determine the headwater level by applying the following energy equation between the tailwater and the headwater

$$Z_u + \frac{V_u^2}{2g} = H_l + Z_d + \frac{V_d^2}{2g} \qquad (10\text{-}14)$$

in which Z = elevation of the water level; H_l = sum of all the head losses between the headpond and the tailwater; and the subscripts u and d refer to the conditions on the upstream and on the downstream sides of the culvert. The head losses include the entrance loss, friction and form losses in the culvert and the exit losses. Normally, the flow velocities in the upstream and in the downstream areas are small and may be neglected. (Note that the velocity heads will cancel if they are equal.) If we express the total losses as KQ^2, then the above equation may be written as

$$Q = \frac{1}{K}\sqrt{Z_u - Z_d} \qquad (10\text{-}15)$$

Several modifications have been proposed to improve the entrance conditions [Harrison et al., 1972] to reduce losses, for vortex control, or for debris control [Reihsen and Harrison, 1971]. A number of innovations for fish passage have also been proposed [Normann et al., 1985].

10-5 Flow Measurement

Since the rate of discharge in a channel depends upon the flow depth as well as on the flow velocity, the value of both are needed to determine the flow rate. For critical flow, however, only the flow depth is sufficient because of a unique relationship between the flow depth and the flow velocity. To measure the rate of discharge in a channel, several methods [Ackers, 1978; World Meteorological Organization, 1971; International Standards Organization, 1977] are available, e.g., gauging structures, velocity-area methods, dilution techniques, slope-area method. In the laboratory or in a small channel, weirs or flumes and the velocity-area methods may be used. In the field or for large channels, dilution technique or the slope-area method may be utilized. A brief description of the weirs and other hydraulic structures was presented in Chapter 7, the flow through culverts in the previous section, and a number of other methods are discussed in the following paragraphs.

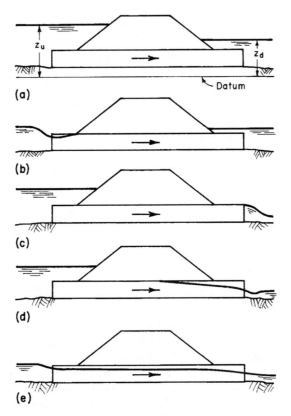

Fig. 10-8   Flow conditions for outlet control (After Normann et al. [1985])

Velocity-area method

In the velocity-area method, we determine the depth-averaged velocity at a number of locations along a line perpendicular to the flow direction and then sum up the product of the flow velocities and the corresponding flow areas. A general procedure [Rantz et al., 1982] for determining the depth-averaged velocity in this method is to average the velocity measured at $0.2d$ and $0.8d$, where d = total depth of flow. Alternately, the velocity measured at $0.6d$ from the channel bottom has also been used for this purpose. For a logarithmic velocity profile, Vanoni [1941] showed that the flow velocity measured at $0.368d$ from the water surface is equal to the theoretical logarithmic profile depth-averaged velocity. Walker [1988] proved that, for typical natural channels, the error introduced by averaging the flow velocities at $0.2d$ and $0.8d$ is less than 2 per cent. By assuming logarithmic velocity distribution, he derived the following equation to determine the average flow velocity at a section

$$\bar{V} = \frac{(1 + \ln d_2)V_1 - (1 + \ln d_1)V_2}{\ln(V_2/V_1)} \tag{10-16}$$

in which $\bar{V}$ = depth-averaged flow velocity; V_1 = velocity measured at depth d_1; and V_2 = velocity measured at depth d_2. This expression is useful for determining the depth-averaged flow velocity in situations where the flow velocities are measured at fixed locations, such as at automatic gauging stations, irrespective of the flow depth.

For natural channels, the flow section is divided into several vertical subsections. The flow velocity is measured at 0.2 and 0.6 of the flow depth in each subsection. The average of these two values is the depth-averaged flow velocity for the vertical section. The discharge through the subsection is computed by multiplying this average velocity and the flow area for the subsection. Then, the sum of the discharges through all the sub-sections is the total discharge in the channel.

Slope-area method

The slope-area method is based on the assumption of uniform flow in a channel reach. Therefore, caution must be exercised in applying this method to determine the rate of discharge during a flood. This is due to the fact that the assumption of uniform flow may be valid only during the periods when the flow is changing at a slow rate with respect to time.

A straight channel reach of length L is selected [Benson and Dalrymple, 1966], and the flow areas, A, and the conveyance factors, K, at the upstream and at the downstream ends of the reach are determined. A representative value for the Manning n for the channel is also selected. Let us refer to the quantities for the upstream and for the downstream ends by the subscripts u and d and the average value for the reach by a bar on the variable name. As discussed in Chapter 4 (Eq. 4-31), the expression for K for the Manning equation is $K = \frac{C_o}{n} AR^{2/3}$. If we assume that the velocity heads at the upstream and at the downstream ends of the selected reach are approximately the same, then the slope of the water surface is equal to the slope of the energy grade line. The geometric mean of the conveyance factors at the upstream and at the downstream ends of the reach may be taken as the average conveyance factor for the reach, i.e., $\bar{K} = \sqrt{K_u K_d}$. Then, the discharge is computed from the following equation

$$Q = \bar{K} S^{1/2} \tag{10-17}$$

in which the slope of the water surface, $S = (Z_u - Z_d)/L$; Z_u = the elevation of the water surface at the upstream end and Z_d = elevation of the water surface at the downstream end.

Flumes

The discharge measurement by flumes [Ackers , et al., 1978] is based on the assumption that critical flow is produced by constricting the width, raising the bottom, or a combination of the two. Then, a single flow depth is sufficient to determine the discharge. In flows carrying sediment or debris, a step rise in the channel bottom results in the deposition of sediment or the accumulation of debris on the upstream side of the raised bottom. To prevent this problem, a step rise is usually avoided. By selecting a suitable value for the throat width and for the bottom level at the throat section and in the downstream part, critical flow may be produced in the throat area followed by supercritical flow. A hydraulic jump is formed downstream of the supercritical flow part. Such flumes are referred to as the *standing wave flumes*. Parshall flume is a common flume of this type since the 1920s. It is available in various standard sizes for a range of discharges and extensive amount of experimental data is available for these designs. Figure 10-9 shows a typical layout and configuration of this flume.

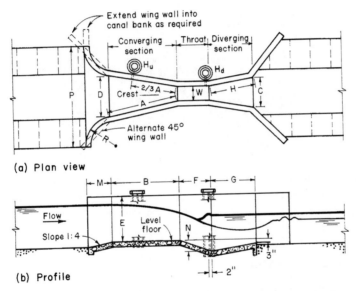

Fig. 10-9   **Parshall flume** (After U. S. Bureau of Reclamation)

10-6 Velocity Measurement*

* The velocity and cross-sectional area may be measured using a variety of methods and equipment. These methods and a number of advanced techniques to measure stage and streamflow are discussed in this section. Typically, these

* This section is authored by M. Elkholy.

methods are non-intrusive, based on remote sensing or image sequence techniques.

Current meter

A current meter may be used to measure the flow velocity. The Price AA current meter shown in Fig. 10-10 is the most commonly used current meter. Six metal cups attached to a wheel rotate around a vertical axis. The meter transmits an electronic signal for each rotation. The revolutions are recorded and timed to measure the flow velocity since the revolving rate of the cups depends upon the flow velocity. The Price AA meter is designed for attachment to a reel system and a cable for measurements in deep water, or to a wading rod in shallow waters.

Fig. 10-10   Current meter for velocity measurement (Courtesy, U.S. Geological Survey)

Acoustic Doppler Current Profiler

An Acoustic Doppler Current Profiler (ADCP) is based on the principle of the Doppler Effect to measure flow velocity. The Doppler Effect is the phenomenon of change in frequency we experience such as when a passing train blows a horn and the horn seems to become less frequent as it passes.

An ADCP monitors the change in the frequency of a sound signal transmitted into the water and receiving it back after being reflected from silt or other particulates being transported in the water, to measure flow velocity. This Doppler Shift or change in frequency is converted into water velocity. The ADCP also employs acoustics to determine water depth by measuring the time it takes for a sound pulse to travel from the river bed to the receiver.

An ADCP is mounted on a boat or small watercraft and its acoustic beams are sent into the water from the water surface for discharge measurements. Then the ADCP is steered across the river surface to measure flow velocity and depth. To measure the channel width, river-bottom tracking capacity of the ADCP acoustic beams or a Global Positioning System (GPS) is employed. The flow area is calculated for the measured channel width and the flow depth and then using the measured velocity the discharge is determined.

Ultrasonic Velocity Profiling (UVP)

The ultrasonic wave has been used for a long time to measure blood flow and flow investigations in industrial processes. Flow meters have been developed by measuring the travel time of an ultrasonic pulse in flowing media. UVP method may be used to determine the flow velocity profile instantaneously. Thus it is possible to obtain profiles with a relatively short sampling period to develop a time-dependent velocity profile and spatiotemporal velocity field.

Figure 10-11 shows a typical UVP configuration for measuring pipe flow. An ultrasonic transducer attached to the wall with Gel-like material emits pulsed ultrasonic wave in the fluid inside the pipe. The contaminants and impurities in the fluid or special seeding particles used in laboratory measurements reflect this wave which is then received by the transducer. Similar configurations may be used to measure discharge in free-surface flows.

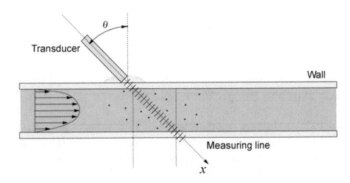

Fig. 10-11   **Ultrasonic Velocity Profiling method (After Takeda, [1995])**

Surface-Velocity Radar (SVR)

Several non-portable Doppler radars have been developed for measuring surface velocity since the early 2000s. Test setups include instruments or devices installed under bridges, on cableways, fixed or van-mounted on banks and aircraft-mounted devices. Typical applications for collecting velocity data continuously in unsteady flow conditions are on large rivers with widths ranging from 100 to 1000 m.

Hand-held radars are economical, and are easy to use to gauge multiple field sites. Welber et al. (2016) used hand-held Decatur Surface Velocity Radar (SVR) to measure velocity in various riverine settings and evaluated the uncertainty of SVR-based discharge measurements. Typically, an SVR has a horizontal beam width of 128, a measuring range of 0.1–9.1 m/s, and an accuracy of 5% according to the manufacturer.

Image-Based Approaches

A promising alternative for surface velocity measurements utilizes images from one camera or more for stereoscopic tracking. A fixed camera mounted on a roof, on a bridge, on a telescopic rod, or on a tripod has been used in several hydrological studies to record the displacement of natural tracers (leaves or foams) or artificial tracers added to the surface allowing for surface flow monitoring at a limited number of locations. Due to their flexibility, they are widely used to collect data using various sensors and platforms. RGB sensors, as well as thermal cameras, have been employed [Muste et al., 2008 and Puleo et al., 2012]. Ran et al. [2016] showed that a low-cost Raspberry Pi camera can be used to observe flash floods. The applicability of crowd-sourced photography for post-flood analysis has been demonstrated by Le Coz et al. [2016] and Guillén et al. [2017]. Continuous recording of river reaches in image-based setups allows the assessment of the flow dynamics.

Several image-based approaches are available; the following are widely used to monitor rivers.

Large-Scale Particle Image Velocimetry

Large-Scale Particle Image Velocimetry (LSPIV) detects features on the water surface created by naturally occurring floating particles or free surface deformations induced by waves or ripples [Fujita et al., 1998], such as those caused by turbulence or wind [Muste et al., 2008]. LSPIV results are shown in Fig. 10-12 [Muste et al., 2004].

LSPIV technique has been successfully used to measure flow velocity during floods where high water velocities can be dangerous. Recording of the free surface from the air or from the shore is a safer alternative (Fig. 10-13). High velocity and turbulence during floods generate several types of seeding. Ripples and boils formed by large-scale turbulent eddies provide natural seeding for LSPIV observations [Fujita et al., 2007]. Large-scale, turbulent eddies entraining silt throughout the flow depth are a second sort of tracer that occurs naturally during floods. Floating debris in the stream, common during floods, is the third tracer type.

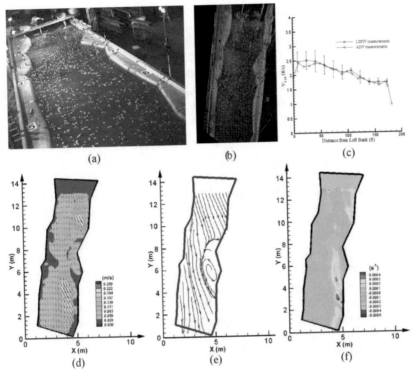

(a)   (b)   (c)

(d)   (e)   (f)

Fig. 10-12   LSPIV results (a) Hydraulic model with seedings, (b) Vector field, (c) Results of LSPIV and ADV velocity measurements, (d) Mean vector field, (e) Streamlines, and (f) Vorticity field (After Muste et al. [2004])

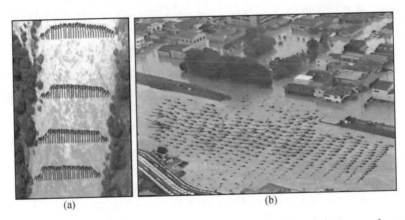

(a)   (b)

Fig. 10-13   Flow distribution (a) Katsura River; (b) Levee breach on the Shinkawa River (After Muste et al. [2008])

Space Temporal Image Velocimetry

By using a Space Temporal Image Velocimetry algorithm (STIV), Fujita et al. [2007] improved the LSPIV approach. Because of 1D tracking rather than 2D, STIV is faster. Profiles are extracted along the primary flow direction, and 15 particle movements are then drawn along the time axis (i.e., changes along the profiles between frames), resulting in a space-time image. The real flow velocity is determined by the angle of the pattern inside that image.

Optical Flow Algorithms

The optical flow algorithms created by the computer vision field has been used in small rivers to measure large floods [Perks et al., 2016; Eltner et al., 2019]. The settings of an affine transformation inside an optimization procedure are adapted to minimize variations in the gray scale value between the template and the search region to track surface-water features from UAVs and to convert to velocities [Perks et al., 2016]. Eltner et al. [2019] present an autonomous method for measuring surface flow velocities in rivers, in which identified particles and accompanying feature tracks are filtered to reduce the impact of outliers and sun glare. The approach is suitable for imagery from various views, including terrestrial and aerial (i.e., UAV) perspectives. They applied it to short river reaches in Saxony, Germany.

Particle Tracking Velocimetry

Particle tracking velocimetry technique (PTV) is similar to LSPIV that employs correlation techniques. Single particles are recognised initially and searched in consecutive images. Elkholy and Chaudhry [2009], Elkholy [2011], and LaRocque et al. [2013] applied this to develop levee-breach closure procedures and to determine the size of sandbag required for the closure on a 1:50 scale hydraulic model (Fig. 10-14) of the 17th Street Canal levee breach in New Orleans caused by Hurricane Katrina in 2005. Figure 10-15 shows the trajectories of hollow plastic balls tracked by a HD video camera utilizing PTV, and Fig. 10-16 shows the resultant surface velocities in the model.

Satellite Observations

Where field data is sparse, satellite observations paired with algorithms designed for river engineering could address large gaps in our knowledge of river flows worldwide. Recent efforts have focused on using in-situ gauge discharge data to calibrate the river width [Smith and Pavelsky, 2008; Pavelsky, 2014], stage [Papa et al., 2012; Tarpanelli et al., 2013], both stage and width [Sichangi et al., 2016], and ice status [Brakenridge et al., 2007]. The

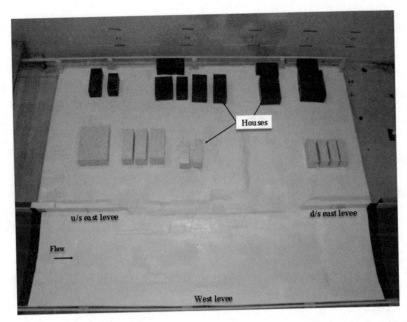

Fig. 10-14   Physical model of the 17th St Canal levee breach (After Elkholy, [2011])

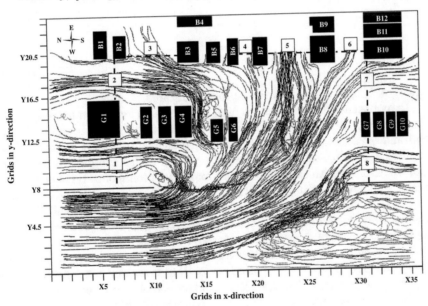

Fig. 10-15   Trajectories of hollow plastic balls using PTV technique (After LaRocque et al. [2013])

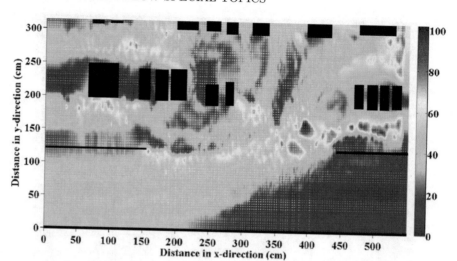

Fig. 10-16 Surface velocity, 17th St Canal Levee breach model (After Elkholy, [2011)

following relationships relating the velocity, v; depth, d; and width, w to discharge have proven to be correct. [Leopold and Maddock, 1953].

$$w = a\, Q^b$$
$$d = c\, Q^f$$
$$v = k\, Q^m \tag{10-18}$$

where, a, b, c, f, k, and m are empirical constants.

Remote-sensing data has been used as surrogates for in-situ measurements for hydrological model calibration [Sun et al., 2010a, b]. At-many-stations hydraulic geometry (AMHG) [Gleason and Smith, 2014] allows the estimation of discharge using remote-sensing data which, unlike ground observations, allows comprehensive coverage of rivers and other bodies of water. The Surface Water and Ocean Topography (SWOT satellite mission, jointly developed by NASA, the UK Space Agency, the Canadian Space Agency, and France's Centre National d'Etudes Spatiales (CNES) (http://swot.jpl.nasa.gov/) to be launched in 2022 aims to assess freshwater exports from rivers. The idea utilizes satellite observations of rivers (widths, slopes, surface height with time) and the application of basic laws of fluid flows and mass conservation to determine the river discharge that produce the observed information.

However, note that SWOT-based discharge will not replace in-situ discharge measurements. SWOT overpasses follow a cycle of 21 days with irregular sampling of mid-latitude locations, three times every cycle. This is different than continuous gauge measurements. While this is sufficient for various global water cycle issues, it is insufficient for many small-scale rivers for which data is required frequently at sub-hourly intervals. In addition, SWOT discharge predictions are unlikely to be as accurate as gauged discharge. However, geographically continuous SWOT measurements allow for data in ungauged basins as well as measurements of spatially scattered phenomena, such as flood wave propagation along rivers [Durand et al., 2009; Pavelsky, 2014; Paiva et al., 2015]. SWOT mission can offer discharge estimates at the continental scale to assist minimizing runoffs errors in existing models.

10-7 Summary

In this chapter, a number of special topics were discussed. First, the analysis of flow in a channel connecting two reservoirs was presented. The reservoir water level may vary in either the upstream, downstream, or both reservoirs and the channel bottom slope may be mild or steep. Salient features of air entrainment in high velocity flow were outlined. Flow through culverts with upstream or downstream control was discussed and a number of flow and velocity measurement methods were outlined.

Problems

10-1 A 10-m wide and 1 km long rectangular channel having a bottom slope of 0.0002 connects lakes B and C. Assume the channel is concrete-lined with $n = 0.013$ and the channel bottom at Lake B entrance is at El. 94.

  i. Plot the delivery curve for the channel if the water level in Lake B is constant at El. 100 and the water level in Lake C is variable;

  ii. Plot a curve between the discharge versus Lake B level if the water level in Lake C is constant at El. 98; and

  iii. Plot a diagram between the water levels in both lakes and the channel discharge if the water levels in both lakes are variable.

10-2 An overflow spillway has a slope of $30°$. For a head of 8 m above the spillway crest, compute the development of boundary layer thickness along the spillway length.

10-3 In the uniform aerated flow region of the spillway of Prob. 10-2, determine the average air concentration, the flow depth corresponding to 99 per cent air concentration, equivalent friction factor and the bulk flow depth.

10-4 Compute the rating curve for a 2-m wide 4-m high box culvert with a bottom slope of 0.005. Assume the downstream end of the culvert remains unsubmerged.

10-5 Plot the rating curve for the culvert of Prob. 10-4 if the bottom slope is 0.001. Assume the tailwater level remains below the culvert top at the downstream end.

10-6 If the culvert of Prob. 10-5 is 100 m long, compute and plot the water surface profile in the culvert if the tailwater level is 4.5 m above the culvert invert at the outlet.

10-7 Assuming a logarithmic velocity distribution, prove that the flow velocity at $0.368d$ (d = flow depth) from the free surface is the depth-averaged flow velocity.

10-8 Show that the average of the flow velocities at 0.2 and 0.8 depth gives a depth-averaged flow velocity with an error of 2 percent. Assume the velocity distribution is logarithmic.

10-9 The water level in the upstream lake of the channel system shown in Fig. 10-17 remains constant at El. 108 m; the water level in the downstream lake may vary between El. 98 and 108 m.

 i. Compute the rates of discharge in the channel for different downstream lake levels and plot a curve between the discharge and the downstream lake level; and

 ii. Compute and plot the water surface profile in the channel for the downstream levels of El. 98, 102, 104, and 106.

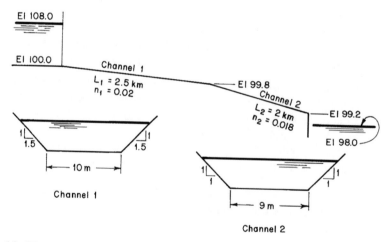

Fig. 10-17 **Channel system of Prob. 10-9**

10-10 A branch channel (channel 3) takes off at the junction of channels 1 and 2 of Fig. 10-17. Channel 3 is 1 km long, has the same cross section as that of channel 2, and Manning n is 0.015. A lake with constant water level at El. 105 is located at the downstream end of the branch channel.

i. Compute and plot the delivery curves for channels 2 and 3 for the different downstream lake levels downstream of channel 2; and

ii. Plot the water surface profiles in all three channels for the water levels of El. 98, 102 and 104 in the lake at the downstream end of channel 2.

10-11 Develop the delivery curve for a channel connecting two large reservoirs, with one reservoir having constant water level. Plot the delivery curve for the upstream reservoir if $x < 12$ and plot the delivery curve for the downstream reservoir if $x > 12$, where x is the number of the first letter of your last name in the English alphabet, (e.g., $x = 10$ for the last name Johnson).

The data for the trapezoidal channel connecting the two reservoirs is as follows: Channel length, in km $= x/(2i)$ where i is an integer such that the channel length is $800m$ or nearest to $800m$; Bottom slope $= 0.000|(|w|)|$ where $|(|w|)|$ means the digital value of w, where w is your date of birth (e.g. $w = 13$ for July 13). For example, for $w = 5$, the bottom slope $= 0.0005$; Manning $n = 0.0|(|x|)|$ where $|(|x|)|$ means the digital value of x, e.g., for $x = 8$, Manning $n = 0.08$; Channel Bottom width, in m $= 10 + 2x$; Channel side slopes $= 0.1x$ horizontal to 1 vertical. The constant water level in the upstream or the downstream reservoir is 5 m above the channel bottom at the upstream end.

Extra Credit:

Plot the delivery curve for both the upstream and the downstream reservoirs for 10 points for extra credit, i.e., plot the delivery curve for the downstream reservoir if $x \leq 12$; plot the delivery curve for the upstream reservoir if $x > 12$.

Graduate students, use any two methods of your choice for the analysis instead of only one for undergraduates.

References

Ackers, P., White, W. R., Perkins, J. A., and Harrison, A. J. M., 1978, *Weirs and Flumes for Flow Measurement,* John Wiley & Sons, New York, NY.

Benson, M. A., and Dalrymple, T., 1966, "General Field and Office Procedures for Indirect Discharge Measurements: U.S. Geological Survey Techniques Water-Resources Inv.," Book 3, Chapter A1.

Bos, M.G. (ed.), 1976, *Discharge Measurement Structures,* Publication No. 20, International Institution for Land Reclamation and Improvement, Delft, The Netherlands.

Brakenridge, G. R., Nghiem, S. V., Anderson, E., and Mic, R., 2007, "Orbital Microwave Measurement of River Discharge and Ice Status," *Water Resources Research,* 43(4).

Cheng, R. T., Gartner, J. W., Mason Jr, R. R., Costa, J. E., Plant, W. J., Spicer, K. R., ... and Hayes, K., 2004, "Evaluating a Radar-Based, Non Contact Streamflow Measurement System in the San Joaquin River at Vernalis, California," *Report No.* 2004-1015.

Chow, V.T., 1959, *Open Channel Hydraulics,* McGraw-Hill Book Co., New York, NY.

Creutin, J. D., Muste, M., Bradley, A. A., Kim, S. C., and Kruger, A., 2003, "River Gauging Using PIV Techniques: A Proof of Concept Experiment on the Iowa River," *Journal of Hydrology,* 277(3-4), 182-194.

Durand, M., Rodriguez, E., Alsdorf, D. E., and Trigg, M., 2009, "Estimating River Depth from Remote Sensing Swath Interferometry Measurements of River Height, Slope, and Width," *IEEE Journal of Selected Topics in Applied Earth Observations and Remote Sensing,* 3(1), 20-31.

Elkholy, M., and Chaudhry, M. H. 2009, August, "Tracking Sandbags Motion During Levee Breach Closure Using DPTV Technique," *In 33rd Congress, International Association of Hydraulic Engineering and Research,* Vancouver, Canada.

Elkholy, M., 2011, "Mechanics of Sandbag Motion in Free-Surface Flow," Ph.D. Thesis, University of South Carolina.

Eltner, A., Sardemann, H., and Grundmann, J., 2019, "Flow Velocity and Discharge Measurement in Rivers Using Terrestrial and UAV Imagery," *Hydrology and Earth System Sciences,* https://doi.org/10.5194/hess-2019-289.

Falvey, H. T., 1980, *Air-Water Flow in Hydraulic Structures,* Monograph No. 41, U. S. Bureau of Reclamation, Denver, CO.

Fujita, I., Muste, M., and Kruger, A., 1998, "Large-Scale Particle Image Velocimetry for Flow Analysis in Hydraulic Engineering Applications," *Journal of hydraulic Research,* 36(3), 397-414.

Fujita, I., Watanabe, H., and Tsubaki, R., 2007, "Development of a Non-Intrusive and Efficient Flow Monitoring Technique: The Space-Time Image Velocimetry (STIV)," *International Journal of River Basin Management,* 5(2), 105-114.

Gleason, C. J., and Smith, L. C., 2014, "Toward Global Mapping of River Discharge Using Satellite Images and At-Many-Stations Hydraulic Geometry," *Proceedings of the National Academy of Sciences,* 111(13), 4788-4791.

Gosling, S. N., and Arnell, N. W., 2011, "Simulating Current Global River Runoff With A Global Hydrological Model: Model Revisions, Validation, and Sensitivity Analysis," *Hydrological Processes,* 25(7), 1129-1145.

Guillén, N. F., Patalano, A., García, C. M., and Bertoni, J. C., 2017, "Use of LSPIV in Assessing Urban Flash Flood Vulnerability," *Natural Hazards,* 87(1), 383-394.

Hager, W. H., 1991, "Uniform Aerated Chute Flow," *Jour. Hydraulic Engineering,* Amer. Soc. Civil Engrs., vol 117, no 4, pp. 528-533.

Harris, J. D., 1982, "Hydraulic Design of Culverts," Chapter D, Drainage Manual, Ontario Ministry of Transportation and Communication, Dawnsview, Ont., Canada.

Harrison, L. J., J. L. Morris, J. M. Normann and F. L. Johnson, 1972, "Hydraulic Design of Improved Inlets for Culverts," HEC No. 13, Bridge Division, Federal Highway Administration, Washington, DC.

Henderson, F. M., 1966, *Open Channel Flow*, MacMillan, New York, NY.

Hopping, P. L., and Hoopes, J. A., 1988, "Development of a Numerical Model to Predict the Behavior of Air/Water Mixtures in Open Channels," in *Model Prototype Correlation of Hydraulic Structures*, Philip Burgi (ed.), pp. 419-428.

International Standards Organization, 1977, *Liquid Flow Measurement in Open Channels*, Geneva, Switzerland.

Jodeau, M., Hauet, A., Paquier, A., Le Coz, J., and Dramais, G., 2008, "Application and Evaluation of LS-PIV Technique for the Monitoring of River Surface Velocities in High Flow Conditions," *Flow Measurement and Instrumentation*, 19(2), 117-127.

Keller, R. J., and Rastogi, A. K., 1975, "Prediction of Flow Development on Spillways," *Jour. Hydraulics Div.*, Amer. Soc. Civil Engrs., vol 101, no 9, pp. 1171-1184.

Keller, R. J., Lai, K. K., Wood, I. R., 1974, "Developing Region in Self Aerating Flows," *Jour. Hydraulics Div.*, Amer. Soc. Civil Engrs., vol 100, no 4, pp. 553-568.

Killen, J. M., and Anderson, A. G., 1969, "A Study of the Air-Water Interface in Air Entrained Flow in Open Channels," *Proc. 13th Congress*, International Association for Hydraulic Research, Japan, vol 2, pp. 339-347.

Kulin, G., and Compton, P. R., 1975, *A Guide to Methods and Standards for the Measurement of Water Flow*, Special Publication No. 421, National Bureau of Standards, US Department of Commerce, Washington, DC.

Lane, E. W., 1939, "Entrainment on Spillway Faces," *Civil Engineering*, Amer. Soc. Civ. Engrs., vol. 9, pp. 89-96.

LaRocque, L. A., Elkholy, M., Chaudhry, M. H., and Imran, J., 2013, "Experiments on Urban Flooding Caused By a Levee Breach," *Journal of Hydraulic Engineering*, 139(9), 960-973.

Le Coz, J., Patalano, A., Collins, D., Guillén, N. F., García, C. M., Smart, G. M., Bind, J., Chiaverini, A., Le Boursicaud, R., Dramais, G., and Braud, I., 2016, "Crowdsourced Data for Flood Hydrology: Feedback from Recent Citizen Science Projects in Argentina, France and New Zealand," *Journal of Hydrology*, 541, 766-777.

Leopold, L. B., and Maddock, T., 1953, "The Hydraulic Geometry of Stream Channels and Some Physiographic Implications," *Geological Survey Professional Paper 252*; US Government Printing Office: Washington, DC, USA, 57p.

McClellan, T. J., 1971, "Fish Passage Through Highway Culverts," PB 204 983, Federal Highway Administration, Region 8, Portland, OR.

Muste, M., Yu, K., and Spasojevic, M., 2004, "Practical Aspects of ADCP Data Use for Quantification of Mean River Flow Characteristics; Part

I: Moving-Vessel Measurements," *Flow measurement and instrumentation,* 15(1), 1-16.

Muste, M., Fujita, I., and Hauet, A., 2008, "Large-Scale Particle Image Velocimetry for Measurements in Riverine Environments," *Water resources research,* 44(4).

Normann, J. M., Houghtalen, R.J., and Johnston, W.J., 1985, *Hydraulic Design of Highway Culverts,* Report no. FHWA-IP-85-15, Federal Highway Administration, McLean, VA.

Parshall, R. L., 1926, "The Improved Venture Flume," *Trans. Amer. Soc. Civil Engrs.,* vol 89, p. 841.

Parshall, R. L., 1953, "Parshall Flumes of Large Size," *Bulletin, Colorado Agricultural Experiment Station,* no. 426A, March.

Paiva, R. C., Durand, M. T., and Hossain, F., 2015, "Spatiotemporal Interpolation of Discharge Across a River Network By Using Synthetic SWOT Satellite Data," *Water Resources Research,* 51(1), 430-449.

Papa, F., Bala, S. K., Pandey, R. K., Durand, F., Gopalakrishna, V. V., Rahman, A., and Rossow, W. B., 2012, "Ganga-Brahmaputra River Discharge from Jason-2 Radar Altimetry: An Update to the Long-Term Satellite-Derived Estimates of Continental Freshwater Forcing Flux into the Bay of Bengal," *Journal of Geophysical Research: Oceans,* 117(C11).

Pavelsky, T. M., 2014, "Using Width-Based Rating Curves from Spatially Discontinuous Satellite Imagery to Monitor River Discharge," *Hydrological Processes,* 28(6), 3035-3040.

Perks, M. T., Russell, A. J., and Large, A. R., 2016, "Advances in Flash Flood Monitoring Using Unmanned Aerial Vehicles (UAVs)," *Hydrology and Earth System Sciences,* 20(10), 4005-4015.

Plant, W. J., Keller, W. C., and Hayes, K., 2005, "Measurement of River Surface Currents With Coherent Microwave Systems," *IEEE Transactions on Geoscience and Remote Sensing,* 43(6), 1242-1257.

Puleo, J. A., McKenna, T. E., Holland, K. T., and Calantoni, J., 2012, "Quantifying Riverine Surface Currents from Time Sequences of Thermal Infrared Imagery," *Water Resources Research,* 48(1).

Rantz, S. E., et al., 1982, "Measurement and Computation of Streamflow: Volume 1. Measurement of Stage and Discharge," *U.S. Geological Survey Water-Supply Paper 2175.*

Ran, Q. H., Li, W., Liao, Q., Tang, H. L., and Wang, M. Y., 2016, "Application of an Automated LSPIV System in a Mountainous Stream for Continuous Flood Flow Measurements," *Hydrological Processes,* 30(17), 3014-3029.

Reihsen, G., and L. J. Harrison, 1971, "Debris Control Structures," HEC No. 9, Bridge Div., Federal Highway Administration, Washington, DC.

Sichangi, A. W., Wang, L., Yang, K., Chen, D., Wang, Z., Li, X., Zhou, J., Liu, W., and Kuria, D., 2016, "Estimating Continental River Basin Discharges Using Multiple Remote Sensing Data Sets," *Remote Sensing of Environment,* 179, 36-53.

Smith, L. C., and Pavelsky, T. M., 2008, "Estimation of River Discharge, Propagation Speed, and Hydraulic Geometry from Space: Lena River, Siberia," *Water Resources Research*, 44(3).

Straub, L. G., and Anderson, A. G., 1958, "Experiments on Self-aerated Flow in Open Channels," *Jour. Hyd. Div.*, Amer. Soc. Civil Engrs., vol. 84, no. 7, pp. 1–35.

Sun, W., Ishidaira, H., and Bastola, S., 2012a, "Calibration of Hydrological Models in Ungauged Basins Based on Satellite Radar Altimetry Observations of River Water Level," *Hydrological Processes*, 26(23), 3524-3537.

Sun, W., Ishidaira, H., and Bastola, S., 2012b, "Prospects for Calibrating Rainfall-Runoff Models Using Satellite Observations of River Hydraulic Variables as Surrogates for In Situ River Discharge Measurements," *Hydrological processes*, 26(6), 872-882.

Takeda, Y., 1995, "Velocity profile measurement by ultrasonic Doppler method," *Experimental thermal and fluid science*, 10(4), 444-453.

Tarpanelli, A., Barbetta, S., Brocca, L., and Moramarco, T., 2013, "River Discharge Estimation by Using Altimetry Data and Simplified Flood Routing Modeling," *Remote Sensing*, 5(9), 4145-4162.

Tauro, F., Grimaldi, S., Petroselli, A., and Porfiri, M., 2012a, "Fluorescent Particle Tracers for Surface Flow Measurements: a Proof of Concept in a Natural Stream," *Water Resources Research*, 48(6).

Tauro, F., Mocio, G., Rapiti, E., Grimaldi, S., and Porfiri, M., 2012b, "Assessment of Fluorescent Particles for Surface Flow Analysis," *Sensors*, 12(11), 15827-15840.

Tauro, F., Olivieri, G., Petroselli, A., Porfiri, M., and Grimaldi, S., 2016, "Flow Monitoring with a Camera: A Case Study on a Flood Event in the Tiber River," *Environmental monitoring and assessment*, 188(2), 118.

U. S. Bureau of Reclamation, 1978, *Design of Small Canal Structures*, U. S. Department of Interior, U.S. Govt. Printing Office.

Vanoni, V. A., 1941, "Velocity Distribution in Open Channels," *Civil Engineering*, Amer. Soc. Civil Engrs., vol. 11, no. 6, pp. 356–57.

Walker, J. F., 1988, "General Two-Point Method for Determining Velocity in Open Channel," *Jour. Hydraulic Engineering*, Amer. Soc. Civil Engrs., vol. 114, no. 7, pp. 801-805.

Welber, M., Le Coz, J., Laronne, J. B., Zolezzi, G., Zamler, D., Dramais, G., ... and Salvaro, M., 2016, " Field Assessment of Noncontact Stream Gauging Using Portable Surface Velocity Radars (SVR)," *Water Resources Research*, 52(2), 1108-1126.

Widén-Nilsson, E., Halldin, S., and Xu, C. Y., 2007, "Global Water-Balance Modelling with WASMOD-M: Parameter Estimation and Regionalisation," *Journal of Hydrology*, 340(1-2), 105-118.

Wood, I. R., (ed.) 1991, *Air Entrainment in Free-Surface Flows*, A.A. Balkema, Rotterdam, Netherlands.

World Meteorological Organization, 1971, *Use of Weirs and Flumes in Stream Gauging*, WMO No. 280, Technical Note 117, Geneva, Switzerland.

11

UNSTEADY FLOW

Propagation of surge wave in Seton Canal produced by the closure of turbine gates at the downstream end (Courtesy, British Columbia Hydro and Power Authority, Canada)

© Springer Nature Switzerland AG 2022
M. H. Chaudhry, *Open-Channel Flow*,
https://doi.org/10.1007/978-3-030-96447-4_11

337

11-1 Introduction

In the previous chapters, we discussed steady flow in open channels. However, the flow conditions in the real-life systems usually vary with time and thus the flows are unsteady. The unsteadiness may be due to natural processes, due to human actions, or due to accidents and incidents. The analysis of unsteady flows is usually more complex than that of steady flows because unsteady-flow conditions vary with respect to both space and time, i.e., they are function of both space and time. Therefore, partial differential equations describe unsteady flows since the dependent variables (flow depth and flow velocity) are functions of more than one independent variables (space and time). A closed-form solution of these equations is not available except for very simplified cases and thus numerical methods are employed for their solution.

Unsteady flow is discussed in Chapters 11 through 15. A brief introduction is presented in this chapter; governing equations are derived in the next chapter, and numerical methods for their solution are presented in Chapters 13 and 14. Two-dimensional unsteady flow is discussed in Chapter 15, and a number of special topics related to unsteady flow are described in Chapter 18.

In this chapter, a number of commonly used terms are first defined. The causes of unsteady flow are then discussed and equations for the velocity of a gravity wave are derived.

11-2 Definitions

A *wave* is defined as a temporal (i.e., with respect to time) or spatial (i.e., with respect to distance) variation of flow depth and rate of discharge. The *wave length, L,* is the distance between two adjacent wave crests or troughs and the *amplitude, z,* of a wave is the height of the maximum water level above the still water level (Fig. 11-1).

Based on different criteria, waves may be classified into several categories. A wave is called *oscillatory wave* if there is no mass transport in the direction of wave travel and it is called *translatory wave* if there is net mass transport. For example, sea waves are oscillatory waves and flood waves are translatory. The translatory waves may be further classified as *solitary* or a *wave train.* A solitary wave has a rising and a falling limb and has a single peak. A wave train is, however, a group of waves in succession. A translatory wave having a steep front is called a *surge.* A *positive wave* has the water depth behind the wave higher than the undisturbed flow depth, and a *negative wave* has the flow depth behind the wave lower than the undisturbed flow depth. A positive wave having a steep front is referred to as a *bore,* or a *shock.* The latter term is borrowed from gas dynamics.

As the wave passes a section, the entire flow depth is disturbed in a *shallow-water wave* while only the top layers, and not the entire section, are affected

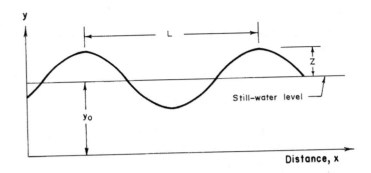

Fig. 11-1   Wave length and amplitude

in a *deep-water wave*. The ratio of wave length to the water depth is greater than 20 for shallow water waves; this ratio is less than 20 for the deep water waves.

The *wave celerity* is defined as the relative velocity of a wave with respect to the fluid in which it is traveling, whereas *absolute wave velocity* is the velocity with respect to a fixed reference frame. Thus, the absolute wave velocity, $\mathbf{V_w}$, is equal to the vectorial sum of the flow velocity, $\mathbf{V}$, and the wave celerity, $\mathbf{c}$, i.e.,

$$\mathbf{V_w} = \mathbf{V} + \mathbf{c} \tag{11-1}$$

In one-dimensional flow, the wave velocity is either in the direction of flow or in the opposite direction. Therefore,

$$V_w = V \pm c \tag{11-2}$$

The plus sign is used if the wave is traveling in the direction of flow and the negative sign is used if the wave is traveling opposite to the flow direction.

By neglecting the viscosity and surface tension, Airy derived the following expression (Henderson, 1966) for the celerity of a small-amplitude wave:

$$c = \sqrt{\frac{gL}{2\pi} \tanh \frac{2\pi y_o}{L}} \tag{11-3}$$

in which y_o = undisturbed flow depth. As we discussed in the previous paragraphs, y_o/L is large in deep-water waves. Thus, $\tanh 2\pi y_o/L \to 1$ and the celerity for the deep-water wave reduces to

$$c = \sqrt{\frac{gL}{2\pi}} \tag{11-4}$$

The ratio y_o/L is very small for shallow-water waves; therefore, $\tanh 2\pi y/L \to 2\pi y_o/L$. Hence, the expression for the celerity of these waves becomes

$$c = \sqrt{gy_o} \tag{11-5}$$

Note that Eq. 11-5 is valid only for small amplitude waves. We will derive a general expression for celerity in Section 11-4 and then deduce Eq. 11-5 from that expression.

11-3 Occurrence of Unsteady Flow

Transients in open channels are produced whenever flow conditions are changed. These changes may be caused by natural processes or due to accidents or planned actions. The flow conditions include the flow depth and flow velocity. If the discharge is varied at a rapid rate or the channel has low friction, a *bore* may develop during the transient conditions.

Typical situations in which unsteady flows occur are as follows:

1. Surges in power canals or tunnels produced by starting or stopping of turbines or due to the opening or closing of the turbine gates to meet the load changes;
2. Surges in upstream or downstream channels produced by starting or stopping of pumps and opening or closing of control gates;
3. Waves in the navigation channels produced by the operation of navigation locks;
4. Flood waves in streams, rivers, and drainage channels due to rain-storms and/or snow-melt or produced by the failure of dams, dykes, levees or other control structures;
5. Tides in estuaries, bays and inlets;
6. Waves generated by landslides and avalanches in rivers, channels, reservoirs, and lakes;
7. Storm runoff in sewers and drainage channels;
8. Circulation in lakes and reservoirs produced by wind or by temperature and density gradients; and
9. Waves in lakes, reservoirs, estuaries, bays, inlets, and oceans produced by wind storms, cyclones, and earthquakes.

11-4 Height and Celerity of a Gravity Wave

In this section, we will derive expressions for determining the celerity and the height of a gravity wave produced by a sudden change in discharge. These expressions are general and may be used for small or large amplitude waves. We will make the following simplifying *assumptions* in the derivations

1. The channel is frictionless and the channel bottom is horizontal (same equations are obtained if the component of the weight of the liquid in the downstream direction is equal to the shear force acting on the channel sides and bottom);

2. The pressure distribution on both sides of the wave front is hydrostatic;
3. The velocity distribution is uniform on both sides of the wave front;
4. The wave is an abrupt discontinuity of negligible length;
5. The wave does not change in shape as it propagates in the channel; and
6. The water surface behind the wave is parallel to the initial water surface.

Let the flow be suddenly increased from Q_1 to Q_2 in a channel which increases the flow depth from y_1 to y_2, as shown in Fig. 11-2. Let this wave be traveling at absolute velocity V_w in the downstream direction. We consider this direction as positive. We are interested in determining the velocity of this wave and the wave height, $y_2 - y_1$. These equations may be derived by using different methods. In the following paragraphs, we will use the control volume approach to the unsteady flow. However, we may convert the unsteady-flow situation to steady flow by applying velocity V_w on the entire system in the upstream direction. Then we may either use the control volume approach as in this section, or we may apply the common continuity and momentum principles (as we did in Section 3-2) to derive these equations.

Let us consider a control volume, as shown in Fig. 11-2, in which the wave front has moved during time interval Δt, as shown. We will apply the Reynolds transport theorem [Roberson and Crowe, 1997] in this section to derive the expressions for the wave height and the wave velocity.

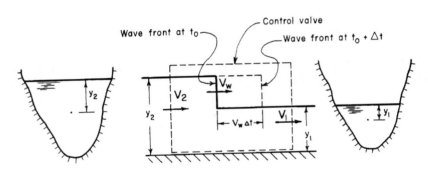

Fig. 11-2   Definition sketch

Continuity equation

Referring to Fig. 11-2

$$\frac{d}{dt} \int_{cv} \rho d\mathcal{V} + \sum_{cs} \rho V A = 0 \qquad (11\text{-}6)$$

in which ρ = mass density of water; $\mathcal{V}$ = volume of the control volume; V = flow velocity; and A = flow area. By using subscripts 1 and 2 to denote quantities for sections 1 and 2 and subscript t_o and $t_o + \Delta t$, for quantities at these times, respectively, we may write

$$\sum_{cs} \rho V A = \rho V_1 A_1 - \rho V_2 A_2 \tag{11-7}$$

Since water may be assumed incompressible

$$\frac{d}{dt} \int_{cv} \rho d\mathcal{V} = \rho \frac{d}{dt} \int_{cv} d\mathcal{V}$$
$$= \rho \times \text{Rate of change of volume of liquid}$$
$$\text{in the control volume}$$
$$= \rho \frac{\mathcal{V}_{t_o + \Delta t} - \mathcal{V}_{t_o}}{\Delta t}$$
$$= \rho \frac{[\mathcal{V}_{t_o} + V_w \Delta t (A_2 - A_1)] - \mathcal{V}_{t_o}}{\Delta t}$$
$$= \rho V_w (A_2 - A_1) \tag{11-8}$$

By substituting Eqs. 11-7 and 11-8 into Eq. 11-6 and simplifying, we obtain

$$A_1(V_1 - V_w) = A_2(V_2 - V_w) \tag{11-9}$$

Momentum equation

The intensive property for the momentum equation is $\beta = V$. Therefore, according to the Reynolds transport theorem

$$\sum F = \frac{d}{dt} \int_{cv} \rho V d\mathcal{V} + \sum_{cs} V \rho V A \tag{11-10}$$

Referring to Fig. 11-2,

$$\sum F = \gamma \bar{y}_2 A_2 - \gamma \bar{y}_1 A_1 \tag{11-11}$$

and

$$\sum_{cs} \rho V^2 A = \rho A_1 V_1^2 - \rho A_2 V_2^2 \tag{11-12}$$

in which $\sum F$ = sum of the external forces on the control volume in the positive x-direction; $\bar{y}$ = depth of the centroid of the flow section; and γ = specific weight of the liquid.

The time rate of change of momentum in the control volume is

$$\frac{d}{dt} \int_{cv} \rho V d\mathcal{V} = \rho V_w (A_2 V_2 - A_1 V_1) \tag{11-13}$$

Substituting Eqs. 11-11 and 11-12 into Eq. 11-9, utilizing Eq. 11-8, and simplifying the resulting equation, we obtain

$$\frac{A_1}{g}(V_1 - V_w)(V_1 - V_2) = \bar{y}_2 A_2 - \bar{y}_1 A_1 \qquad (11\text{-}14)$$

Elimination of V_2 from Eqs. 11-9 and 11-14 and the rearrangement of the resulting equation yield

$$(V_1 - V_w)^2 = \frac{gA_2}{A_1(A_2 - A_1)}(A_2\bar{y}_2 - A_1\bar{y}_1) \qquad (11\text{-}15)$$

The velocity of the surge wave, V_w, must be greater than the undisturbed flow velocity, V_1, for the wave to propagate in the downstream direction. Therefore, it follows from the above equation that

$$V_w = V_1 + \sqrt{\frac{gA_2}{A_1(A_2 - A_1)}(A_2\bar{y}_2 - A_1\bar{y}_1)} \qquad (11\text{-}16)$$

Transposing V_1 to the left-hand side

$$V_w - V_1 = \sqrt{\frac{gA_2}{A_1(A_2 - A_1)}(A_2\bar{y}_2 - A_1\bar{y}_1)} \qquad (11\text{-}17)$$

As we defined earlier, the celerity, c, of a wave is the velocity relative to the fluid in which it is traveling, i.e., $c = V_w - V_1$. Thus, the left-hand side of Eq. 11-17 is the celerity of the surge wave.

$$c = \sqrt{\frac{gA_2}{A_1(A_2 - A_1)}(A_2\bar{y}_2 - A_1\bar{y}_1)} \qquad (11\text{-}18)$$

The height of the surge wave, $y_2 - y_1$, produced by a sudden change in discharge may be determined from the following relationship between the flow depths and flow velocities at sections 1 and 2 (Fig. 11-1), derived by eliminating V_w from Eqs. 11-9 and 11-15:

$$A_2\bar{y}_2 - A_1\bar{y}_1 = \frac{A_1 A_2}{g(A_2 - A_1)}(V_1 - V_2)^2 \qquad (11\text{-}19)$$

Let us assume that we know the values of y_1 and V_1, or Q_1. Then, for a specified change in discharge from Q_1 to Q_2, we can determine by trial and error the values of y_2 and V_2 from $Q_2 = V_2 A_2$ and Eq. 11-19. The value of the wave velocity, V_w, may then be determined from Eq. 11-16 for a wave traveling in the downstream direction. For a wave traveling in the upstream direction, use a negative sign with the radical term.

The preceding equations may be used for any cross section provided the simplifying assumptions we listed at the beginning of this section are valid.

For a *rectangular channel,* the above equations are simplified as follows. For a channel of width B, $A_1 = By_1$; $A_2 = By_2$; $\bar{y}_1 = \frac{1}{2}y_1$, and $\bar{y}_2 = \frac{1}{2}y_2$. Substituting these relationships into Eq. 11-18 and simplifying the resulting equation, we obtain

$$c = \sqrt{\frac{gy_2}{2y_1}(y_1 + y_2)} \qquad (11\text{-}20)$$

For waves of small height, $y_1 \simeq y_2 = y$ (say). Then, Eq. 11-20 becomes

$$c = \sqrt{gy} \qquad (11\text{-}21)$$

Note that this is the same expression as that we derived in Chapter 3 (Eq. 3-30). Substituting $A_1 = By_1$ and $A_2 = By_2$ into Eq. 11-9, and simplifying, we obtain

$$V_w = \frac{V_2 y_2 - V_1 y_1}{y_2 - y_1} \qquad (11\text{-}22)$$

Substitution of Eq. 11-20 into Eq. 11-2 for a wave traveling in the downstream direction and elimination of V_w from the resulting equation and Eq. 11-22 yield

$$(V_1 - V_2)^2 = \frac{g(y_1 - y_2)}{2y_1 y_2}(y_1^2 - y_2^2) \qquad (11\text{-}23)$$

Example 11-1

A 1-m wide rectangular channel is carrying a flow of 5 m³/s at a flow depth of 2 m. Determine the height of a surge wave and its velocity if the discharge is suddenly increased to 10 m³/s at the upstream end.

Given:

$Q_1 = 5$ m³/s;
$Q_2 = 10$ m³/s;
$y_1 = 2$ m;
$B = 1$ m.

Determine:

$y_2 = ?$
$V_w = ?$

Solution:

The flow velocity at section 1,

$$V_1 = Q_1/(By_1) = 5/(1.\text{x}2.) = 2.5\,\text{m/s}$$

Now, $Q_2 = By_2V_2$ or $V_2y_2 = 10/1 = 10$. By substituting the values of y_1 and V_1 and $V_2y_2 = 10$ into Eq. 11-22, and simplifying, we obtain

$$(1 - \frac{4}{y_2})^2 = \frac{2.452(y_2 - 2)}{y_2}(y_2^2 - 4)$$

Solution of this equation by trial and error gives $y_2 = 2.334$ m. Thus, the height of the surge $= 2.334-2. = 0.334$ m.

Substituting values of y_1, y_2, V_1 and $V_2y_2 = 4$ into Eq. 11-21, we obtain

$$\begin{aligned}
V_w &= \frac{2 - 4}{2.0 - 2.334} \\
&= \frac{2}{0.334} \\
&= 5.99\,\text{m/s}
\end{aligned}$$

11-5 Summary

In this chapter, a brief introduction on the unsteady flow was presented. Commonly used terms were defined and causes that produce unsteady flow were discussed. Expressions were derived for determining the celerity and height of a gravity wave produced by a sudden change in discharge.

Problems

11-1 Consider a small wave of height, Δy which changes the flow velocity from V to $V + \Delta V$. By applying the continuity and momentum principles, derive Eq. 11-4b.

11-2 Determine the speed and height of a surge wave produced by instantaneously closing a downstream control gate in a 5-m wide rectangular channel carrying a flow of 7.5 m^3/s at a flow depth of 1.5 m.

11-3 If the width of the channel of Problem 11-2 is reduced to 4 m at a distance of 500 m upstream of the control gate, determine the height of the wave in the constricted channel.

11-4 What will be the wave height if the width of the channel of Problem 11-3 is increased to 7.5 m instead of reduced to 4 m?

11-5 By assuming that the shape of the wave does not change as it travels, the unsteady flow of Fig. 11-2 may be converted to steady flow by superimposing velocity V_w in the upstream direction on the entire system. Apply the continuity and momentum principles to the steady flow to derive Eqs. 11-8 and 11-17.

11-6 By utilizing the expression for the celerity of a small wave, show that a positive wave steepens and a negative wave flattens as they travel in a channel having negligible friction.

11-7 Derive expressions for the absolute wave velocities and heights of the reflected waves after two positive waves meet.

11-8 Derive the expressions for the wave velocities and the wave heights of the reflected and transmitted waves at a step rise in the channel bottom.

11-9 A monoclical wave does not change shape as it propagates in a channel, and the flow is assumed to be uniform upstream and downstream sides of the wave. If the subscripts u and d refer to the variables for the upstream and downstream sides, prove that the velocity of this wave is

$$V_w = \frac{Q_u - Q_d}{A_u - A_d}$$

References

Abbott, M. B., 1979, *Computational Hydraulics: Elements of the Theory of Free Surface Flows,* Pitman, London.

Basco, D. R., 1983, *Computation of Rapidly Varied, Unsteady Free-Surface Flow,* U. S. Geological Survey Report WRI 83-4284.

Chow, V. T., 1959, *Open-Channel Hydraulics,* McGraw-Hill, New York, NY.

Cunge, J. A., Holly, F. M., and Verwey, A., 1980, *Practical Aspects of Computational River Hydraulics,* Pitman, London.

Dronkers, J. J., 1964, *Tidal Computations in Rivers and Coastal Waters,* North-Holland Publ., Amsterdam, The Netherlands.

Henderson, F. M., 1966, *Open Channel Flow,* MacMillian, New York, NY.

Lai, C., 1986, "Numerical Modeling of Unsteady Open-Channel Flow," *Advances in Hydroscience,* vol. 14, pp. 161-333.

Le Mehaste, B., 1976, *Hydrodynamics,* Springer Verlag, New York, NY.

Mahmood, K., and Yevjevich, V., eds., 1975, *Unsteady Flow in Open Channels,* vol. 1, Water Resources Publications, Fort Collins, CO.

Roberson, J. A., and Crowe, C. T., 1997, *Engineering Fluid Mechanics,* Sixth ed., John Wiley & Sons, Inc., New York, NY.

Stoker, J. J., 1957, *Water Waves,* Wiley (Interscience), New York, NY.

12

GOVERNING EQUATIONS FOR ONE-DIMENSIONAL FLOW

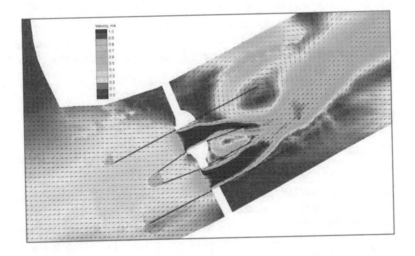

Lake Borgne surge barrier modeling AdH (Courtesy ERDC, US Army Corps of Engineers)

© Springer Nature Switzerland AG 2022
M. H. Chaudhry, *Open-Channel Flow*,
https://doi.org/10.1007/978-3-030-96447-4_12

12-1 Introduction

Three conservation laws – mass, momentum, and energy – are used to describe open-channel flows. Two flow variables, such as the flow depth and velocity, or the flow depth and rate of discharge, are sufficient to define the flow conditions at a channel cross section. Therefore, two governing equations may be used to analyze a typical flow situation. The continuity equation and the momentum or energy equation are used for this purpose. Except for the velocity head coefficient, α, and the momentum coefficient, β, the momentum and energy equations are equivalent [Cunge, et al., 1980] provided the flow depth and velocity are continuous, i.e., there are no discontinuities, such as a jump or a bore. However, the momentum equation should be used if the flow has discontinuities, since, unlike the energy equation, it is not necessary to know the amount of losses in the discontinuities in the application of the momentum equation.

In this chapter, we will derive the continuity and momentum equations, usually referred to as de Saint Venant equations. Several investigators [Stoker, 1957; Chow, 1959; Dronkers, 1964; Henderson, 1966; Strelkoff, 1969; Yen, 1973; Liggett, 1975; Cunge et al., 1980; Lai, 1986; and Abbott and Basco, 1990] derived these equations by using different procedures. For illustration purposes, we will use two different procedures in our derivations. We will use the Reynolds transport theorem for the prismatic channels having lateral inflows or outflows. The type of the governing equations is then discussed. The equations describing flows having non-hydrostatic pressure distribution are derived by integrating the continuity and momentum equations for two-dimensional flows. The chapter concludes by presenting integral forms of the governing equations.

12-2 St. Venant Equations

We will make the following *assumptions* in the derivation of the governing equations.

1. The pressure distribution is hydrostatic. This is a valid assumption if the streamlines do not have sharp curvatures;
2. The channel bottom slope is small so that the flow depth measured normal to the channel bottom or measured vertically are approximately the same;
3. The flow velocity over the entire channel cross section is uniform;
4. The channel is prismatic, – i.e., the channel cross section and the channel bottom slope do not change with distance. The variations in the cross section or bottom slope may be taken into consideration by approximating the channel into several prismatic reaches;
5. The head losses in unsteady flow may be simulated by using the steady-state resistance laws, such as the Manning or Chezy equation, i.e., head

losses for a given flow velocity during unsteady flow are the same as that during steady flow.

Continuity Equation

In open-channel flows, we mostly deal with the flow of water which may be assumed to be incompressible and to have constant mass density. Therefore, the law of conservation of mass is the same as the continuity equation.

Let us consider a control volume having fixed boundaries, as shown in Fig. 12-1. If the flow between section 1 and 2 is unsteady and nonuniform, then the rate of discharge, Q, flow velocity, V, and flow depth, y, are functions of distance x (measured positive in the downstream direction), and time, t. For applying the Reynolds transport theorem to the control volume, the extensive property, $\mathcal{B} = $ mass, M, and the intensive property, $\beta = \Delta m / \Delta m = 1$.

Referring to Fig. 12-1, substituting $\mathcal{B} = M$ and $\beta = 1$ into Eq. 1-26, using subscripts 1 and 2 to indicate flow variables at sections 1 and 2, respectively, and noting that $dM/dt = 0$ (law of conservation of mass), we obtain

$$\frac{dM}{dt} = \frac{d}{dt} \int_{x_1}^{x_2} \rho A \, dx + \rho A_2 V_2 - \rho A_1 V_1 - \rho q_l(x_2 - x_1) = 0 \qquad (12\text{-}1)$$

in which $A = $ flow area; $\rho = $ mass density of water; and $q_l = $ volumetric rate of lateral inflow or outflow per unit length of the channel between sections 1 and 2. The lateral inflow, q_l, into the channel is considered positive, whereas the outflow is considered negative. These lateral inflow or outflow may be due to infiltration, evaporation, or flows to or from the channel banks. Since water may be assumed incompressible, mass density, ρ, is constant. Therefore, Eq. 12-1 may be written as

$$\frac{d}{dt} \int_{x_1}^{x_2} A \, dx + A_2 V_2 - A_1 V_1 - q_l(x_2 - x_1) = 0 \qquad (12\text{-}2)$$

We may derive the differential or integral form of the continuity equation from Eq. 12-2. The differential form requires that the flow variables be continuous, but there is no such restriction on the integral form. We derive the differential form in the following paragraphs and the integral form in Section 12-5.

Let us apply Leibnitz's rule* to the first term on the left-hand side of Eq. 12-1. To do this, it is necessary that both A and $\partial A/\partial t$ are continuous with respect to both x and t. In other words, between x_1 and x_2, there are no abrupt discontinuities in the channel cross section (i.e., the channel width

---

* According to this rule,

$$\frac{d}{dt} \int_{f_1(t)}^{f_2(t)} F(x, t) \, dx = \int_{f_1(t)}^{f_2(t)} \frac{\partial}{\partial t} F(x, t) \, dx + F(f_2(t), t) \frac{df_2}{dt} - F(f_1(t), t) \frac{df_1}{dt}.$$

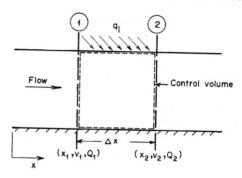

Fig. 12-1 Definition sketch for continuity equation

and the channel bottom do not have step changes) and in the flow depth (i.e., there is no bore or jump). By applying this rule, noting that $dx_1/dt = 0$ and $dx_2/dt = 0$, since the boundaries of the control volume are fixed, and $Q_1 = A_1 V_1$ and $Q_2 = A_2 V_2$, we obtain

$$\int_{x_1}^{x_2} \frac{\partial A}{\partial t}\, dx + Q_2 - Q_1 - q_l(x_2 - x_1) = 0 \tag{12-3}$$

Based on the mean value theorem [†] (to apply this theorem, it is necessary that both Q and $\partial Q/\partial x$ are continuous), Eq. 12-3 takes the form

$$\frac{\partial A}{\partial t} + \frac{\partial Q}{\partial x} = q_l \tag{12-4}$$

Equation 12-4 is referred to as the continuity equation in the *conservation* or *divergent form*. According to this equation, mass is conserved along any closed contour in the x-t plane if the right-hand side of this equation is zero. If the term on the right-hand side is not zero, then this term acts like a source or a sink depending upon the sign of q_l.

For a channel having a regular cross section (i.e., top water-surface width, B, is a continuous function of the flow depth y), the change in flow area, ΔA, for a small change in flow depth, Δy, may be approximated as $B\Delta y$. In the limit as $\Delta y \to 0, dA/dy = B$. Hence, Eq. 12-4 may be written as

$$B\frac{\partial y}{\partial t} + \frac{\partial Q}{\partial x} = q_l \tag{12-5}$$

Similarly, we may write $Q = VA$. Substituting this equation into Eq. 12-5, expanding the second term, and noting that $\partial A/\partial x = B\partial y/\partial x$ and the hydraulic depth, $D = A/B$, this equation becomes

---

[†] According to this theorem,

$$\int_{x_1}^{x_2} F(x)\, dx = (x_2 - x_1)F(\zeta) \quad x_1 < \zeta < x_2.$$

$$\frac{\partial y}{\partial t} + D\frac{\partial V}{\partial x} + V\frac{\partial y}{\partial x} - \frac{q_l}{B} = 0 \tag{12-6}$$

Momentum Equation

For the momentum equation, the extensive property, $\mathcal{B} = $ momentum of water in the control volume $= mV$ and the intensive property $\beta = V\Delta m/\Delta m = V$. In addition, according to the Newton's second law of motion, the rate of change of momentum is equal to the resultant force acting on the control volume, i.e., $\sum F = d\mathcal{B}/dt$. Substitution of these relationships into Eq. 1-26 yields

$$\sum F = \frac{d}{dt}\int_{x_1}^{x_2} V\rho A dx + V_2\rho A_2 V_2 - V_1\rho A_1 V_1 - V_x\rho q_l(x_2 - x_1) \tag{12-7}$$

in which V_x is the component of the velocity of lateral inflow in the x-direction. Note that q_l is positive for lateral inflow and it is negative for lateral outflow.

By applying Leibnitz's rule and writing $Q = VA$, Eq. 12-7 becomes

$$\sum F = \int_{x_1}^{x_2} \rho\frac{\partial Q}{\partial t}\, dx + \rho Q_2 V_2 - \rho Q_1 V_1 - V_x\rho q_l(x_2 - x_1) \tag{12-8}$$

By dividing throughout by $\rho(x_2 - x_1)$ and applying the mean value theorem, we obtain

$$\frac{\sum F}{\rho(x_2 - x_1)} = \frac{\partial Q}{\partial t} + \frac{\partial(QV)}{\partial x} - V_x q_l \tag{12-9}$$

For simplicity, let us neglect the shear stresses on the flow surface due to wind and neglect the effects of the Coriolis acceleration. These are valid assumptions for typical hydraulic engineering applications. Since the channel is assumed to be prismatic, there are no external forces acting on the control volume due to changes in the channel cross section. Hence, referring to Fig. 12-2, the following forces are acting on the control volume:

Pressure force, F_1, acting on the upstream end, $= \rho g A_1 \bar{y}_1$ $\qquad$ (12-10a)

in which $\bar{y}_1 = $ depth of the centroid of flow area A_1. Similarly,

Pressure force, F_2, acting on the downstream end $= \rho g A_2 \bar{y}_2$ $\qquad$ (12-10b)

The component of weight of water in the control volume in the x-direction, F_3, may be written as

$$F_3 = \rho g \int_{x_1}^{x_2} AS_o dx \tag{12-10c}$$

in which $S_o = $ channel bottom slope, considered positive sloping downwards.

The frictional force, F_4, is due to shear between water and the channel sides and the channel bottom. This may be expressed in terms of the friction slope, S_f, or the energy gradient needed to overcome friction as

$$F_4 = \rho g \int_{x_1}^{x_2} A S_f dx \qquad (12\text{-}10\text{d})$$

For any exponential formula for friction losses, the expression for the friction slope may be written as

$$S_f = \frac{C V |V|^{m-1}}{R^p} \qquad (12\text{-}10\text{e})$$

in which C and p are coefficients that depend upon the formula employed, $R =$ hydraulic radius, and m depends upon the flow type. For example, $m = 1$ for laminar flow; $m = 1.75$ for smooth turbulent flow; and $m = 2$ for fully rough turbulent flow. Hence, the resultant force acting on the control volume is

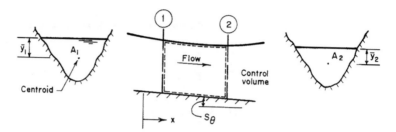

Fig. 12-2   Definition sketch for momentum equation

$$\sum F = F_1 - F_2 + F_3 - F_4 \qquad (12\text{-}11)$$

Substitution of expressions for F_1 to F_4 from Eq. 12-10 into Eq. 12-11 and division by $\rho(x_2 - x_1)$ yield

$$\frac{\sum F}{\rho(x_2 - x_1)} = \frac{g(A_1 \bar{y}_1 - A_2 \bar{y}_2)}{x_2 - x_1} + \frac{g}{x_2 - x_1} \int_{x_1}^{x_2} A(S_o - S_f) dx \qquad (12\text{-}12)$$

Equating the right-hand sides of Eqs. 12-9 and 12-12 and applying the mean value theorem, we obtain

$$\frac{\partial Q}{\partial t} + \frac{\partial (QV)}{\partial x} - V_x q_l = -g \frac{\partial}{\partial x}(A\bar{y}) + gA(S_o - S_f) \qquad (12\text{-}13)$$

This equation may be written as

$$\frac{\partial Q}{\partial t} + \frac{\partial}{\partial x}(QV + gA\bar{y}) = gA(S_o - S_f) + V_x q_l \qquad (12\text{-}14)$$

This equation is referred to as the momentum equation in the conservation form. This means that the momentum along any closed contour in the x-t plane is conserved if the right-hand side of Eq. 12-14 is zero [Cunge et al., 1980]. Nonzero terms on the right-hand side act as sources or sinks.

Now, $\Delta(A\bar{y}) = [A(\bar{y} + \Delta y) - \frac{1}{2}B(\Delta y)^2] - A\bar{y}$. Thus, by neglecting the higher-order terms and letting $\Delta y \to 0$, we obtain $\frac{\partial}{\partial y}(A\bar{y}) = A$. By utilizing this expression, we may write

$$\frac{\partial}{\partial x}(gA\bar{y}) = g\frac{\partial}{\partial y}(A\bar{y})\frac{\partial y}{\partial x} = gA\frac{\partial y}{\partial x}$$

Hence, Eq. 12-14 becomes

$$\frac{\partial Q}{\partial t} + \frac{\partial(QV)}{\partial x} + gA\frac{\partial y}{\partial x} = V_x q_l + gA(S_o - S_f) \qquad (12\text{-}15)$$

By expanding the first two terms on the left-hand side and rearranging, we get

$$V\left[B\frac{\partial y}{\partial t} + A\frac{\partial V}{\partial x} + BV\frac{\partial y}{\partial x} - \frac{V_x}{V}q_l\right]$$
$$+ A\left(\frac{\partial V}{\partial t} + V\frac{\partial V}{\partial x} + g\frac{\partial y}{\partial x} + gS_f - gS_o\right) = 0 \qquad (12\text{-}16)$$

According to Eq. 12-6, the sum of the terms within the brackets is zero if $V_x = 0$ or if $V_x = V$. Hence,

$$\frac{\partial V}{\partial t} + g\frac{\partial}{\partial x}\left(\frac{V^2}{2g} + y\right) = g(S_o - S_f) \qquad (12\text{-}17)$$

This equation has been referred to in the literature as the momentum equation, equation of motion, and dynamic equation. Since it does not truly describe the conservation of momentum, we refer to it as the *dynamic equation*.

We may rearrange the terms of Eq. 12-17 as

$$\underbrace{\underbrace{\underbrace{S_f = S_o}_{Steady, uniform} \quad -\frac{\partial}{\partial x}(\frac{V^2}{2g} + y)}_{Steady, nonuniform} \quad -\frac{1}{g}\frac{\partial V}{\partial t}}_{Unsteady, nonuniform}$$

The significance of each term is clear from this equation. For steady-uniform flow, the slope of the energy grade line is the same as the channel-bottom slope. The equation for steady, gradually varied flow is obtained by including the variation of the flow depth and the velocity head, i.e., by including the derivative with respect to distance, x. The unsteadiness or the local acceleration term is added to make the equation valid for unsteady, nonuniform flow.

12-3 General Remarks

The continuity and momentum equations form a set of nonlinear partial differential equations. A closed form solution of these equations is not available except for very simplified cases. Therefore, numerical methods are used for their integration. To select a numerical scheme, it is helpful if we know the type of these equations, i.e., whether they are hyperbolic, parabolic, or elliptic. We may determine the type of these equations as follows.

Let us write Eqs. 12-6 and 12-17 in vector form as

$$\frac{\partial \mathbf{U}}{\partial \mathbf{t}} + \mathbf{A}\frac{\partial \mathbf{U}}{\partial \mathbf{x}} = \mathbf{S} \tag{12-18}$$

in which

$$\mathbf{U} = \begin{pmatrix} y \\ V \end{pmatrix}; \quad \mathbf{A} = \begin{pmatrix} V & D \\ g & V \end{pmatrix}; \quad \mathbf{S} = \begin{pmatrix} -\frac{q_l}{B} \\ g(S_o - S_f) \end{pmatrix} \tag{12-19}$$

The eigenvalues of matrix $\mathbf{A}$ define the type of Eq. 12-18. These eigenvalues are determined from the equation

$$\begin{vmatrix} V - \lambda & D \\ g & V - \lambda \end{vmatrix} = 0 \tag{12-20}$$

It follows from this equation that

$$(V - \lambda)^2 - gD = 0 \tag{12-21}$$

Two roots of this equation are

$$\lambda_1 = V + \sqrt{\frac{gA}{B}}$$

$$\lambda_2 = V - \sqrt{\frac{gA}{B}} \tag{12-22}$$

The radical term of Eq. 12-22 is an expression for the wave celerity. Hence, the eigenvalues of $\mathbf{A}$ represent absolute wave velocities, $V + c$ and $V - c$.

The expressions for the eigenvalues, λ_1 and λ_2, show that both of them are real and distinct for sub- and supercritical flows. Therefore, Eqs. 12-18 are a set of *hyperbolic partial differential equations*. This type of equation represents the propagation of waves in different media. Computational procedures, referred to as the marching procedure, are suitable for the numerical integration of these equations.

12-4 Boussinesq Equations

If the streamlines in a flow have sharp curvatures, then the pressure distribution is not hydrostatic because of acceleration. In this section, we derive the equations to describe these flows by integrating the two-dimensional flow equations in the vertical direction and by utilizing the Boussinesq assumption. These equations are referred to as the *Boussinesq equations*. According to the Boussinesq assumption, the flow velocity, w, in the vertical direction varies linearly from zero at the channel bottom to the maximum value at the flow surface.

In the derivation, we assume that the flow velocity in the lateral direction is zero; channel bottom slope is small; flow velocity, u, in the x-direction is uniform over the flow depth, and the datum lies along the channel bottom. In addition, we assume for simplicity that the channel is frictionless.

Continuity equation

The continuity equation for a two-dimensional flow may be written as

$$\frac{\partial u}{\partial x} + \frac{\partial w}{\partial z} = 0 \tag{12-23}$$

in which $u = $ flow velocity in the x-direction, and $w = $ flow velocity in the z-direction. By multiplying this equation throughout by dz and integrating it from the channel bottom $(z = 0)$ to the free surface $(z = y)$, we obtain

$$\int_0^y \frac{\partial u}{\partial x} dz + \int_0^y \frac{\partial w}{\partial z} dz = 0 \tag{12-24}$$

Based on the Leibnitz rule, this equation becomes

$$\frac{\partial}{\partial x} \int_0^y u \, dz + \left. \left(u\frac{\partial z}{\partial x}\right)\right|_{z=0} - \left. \left(u\frac{\partial z}{\partial x}\right)\right|_{z=y} + w|_{z=y} - w|_{z=0} = 0 \tag{12-25}$$

Let us refer to quantities for the channel bottom $(z = 0)$ and for the water surface $(z = y)$ by subscripts "b" and "s" respectively. Then,

$$\frac{\partial}{\partial x} \int_0^y u \, dz + u_b \frac{\partial z_b}{\partial x} - u_s \frac{\partial z_s}{\partial x} + w_s - w_b = 0 \tag{12-26}$$

For the velocity of the water surface, w_s, we may write

$$w_s = \frac{dy}{dt} = \frac{\partial y}{\partial t} + u_s \frac{\partial y}{\partial x} \tag{12-27}$$

Since the datum is along the channel bottom, $\partial z_b/\partial x = 0$. In addition, for a rigid channel bottom, $w_b = 0$. Hence, based on Eq. 12-27, Eq. 12-26 may be written as

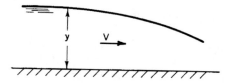

Fig. 12-3   Notation

$$\frac{\partial y}{\partial t} + \frac{\partial(uy)}{\partial x} = 0 \tag{12-28}$$

This equation may be expanded as

$$\frac{\partial y}{\partial t} + u\frac{\partial y}{\partial x} + y\frac{\partial u}{\partial x} = 0 \tag{12-29}$$

or

$$\frac{\partial u}{\partial x} = -\frac{1}{y}\left(\frac{\partial y}{\partial t} + u\frac{\partial y}{\partial x}\right) \tag{12-30}$$

Now, it follows from Eq. 12-23 that

$$\frac{\partial w}{\partial z} = -\frac{\partial u}{\partial x} \tag{12-31}$$

Since u is assumed to be constant across the flow depth y, we can integrate Eq. 12-31 as

$$w = -\frac{\partial u}{\partial x}z \tag{12-32}$$

Substitution of Eq. 12-30 into Eq. 12-32 yields

$$w = \frac{1}{y}\left(\frac{\partial y}{\partial t} + u\frac{\partial y}{\partial x}\right)z \tag{12-33}$$

Momentum Equation in z-direction

The momentum equation in the z-direction may be written as

$$\frac{\partial w}{\partial t} + u\frac{\partial w}{\partial x} + w\frac{\partial w}{\partial z} = -\frac{1}{\rho}\frac{\partial p}{\partial z} - g \tag{12-34}$$

By multiplying Eq. 12-34 by z and rearranging the terms, we obtain

$$\frac{p}{\rho} = \frac{\partial(zw)}{\partial t} + \frac{\partial(uzw)}{\partial x} + \frac{\partial(w^2 z)}{\partial z} + gz + \frac{1}{\rho}\frac{\partial(pz)}{\partial z} - wz\left(\frac{\partial u}{\partial x} + \frac{\partial w}{\partial z}\right) - w^2 \tag{12-35}$$

Based on the continuity equation (Eq. 12-23), the sum of the terms in the parenthesis on the right-hand side is zero. Hence, Eq. 12-35 becomes

$$\frac{p}{\rho} = \frac{\partial(zw)}{\partial t} + \frac{\partial(uzw)}{\partial x} + \frac{\partial(w^2 z)}{\partial z} + gz + \frac{1}{\rho}\frac{\partial(pz)}{\partial z} - w^2 \tag{12-36}$$

Integration of this equation over the flow depth, y, yields

$$\int_0^y \frac{p}{\rho}\, dz = \int_0^y \frac{\partial(zw)}{\partial t}\, dz + \int_0^y \frac{\partial(uzw)}{\partial x}\, dz + \int_0^y \frac{\partial(w^2 z)}{\partial z}\, dz$$

$$+ \int_0^y gz\, dz + \int_0^y \frac{1}{\rho}\frac{\partial(pz)}{\partial z}\, dz - \int_0^y w^2\, dz \qquad (12\text{-}37)$$

Applying the Leibnitz rule and referring to the quantities for the free surface and for the channel bottom by the subscripts s and b respectively, this equation may be written as

$$\int_0^y \frac{p}{\rho}\, dz = \frac{\partial}{\partial t}\int_0^y wz\, dz - (wz)_s \frac{\partial y}{\partial t} + (wz)_b \frac{\partial z_b}{\partial t}$$

$$+ \frac{\partial}{\partial x}\int_0^y uwz\, dz - (uwz)_s \frac{\partial h}{\partial x}$$

$$+ (uwz)_b \frac{\partial z_b}{\partial x} + (w^2 z)_s - (w^2 z)_b$$

$$+ \frac{1}{2}gy^2 - \int_0^y w^2\, dz + \frac{1}{\rho}[(pz)_s - (pz)_b] \qquad (12\text{-}38)$$

The flow conditions at the free surface and at the channel bottom are described by the following equations:

Free surface $(z = z_s)$:

i. $w_s = \frac{\partial y}{\partial t} + u_s \frac{\partial y}{\partial x}$ (This equation follows from Eq. 12-27);
ii. $p_s = 0$, since the pressure at the free surface is atmospheric.

Channel bottom $(z = z_b)$:

i. $w_b = 0$, since $w = 0$ at the channel bottom;
ii. $p_b z_b = 0$, since the channel bottom is used as datum.

Based on these conditions, Eq. 12-38 may be written as

$$\int_0^y \frac{p}{\rho}\, dz = \frac{\partial}{\partial t}\int_0^y wz\, dz + \frac{\partial}{\partial x}\int_0^y uwz\, dz + \frac{1}{2}gy^2 - \int_0^y w^2\, dz \qquad (12\text{-}39)$$

Momentum Equation in x-direction

The momentum equation in the x-direction may be written as

$$\frac{\partial u}{\partial t} + u\frac{\partial u}{\partial x} + w\frac{\partial u}{\partial z} = -\frac{1}{\rho}\frac{\partial p}{\partial x} \qquad (12\text{-}40)$$

Based on Eq. 12-24 and rearranging the terms, Eq. 12-40 may be written as

$$\frac{\partial u}{\partial t} + \frac{\partial u^2}{\partial x} + \frac{\partial(uw)}{\partial z} + \frac{1}{\rho}\frac{\partial p}{\partial x} = 0 \qquad (12\text{-}41)$$

By multiplying Eq. 12-41 throughout by dz, integrating from 0 to y, and applying the conditions at the free surface and at the bottom, we obtain

$$\frac{\partial(yu)}{\partial t} + \frac{\partial(yu^2)}{\partial x} + \int_0^y \frac{1}{\rho}\frac{\partial p}{\partial x}\, dz = 0 \tag{12-42}$$

By applying the Leibnitz rule, this equation becomes

$$\frac{\partial(yu)}{\partial t} + \frac{\partial(yu^2)}{\partial x} + \frac{\partial}{\partial x}\int_0^y \frac{p}{\rho}\, dz - (\frac{p}{\rho})_s\frac{\partial y}{\partial x} + (\frac{p}{\rho})_b\frac{\partial z_b}{\partial x} = 0 \tag{12-43}$$

As discussed in the previous paragraphs, $p_s = 0$ and $\partial z_b/\partial x = 0$. Substitution of these relationships and Eq. 12-39 into Eq. 12-43 and rearrangement of the terms of the resulting equation yield

$$\frac{\partial(yu)}{\partial t} + \frac{\partial}{\partial x}(yu^2 + \frac{1}{2}gy^2) + \frac{\partial}{\partial x}\left(\frac{\partial}{\partial t}\int_0^y wz\, dz + \frac{\partial}{\partial x}\int_0^y uwz\, dz \right.$$
$$\left. + \int_0^y w^2\, dz\right) = 0 \tag{12-44}$$

Let us simplify the last term of this equation. By substituting Eq. 12-32 into the first expression of the last term we obtain

$$\frac{\partial}{\partial t}\int_0^y wz\, dz = \frac{\partial}{\partial t}\int_0^y (-\frac{\partial u}{\partial x}z^2)\, dz$$

or

$$\frac{\partial}{\partial t}\int_0^y wz\, dz = \frac{\partial}{\partial t}\left(-\frac{\partial u}{\partial x}\frac{y^3}{3}\right) \tag{12-45}$$

Similarly

$$\frac{\partial}{\partial x}\int_0^y uwz\, dz = \frac{\partial}{\partial x}\left(-u\frac{\partial u}{\partial x}\frac{y^3}{3}\right) \tag{12-46}$$

and

$$\int_0^y w^2\, dz = \left(\frac{\partial u}{\partial x}\right)^2 \frac{y^3}{3} \tag{12-47}$$

By substituting Eq. 12-45 through 12-47 into the third term of Eq. 12-44, and expanding and simplifying the resulting expression, we get

$$\frac{\partial}{\partial x}\left(\frac{\partial}{\partial t}\int_0^y wz\, dz + \frac{\partial}{\partial x}\int_0^y uwz\, dz - \int_0^y w^2\, dz\right)$$
$$= -\frac{\partial}{\partial x}[y^2\frac{\partial u}{\partial x}(\frac{\partial y}{\partial t} + u\frac{\partial y}{\partial x})$$
$$+ \frac{y^3}{3}(\frac{\partial^2 u}{\partial x\partial t} + 2(\frac{\partial u}{\partial x})^2 + u\frac{\partial^2 u}{\partial x^2})] \tag{12-48}$$

Based on Eq. 12-29, this equation may be written as

$$\frac{\partial}{\partial x}\left(\frac{\partial}{\partial t}\int_0^y wz\,dz + \frac{\partial}{\partial x}\int_0^y uwz\,dz - \int_0^y w^2\,dz\right)$$

$$= -\frac{1}{3}\frac{\partial}{\partial x}\left\{y^3\left(\frac{\partial^2 u}{\partial x\partial t} + u\frac{\partial^2 u}{\partial x^2} - (\frac{\partial u}{\partial x})^2\right)\right\} \qquad (12\text{-}49)$$

Hence, the momentum equation in the x-direction is

$$\frac{\partial yu}{\partial t} + \frac{\partial}{\partial x}\left[yu^2 + \frac{1}{2}gy^2 - \frac{y^3}{3}\left(\frac{\partial^2 u}{\partial x\partial t}\right.\right.$$

$$\left.\left. +u\frac{\partial^2 u}{\partial x^2} - (\frac{\partial u}{\partial x})^2\right)\right] = 0 \qquad (12\text{-}50)$$

By expanding various terms of this equation, we obtain

$$\frac{\partial yu}{\partial t} + \frac{\partial}{\partial x}\left(yu^2 + \frac{1}{2}gy^2\right)$$

$$+ y^2\frac{\partial y}{\partial x}\left\{(\frac{\partial u}{\partial x})^2 - u\frac{\partial^2 u}{\partial x^2} - \frac{\partial^2 u}{\partial x\partial t}\right\}$$

$$+ \frac{y^3}{3}\left\{\frac{\partial u}{\partial x}\frac{\partial^2 u}{\partial x^2} - \frac{\partial^3 u}{\partial x^2\partial t} - u\frac{\partial^3 u}{\partial x^3}\right\} = 0 \qquad (12\text{-}51)$$

Usually, the terms containing the product of derivatives and the third space derivatives are neglected [Basco, 1983]. Therefore, the preceding equation simplifies to

$$\frac{\partial yu}{\partial t} + \frac{\partial}{\partial x}\left(yu^2 + \frac{1}{2}gy^2\right) - \frac{1}{3}\frac{\partial^3 u}{\partial x^2\partial t} = 0 \qquad (12\text{-}52)$$

Equations 12-28 and 12-52, referred to as the Boussinesq equations, describe flows if the pressure distribution is not hydrostatic. Note that if the additional terms accounting for the nonhydrostatic pressure distribution are neglected, these equations reduce to the St. Venant equations.

12-5 Integral Forms

In Section 12-2, we developed differential form of the governing equations. For the derivation of these equations, we had to assume that the flow variables and the parameters of the channel section are continuous both in space and time. This was needed for the derivatives of various variables to exist. In this section, we will derive the integral form of the equations for which we do not have to make this assumption. Thus, we can analyze flows having discontinuities such as bores and shocks.

By multiplying Eq. 12-2 by dt and integrating from t_1 to t_2, we obtain

$$\int_{x_1}^{x_2}(A_{t_2} - A_{t_1})\,dx + \int_{t_1}^{t_2}(Q_2 - Q_1)\,dt + \int_{t_1}^{t_2}\int_{x_1}^{x_2} q_l\,dxdt = 0 \qquad (12\text{-}53)$$

This is the continuity equation in the integral form.

Multiplying Eq. 12-7 by dt, integrating from t_1 to t_2, substituting expressions for $\sum F$ from Eq. 12-12 and noting that ρ is constant, we obtain

$$\int_{x_1}^{x_2} (Q_{t_2} - Q_{t_1})\, dx + \int_{t_1}^{t_2} Q_2 V_2\, dt - \int_{t_1}^{t_2} Q_1 V_1\, dt$$

$$- \int_{t_1}^{t_2} \int_{x_1}^{x_2} V_x q_l\, dx dt = g \int_{t_1}^{t_2} (A_1 \bar{y}_1 - A_2 \bar{y}_2)\, dt$$

$$+ g \int_{t_1}^{t_2} \int_{x_1}^{x_2} A(S_o - S_f)\, dx\, dt \qquad (12\text{-}54)$$

This is the momentum equation in the integral form.

12-6 Summary

In this chapter, we derived the continuity and momentum equations describing unsteady flow in open channels and presented different formulations of these equations. It was proved that these partial differential equations are hyperbolic. Then, these equations were derived for flows having nonhydrostatic pressure distribution based on the assumption that the vertical flow velocity varies from zero at the channel bottom to maximum at the water surface. The chapter concluded with the derivation of the integral form of the governing equations.

Problems

12-1 Figure 12-4 shows an infinitesimal length of a channel having a small bottom slope and no lateral inflow or outflow. By applying the law of conservation of mass between sections 1 and 2, derive the continuity equation for one-dimensional unsteady flow.

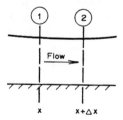

Fig. 12-4   Channel segment

12-2 By applying the Newton's second law of motion to the volume of water between sections 1 and 2 of Fig. 12-4, derive the momentum equation.

12-3 Deduce Eq. 5-5 from Eq. 12-17. Assume $\alpha = 1$.

12-4 If we use stage (elevation above a specified datum), Z, instead of flow depth y, show that the continuity and momentum equation for a prismatic channel become

$$\frac{\partial Z}{\partial t} + V\frac{\partial Z}{\partial x} + \frac{A}{B}\frac{\partial V}{\partial x} + VS_o = 0$$

$$\frac{\partial V}{\partial t} + V\frac{\partial V}{\partial x} + g\frac{\partial Z}{\partial x} + gS_f = 0$$

12-5 Starting with the momentum equation, show that the governing equation for steady, uniform flow is $S_f = S_o$.

12-6 If wind stress on the flow surface is included, prove that the continuity and momentum equations become

$$B\frac{\partial y}{\partial t} + \frac{\partial Q}{\partial x} - q_l = 0$$

$$\frac{\partial Q}{\partial t} + \frac{Q}{A}\frac{\partial Q}{\partial x} + Q\frac{\partial}{\partial x}\left(\frac{Q}{A}\right) + gA\frac{\partial y}{\partial x}$$
$$- gA(S_o - S_f) - q_l u - k_w BV_w^2 \cos\theta = 0$$

in which u = velocity component of the lateral flow in the positive x-direction; V_w = wind velocity, and k = dimensionless wind stress coefficient.

References

Abbott, M. B., 1979, *Computational Hydraulics: Elements of the Theory of Free Surface Flows*, Pitman, London.

Abbott, M. B., and Basco, D. R., 1990, *Computational Fluid Dynamics*, John Wiley, New York, NY.

Basco, D. R., 1983, *Computation of Rapidly Varied, Unsteady Free-Surface Flow*, U. S. Geological Survey Report WRI 83-4284.

Boussinesq, J., 1877, "Essais sur la theorie des eaux courantes," *Memoires presertes par divers Savants a l'Academie des Sciences de l'Institut de France*, vol. 23, pp. 1-680; vol. 24, 1-64.

Chow, V. T., 1959, *Open-Channel Hydraulics*, McGraw-Hill, New York, NY.

Cunge , J. A., Holly , F. M., and Verwey , A., 1980, *Practical Aspects of Computational River Hydraulics*, Pitman, London.

de Saint-Venant, B., 1871, "Theorie du mouvement non permanent des eaux, avec application aux crues de rivieras et a l'introduction des marces dans leur lit,' *Comptes Rendus de l'Academic des Sciences*, vol. 73, Paris, pp. 147-154, 237-240.

Dronkers, J. J., 1964, *Tidal Computations in Rivers and Coastal Waters*, North-Holland Publ., Amsterdam, Netherlands.

Henderson, F. M., 1966, *Open Channel Flow*, MacMillian, New York, NY.

Lai , C., 1986, "Numerical Modeling of Unsteady Open-Channel Flow," *Advances in Hydroscience*, vol. 14, pp. 161-333.

Liggett, J. A., 1975, "Basic Equations of Unsteady Flow," in *Unsteady Flow in Open Channels,* (Mahmood, K., and Yevjevich, V., eds.), vol. 1, Chap. 2, Water Resources Publications, Fort Collins, CO.

Stoker, J. J., 1957, *Water Waves,* Wiley (Interscience), New York, NY.

Strelkoff, T., 1969, "One-dimensional Equations of Open-channel Flow," *Jour. Hyd. Div.,* Amer. Soc. Civ. Engs., vol. 95, HY3, pp. 861-866.

Yen, B. C., 1973, "Open-channel Flow Equations Revisited," *Jour. Engineering Mechanics Div,* Amer. Soc. Civil Engrs., vol. 99, no. 5, pp. 979–1009.

13

NUMERICAL METHODS

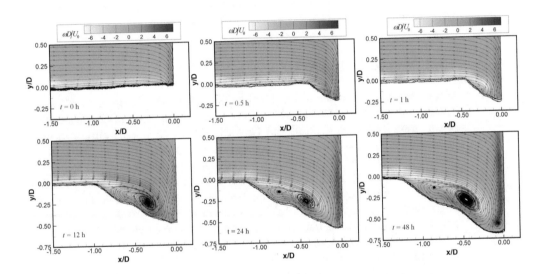

Average streamlines and vorticity distributions in a developing scour hole (Courtesy, D. Guan, Y. M. Chiew and M. Wei [2019])

© Springer Nature Switzerland AG 2022
M. H. Chaudhry, *Open-Channel Flow*,
https://doi.org/10.1007/978-3-030-96447-4_13

13-1 Introduction

In Section 12-3, we showed that the unsteady flow in open channels is described by a set of hyperbolic partial differential equations. These equations describe the conservation of mass and momentum in terms of the partial derivatives of dependent variables, flow velocity, V, and flow depth, y. However, for practical applications, we need to know the values of these variables instead of the values of their derivatives. Therefore, we integrate the governing equations. Because of the presence of nonlinear terms, a closed-form solution of these equations is not available, except for very simplified cases. Therefore, they are integrated numerically for which several numerical methods have been presented.

In this chapter, we introduce the method of characteristics and discuss necessary boundary and initial conditions for the numerical solution of governing equations. Various available numerical methods are presented and their advantages and disadvantages are briefly discussed.

13-2 Method of characteristics

Monge in 1789 developed a graphical procedure for the integration of partial differential equations. He called this procedure the method of characteristics. It was used by Massau [1889] and Craya [1946] for analyzing surges in open channels and it was utilized to investigate the propagation of flood waves [Isaacson et al., 1954] and other unsteady flow problems. Although it has become a standard method for the analysis of transients in closed conduits, its applications to open channels has become almost negligible. The concept of characteristic curves is helpful in understanding the propagation of waves and the development of boundary conditions for the explicit finite difference methods. We present a brief development of the method; readers interested in details should see Stoker [1957]; Abbott [1966, 1979]; and Lai [1986].

Let us rewrite Eqs. 12-6 and 12-17 for prismatic channels having no lateral inflow or outflow:

$$\frac{\partial y}{\partial t} + D_h \frac{\partial V}{\partial x} + V \frac{\partial y}{\partial x} = 0 \tag{13-1}$$

$$\frac{\partial V}{\partial t} + V \frac{\partial V}{\partial x} + g \frac{\partial y}{\partial x} = g(S_o - S_f) \tag{13-2}$$

in which V = flow velocity; y = flow depth; $D_h = A/B$ = hydraulic depth*; A = flow area; B = top water surface width; S_o = channel bottom slope; S_f = slope of the energy grade line; x = distance along the channel length; t = time; and g = acceleration due to gravity.

---

* Only in this section, we will use D_h for the hydraulic depth instead of D in order to differentiate between the symbols for the hydraulic depth and for the total derivatives.

By multiplying Eq. 13-1 by an unknown multiplier, λ, adding it to Eq. 13-2, and rearranging the terms of the resulting equation, we obtain

$$\left[\frac{\partial V}{\partial t} + (V + \lambda D)\frac{\partial V}{\partial x}\right] + \lambda\left[\frac{\partial y}{\partial t} + (V + \frac{g}{\lambda})\frac{\partial y}{\partial x}\right] = g(S_o - S_f) \qquad (13\text{-}3)$$

Since $V = V(x,t)$ and $y = y(x,t)$, the total derivatives

$$\frac{DV}{Dt} = \frac{\partial V}{\partial t} + \frac{\partial V}{\partial x}\frac{dx}{dt}$$
$$\frac{Dy}{Dt} = \frac{\partial y}{\partial t} + \frac{\partial y}{\partial x}\frac{dx}{dt} \qquad (13\text{-}4)$$

A comparison of Eqs. 13-3 and 13-4 shows that we can write the terms inside the brackets as total derivatives if we define the unknown multiplier λ such that

$$V + \lambda D_h = \frac{dx}{dt} = V + \frac{g}{\lambda} \qquad (13\text{-}5)$$

or

$$\lambda_{1,2} = \pm\sqrt{\frac{g}{D_h}} = \pm\sqrt{\frac{gB}{A}} \qquad (13\text{-}6)$$

In Chapter 12, we derived that the *celerity* of a gravity wave, $c = \sqrt{gA/B}$. Thus, by defining $\lambda_1 = g/c$ and utilizing the expressions of Eq. 13-4 for the total derivatives, we can write Eq. 13-3 as

$$\frac{DV}{Dt} + \frac{g}{c}\frac{Dy}{Dt} = g(S_o - S_f) \qquad (13\text{-}7)$$

if

$$\frac{dx}{dt} = V + c \qquad (13\text{-}8)$$

Similarly, by defining $\lambda_2 = -g/c$, we can write Eq. 13-3 as

$$\frac{DV}{Dt} - \frac{g}{c}\frac{Dy}{Dt} = g(S_o - S_f) \qquad (13\text{-}9)$$

if

$$\frac{dx}{dt} = V - c \qquad (13\text{-}10)$$

Note that Eq. 13-7 is valid if Eq. 13-8 is satisfied and Eq. 13-9 is valid if Eq. 13-10 is satisfied. Equation 13-8 plots as a curve in the x-t plane (Fig. 13-1). This curve is referred to as the *positive characteristic*, C^+. Similarly, Eq. 13-10 plots as the *negative characteristic, C.⁻* In other words, Eq. 13-7 is valid along the positive characteristic AP and Eq. 13-9 is valid along the negative characteristic BP. Equations 13-7 and 13-9 are called the *compatibility equations*. Thus, by these simple algebraic manipulations, we have eliminated the space variable, x, from the governing equations and converted them into

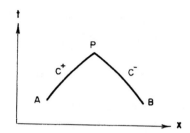

Fig. 13-1   Positive and negative characteristics

ordinary differential equations. However, we had to pay a price in transforming the partial differential equation into ordinary differential equations, i.e., Equations 13-1 and 13-2 are valid for any values of x and t; however, the transformed equations are valid only along the characteristics.

By multiplying Eqs. 13-7 and 13-9 by dt and integrating along the characteristics AP and BP, we obtain

$$\int_A^P dV + \int_A^P \frac{g}{c} \, dy = g \int_A^P (S_o - S_f) \, dt \qquad (13\text{-}11)$$

and

$$\int_B^P dV - \int_B^P \frac{g}{c} \, dy = g \int_B^P (S_o - S_f) \, dt \qquad (13\text{-}12)$$

To evaluate these integrals, c and S_f along the characteristics should be known, i. e., y and V along the characteristics should be known, since c and S_f are functions of y and V. However, y and V are the unknowns we want to determine. Therefore, we have to make some approximations to evaluate these integrals. For this purpose, let us assume that the values of c and S_f computed by using V and y at A and B are valid along the entire characteristics AP and BP, respectively. Based on this approximation, Eqs. 13-11 and 13-12 may be written as

$$V_P - V_A + \left(\frac{g}{c}\right)_A (y_P - y_A) = g(S_o - S_f)_A (t_P - t_A) \qquad (13\text{-}13)$$

and

$$V_P - V_B - \left(\frac{g}{c}\right)_B (y_P - y_B) = g(S_o - S_f)_B (t_P - t_B) \qquad (13\text{-}14)$$

in which the subscripts A, B, and P refer to the quantities at points A, B, and P, respectively. If we know the values of V and y at points A and B, then their values at point P may be determined by solving Eqs. 13-13 and 13-14 simultaneously. We may write these equations in a slightly different form by combining the known and unknown quantities as follows:

$$V_P = C_p - C_A \, y_P \qquad (13\text{-}15)$$

and

$$V_P = C_n + C_B \, y_P \tag{13-16}$$

in which we have combined the known quantities in the following two constants, C_p and C_n,

$$C_p = V_A + C_A y_A + g(S_o - S_f)_A (t_P - t_A)$$
$$C_n = V_B - C_B y_B + g(S_o - S_f)_B (t_P - t_B)$$
$$C = \frac{g}{c} \tag{13-17}$$

Note that C_p and C_n are constants during the time interval $t_P - t_A$ and $t_P - t_B$, respectively, although they may vary from one time interval to another.

Characteristics

In the previous paragraphs, we defined the characteristics as the curves that are plots of $dx/dt = V \pm c$. In this section, we discuss their physical significance.

As we mentioned in Chapter 3, a flow disturbance (depth and/or velocity) propagates in two directions if the flow is subcritical and only in the downstream direction if the flow is supercritical. The absolute velocity at which this disturbance travels is $V \pm c$. If we plot this propagation in the x-t plane assuming the disturbance is at point C at time $t = 0$ and distance $x = x_o$, then its influence will be felt in the shaded region shown in Fig. 13-2. This region is referred to as the *zone of influence*. The curve defining the propagation in the downstream direction is called as C^+ and that in the upstream direction as C^-. Any point outside the shaded region is not affected at all by the propagation of the disturbance.

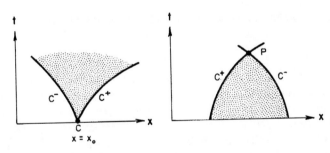

Fig. 13-2   Zones of influence and dependence

Now, let us discuss the conditions that affect the flow conditions at point P. To understand this, let us draw the characteristics in the backward direction through point P (Fig. 13-2). Then, only the conditions within the shaded region affect the flow conditions at point P. In other words, we may say that the flow conditions at point P depend upon the flow conditions in the shaded region. This is called the *zone of dependence*.

Depending upon the relative magnitude of the flow velocity and the celerity, a disturbance may or may not travel in the upstream direction. If the flow is subcritical, then the characteristic directions are both positive and negative. For critical flow, one of the characteristic direction is zero, whereas the second is positive. In supercritical flows, both characteristic directions are positive. Figure 13-3 shows the characteristics for different types of flow. Mathematically speaking [Cunge et al., 1980], the discontinuities in the first

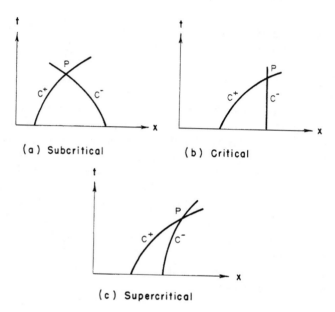

(a) Subcritical   (b) Critical

(c) Supercritical

Fig. 13-3   Characteristics for subcritical, critical, and supercritical flow

and higher derivatives of the dependent variables and the physical parameters that appear in the governing equations propagate along the characteristics. Thus the discontinuities in the slope of the water surface $\partial y/\partial x$ or in the velocity gradient $\partial V/\partial x$ propagate along the characteristics at the velocity of shallow-water waves. However, this is not the case for the discontinuities in the flow variables themselves, i. e., a bore does not propagate at the velocity of shallow-water waves.

Let us now reformulate Eqs. 13-7 (valid along the positive characteristics) and 13-9 (valid along the negative characteristics). By assuming B to be constant, we can write

$$\frac{dc}{dt} = \frac{d}{dt}\left(\frac{gA}{B}\right)^{\frac{1}{2}}$$
$$= \frac{1}{2}\frac{g}{c}\frac{dy}{dt} \tag{13-18}$$

Thus, Eqs. 13-7 and 13-9 may be written as

$$\frac{d}{dt}(V + 2c) = S \tag{13-19}$$

in which $S = g(S_o - S_f)$. This equation is valid along the positive characteristic, $dx/dt = V + c$. Similarly, Eq. 13-9, which is valid along the negative characteristic, may be written as

$$\frac{d}{dt}(V - 2c) = S \tag{13-20}$$

The term on the right-hand side of Eqs. 13-19 and 13-20 is referred to as the *source term*. The source term is zero if the channel is frictionless and the channel bottom slope is zero, or if the flow is uniform (i.e., $S_f = S_o$). If $S = 0$, then it follows from Eqs. 13-19 and 13-20 that

$$V + 2c = J^+ \tag{13-21}$$

and

$$V - 2c = J^- \tag{13-22}$$

The constants J^+ and J^- are called *Riemann invariants*. Note that their values may vary from one characteristic to another.

For a sloping prismatic channel, including the friction losses, integration of Eqs. 13-19 and 13-20 along the characteristics (Fig. 13-4) yields

$$[V + 2c]_1^2 = \int_{t_1}^{t_2} S\,dt \tag{13-23}$$

and

$$[V - 2c]_3^4 = \int_{t_3}^{t_4} S\,dt \tag{13-24}$$

If the right-hand sides of Eqs. 13-23 and 13-24 are nonzero but are sufficiently small, then the left-hand sides are called *Riemann quasi-invariants*. By a proper selection of t_1 and t_2, the right-hand sides may be made small [Abbott, 1979]. Riemann invariants in an analogous form cannot be derived for nonprismatic channels [Cunge et al., 1980] because of the additional terms introduced in the source term S.

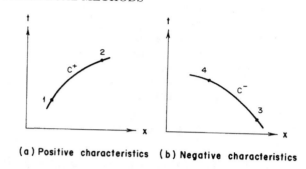

(a) Positive characteristics   (b) Negative characteristics

Fig. 13-4   Positive and negative characteristics

13-3 Initial and Boundary Conditions

In the analysis of unsteady flow in open channels, we usually start our calculations at a specified time. The flow conditions at this starting time are referred to as the *initial conditions*. Since the boundaries of all physical systems are located at finite distances, we have to specify in our calculations some particular conditions at the limits or boundaries of the physical system. These conditions are called the *boundary conditions*. In this section, we closely follow Cunge et al. [1980], in our discussion on how to specify these conditions.

Figure 13-5 shows the computational domain for a one-dimensional system. The computations start at time $t = t_o$; the upstream end of the system is at $x = x_o$; and the downstream end is at $x = x_1$. We call the upstream and downstream flow directions with reference to the initial steady-state flow directions. To facilitate understanding, let us consider the channel to be horizontal and frictionless. Thus the Riemann invariants are given by Eqs. 13-21 and 13-22.

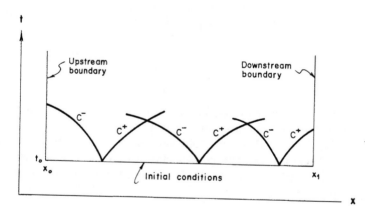

Fig. 13-5   Computational domain for a one-dimensional system

If we specify the initial steady-state flow depth and flow velocity at $t = t_o$, then the values of Riemann invariants are defined along the characteristics, since c is a function of the flow depth. We may independently specify either V or y anywhere on this characteristics, the other is then determined from the Riemann invariant. At any other point in the computational domain where the positive and negative characteristics intersect, V and y are determined from the values of J^+ and J^- for the characteristics passing through point P. Thus, we specify the *flow depth* and the *flow velocity* at the computational nodes for the initial conditions. At the boundaries, however, we specify the flow depth or the flow velocity, or a relationship between the flow velocity and flow depth (e.g., a rating curve). The latter condition should be independent from the compatibility conditions.

Figure 13-6 shows the characteristics at the limits of the computational domain for the subcritical and supercritical flows. The following rule [Cunge et al., 1980] may be followed for specifying the conditions at the boundaries (the initial conditions may be considered as a special case of the boundary condition in time): One condition has to be specified for each characteristic entering the computational domain at its limits. For example, two characteristics enter the computational domain at time $t = t_o$. Thus, two conditions have to be given as the initial conditions. At the upstream boundary, one condition is needed for subcritical flows and two conditions are needed for the supercritical flows. At the downstream boundary, we have to specify one condition for the subcritical flow and none for the supercritical flows. These additional conditions at the boundaries should be independent from the governing equations and of the Riemann invariants. In addition, if two conditions are specified, they have to be independent from one another. For example, we may not specify y and $\partial y / \partial x$ independently as the initial conditions. Similarly, we may not specify the flow variables y and V independently such that the continuity or momentum equations are not satisfied.

In the previous paragraphs, we discussed the zone of dependence and the zone of influence. By referring to them, we can see how the initial and boundary conditions affect the solution. For channels having low friction losses, the effect of initial conditions may persist for a long time. In such situations, these conditions should be specified accurately. Similarly, the effect of boundary conditions are carried through for some time. In certain situations, it may be difficult to define proper initial conditions – e.g., tidal flow in an estuary [Chaudhry, 2014]. In such cases, we start the computations with assumed initial conditions and compute the flow conditions for two or three tidal cycles. Usually, after two tidal cycles, conditions become periodic. However, to investigate a new system, the flow conditions between any two consecutive tidal cycles should be compared; the computations should be carried out for one more cycle if they are significantly different. However, they may be assumed correct if they are close.

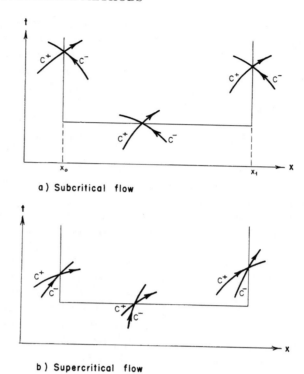

a) Subcritical flow

b) Supercritical flow

Fig. 13-6   **Characteristics at the limits of computational domain**
(After Cunge, et al. [1980])

13-4 Characteristic Grid Method

In Section 13-2, we derived equations for the characteristic method in which the time and space interval are specified. The method is, therefore, referred to as the *method of specified intervals*. To use specified intervals, interpolations become necessary either in time or in space, since all characteristics do not pass through the grid points. In this section, we integrate the characteristic form of the equations so that interpolations are not necessary, since we integrate along the characteristics and we do not use specified intervals for time and distance. Instead, the locations of the computational nodes after the initial conditions depend upon the computed values of V and y and are determined by the intersection of the characteristics.

Let us rewrite Eqs. 13-19 and 13-20 as well as the equations of characteristics along which they are valid. The following equation

$$\frac{d}{dt}(V + 2c) = g(S_o - S_f) \tag{13-25}$$

is valid if

$$\frac{dx}{dt} = V + c \tag{13-26}$$

Similarly, the equation

$$\frac{d}{dt}(V - 2c) = g(S_o - S_f) \tag{13-27}$$

is valid if

$$\frac{dx}{dt} = V - c \tag{13-28}$$

Let us assume that the flow depth (and hence c) and the flow velocity are known at points L and R (Fig. 13-7). Note that L and R may not be at the same time level. The characteristics passing through L and R intersect at point P. The location of this point is not known *a priori* but is determined from Eqs. 13-26 and 13-28. By multiplying Eqs. 13-25 to 13-28 by dt, integrating

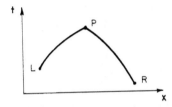

Fig. 13-7   Definition sketch

and applying the limits, we get

$$(V + 2c)_P = (V + 2c)_L + g \int_L^P (S_o - S_f)\, dt \tag{13-29}$$

$$x_P = x_L + \int_L^P (V + c)\, dt \tag{13-30}$$

$$(V - 2c)_P = (V - 2c)_R + g \int_R^P (S_o - S_f)\, dt \tag{13-31}$$

$$x_P = x_R + \int_R^P (V - c)\, dt \tag{13-32}$$

Note that we have not made any approximation in the integration of Eqs. 13-29 through 13-32. However, approximations become necessary while evaluating the integrals numerically, since we do not know the values of V, c, and S_f along the characteristics. By using the trapezoidal rule to evaluate the integrals, Eqs. 13-29 through 13-32 may be written as

$$x_P = x_L + (t_P - t_L)\left[\frac{1}{2}(V_P + c_P) + \frac{1}{2}(V_L + c_L)\right] \tag{13-33}$$

$$V_P = V_L + 2(c_L - c_P)$$
$$+ \frac{1}{2}g(t_P - t_L)\left((S_o - S_f)_P + (S_o - S_f)_L\right) \tag{13-34}$$

$$x_P = x_R + (t_P - t_R)\left[\frac{1}{2}(V_P - c_P) + \frac{1}{2}(V_R - c_R)\right] \tag{13-35}$$

$$V_P = V_R + 2(c_P - c_R)$$
$$+ \frac{1}{2}g(t_P - t_R)\left((S_o - S_f)_P + (S_o - S_f)_R\right) \tag{13-36}$$

Equations 13-33 through 13-36 are four nonlinear algebraic equations in four unknowns, namely, V_P, y_P, x_P, and t_P; the corresponding values of these variables and the coefficients at points L and R are known. These equations may be solved simultaneously to determine the values of these unknowns. By proceeding similarly, we determine V and y at later times.

At time $t = 0$, we compute the initial conditions at discrete grid points along the channel length; the distance between these points may not be equal. The distance and time where the conditions after the initial conditions will be computed are determined from Eqs. 13-33 through 13-36.

Note that we can use Eqs. 13-33 through 13-36 only at the interior points. At the upstream end, we cannot use Eqs. 13-33 and 13-34, since there is no grid point upstream of the upstream end. However, the distance at the upstream end is fixed, say $x = 0$. Still we need one more equation to have a unique solution for V, y, and t. This equation is provided by the condition imposed by the boundary. This condition might be in the form of specifying the flow depth, flow velocity, or a relationship between these two variables that is independent of the characteristic equations.

Similarly, at the downstream end, we cannot write Eqs. 13-35 and 13-36, since there is no grid point downstream of the downstream boundary. We may again specify the distance x (for example, $x = L$). The additional equation needed for a unique solution is again provided by specifying the flow depth, flow velocity, or a relationship between V and y.

By proceeding in a similar manner, we can compute the flow conditions in the channel for any desired length of time. If the conditions are needed at a specified location, they may be computed by interpolation from the values computed at the neighboring points.

For continuous flows, the procedure gives satisfactory results although it may be necessary to interpolate the channel parameters or the computed values. In addition, once a bore is formed in the solution, it is possible to have multiple values for the variables, and the procedure fails. If we plot the characteristics on the x-t plane as the computations progress, then the formation of the bore or steep-fronted waves is indicated by the convergence of the characteristics.

13-5 Method of Specified Intervals

In the method of specified intervals, we specify the size of the spatial and time grids [Lister, 1969; Wylie and Streeter, 1983; Lai, 1986; Chaudhry, 2014]. If we draw characteristics through a grid point, then they do not pass through the neighboring grid point. For example, the characteristics through point P do not pass through points A and B, instead they intersect at points R and S (Fig. 13-8). To compute conditions at point P, the conditions at points R and S should be known. These may be determined by interpolation from the known values at points A, B, and C as follows. We are using linear interpolations in the following derivation; higher-order interpolations may be utilized, if necessary, although they may not improve the results significantly.

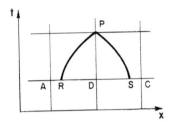

Fig. 13-8   Interpolation

Referring to Fig. 13-8, we may write

$$\frac{V_C - V_R}{V_C - V_A} = \frac{x_C - x_R}{x_C - x_A} = \frac{x_P - x_R}{x_C - x_A} = \frac{(V_R + c_R)\Delta t}{\Delta x} \tag{13-37}$$

Similarly, we may write

$$\frac{c_C - c_R}{c_C - c_A} = \frac{(V_R + c_R)\Delta t}{\Delta x} \tag{13-38}$$

Elimination of c_R from these two equations yields

$$V_R = \frac{V_C - \frac{\Delta t}{\Delta x}(c_A V_C - c_C V_A)}{1 + (V_C - V_A + c_C - c_A)\frac{\Delta t}{\Delta x}} \tag{13-39}$$

Now c_R and y_R may be determined from

$$c_R = \frac{c_C - V_R\frac{\Delta t}{\Delta x}(c_C - c_A)}{1 + \frac{\Delta t}{\Delta x}(c_C - c_A)}$$

$$y_R = y_C - \frac{\Delta t}{\Delta x}(V_R + c_R)(y_C - y_A) \tag{13-40}$$

For subcritical flow, point S lies between C and B. Proceeding similarly as before

$$V_S = \frac{V_C - \frac{\Delta t}{\Delta x}(c_B V_C - c_C V_B)}{1 - \frac{\Delta t}{\Delta x}(V_C - V_B - c_C + c_B)}$$

$$c_S = \frac{c_C + V_S \frac{\Delta t}{\Delta x}(c_C - c_B)}{1 + \frac{\Delta t}{\Delta x}(c_C - c_B)}$$

$$y_S = y_C + \frac{\Delta t}{\Delta x}(V_S - c_S)(y_C - y_B) \tag{13-41}$$

By using these interpolated values, y_P and V_P may be determined from Eqs. 13-15 and 13-16, where

$$C_p = V_R + \frac{g}{c_R}y_R + g(S_o - S_f)_R \Delta t$$

$$C_n = V_S - \frac{g}{c_S}y_S + g(S_o - S_f)_S \Delta t \tag{13-42}$$

13-6 Other Numerical Methods

In addition to the characteristic method, the following methods have been used for the numerical integration of hyperbolic partial differential equations:

1. Finite-difference methods;
2. Finite-element method;
3. Finite-volume method;
4. Spectral method;
5. Boundary-element method.

The method of characteristics, finite-difference methods, and finite-element method have been employed for the analysis of unsteady open-channel flow. We presented the details of the method of characteristics in the previous two sections. The finite-difference methods are discussed in the next two chapters. The finite-element method [Baker, 1983] has been used only to a limited extent for open-channel analysis [Katopodes, 1984]. It does not offer any significant advantage as compared to the other methods for one-dimensional flow problems, and several difficulties have to be overcome if a shock or bore is formed in the solution.

A number of problems arise in the application of the spectral method if the boundary conditions are nonperiodic [Canuto, et al., 1988]. As compared with the other methods, the boundary-element method [Liggett, 1984; Brebbia and Domingues, 1989] has not proven to be very successful for time dependent problems. We will not discuss these two methods. Finite volume methods have also been used to solve the Saint-Venant equations numerically and interested readers should see Crossley et al. [2003], and Ying et al. [2004].

In addition, some numerical methods have been presented for modeling the open-channel flow dynamics with emphasis on specific design problems. Litrico and Fromion [2004] proposed a scheme to design controllers with the classical automatic control techniques for water distribution in an open-channel system. They computed the frequency response of the Saint-Venant transfer matrix linearized around the stationary regimes, including backwater curves and reported the possibility of using this method to validate the finite-difference models in the frequency domain. Dulhoste et al. [2004] proposed a collocation model for the irrigation canals or for the dam-river systems for the nonlinear control of open-channel flow to control water levels along reaches. In their work, an orthogonal collocation method has been used, together with the Lagrange polynomials bases approximations.

13-7 Summary

In this chapter, we described the concept of characteristics and discussed the inclusion of initial and boundary conditions in the numerical computations. The characteristic method was introduced and its two different formulations were presented. These formulations are the method of specified intervals and the characteristic grid method. A brief introduction to the other available numerical methods for the integration of the shallow-water equations was presented.

Problems

13-1 A trapezoidal channel having a bottom width of 6.1 m and side slopes of 1.5H : 1V is carrying a flow of 126 m^3/s at a flow depth of 5.79 m. The bottom slope is 0.00008, Manning n = 0.013 and the channel length is 5 km. There is constant-level reservoir at the upstream end of the channel. A sluice gate at the downstream end is suddenly closed at time $t = 0$.

Compute the transient conditions until $t = 2000$ s by using the (a) grid of characteristics and (b) method of specified intervals.

Plot the computed flow depth in the channel at $t = 0$, 500, 1000, 1500 and 2000 s; and plot the variation of the flow depth with time at 1.5, 2.5, 3 and 5 km from the reservoir.

13-2 By using different values of $\Delta t/\Delta x$, investigate the effect of interpolation error on the computed wave height and wave shape. Use data of Problem 13-1.

13-3 Show that the method of characteristic grid fails when a shock or bore is formed. Use data for Problem 13-1 except let the gate close in 10 s and let the wave propagate in the channel.

References

Abbott, M. B., 1966, *An Introduction to the Method of Characteristics*, Thames and Hudson, London, and American Elsevier, New York, NY.

Abbott, M. B., 1975, "Method of Characteristics," Chapter 3 and "Weak Solution of the equations of Open Channel Flow," Chapter 7 of *Unsteady Open Channel Flow*, (Mahmood, K., and Yevjevich, V. eds.), Water Resources Publications, Fort Collins, CO.

Abbott, M. B., 1979, *Computational Hydraulics; Elements of the Theory of Free Surface Flows*, Pitman Publishing Ltd., London.

Abbott, M. B., and Verwey, A., 1970, "Four-Point Method of Characteristics," *Jour. Hyd. Div.*, Amer. Soc. Civ. Engrs., vol 96, Dec., pp. 2549-2564.

Amein, M., and Fang, C. S., 1970, "Implicit Flood Routing in Natural Channels," *Jour. Hyd. Div.*, Amer. Soc. Civ. Engrs., vol. 96, Dec., pp. 2481-2500.

Anderson, D. A., Tannehill, J. C., and Pletcher, R. H., 1984, *Computational Fluid Mechanics and Heat Transfer*, McGraw Hill, New York, NY.

Baker, J. A., 1983, *Finite-Element Computational Fluid Dynamics*, McGraw-Hill, New York, NY.

Brebbia, C. A., and Dominguez, J., 1989, *Boundary Elements, An Introductory Course*, Computational Mechanics Publications, London.

Canuto, C., Hussaini, M. Y., Quarteroni, A., and Zang, T. A., 1988, *Spectral Methods in Fluid Dynamics*, Springer-Verlag, New York, NY.

Chaudhry, M. H., 2014, *Applied Hydraulic Transients*, 3rd ed., Springer, New York, NY.

Craya, A., 1946, "Calcul graphique des regimes variables dans les canaux," *La Houille Blanche*, no. 1, Nov. 1945-Jan 1946, pp. 79-138, and no. 2, Mar 1946, pp. 117-130.

Crossley, A. J., Wright, N. G., and Whitlow, C. D., 2003, "Local Time Stepping for Modeling Open Channel Flows," *Jour. Hyd. Engineering*, Amer. Soc. Civ. Engrs., vol. 129, no. 6, pp.455-462.

Cunge, J., Holly, F. M., and Verwey, A., 1980, *Practical Aspects of Computational River Hydraulics*, Pitman, London.

Dulhoste, J.F., Georges, D., and Besancon, G, 2004, "Nonlinear Control of Open-Channel Water Flow based on Collocation Control Model," *Jour. Hyd. Engineering*, Amer. Soc. Civ. Engrs., vol. 130, no. 3, pp. 254-266.

Fread, D. L., and Harbaugh, T. E., 1973, "Transient Simulation of Breached Earth Dams," *Jour. Hyd. Div.*, Amer. Soc. Civil Engrs., Jan., pp. 139-154.

Guan, D., Chiew, Y.-M., Wei, M., and Hsieh, S.-C. ,2019, "Characterization of horseshoe vortex in a developing scour hole at a cylindrical bridge pier," *Int. J. Sediment Res.*, 34(2), 118-124.

Katopodes, N., 1984, "A Dissipative Galerkin Scheme for Open-Channel Flow," *Jour. Hydraulic Engineering*, Amer. Soc. Civ. Engrs., vol. 110, April, pp. 450-466.

Isaacson, E., Stoker, J. J., and Troesch, B. A., 1954, "Numerical Solution of Flood Prediction and River Regulation Problems (Ohio-Mississippi Floods)," *Report II*, Inst. Math. Sci. Rept. IMM-NYU-205, New York University.

Lai, C., 1986, "Numerical Modeling of Unsteady Open-Channel Flow," in *Advances in Hydroscience*, vol. 14, Academic Press, New York., pp. 161-333.

Lai, C., 1988, "Comprehensive Method of Characteristics Models for Flow Simulation," *Jour. Hydraulic Engineering*, Amer. Soc. Civil Engrs., vol. 114, no. 9, pp. 1074-1097.

Lax, P. D., 1954, "Weak Solutions of Nonlinear Hyperbolic Partial Differential Equations and Their Numerical Computation," *Communications on Pure and Applied Mathematics,"* vol. 7, pp. 159-163.

Leendertse, J. J., 1967, "Aspects of a Computational Model for Long Period Water-Wave Propagation," *Memo* RM-5294-PR, Rand Corporation, May.

Liggett, J. A., 1984, "The Boundary Element Method – Some Fluid Applications," in *Multi-Dimensional Fluid Transients,* (Chaudhry, M. H., and Martin, C. S. eds.), Amer. Soc. Mech. Engrs., Dec., New York, NY, pp. 1-8.

Litrico, X., and Fromion V., 2004, "Frequency Modeling of Open-Channel Flow," *Jour. Hyd. Engineering*, Amer. Soc. Civ. Engrs., vol. 130, no. 8, pp. 806-815.

Massau, J., 1889, "L'integration graphique and Appendice au memoire sur l'integration graphique," *Assoc. des Ingenieurs sortis des Ecoles Speciales de Gand*, Belgium, Annales, vol. 12, pp. 185-444.

Monge, G., "Graphical integration," *Ann. des Ing. Sorits des Ecoles de Grand,* 1789

Price, R. K., 1974, "Comparison of Four Numerical Flood Routing Methods," *Jour. Hyd. Div.*, Amer. Soc. Civ. Engrs., vol. 100, July, pp. 879-899.

Richtmyer, R. D., and Morton, K. W., 1967, *Difference Methods for Initial Value Problems,* 2nd. ed., Interscience, New York, NY.

Stoker, J. J., 1957, *Water Waves*, Interscience, New York, NY.

Strelkoff, T., 1970, "Numerical Solution of St. Venant Equations," *Jour. Hyd. Div.*, Amer. Soc. Civ. Engrs., vol. 96, January, pp. 223-252.

Terzidis, G., and Strelkoff, T., 1970, "Computation of Open Channel Surges and Shocks," *Jour. Hyd. Div.*, Amer. Soc. Civ. Engrs., vol. 96, Dec., pp. 2581-2610.

Ying, X., Khan, A. A., and Wang, S. S. Y., 2004, "Upwind Conservative Scheme for the Saint Venant Equations," *Jour. Hyd. Engineering*, Amer. Soc. Civ. Engrs., vol. 130, no. 10, pp. 977-987.

14

FINITE-DIFFERENCE METHODS

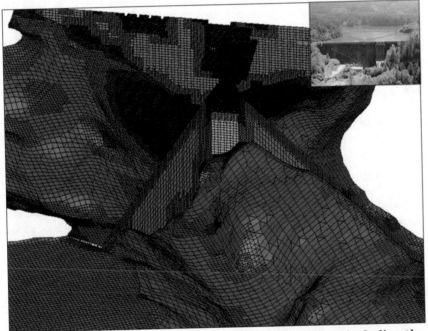

CFD Model Grid of Upper Baker Hydro Facility Forbay including the proposed floating Surface Collector structures (Courtesy, Puget Sound Energy, WA)

Supplementary Information The online version contains supplementary material available at (https://doi.org/10.1007/978-3-030-96447-4_14)

14-1 Introduction

We discussed in Chapter 12 that de Saint Venant equations are nonlinear partial differential equations for which a closed form solution is not available except for very simplified cases. In Chapter 13, we presented several numerical methods that may be used for their integration. Of these methods, the finite-difference methods have been utilized extensively; details of some of these methods are outlined in this chapter. Either a conservation or nonconservation form of the governing equations may be used in some methods whereas only one of these forms may be used in others. A conservation form should be preferred, since it conserves various quantities better and it simulates the celerity of wave propagation more accurately than the nonconservation form [Cunge et al., 1980; Miller and Chaudhry, 1989].

We first discuss a number of commonly used terms. Then, a number of explicit and implicit finite-difference methods are presented and the inclusion of boundary conditions in these methods is outlined. The consistency of a numerical scheme is briefly discussed and the stability conditions are then derived. The results computed by different schemes are compared.

14-2 Terminology

In this section, we briefly introduce a number of terms commonly used in finite difference applications.

Finite-difference approximations

To simplify presentation, let us first consider a function $f(x)$ of one independent variable, x. Let us assume that the value of this function, $f(x_o)$, at x_o is known. Then, by using a Taylor series expansion, the function $f(x_o + \Delta x)$ may be written as

$$f(x_o + \Delta x) = f(x_o) + \Delta x f'(x_o) + \frac{(\Delta x)^2}{2!} f''(x_o) + O(\Delta x)^3 \qquad (14\text{-}1)$$

in which a prime $'$ refers to derivative with respect to x, e. g., $f'(x_o) = dy/dx$ evaluated at $x = x_o$, and $O(\Delta x)^3$ indicates terms of third- or higher-order of Δx. Similarly $f(x_o - \Delta x)$ may be expanded as

$$f(x_o - \Delta x) = f(x_o) - \Delta x f'(x_o) + \frac{(\Delta x)^2}{2!} f''(x_o) + O(\Delta x)^3 \qquad (14\text{-}2)$$

Equation 14-1 may be written as

$$f(x_o + \Delta x) = f(x_o) + \Delta x f'(x_o) + O(\Delta x)^2 \qquad (14\text{-}3)$$

It follows from Eq. 14-3 that

$$\frac{df}{dx}\bigg|_{x=x_o} = \frac{f(x_o + \Delta x) - f(x_o)}{\Delta x} + O(\Delta x) \tag{14-4}$$

Similarly, it follows from Eq. 14-2 that

$$\frac{df}{dx}\bigg|_{x=x_o} = \frac{f(x_o) - f(x_o - \Delta x)}{\Delta x} + O(\Delta x) \tag{14-5}$$

Neglecting the $O(\Delta x)$ terms in Eqs. 14-4 and 14-5, we obtain

$$\frac{df}{dx}\bigg|_{x=x_o} = \frac{f(x_o + \Delta x) - f(x_o)}{\Delta x} \tag{14-6}$$

and

$$\frac{df}{dx}\bigg|_{x=x_o} = \frac{f(x_o) - f(x_o - \Delta x)}{\Delta x} \tag{14-7}$$

The finite-difference approximation of Eq. 14-6 is referred to as the *forward finite-difference* and that of Eq. 14-7 as the *backward finite difference*. Note that in both of these cases, the discarded terms are of the first order of Δx. Therefore, both forward and backward finite-difference approximations are referred to as *first-order accurate*.

Now, let us subtract Eq. 14-2 from Eq. 14-1, rearrange the terms and divide by Δx.

$$\frac{df}{dx}\bigg|_{x=x_o} = \frac{f(x_o + \Delta x) - f(x_o - \Delta x)}{2\Delta x} + O(\Delta x)^2 \tag{14-8}$$

By neglecting the last term, we obtain

$$\frac{df}{dx}\bigg|_{x=x_o} = \frac{f(x_o + \Delta x) - f(x_o - \Delta x)}{2\Delta x} \tag{14-9}$$

This approximation is referred to as the *central finite-difference approximation*. Note that the neglected term in this case is of the order of $(\Delta x)_j^2$; therefore, it is referred to as *second-order accurate*.

Figure 14-1 shows a geometrical representation of the forward, backward, and central finite-difference approximations. The forward finite-difference approximation replaces the slope of the tangent to the curve at B by the slope of line BC, the backward finite-difference approximation replaces this slope by the slope of line AB and the central finite difference approximation replaces it by the slope of the chord line AC. It is clear from this figure that the central finite-difference approximation is more accurate than the forward or backward finite-difference approximations. Now, let us consider finite-difference approximations for a partial derivative. Let us consider a function $f(x,t)$ with two independent variables: x and t. We may divide the x-t plane into a grid as shown in Fig. 14-2. The grid interval along the x-axis is Δx and the grid

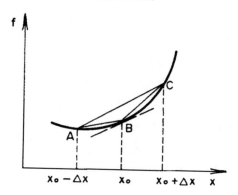

Fig. 14-1   **Finite-difference approximations**

interval along the t-axis is Δt. In this figure, we are assuming the grid size is uniform along each axis, although it is not necessary to do so. For brevity, we will call the $i\Delta x$ grid point i and the $(i+1)\Delta x$ grid point $i+1$. For the time axis, we will use k for $k\Delta t$ grid point and $k+1$ for the $(k+1)\Delta t$ grid point. To refer to different variables at these grid points, we will use the number of the spatial grid as a subscript and that of the time grid as a superscript. For example, the flow depth, y, at the ith spatial grid point and kth time grid point will be denoted by y_i^k. We will denote the known time level by superscript k and the unknown time level by $k+1$. By the known time level we mean that the values of different dependent variables are known at this time and we want to compute their values at the unknown time level. The known conditions may be the specified values as the *initial conditions* or they may have been computed during the previous time step. If the computations progress from one step to the next, then the procedure is referred to as a *marching procedure*. Most of the phenomenon described by hyperbolic partial differential equations are solved by using the marching procedures. The conditions specified at time $t = 0$ are referred to as the *initial conditions*. The conditions specified at the channel ends or at the interior boundaries are called the *boundary conditions*. The conditions at the channel ends are referred to as the *end conditions*.

The partial derivatives may be approximated in several different ways. The spatial partial derivatives replaced in terms of the variables at the known time level are referred to as the *explicit* finite differences whereas those in terms of the variables at the unknown time level are called *implicit* finite differences. Thus, referring to Fig. 14-2, if k is the known time level and $k + 1$ is the unknown time level, then some typical finite difference approximations for the spatial partial derivative, $\partial f/\partial x$, at the grid point (i, k) are as follows:

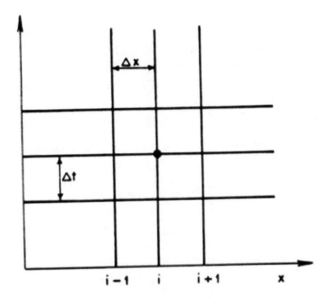

Fig. 14-2   Finite-difference grid

Explicit finite differences

Backward:
$$\frac{\partial f}{\partial x} = \frac{f_i^k - f_{i-1}^k}{\Delta x}$$

Forward:
$$\frac{\partial f}{\partial x} = \frac{f_{i+1}^k - f_i^k}{\Delta x}$$

Central:
$$\frac{\partial f}{\partial x} = \frac{f_{i+1}^k - f_{i-1}^k}{2\Delta x} \qquad (14\text{-}10)$$

Implicit finite differences

Backward:
$$\frac{\partial f}{\partial x} = \frac{f_i^{k+1} - f_{i-1}^{k+1}}{\Delta x}$$

Forward:
$$\frac{\partial f}{\partial x} = \frac{f_{i+1}^{k+1} - f_i^{k+1}}{\Delta x}$$

Central:
$$\frac{\partial f}{\partial x} = \frac{f_{i+1}^{k+1} - f_{i-1}^{k+1}}{2\Delta x} \qquad (14\text{-}11)$$

Many other finite-difference approximations are possible if the values at three or more grid points, instead of just two grid points, are used as in Eqs. 14-10 and 14-11.

14-3 Explicit Finite-Difference Schemes

Several explicit finite-difference schemes have been proposed for the solution of hyperbolic partial differential equations In this section, we will present a number of typical schemes that have been employed in hydraulic engineering.

Unstable scheme

To solve de Saint Venant equations, we may select the following finite-difference approximations:

$$\frac{\partial f}{\partial x} = \frac{f_{i+1}^k - f_{i-1}^k}{2\Delta x}$$

$$\frac{\partial f}{\partial t} = \frac{f_i^{k+1} - f_i^k}{\Delta t} \tag{14-12}$$

in which, for brevity, f refers to both dependent variables y and V.

This finite-difference scheme is inherently unstable; i.e., computations become unstable irrespective of the size of grid spacing. This may be proved analytically following the stability analysis procedure presented in Section 14-6.

French [1985] used this scheme to solve the shallow-water equations. The friction losses were increased by a large amount to make the scheme stable. However, because of arbitrarily increase in the head losses, the accuracy of the computed results may be questionable.

Diffusive scheme

Lax presented [1954] a slight variation of the unstable scheme discussed in the previous section. This scheme is one of the simplest to program and yields satisfactory results [Chaudhry, 2014] for typical engineering applications.

General formulation

In this scheme, the partial derivatives and other variables are approximated as follows:

$$\frac{\partial f}{\partial x} = \frac{f_{i+1}^k - f_{i-1}^k}{2\Delta x}$$

$$\frac{\partial f}{\partial t} = \frac{f_i^{k+1} - f^*}{\Delta t}$$

$$f^* = \frac{1}{2}(f_{i-1}^k + f_{i+1}^k)$$

$$D^* = \frac{1}{2}(D_{i-1}^k + D_{i+1}^k)$$

$$S_f^* = \frac{1}{2}(S_{f_{i-1}}^k + S_{f_{i+1}}^k) \tag{14-13}$$

in which, for brevity, we are using f for both dependent variables, y and V. We use these approximations in the conservation and nonconservation forms of the governing equations as follows.

Nonconservation form

Substitution of these expressions for the partial derivatives and for the coefficients of Eqs. 12-6 and 12-17 and simplification of the resulting equations yield

$$y_i^{k+1} = \frac{1}{2}(y_{i-1}^k + y_{i+1}^k) - \frac{1}{2}\frac{\Delta t}{\Delta x}D_i^*(V_{i+1}^k - V_{i-1}^k)$$
$$- \frac{1}{2}\frac{\Delta t}{\Delta x}V_i^*(y_{i+1}^k - y_{i-1}^k)$$
$$V_i^{k+1} = \frac{1}{2}(V_{i-1}^k + V_{i+1}^k) - \frac{1}{2}\frac{\Delta t}{\Delta x}g(y_{i+1}^k - y_{i-1}^k)$$
$$- \frac{1}{2}\frac{\Delta t}{\Delta x}V_i^*(V_{i+1}^k - V_{i-1}^k) + g\Delta t(S_o - S_f^*) \qquad (14\text{-}14)$$

Conservation form

The conservation form of the governing equations in the matrix form may be written as
$$\mathbf{U}_t + \mathbf{F}_x + \mathbf{S} = 0 \qquad (14\text{-}15)$$

in which

$$\mathbf{U} = \begin{pmatrix} A \\ VA \end{pmatrix}; \quad \mathbf{F} = \begin{pmatrix} VA \\ V^2A + gA\bar{y} \end{pmatrix}; \quad \mathbf{S} = \begin{pmatrix} 0 \\ -gA(S_o - S_f) \end{pmatrix} \qquad (14\text{-}16)$$

and $A\bar{y}$ = moment of flow area about the free surface. This moment may be computed from $\int_0^{y(x)} [y(x) - \eta]\,\sigma(\eta)\,d\eta$ in which σ is the water-surface width at depth η. Substitution of the finite-difference approximations of Eq. 14-13 into Eq. 14-15 yields

$$\mathbf{U}_i^{k+1} = \frac{1}{2}(\mathbf{U}_{i+1}^k + \mathbf{U}_{i-1}^k) - \frac{1}{2}\frac{\Delta t}{\Delta x}(\mathbf{F}_{i+1}^k - \mathbf{F}_{i-1}^k) - \mathbf{S}^*\Delta t \qquad (14\text{-}17)$$

Once the values of A and VA have been determined at the $(k+1)$ time level, we determine the values of variables of interest, y and V, and then proceed to the next time step.

Boundary Conditions

Equations 14-14 or 14-17 may be used at the interior grid points to compute the unsteady flow depth and flow velocity. At the boundaries, however, we cannot use these equations, since there is no grid point outside the flow domain. Therefore, different procedures have been proposed (for details, see Liggett and Cunge, 1975) to include the boundaries in the computations. For one-dimensional flows, the method of specified interval presented in Section 13-5 gives acceptable results and is employed here.

In this procedure, we solve the positive characteristic equation (Eq. 13-15) simultaneously with the condition imposed by the boundary for the downstream-end condition and the negative characteristic equation (Eq. 13-16) with the upstream-end condition for the upstream boundary. The end condition may specify time variation of flow depth, flow velocity (or discharge), or a combination of these variables. For example, the flow depth remains constant for a constant-level reservoir; flow velocity is always zero for a dead-end or completely closed gate; and a relationship between the flow depth and discharge is specified for a rating curve. Similarly, discharge through a partially open gate is a function of the flow depth upstream of the gate.

For an intermediate boundary, e.g., a junction of two channels, we solve the positive and negative characteristic equations simultaneously with the continuity and energy equations for the junction. For example, for a junction of channels i and $i + 1$ (Fig. 14-3), we have two nodes at the junction. One of these nodes is the last node on channel i and the second node is the first node on channel $i + 1$. We consider these nodes to be close to each other, so that the channel length between them may be neglected. However, they are separate nodes in the sense that they may have different flow depths and flow velocities, and the head losses at the junction are assumed to be concentrated between these two nodes. For such a boundary, we need four equations, since we have two unknowns (flow depth and flow velocity) for each computational node at the boundary. We use the positive characteristic equation for the last node on channel i and the negative characteristic equation for the first node on channel $i + 1$. The remaining two equations are the continuity and energy equations between the two nodes at the channel junction. For a junction of more than two channels, we utilize the positive characteristics equations for the channels upstream of the junction and the negative characteristics for the channels downstream of the junction. The additional equations are provided by the continuity and energy equations between channels nodes at the junction.

Referring to Fig. 14-3, the following equations describe the conditions imposed by different boundaries. These equations are solved simultaneously with the positive and/or negative characteristic equations to determine the flow conditions at the boundary nodes.

Upstream reservoir

$$y_i^{k+1} = y_{res} - (1 + K_e)\frac{(V_{i,1}^{k+1})^2}{2g}$$

Downstream reservoir

$$y_{i,n+1}^{k+1} = y_{res} - (1 - K_d)\frac{(V_{i,n+1}^{k+1})^2}{2g}$$

Sluice gate

$$Q_{i,n+1}^{k+1} = C_d A_g \sqrt{2g y_{i,n+1}^{k+1}}$$

Rating curve

$$Q = f(y_{i,n+1}^{k+1})$$

Series junction

$$V_{i,n+1}^{k+1} A_{i,n+1}^{k+1} = V_{i+1,1}^{k+1} A_{i+1,1}^{k+1}$$

$$z_i + y_{i,n+1}^{k+1} + \frac{(V_{i,n+1}^{k+1})^2}{2g} = z_{i+1} + y_{i+1,1}^{k+1} + (1 + K)\frac{(V_{i+1,1}^{k+1})^2}{2g}$$

$$(14\text{-}18)$$

in which K_e, K_d, and K are the head-loss coefficients at the entrance, at the exit, and at the junction, respectively; y_{res} = reservoir depth at the channel entrance; Q = rate of discharge; C_d = discharge coefficient for the sluice gate; A_g = area of the gate opening; and z = elevation of the channel bottom above a specified datum.

For completeness, we have included the velocity head and the head-loss terms in Eq. 14-18. These terms are usually very small and may be neglected. If these terms are neglected, then these conditions are simplified considerably, i.e., the flow depth in the reservoir is equal to that in the channel at the channel entrance and at the exit, and the flow depths in the channels at the junction are equal to each other. In addition, this simplifies the algorithm in case there is a possibility of flow reversal.

Stability

For the stability* of the scheme, it is necessary that the Courant number, C_n, is less than or equal to 1, where

$$C_n = \frac{\text{Actual wave velocity}}{\text{Numerical wave velocity}} = \frac{|V| \pm c}{\Delta x / \Delta t} \qquad (14\text{-}19)$$

Thus, the computational time interval depends upon the spatial grid spacing, flow velocity, and celerity, which are functions of the flow depth. Since the flow

---

* We present a detailed discussion in Section 14-6.

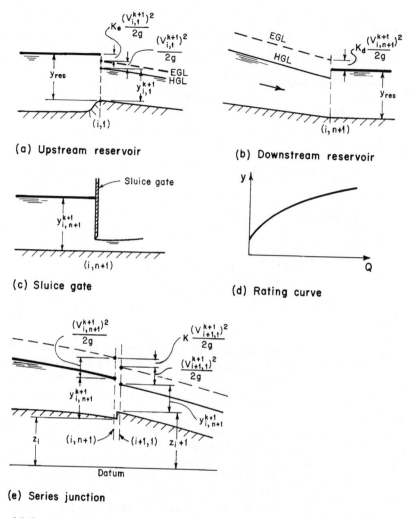

(a) Upstream reservoir

(b) Downstream reservoir

(c) Sluice gate

(d) Rating curve

(e) Series junction

Fig. 14-3 Typical boundaries

depth and the flow velocity may change significantly during the computations, it may be necessary to reduce the size of computational time interval for stability. The time interval should be such that C_n is as close to 1 as possible. If it is significantly less than unity, then the interval size should be increased to improve accuracy and to prevent the smearing of bores and steep waves.

MacCormack Scheme

The MacCormack scheme is an explicit, two-step predictor-corrector scheme [MacCormacK, 1969; Anderson et al., 1984] that is second-order accurate in space and time and is capable of capturing the shocks without isolating them. This scheme has been applied for analyzing one-dimensional, unsteady, open-channel flows by Fennema and Chaudhry [1986, 1987] and Dammuller, Bhallamudi, and Chaudhry [1989].

General formulation

For one-dimensional flow, two alternatives of this scheme are possible. In one alternative, backward finite-differences are used to approximate the spatial partial derivatives in the predictor part and forward finite-differences are utilized in the corrector part. The values of variables determined during the predictor part are used during the corrector part. In the second alternative, forward finite-differences are used in the predictor part and backward finite-difference are used in the corrector part. A general recommended procedure is to alternate the direction of differencing from one time step to the next; i.e., use alternative 1 during one time step, alternative 2 during the next time step, and alternate this sequence thereafter. Recent investigations show that the computed results are better if the direction of differencing in the predictor step is the same as that of the propagation of the wave front in the system.

   The finite-difference approximations for the *first alternative* of this scheme are given in the following paragraphs; equations for the second alternative of the scheme may be written similarly by reversing the direction of the spatial finite-difference approximations, as discussed in the previous paragraphs.

Predictor

$$\frac{\partial \mathbf{U}}{\partial \mathbf{t}} = \frac{\mathbf{U}_i^* - \mathbf{U}_i^k}{\Delta t}$$
$$\frac{\partial \mathbf{F}}{\partial \mathbf{x}} = \frac{\mathbf{F}_i^k - \mathbf{F}_{i-1}^k}{\Delta x} \tag{14-20}$$

in which notation of Fig. 14-2 is used and the superscript* refers to the variables computed during the predictor part. Substitution of these finite differences into Eq. 14-15 and simplification of the resulting equation yield

$$\mathbf{U}_i^* = \mathbf{U}_i^k - \frac{\Delta t}{\Delta x} \left(\mathbf{F}_i^k - \mathbf{F}_{i-1}^k \right) - \mathbf{S}_i^k \Delta t \tag{14-21}$$

The computed value of $\mathbf{U}_i^*$ gives A^* and Q^*, from which we determine the values of V^* and y^*. We compute these for all the computational nodes. These values are then used in the corrector part to compute $\mathbf{F}^*$ and $\mathbf{S}^*$.

Corrector

$$\frac{\partial \mathbf{U}}{\partial \mathbf{t}} = \frac{\mathbf{U}_i^{**} - \mathbf{U}_i^k}{\Delta t}$$

$$\frac{\partial \mathbf{F}}{\partial \mathbf{x}} = \frac{\mathbf{F}_{i+1}^* - \mathbf{F}_i^*}{\Delta x} \tag{14-22}$$

Substituting these finite differences and $\mathbf{S} = \mathbf{S}_i^*$ into Eq. 14-15, we obtain

$$\mathbf{U}_i^{**} = \mathbf{U}_i^k - \frac{\Delta t}{\Delta x}(\mathbf{F}_{i+1}^* - \mathbf{F}_i^*) - \mathbf{S}_i^* \Delta t \tag{14-23}$$

in which the superscript $**$ refers to the values of the variables after the corrector step. The value of $\mathbf{U}_i$ at the unknown time level $k + 1$ is given by

$$\mathbf{U}_i^{k+1} = \frac{1}{2}(\mathbf{U}_i^* + \mathbf{U}_i^{**}) \tag{14-24}$$

Boundary conditions

The preceding equations are for the interior nodes. The boundary nodes may be included in the analysis in the same manner as that we discussed for the Lax scheme. The simulation of the boundary nodes is, therefore, first-order accurate. Whether this first-order simulation of boundaries in an otherwise second-order scheme affects the overall accuracy of computed results is controversial. MacCormack heuristically showed that if the order of accuracy of the end conditions is one less than that of the interior nodes, then the overall accuracy of the computed results is not impaired. However, the validity of this statement is considered questionable.

Stability

The MacCormack scheme is stable if $C_n \leq 1$. For the computations to be stable, this condition must be satisfied at each grid point during every computational interval.

Artificial viscosity

The solution obtained by a finite-difference scheme has dissipative errors if the leading term of the truncation error in the scheme has even derivatives, and the solution has dispersive errors if the leading term has odd derivatives. The dispersive errors usually produce oscillations in the computed results in the vicinity of steep wave fronts. These oscillations are purely due to numerical errors and have nothing to do with the physical phenomenon being simulated. To smooth these oscillations, artificial viscosity is added to the scheme, for which several procedures have been reported. In this section, we summarize a procedure presented by Jameson, Schmidt, and Turkel [1981]. This procedure smoothes the oscillations where large gradients are present; however, it leaves the relatively smooth areas undisturbed.

To apply this procedure to open-channel flows, we first compute the following parameter based on the normalized form of the computed water-surface gradients:

$$\nu_i = \frac{|y_{i+1} - 2y_i + y_{i-1}|}{|y_{i+1}| + 2|y_i| + |y_{i-1}|}$$

$$\nu_{i+\frac{1}{2}} = \kappa \max(\nu_{i+1}, \nu_i) \tag{14-25}$$

in which κ is used to regulate the amount of artificial viscosity and a suitable value may be selected for a particular application by trial and error. At the grid points where y_{i+1} and y_{i-1} do not exist, we use the following expressions instead of that of Eq. 14-25:

$$\nu_i = \frac{|y_i - y_{i-1}|}{|y_i| + |y_{i-1}|}$$

$$\nu_i = \frac{|y_{i+1} - y_i|}{|y_{i+1}| + |y_i|} \tag{14-26}$$

The computed dependent variable f is then modified as

$$f_i^{k+1} = f_i^{k+1} + \nu_{i+\frac{1}{2}}(f_{i+1}^{k+1} - f_i^{k+1}) - \nu_{i-\frac{1}{2}}(f_i^{k+1} - f_{i-1}^{k+1}) \tag{14-27}$$

in which f refers to both y and V. This equation should be considered as a replacement statement.

14-4 Implicit Finite-Difference Schemes

In the implicit finite-difference schemes, the spatial partial derivatives and/or the coefficients are replaced in terms of the values at the unknown time level. The unknown variables, therefore, appear implicitly in the algebraic equations and the methods are called *implicit* methods. The algebraic equations for the entire system have to be solved simultaneously in these methods.

Several implicit finite-difference schemes have been used for the analysis of unsteady open-channel flows [Amein and Fang, 1970; Strelkoff, 1970; Terzidis and Strelkoff, 1970; Liggett and Cunge, 1975; Abbott, 1979; and Cunge, Holly and Verwey, 1980].   We present details of three of these schemes.

Preissmann Scheme

The Preissmann scheme has been extensively used since the early 1960s [Preissmann and Cunge, 1961; Liggett and Cunge, 1975; and Cunge et al., 1980].   It has the advantages that a variable spatial grid may be used; steep wave fronts may be properly simulated by varying the weighting coefficient; and the scheme yields exact solution of linearized form of the governing equations for a particular value of Δx and Δt.

General formulation

The partial derivatives and other coefficients are approximated as follows:

$$\frac{\partial f}{\partial t} = \frac{(f_i^{k+1} + f_{i+1}^{k+1}) - (f_i^k + f_{i+1}^k)}{2\Delta t}$$

$$\frac{\partial f}{\partial x} = \frac{\alpha(f_{i+1}^{k+1} - f_i^{k+1})}{\Delta x} + \frac{(1-\alpha)(f_{i+1}^k - f_i^k)}{\Delta x}$$

$$f = \frac{1}{2}\alpha(f_{i+1}^{k+1} + f_i^{k+1}) + \frac{1}{2}(1-\alpha)(f_{i+1}^k + f_i^k) \qquad (14\text{-}28)$$

in which α is a weighting coefficient; f refers to both V and y in the partial derivatives, and f stands for S_f, and V as a coefficient. By selecting a suitable value for α, the scheme may be made totally explicit ($\alpha = 0$) or totally implicit ($\alpha = 1$). The scheme is stable if $.55 < \alpha \leq 1$. Steep wave fronts are properly simulated for low values of α, but there are oscillations behind the wave front. These oscillations are eliminated for α close to unity; however, steep wave fronts are somewhat smeared (Fig. 14-4). For typical applications, α value between 0.6 and 0.7 may be used.

In a more general form of the Preissmann scheme [1961], known as the four-point scheme, the dependent variables as well as the time and space derivatives are weighted both in space and time computational grids [e.g., Evans, 1977, Lyn and Goodwin, 1987, Samuels and Skeels, 1990, Meselhe and Holly, 1997, and Venutelli, 2002].

By substituting the above finite-difference approximations and the coefficients into Eqs. 14-15, and rearranging the terms of the resulting equation, we obtain

$$\mathbf{U}_i^{k+1} + \mathbf{U}_{i+1}^{k+1} + 2\frac{\Delta t}{\Delta x}[\alpha(\mathbf{F}_{i+1}^{k+1} - \mathbf{F}_i^{k+1}) + (1-\alpha)(\mathbf{F}_{i+1}^k - \mathbf{F}_i^k)]$$

$$+ \Delta t[\alpha(\mathbf{S}_i^{k+1} + \mathbf{S}_{i+1}^{k+1}) + (1-\alpha)(\mathbf{S}_{i+1}^k + \mathbf{S}_i^k)] = \mathbf{U}_i^k + \mathbf{U}_{i+1}^k$$

$$(14\text{-}29)$$

In Eqs. 14-29, we have four unknowns, namely, $V_i^{k+1}, y_i^{k+1}, V_{i+1}^{k+1}$, and y_{i+1}^{k+1}. If we write these two equations for each grid point, we have $2N$ equations ($N =$ number of reaches on the channel). We cannot write these equations for the downstream end. However, we have $2(N+1)$ unknowns, i.e., two unknowns for each grid point. Thus, for a unique solution we need two more equations. These are provided by the *end conditions,* as discussed in the following paragraphs.

Boundary Conditions

Unlike the explicit schemes, we include directly in the system of equations the equations describing the end conditions (equations describing a number of typical boundaries are listed in Eq. 14-18). In other words, we do not have

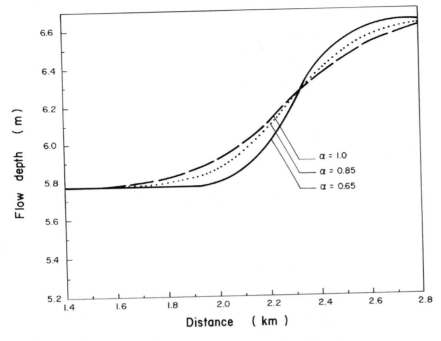

Fig. 14-4   Effect of the variation of α on wave front

to use the characteristic equations or the reflection procedures. This is one of the main advantage of the implicit schemes. For example, for an upstream constant-level reservoir, we include in the system of governing equations an equation of the form $y_1^{k+1} = y_{res}$ (entrance losses and the velocity head are neglected in this equation). A similar equation may be written for a downstream reservoir. If a rating curve is specified at the downstream end, then an equation relating the flow depth and the rate of discharge is input in the system of equations. Similarly, other boundary conditions may be incorporated.

Stability

The scheme is unconditionally stable provided $\alpha > 0.5$, i.e., the flow variables are weighted towards the $k + 1$ time level. An unconditional stability means that there is no restriction on the size of Δx and Δt for stability. However, for accuracy C_n should be close to 1. The preceding unconditional stability criterion is derived assuming a frictionless channel. If a linearized form of the friction-loss term is included in the analysis [Samuels and Skeels, 1990] then Vedernikov number, $\mathbf{V}_n \leq 1$, for the scheme to be stable, where

$$\mathbf{V}_n = \frac{p A \mathbf{F}_r}{m R} \frac{dR}{dA} \qquad (14\text{-}30)$$

The exponents m and p are as defined in Eq. 12-10e and $\mathbf{F}_r$ = Froude number. Through numerical experimentation, Evans [1977] and Yen and Lin [1986] investigated the stability of this scheme for full St. Venant equations for specific flow parameters. These studies have the advantage that they are not based on linearized analysis; however, they are not general, since the conclusions are based on a specific range of parameters.

Solution procedure

Expansion of Eq. 14-29 yields

$$
\begin{aligned}
A_i^{k+1} + A_{i+1}^{k+1} + 2\frac{\Delta t}{\Delta x}\Big\{\alpha\big[(VA)_{i+1}^{k+1} - (VA)_i^{k+1}\big] \\
+ (1-\alpha)\big[(VA)_{i+1}^k - (VA)_i^k\big]\Big\} = A_i^k + A_{i+1}^k
\end{aligned}
\tag{14-31}
$$

$$
\begin{aligned}
(VA)_i^{k+1} + (VA)_{i+1}^{k+1} + 2\frac{\Delta t}{\Delta x}\Big\{\alpha\big[(V^2A)_{i+1}^{k+1} - (V^2A + gA\bar{y})_i^{k+1}\big]\Big\} \\
- g\Delta t\Big\{\alpha\big[A_{i+1}^{k+1}(S_o - S_f)_{i+1}^{k+1} + A_i^{k+1}(S_o - S_f)_i^{k+1}\big]\Big\} \\
= g\Delta t\Big\{(1-\alpha)\big[A_{i+1}^k(S_o - S_f)_{i+1}^k + A_i^k(S_o - S_f)_i^k\big]\Big\} \\
+ (VA)_i^k + (VA)_{i+1}^k - (1-\alpha)\big[(V^2A + gA\bar{y})_{i+1}^k \\
- (V^2A + gA\bar{y})_i^k\big]
\end{aligned}
\tag{14-32}
$$

The system of equations (Eqs. 14-31 and 14-32) for each node and the boundary conditions) are a set of nonlinear algebraic equations. These may be solved by an iterative technique. The solution by the Newton-Raphson method is discussed in this section; readers interested in the solution by double sweep method, should see Cunge, Holly and Verwey [1980].

In the Newton-Raphson method, we estimate values for the unknown variables V and y at each node and then iterate to refine the solution as we discussed in Chapter 6. To determine the corrections for each iteration, we need the partial derivatives of Eqs. 14-31 and 14-32 and of the boundary conditions with respect to $y_i^{k+1}, y_{i+1}^{k+1}, V_i^{k+1}$, and V_{i+1}^{k+1}. If we derive these partial derivatives and arrange the governing equations, we obtain

$$
\mathbf{Ax} = \mathbf{b}
\tag{14-33}
$$

in which $\mathbf{A}$ is a matrix comprising of the partial derivatives and $\mathbf{x}$ is a column vector comprising the corrections Δy_i and ΔV_i ($i = 1, 2, \cdots, N+1$). Thus, we have $2(n+1)$ equations in $2(n+1)$ unknowns. These unknowns are the corrections. Now, we check that $\sum_i^{N+1} |\Delta y_i| + |\Delta V_i| \le \epsilon$, where ϵ is the specified tolerance. If the sum of the corrections is less than the specified tolerance, then we proceed to the next step after applying the correction. Otherwise, we apply the correction and iterate the procedure.

Note that matrix $\mathbf{A}$ is banded with a bandwidth of 4. We may utilize this fact while solving Eq. 14-33, since a banded-matrix solution routine requires less storage and gives more accurate results. For branching or parallel systems, the nodes may be numbered [Chaudhry, 2014] such that the resulting matrix is banded.

Beam and Warming scheme

Several different formulations of the Beam and Warming scheme have been presented [Anderson et al., 1984]. However, we present two of these formulations, which are based on splitting the coefficients. Thus, a correct signal transmission can be forced, thereby allowing the analysis of supercritical flows. Fennema and Chaudhry [1987] used these schemes for the simulation of dam-break flows. A general description of these schemes for application to two-dimensional flows is presented in the next chapter; details for one-dimensional applications are given in this section.

By using trapezoidal rule, the time difference may be approximated as

$$\mathbf{U}^{k+1} = \mathbf{U}^k + \frac{\Delta t}{2}\left[\left(\frac{\partial \mathbf{U}}{\partial \mathbf{t}}\right)^{k+1} + \left(\frac{\partial \mathbf{U}}{\partial \mathbf{t}}\right)^k\right] \tag{14-34}$$

By substituting the value of $\mathbf{U}_t$ from Eq. 14-15 into this equation

$$\mathbf{U}^{k+1} = \mathbf{U}^k - \frac{\Delta t}{2}\left[\left(\frac{\partial \mathbf{F}}{\partial \mathbf{x}} + \mathbf{S}\right)^{k+1} + \left(\frac{\partial \mathbf{F}}{\partial \mathbf{x}} + \mathbf{S}\right)^k\right] \tag{14-35}$$

The terms $\mathbf{F}^{k+1}$ and $\mathbf{S}^{k+1}$ in Eq. 14-35 are nonlinear and may be linearized as follows. The Taylor series expansion of $\mathbf{F}^{k+1}$ may be written as

$$\begin{aligned}
\mathbf{F}^{k+1} &= \mathbf{F}^k + \Delta t\left(\frac{\partial \mathbf{F}}{\partial \mathbf{t}}\right)^k + O(\Delta t)^2 \\
&= \mathbf{F}^k + \Delta t\left(\frac{\partial \mathbf{F}}{\partial \mathbf{U}}\frac{\partial \mathbf{U}}{\partial \mathbf{t}}\right)^k + O(\Delta t)^2 \\
&= \mathbf{F}^k + \mathbf{A}^k\frac{\partial \mathbf{U}}{\partial \mathbf{t}}\Delta t \\
&= \mathbf{F}^k + \mathbf{A}^k(\mathbf{U}^{k+1} - \mathbf{U}^k)
\end{aligned} \tag{14-36}$$

Similarly

$$\mathbf{S}^{k+1} = \mathbf{S}^k + \mathbf{B}^k(\mathbf{U}^{k+1} - \mathbf{U}^k) \tag{14-37}$$

where $\mathbf{A}$ and $\mathbf{B}$ are the Jacobians of $\mathbf{F}$ and $\mathbf{S}$ respectively and may be expressed as

$$\mathbf{A} = \begin{pmatrix} 0 & 1 \\ gD - V^2 & 2V \end{pmatrix}$$

$$\mathbf{B} = \begin{pmatrix} 0 & 0 \\ (-gS_o - 1.33gn^2V|V|)/R^{1.33} & gn^2|V|/R^{1.33} \end{pmatrix} \tag{14-38}$$

Substituting Eqs. 14-36 and 14-37 into Eq. 14-35 and simplifying

$$\left[\mathbf{I} + \frac{\Delta t}{2}(\frac{\partial \mathbf{A}}{\partial \mathbf{x}} + \mathbf{B})^\mathbf{k}\right]\boldsymbol{\Delta}_\mathbf{t}\mathbf{U}^{\mathbf{k+1}} = -\boldsymbol{\Delta}\mathbf{t}(\frac{\partial \mathbf{F}}{\partial \mathbf{x}} + \mathbf{S})^\mathbf{k} \tag{14-39}$$

For correct signal transmission, the matrices $\mathbf{A}$ and $\mathbf{F}$ may be split as

$$\mathbf{A} = \mathbf{A}^+ + \mathbf{A}^-$$
$$\mathbf{F} = \mathbf{F}^+ + \mathbf{F}^- \tag{14-40}$$

in which $\mathbf{A}^+ = \mathbf{MD}^+\mathbf{M}^{-1}$; $\mathbf{A}^- = \mathbf{MD}^-\mathbf{M}^{-1}$; $\mathbf{F}_x^+ = \mathbf{A}^+\mathbf{U}_x$; $\mathbf{F}_x^- = \mathbf{A}^-\mathbf{U}_x$; $\mathbf{D}$ is the diagonal matrix of eigenvalues of $\mathbf{A}$ and

$$\mathbf{M} = \begin{pmatrix} 1/(2c) & -1/(2c) \\ (V+c)/(2c) & -(V-c)/(2c) \end{pmatrix}$$

By substituting Eq. 14-40 into 14-39, we obtain

$$\left[\mathbf{I} + \frac{\Delta t}{2}\left\{\frac{\partial}{\partial x}(\mathbf{A}^+ + \mathbf{A}^-) + \mathbf{B}\right\}^\mathbf{k}\right]\boldsymbol{\Delta}_\mathbf{t}\mathbf{U}^{\mathbf{k+1}} = -\boldsymbol{\Delta}\mathbf{t}\left[\frac{\partial}{\partial \mathbf{x}}(\mathbf{F}^+ + \mathbf{F}^-) + \mathbf{S}\right]^\mathbf{k}$$

This equation in finite-difference form may be written as

$$\left[\mathbf{I} + \frac{1}{2}\frac{\Delta t}{\Delta x}\left\{(\nabla_x\mathbf{A}^+ + \Delta_x\mathbf{A}^-) + \frac{\Delta t}{2}\mathbf{B}_i\right\}^k\right]\Delta_t\mathbf{U}^{k+1}$$
$$= -\frac{\Delta t}{\Delta x}\left[\mathbf{A}_i^+\nabla_x\mathbf{U} + \mathbf{A}_i^-\Delta_x\mathbf{U}\right]^k - \Delta t\mathbf{S}_i^k \tag{14-41}$$

in which $\nabla_x\mathbf{A}^+ = \mathbf{A}_i^+ - \mathbf{A}_{i-1}^+$ and $\Delta_x\mathbf{A}^- = \mathbf{A}_{i+1}^- - \mathbf{A}_i^-$. The left-hand side of Eq. 14-41 constitutes a block tridiagonal system for each time step and may be solved by using a special algorithm.

Vasiliev Scheme

This scheme was developed by a team of researchers at the Institute of Hydrodynamics, Novosibirsk, Russia [Vasiliev et al., 1965]. It uses the following finite-difference approximations

$$\frac{\partial f}{\partial t} = \frac{f_i^{k+1} - f_i^k}{\Delta t}$$
$$\frac{\partial f}{\partial x} = \frac{f_{i+1}^{k+1} - f_{i-1}^{k+1}}{2\Delta x} \tag{14-42}$$

Vasiliev et al. [1965] applied this discretization to the continuity equation (Eq. 12-4) and to the following form of the momentum equation

$$\frac{\partial Q}{\partial t} + 2V\frac{\partial Q}{\partial x} + (c^2 - V^2)B\frac{\partial y}{\partial x} = gA(S_o - S_f) \qquad (14\text{-}43)$$

If we substitute these finite differences in the governing equations, we obtain $2N - 2$ equations for a channel divided into N reaches, since we cannot write these equations for the boundary nodes. Two equations are provided by the end conditions. Thus, we need two more equations for a unique solution of the resulting system of equations. The characteristic equations for the upstream and the downstream end provide these equations.

Now, we discuss consistency and stability of finite-difference schemes in the next two sections.

14-5 Consistency

A finite-difference scheme is said to be *consistent* if the finite-difference form of the equation tends to the original differential equation as Δx and Δt tend to zero.

To check the consistency of the Lax scheme, let us expand $V_i^{k+1}, V_{i+1}^k, V_{i-1}^k$ in Taylor series about values at grid point (i, k):

$$V_i^{k+1} = V_i^k + \frac{\partial V}{\partial t}\Delta t + \frac{1}{2!}\frac{\partial^2 V}{\partial t^2}(\Delta t)^2 + \frac{1}{3!}\frac{\partial^3 V}{\partial t^3}(\Delta t)^3 + \cdots$$

$$V_{i+1}^k = V_i^k + \frac{\partial V}{\partial x}\Delta x + \frac{1}{2!}\frac{\partial^2 V}{\partial x^2}(\Delta x)^2 + \frac{1}{3!}\frac{\partial^3 V}{\partial x^3}(\Delta x)^3 + \cdots$$

$$V_{i-1}^k = V_i^k - \frac{\partial V}{\partial x}\Delta x + \frac{1}{2!}\frac{\partial^2 V}{\partial x^2}(\Delta x)^2 - \frac{1}{3!}\frac{\partial^3 V}{\partial x^3}(\Delta x)^3 + \cdots \qquad (14\text{-}44)$$

Similarly, we can expand y_i^{k+1}, y_{i+1}^k, and y_{i-1}^k in terms of y_i^k. Substituting these expansions into Eq. 14-14, we obtain

$$V_i^k + \frac{\partial V}{\partial t}\Delta t + \frac{1}{2}\frac{\partial^2 V}{\partial t^2}(\Delta t)^2 + \frac{1}{3!}\frac{\partial^3 V}{\partial t^3}(\Delta t)^3 + \cdots$$

$$-\frac{1}{2}[V_i^k - \frac{\partial V}{\partial x}\Delta x + \frac{1}{2}\frac{\partial^2 V}{\partial x^2}(\Delta x)^2 - \frac{1}{3!}\frac{\partial^3 V}{\partial x^3}(\Delta x)^3 + \cdots$$

$$+V_i^k - \frac{\partial V}{\partial x}\Delta x + \frac{1}{2}\frac{\partial^2 V}{\partial x^2}(\Delta x)^2 - \frac{1}{3!}\frac{\partial^3 V}{\partial x^3}(\Delta x)^3 + \cdots]$$

$$+\frac{1}{2}g\frac{\Delta t}{\Delta x}[y_i^k + \frac{\partial y}{\partial x}\Delta x + \frac{1}{2}\frac{\partial^2 y}{\partial x^2}(\Delta x)^2 + \frac{1}{3!}\frac{\partial^3 y}{\partial x^3}(\Delta x)^2 + \cdots$$

$$-y_i^k + \frac{\partial y}{\partial x}\Delta x - \frac{1}{2}\frac{\partial^2 y}{\partial x^2}(\Delta x)^2 + \frac{1}{3!}\frac{\partial^3 y}{\partial x^3}(\Delta x)^3 + \cdots]$$

$$+\frac{1}{2}V_i^k\frac{\Delta t}{\Delta x}[V_i^k + \frac{\partial V}{\partial x}\Delta x + \frac{1}{2}\frac{\partial^2 V}{\partial x^2}(\Delta x)^2$$

$$+\frac{1}{3!}\frac{\partial^3 V}{\partial x^3}(\Delta x)^3 + \cdots - V_i^k + \frac{\partial V}{\partial x}\Delta x - \frac{1}{2}\frac{\partial^2 V}{\partial x^2}(\Delta x)^2$$

$$+\frac{1}{3!}\frac{\partial^3 V}{\partial x^3}(\Delta x)^3 + \cdots] = 0 \tag{14-45}$$

Simplifying and dividing throughout by Δt, this equation becomes

$$\frac{\partial V}{\partial t} + g\frac{\partial y}{\partial x} + V\frac{\partial V}{\partial x} + \frac{1}{2}\frac{\partial^2 V}{\partial t^2}\Delta t + \frac{1}{3!}\frac{\partial^3 V}{\partial t^3}(\Delta t)^2$$

$$-\frac{1}{2}\frac{\partial^2 V}{\partial x^2}\frac{(\Delta x)^2}{\Delta t} + \frac{1}{3!}gV_i^k\frac{\partial^3 V}{\partial x^3}(\Delta x)^2$$

$$+\frac{1}{3!}\frac{\partial^3 y}{\partial x^3}(\Delta x)^2 = 0 \tag{14-46}$$

In order to have the space and time derivatives converge uniformly as Δx and Δt become smaller so that both of these derivatives have similar errors, the first error term of time derivatives, $\frac{1}{2}\partial^2 V/\partial t^2 \Delta t$ should be equivalent to the first error term of the space derivative, $\frac{1}{6}\partial^2 V/\partial x^2(\Delta x)^2$. Thus, the computational grid has to be constructed so that $(\Delta x)^2 = k\Delta t$, where k is a constant [Liggett and Cunge, 1975].

Keeping this ratio between Δx and Δt constant, letting Δt and Δx approach zero, and neglecting the third- and higher-order terms, this equation takes the form

$$\frac{\partial V}{\partial t} + g\frac{\partial y}{\partial x} + V\frac{\partial V}{\partial x} - \frac{1}{2}k\frac{\partial^2 V}{\partial x^2} = 0 \tag{14-47}$$

In this equation, there is an additional diffusion-like term, $\frac{1}{2}k\partial^2 V/\partial x^2$, which is not present in the original equation. Proceeding similarly, we can show that an additional term $\frac{1}{2}k\partial^2 y/\partial x^2$ is introduced in the continuity equation. Therefore, the finite-difference equations in the Lax scheme are not consistent with the original governing equations.

14-6 Stability

We investigate the stability of a numerical scheme by studying whether an error grows or decays as the solution progresses in a marching procedure. Rigorous procedures are not presently available to determine the stability of nonlinear equations. However, by neglecting or linearizing the nonlinear terms, stability may be studied. If the nonlinearities are not strong, then the criteria developed for the linear equations may be assumed to be valid for the nonlinear equations as well. The effects of boundaries on the stability of a scheme are not included in such analyses.

For illustration purposes, we analyze the stability of the Lax scheme in the following paragraphs. This analysis procedure is referred to as *von Neumann or Fourier analysis.*

To do this analysis, let us linearize the governing equations, Eqs. 13-1 and 13-2, by neglecting the friction term and by using the coefficient V_o and D_o, in which the subscript o indicates steady-state value. In addition, let us drop S_o for simplicity and replace y by Y. Then, the linearized set of governing equations becomes

$$\frac{\partial V}{\partial t} + V_o \frac{\partial V}{\partial x} + g \frac{\partial Y}{\partial x} = 0 \tag{14-48}$$

$$\frac{\partial Y}{\partial t} + D_o \frac{\partial V}{\partial x} + V_o \frac{\partial Y}{\partial x} = 0 \tag{14-49}$$

Substitution of the finite-difference approximations of Eq. 14-13 into these equations and the rearrangement of the resulting equations yield

$$Y_i^{k+1} = \frac{1}{2}(Y_{i-1}^k + Y_{i+1}^k - \frac{1}{2}\frac{\Delta t}{\Delta x}D_o(V_{i+1}^k - V_{i-1}^k)$$
$$- \frac{1}{2}\frac{\Delta t}{\Delta x}V_o(Y_{i+1}^k - Y_{i-1}^k)$$
$$V_i^{k+1} = \frac{1}{2}(V_{i-1}^k + V_{i+1}^k) - \frac{1}{2}\frac{\Delta t}{\Delta x}g(Y_{i+1}^k - Y_{i-1}^k)$$
$$- \frac{1}{2}\frac{\Delta t}{\Delta x}V_o(V_{i+1}^k - V_{i-1}^k) \tag{14-50}$$

Let the exact solution of these equations be V_{exac} and Y_{exac}. Such a solution will be obtained by a computer having an infinite accuracy. However, real machines have only finite accuracy. Let us call the solution obtained on a real machine as V_{comp} and Y_{comp} in which roundoff errors v and y have been introduced. By substituting $V_{exac} = V_{comp} + v$ and $Y_{exac} = Y_{comp} + y$ into Eqs. 14-50 and noting that V_{exac} and Y_{exac} must also satisfy these equations, we obtain

$$v_i^{k+1} - \frac{1}{2}(v_{i-1}^k + v_{i+1}^k) + \frac{1}{2}V_o\frac{\Delta t}{\Delta x}(v_{i+1}^k - v_{i-1}^k)$$
$$+ \frac{1}{2}g\frac{\Delta t}{\Delta x}(y_{i+1}^k - y_{i-1}^k) = 0 \tag{14-51}$$

$$y_i^{k+1} - \frac{1}{2}(y_{i+1}^k + y_{i-1}^k) + \frac{1}{2}D_o\frac{\Delta t}{\Delta x}(v_{i+1}^k - v_{i-1}^k)$$
$$+ \frac{1}{2}\frac{\Delta t}{\Delta x}V_o(y_{i+1}^k - y_{i-1}^k) = 0 \qquad (14\text{-}52)$$

For the scheme to be stable, these errors must decay as the solution progresses from one time step to the next. To investigate this, let us express v_i^k and y_i^k in Fourier series as

$$v_i^k = \sum_{n=0}^{N} A'_n e^{j\theta_n x_i}$$

$$y_i^k = \sum_{n=0}^{N} B'_n e^{j\theta_n x_i} \qquad (14\text{-}53)$$

in which $j = \sqrt{-1}$, the wave number, $\theta_n = n\pi/L$ $(n = 0, 1, 2, \cdots, N)$ and the interval of interest is of length L. Now, $x_{i+1} = x_i + \Delta x$ and $x_{i-1} = x_i - \Delta x$. Hence, we may write

$$v_{i-1}^k = \sum_{n=0}^{N} A'_n e^{j\theta_n x_i} e^{-j\theta_n \Delta x}$$

$$v_{i+1}^k = \sum_{n=0}^{N} A'_n e^{j\theta_n x_i} e^{j\theta_n \Delta x}$$

$$y_{i-1}^k = \sum_{n=0}^{N} B'_n e^{j\theta_n x_i} e^{-j\theta_n \Delta x}$$

$$y_{i+1}^k = \sum_{n=0}^{N} B'_n e^{j\theta_n x_i} e^{j\theta_n \Delta x} \qquad (14\text{-}54)$$

Since the system is linear, we may consider only one term of the series instead of the sum of N terms. In addition, we may write

$$v_i^{k+1} = e^{\alpha t}v_i^k = \xi v_i^k = \xi A' e^{j\theta x_i} \qquad (14\text{-}55)$$

in which ξ is called the *amplification factor*. Depending upon the value of ξ, an error introduced at any time grows or decays as the computations progress in time. If $|\xi| < 1$, the error decays as the computations progress and the scheme is called *stable*; however, if $|\xi| > 1$, then the error grows with time and the scheme is called *unstable*. The scheme is said to be *neutrally stable* if $|\xi| = 1$.

Similarly, we may express the error in the flow depth as

$$y_i^{k+1} = \xi B' e^{j\theta x_i} \qquad (14\text{-}56)$$

Substituting these expressions into Eqs. 14-51 and letting $r = \Delta t/\Delta x$, we obtain

$$\xi A' e^{j\theta x_i} - \frac{1}{2}(A' e^{j\theta x_i} e^{-j\theta \Delta x} + A' e^{j\theta x_i} e^{j\theta \Delta x})$$

$$+ \frac{1}{2} r V_o (A' e^{j\theta x_i} e^{j\theta \Delta x} - A' e^{j\theta x_i} e^{-j\theta \Delta x})$$

$$+ \frac{1}{2} r g (B' e^{j\theta x_i} e^{j\theta \Delta x} - B' e^{j\theta x_i} e^{-j\theta \Delta x}) = 0 \qquad (14\text{-}57)$$

Cancelling out $e^{j\theta x_i}$ and rearranging the terms, we obtain

$$[\xi - \frac{1}{2}(e^{j\theta \Delta x} + e^{-j\theta \Delta x}) + \frac{1}{2} r V_o (e^{j\theta \Delta x} - e^{-j\theta \Delta x})]A'$$

$$+ \frac{1}{2} r g (e^{j\theta \Delta x} - e^{-j\theta \Delta x})B' = 0 \qquad (14\text{-}58)$$

Let $\delta = \theta \Delta x$. Then the preceding equation may be written as

$$(\xi - \cos\delta + jr V_o \sin\delta)A' + jrg \sin\delta B' = 0 \qquad (14\text{-}59)$$

Proceeding similarly, Eq. 14-52 becomes

$$(\xi - \cos\delta + jr V_o \sin\delta)B' + jD_o r \sin\delta A' = 0 \qquad (14\text{-}60)$$

For a nontrivial solution of A' and B', it follows from Eqs. 14-59 and 14-60 that

$$\begin{vmatrix} \xi - \cos\delta + jr V_o \sin\delta & jrg \sin\delta \\ jD_o r \sin\delta & \xi - \cos\delta + jV_o \sin\delta \end{vmatrix} = 0 \qquad (14\text{-}61)$$

It follows from this equation that

$$(\xi - \cos\theta + jr V_o \sin\delta)^2 = -D_o g r^2 \sin^2\delta = 0 \qquad (14\text{-}62)$$

Taking the square root of both sides

$$\xi - \cos\delta + jr V_o \sin\delta = \pm j\sqrt{D_o g}\, r \sin\delta \qquad (14\text{-}63)$$

or

$$\xi = \cos\delta - j(V_o \pm \sqrt{D_o g})r \sin\delta \qquad (14\text{-}64)$$

For the error to decay, $|\xi| < 1$, i.e.,

$$\left(\cos^2\delta + (V_o \pm \sqrt{D_o g})^2 r^2 \sin^2\delta\right)^{\frac{1}{2}} < 1$$

or

$$\cos^2\delta + (V_o \pm \sqrt{D_o g})^2 r^2 \sin^2\delta < 1 \qquad (14\text{-}65)$$

or

$$(1 - \sin^2\delta) + (V_o \pm \sqrt{g D_o})^2 r^2 \sin^2\delta < 1 \qquad (14\text{-}66)$$

or

$$(V_o \pm \sqrt{gD_o})^2 r^2 < 1 \qquad (14\text{-}67)$$

Noting that $c = \sqrt{gD_o}$, it follows from this equation that

$$\frac{\Delta t}{\Delta x} < \frac{1}{V_o \pm c} \qquad (14\text{-}68)$$

It is clear from Eq. 14-63 that the amplification factor depends upon the mesh size and the wave number or frequency. Based on this equation, we may write

$$\xi = |\xi| e^{j\phi} \qquad (14\text{-}69)$$

in which $|\xi|$ = amplitude of the amplification factor and ϕ = phase angle. The expressions for the amplitude and the phase angle are

$$|\xi| = \sqrt{\cos^2 \delta + C_n^2 \sin^2 \delta} \qquad (14\text{-}70)$$

and

$$\phi = \tan^{-1}(-C_n \tan \delta) \qquad (14\text{-}71)$$

in which C_n = Courant number = $(V_o + \sqrt{gD_o})\Delta t/\Delta x$.

Figure 14-5 [Anderson et al., 1984] shows the amplitude and the phase angle for the amplification factor for the Lax scheme for several values of C_n. It is clear from this figure that all frequency components are propagated without attenuation for $C_n = 1$. However, if $C_n < 1$, the attenuation is small for low- and high-frequency components and the attenuation is severe for midrange frequency components. Note that the Courant condition is satisfied

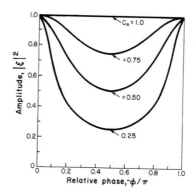

Fig. 14-5   **Amplitude-phase plot for the amplification factor** (After Anderson et al. [1984])

even for larger Δt if Δx is large. However, for the Lax diffusive scheme, Koren [1967] showed that the following additional stability criterion must be satisfied for a small amplitude disturbance applied to an initial steady-state flow to damp out:

$$\Delta t \leq \frac{\sqrt{1 + 2|\frac{V_o}{c_o}|} - 1}{|\frac{V_o}{c_o}|\frac{gS_o}{V_o}} \qquad (14\text{-}72)$$

in which V_o and c_o are the initial velocity and the celerity of a shallow water wave, respectively. Numerical experimentation by Huang and Song [1985] confirm the validity of this criterion in addition to showing that it is applicable to the method of characteristics as well.

Example 14-1

A trapezoidal channel having a bottom width of 6.1 m and side slopes of 1.5H : 1V is carrying a flow of 126 m³/s at a flow depth of 5.79 m. The bottom slope is 0.00008, Manning n = 0.013 and the channel length is 5 km. There is a constant-level reservoir at the upstream end of the channel. If a sluice gate at the downstream end is suddenly closed at time t = 0:

1. *Compute the transient conditions until t = 2000 s by using:*
   - a. *Method of specified intervals;*
   - b. *Lax diffusive scheme;*
   - c. *MacCormack scheme;*
   - d. *Preissmann scheme.*
2. *Plot the computed flow depth in the channel at t = 0, 500, 1000, 1500 and 2000 s.*
3. *Plot the variation of the flow depth with time at distance of 1.5, 2.5, 3 and 5 km from the reservoir.*

Solution

A computer program was written based on these finite difference schemes for the interior grid points. In the first four schemes, the positive characteristic equation and the condition that flow velocity at the downstream end is always zero for $t > 0$ were utilized to simulate the downstream end condition. At the upstream end, the negative characteristic equation was used along with the condition that the flow depth is equal to the reservoir depth. In the Preissmann scheme, the upstream- and downstream-boundary conditions were directly incorporated into the solution. The downstream boundary condition specified the flow velocity at the downstream end to be always zero following the gate closure, and the upstream end condition specified the flow depth to be constant and equal to the reservoir depth at the channel entrance.

The channel length was divided into 50 equal-length reaches and the computational time interval, Δt, was selected so that the stability condition was always satisfied in the explicit schemes at every grid point. If this was not the case, then the computational time interval was reduced by 20 percent and the flow conditions were recalculated. However, if the time step was considerably smaller than that required by the Courant condition, then the time step for the next interval was increased by 15 percent.

The computed flow depths in the channel at different times by using these schemes are shown in Fig. 14-6. The variation of flow depth with respect to time at different locations is plotted in Fig. 14-7.

14-7 Summary

In this chapter, we presented several explicit and implicit finite-difference methods for numerical integration of shallow-water equations. The specification of the initial and boundary conditions was outlined. The consistency and stability of a numerical schemes were discussed. The Von Neumann stability analysis was demonstrated by applying it to the Lax scheme.

Problems

14-1 Based on the finite-difference schemes of this chapter, write a computer program to analyze unsteady flows in a trapezoidal channel having a constant-level reservoir at the upstream end, bottom width of 10 m, side slopes of 2H : 1V and bottom slope of 0.0001. Initially, the flow is uniform at a flow depth of 3 m. At t=0, a control gate is instantaneously closed at the downstream end. Assume Manning n for the channel surface is 0.010.

14-2 Write a computer program by using the conservation and nonconservation forms of the governing equations and the Lax and MacCormack schemes. For the data of Problem 14-1, compare the velocity of propagation of the surge wave and the conservation of mass at different times for the different form of the governing equations and the numerical schemes.

14-3 By using von Neumann analysis, show that the finite-difference scheme given by Eq. 14-12 is unstable irrespective of the size of time and spatial intervals.

14-4 A 3-m wide rectangular channel is carrying a discharge of 30 m^3/s at a flow depth of 1 m. There is a constant-level reservoir at the upper end and the flow is uniform during the initial conditions. The downstream reservoir level is suddenly raised to 6.5 m at $t = 0$. Apply different numerical schemes of this chapter to compute the final steady-state water-surface profile in the channel. (*Hint:* First specify the initial conditions for the supercritical flow. Then, change the downstream end condition and continue the computations till they converge to a steady state.)

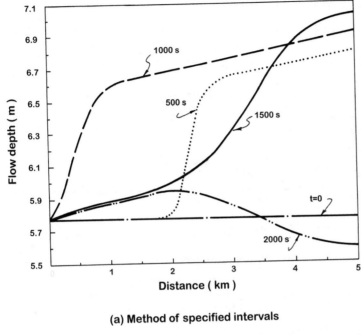

(a) Method of specified intervals

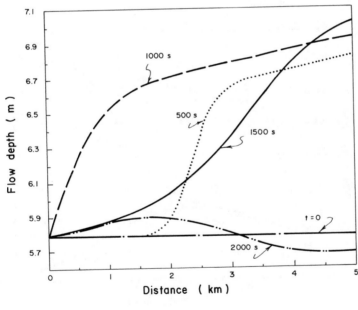

(b) Lax diffusive scheme

Fig. 14-6 Computed flow depths at different times

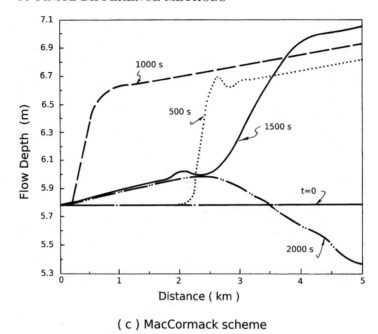

(c) MacCormack scheme

Fig. 14-6   (Continued)

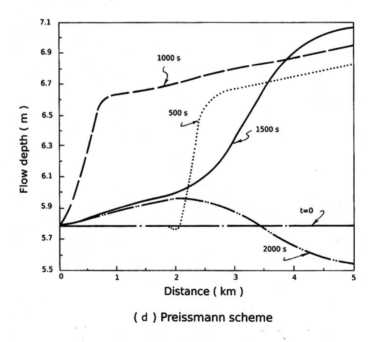

(d) Preissmann scheme

Fig. 14-6   (Concluded)

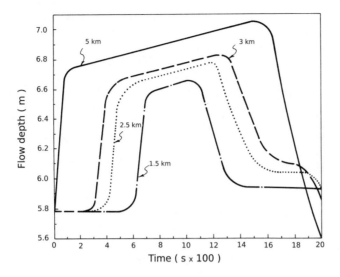

(a) Method of specified intervals

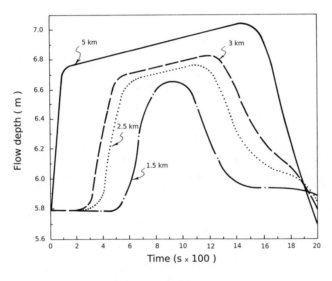

(b) Lax diffusive scheme

Fig. 14-7 Variation of flow depth with time

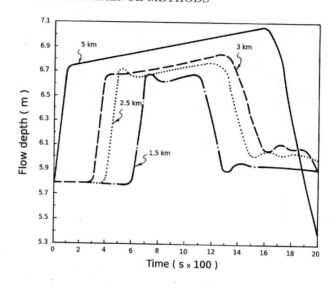

(c) MacCormack scheme

Fig. 14-7   (Continued)

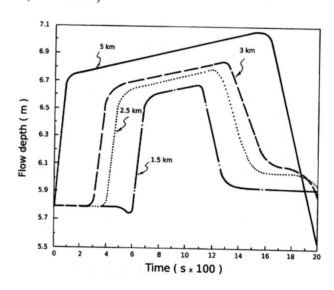

(d) Preissmann scheme

Fig. 14-7   (Concluded)

14-5 Set up a computational procedure to solve the governing equations at a series junction. Derive the stability conditions for the MacCormack scheme.

14-6 Show that the Preissmann scheme is unconditionally stable if the friction-loss term is neglected. Derive the stability condition if a linearized form of the friction term is included.

References

Abbott, M. B., 1979, *Computational Hydraulics; Elements of the Theory of Free Surface Flows*, Pitman, London.

Amein, M., and Fang, C. S., 1970, "Implicit Flood Routing in Natural Channels," *Jour. Hyd. Div.*, Amer. Soc. of Civ. Engrs., vol. 96, no. 12, pp. 2481–2500.

Anderson, D. A., Tannehill, J. C., and Pletcher, R. H., 1984, *Computational Fluid Mechanics and Heat Transfer*, McGraw Hill, New York.

Beam, R.M., and Warming, R.F., 1976 "An Implicit Finite- Difference Algorithm for Hyperbolic Systems in Conservation-Law Form," *Jour. Comp. Phys.*, vol. 22, pp. 87–110.

Chaudhry, M. H., 2014, *Applied Hydraulic Transients*, 3rd ed., Springer, New York, NY.

Cunge, J., Holly, F. M., and Verwey, A., 1980, *Practical Aspects of Computational River Hydraulics*, Pitman, London.

Dammuller, D., Bhallamudi, S. M., and Chaudhry, M. H., 1989, "Modeling of Unsteady Flow in Curved Channels," *Jour Hyd. Engineering*, Amer Soc Civ. Engrs, vol 115, no. 11, pp. 1479–1495.

Evans, E.P., 1977, "The Behaviour of a Mathematical Model of Open-Channel Flow," Paper A97, *Proc. 17ʰh Congress*, Inter. Assoc. for Hyd. Research, Baden Baden, Germany.

Fennema, R. J., and Chaudhry, M. H., 1986, "Explicit Numerical Schemes for Unsteady Free-Surface Flows with Shocks," *Water Resources Research*, vol. 22, no. 13, pp. 1923–1930.

Fennema, R. J., and Chaudhry, M. H., 1987, "Simulation of One-dimensional Dam-break Flows," *Jour. Hyd. Research*, vol. 25, no. 1, pp. 41–51.

Fread, D. L., and Harbaugh, T. E., 1973, "Transient Simulation of Breached Earth Dams," *Jour. Hyd. Div.*, Amer. Soc. Civil Engrs., no. 1, pp. 139–154.

French, R. H., 1985, *Open-Channel Hydraulics*, McGraw-Hill, New York, NY.

Galperin, B., Blumberg, A. F. and Weisberg, R. H., 1992, "The Importance of Density Driven Circulation in Well-mixed Estuaries: The Tampa Bay Experience," *Proc. Water Forum*, Amer. Soc. of Civil Engrs., 11 pp.

Huang, J., and C.C.S. Song, 1985, "Stability of Dynamic Flood Routing Schemes," *Jour. Hydraulic Engineering*, Amer. Soc. Civil Engrs., vol. 111, no. 12, pp.1497–1505.

Isaacson, E., Stoker, J. J., and Troesch , B. A., 1954, "Numerical Solution of Flood Prediction and River Regulation Problems (Ohio-Mississippi Floods)," *Report II,* Inst. Math. Sci. Rept. IMM-NYU-205, New York University.

Jameson, A., Schmidt, W., and Turkel, E., 1981, "Numerical Solutions of the Euler equations by Finite Volume Methods Using Runge-Kutta Schemes," Time-Stepping *AIAA* 14th *Fluid And Plasma Dynamics Conference,* Palo Alto, California, AIAA-81-1259.

Katopodes, N. and Wu, C-T., 1986, "Explicit Computation of Discontinous Channel Flow," *Jour. Hyd. Engineering,* Amer. Soc. Civ. Engrs., vol. 112, no. 6, pp. 456–475.

Koren, V.I., 1967, "The Analysis of Stability of Some Explicit Finite Difference Schemes for the Integration of Saint-Venant Equations," *Meterologiya i Gidrologiya,* no. 1 (in Russian).

Lax, P. D., 1954, "Weak Solutions of Nonlinear Hyperbolic Partial Differential Equations and Their Numerical Computation," *Communications on Pure and Applied Mathematics,* vol. 7, pp. 159–163.

Lai, C., 1986, "Numerical Modeling of Unsteady Open-Channel Flow," in *Advances in Hydroscience,* vol. 14, Academic Press, New York., pp. 161–333.

Leendertse, J. J., 1967, "Aspects of a Computational Model for Long Period Water-Wave Propagation," *Memo* RM-5294-PR, Rand Corporation, Santa Monica, CA, May.

Liggett, J. A., and Woolhiser, D. A., 1967, "Difference Solutions of Shallow-water equations," *Jour. Engineering Mech. Div.,* Amer. Soc. Civil Engrs., vol. 93, no. EM2, pp. 39–71.

Liggett, J. A., and Cunge, J. A., 1975, "Numerical Methods of Solution of Unsteady Flow Equations," in *Unsteady Open Channel Flow,* Mahmood, K. and Yevjevich, V. (eds.), Water Resources Publications, Fort Collins.

Lyn, D.A. and Goodwin, P., 1987, "Stability of a General Preissmann Scheme," *Jour. Hydraulic Engineering,* Amer. Soc. Civil Engrs., vol. 113, no. 1, pp. 16–27.

MacCormacK, R.W., 1969, "The Effect of Viscosity in Hypervelocity Impact Cratering," *Amer. Inst. Aero. Astro.,* Paper 69–354, Cincinnati, Ohio.

Meselhe, E. A., and Holly, F. M., Jr., 1997, "Invalidity of Preissmann Scheme for Transcritical Flow," *Jour. Hydraulic Engineering,* vol. 123, no. 7, pp. 652–655.

Miller, S., and Chaudhry, M. H., 1989, "Dam-break Flows in Curved Channel," *Jour. Hyd. Engineering,* Amer. Soc. Civil Engrs., vol. 115, no. 11, pp. 1465–1478.

Pandolfi, M., 1975, "Numerical Experiments on Free Surface Water Motion with Bores," *Proc.* 4th *Int. Conf. on Numerical Methods in Fluid Dynamics,* Lecture Notes in Physics No. 35, Springer-Verlag, pp. 304–312.

Price, R. K., 1974, "Comparison of Four Numerical Flood Routing Methods," *Jour., Hyd. Div.,* Amer. Soc. Civ. Engrs., vol. 100, no. 7, pp. 879–899.

Preissmann, A., and Cunge, J., 1961, "Calcul du mascaret sur machines electroniques," *La Houille Blanche,* no. 5, pp. 588–596.

Richtmyer, R. D., and Morton, K. W., 1967, *Difference Methods for Initial Value Problems,* 2nd. ed., Interscience, New York.

Roache, P. J., 1972, *Computational Fluid Dynamics,* Hermosa Publishers.

Samuels, P.G., and Skeels, C.P., 1990, "Stability Limits for Preissmann's Scheme," *Jour. Hydraulic Engineering,* Amer. Soc. Civil Engrs., vol 116, no. 8, pp. 997–1012.

Stoker, J. J., 1957, *Water Waves,* Interscience, New York.

Strelkoff, T., 1970, "Numerical Solution of St. Venant Equations," *Jour. Hyd. Div.,* Amer. Soc. Civ. Engrs., vol. 96, no. 1, pp. 223–252.

Terzidis, G., and Strelkoff, T., 1970, "Computation of Open Channel Surges and Shocks," *Jour. Hydraulics Div.,* Amer. Soc. Civ. Engrs., vol. 96, no. 12, pp. 2581–2610.

Vasiliev, O. F., Gladyshev, M. T., Pritvits, N. A., and Sudobiocher, V. G., 1965, "Numerical Method for the Calculation of Shock Wave Propagation in Open Channels," *Proc., 11th Congress,* Inter. Assoc. for Hydraulic Research, vol. 3, paper 3.44, 14pp.

Venutelli, M. 2002, "Stability and Accuracy of Weighted Four-Point Implicit Finite Difference Schemes for Open Channel Flow," *Jour. Hydraulic Engineering,* Amer. Soc. Civ. Engrs., vol. 128, no. 3, pp. 281–288.

Verboom, G. K., Stelling, G.S. and Officier, M. J., 1982, "Boundary Conditions for the Shallow Water Equations," *Engineering Applications of Computational Hydraulics,* vol. 1, (Abbott, M. B. and Cunge, J. A., eds.) Pitman, Boston.

Yen, C-L., and Lin, C.-H., 1986, "Numerical Stability in Unsteady Supercritical Flow Simulation," *Proc. 5th Congress Asian and Pacific Regional Div.,* Int. Assoc. Hyd. Research, Aug.

Younus, M., and Chaudhry, M.H., 1994, "A depth-averaged k-ϵ turbulence model for the computation of free-surface flow," *Jour. Hyd. Research,* vol. 32, no. 3, pp. 415–444.

15

TWO-DIMENSIONAL FLOW

(a)

(b)

(c)

Photographs of the Teton Dam failure at different times on June 5, 1976
(Courtesy, U. S. Geological Survey, photographs by Mrs. E. Olson)

© Springer Nature Switzerland AG 2022
M. H. Chaudhry, *Open-Channel Flow*,
https://doi.org/10.1007/978-3-030-96447-4_15

15-1 Introduction

In the previous chapters, we considered one-dimensional flows. However, the assumption of one-dimensional flow may not be valid in many situations – e.g., flow in a non-prismatic channel (i.e., channel with varying cross section and alignment), flow downstream of a partially breached dam, or lateral flow from a failed dyke. Although flow in these situations is three-dimensional, we may simplify their analysis by considering them as two-dimensional flows by using vertically averaged quantities. Such an assumption not only simplifies the analysis considerably but yields results of reasonable accuracy.

In this chapter, we discuss the analysis of two-dimensional flows. First, we derive the equations describing unsteady two-dimensional flows. Then, we present explicit and implicit finite difference methods for their solution.

15-2 Governing Equations

We will derive the governing equations [Jimenez, 1987] by integrating the Navier-Stokes equations for an incompressible fluid over the flow depth [Lai, 1986].

Except for one-dimensional flow, the same assumptions as those given in Section 12-2 are used. The effects of large bottom slope are included, and it is assumed that the bottom of the channel is an inclined plane. A Cartesian orthogonal coordinate system with the x-y plane parallel to the plane of the channel bottom is considered. For a right-hand system, the positive z-direction points upward and is perpendicular to the x-y plane.

The Navier-Stokes equations for an incompressible fluid are as follows:

Continuity equation

$$\frac{\partial u}{\partial x} + \frac{\partial v}{\partial y} + \frac{\partial w}{\partial z} = 0 \tag{15-1}$$

Momentum equation

$$\frac{\partial u}{\partial t} + u\frac{\partial u}{\partial x} + v\frac{\partial u}{\partial y} + w\frac{\partial u}{\partial z} = g_x - \frac{1}{\rho}\frac{\partial p}{\partial x} + \frac{\mu}{\rho}\nabla^2 u \tag{15-2}$$

$$\frac{\partial v}{\partial t} + u\frac{\partial v}{\partial x} + v\frac{\partial v}{\partial y} + w\frac{\partial v}{\partial z} = g_y - \frac{1}{\rho}\frac{\partial p}{\partial y} + \frac{\mu}{\rho}\nabla^2 v \tag{15-3}$$

$$\frac{\partial w}{\partial t} + u\frac{\partial w}{\partial x} + v\frac{\partial w}{\partial y} + w\frac{\partial w}{\partial z} = g_z - \frac{1}{\rho}\frac{\partial p}{\partial z} + \frac{\mu}{\rho}\nabla^2 w \tag{15-4}$$

in which u, v and w are the components of the velocity along the x, y and z directions; $\mathbf{g} = (g_x, g_y, g_z)^T$ is the gravitational force per unit mass; μ is the dynamic viscosity; p is pressure; and the symbol ∇^2 stands for the Laplace operator,

$$\nabla^2 = \frac{\partial^2}{\partial x^2} + \frac{\partial^2}{\partial y^2} + \frac{\partial^2}{\partial z^2}$$

We will integrate these equations over the flow depth to obtain the depth-averaged equations.

Continuity equation

The depth-averaged continuity equation for two-dimensional flow may be obtained by integrating Eq. 15-1 over the flow depth, i.e.,

$$\int_{Z_b}^{Z} \frac{\partial u}{\partial x} dz + \int_{Z_b}^{Z} \frac{\partial v}{\partial y} dz + w(Z) - w(Z_b) = 0 \tag{15-5}$$

in which Z and Z_b are the z-coordinates of the water surface and the channel bottom, respectively (these are measured perpendicular to the plane of the channel bottom). The integrals of Eq. 15-5 may be evaluated by using the Leibnitz rule

$$\int_{Z_b}^{Z} \frac{\partial u}{\partial x} dz = \frac{\partial}{\partial x} \int_{Z_b}^{Z} u \, dz - u(Z) \frac{\partial Z}{\partial x} + u(Z_b) \frac{\partial Z_b}{\partial x}$$

$$\int_{Z_b}^{Z} \frac{\partial v}{\partial y} dz = \frac{\partial}{\partial y} \int_{Z_b}^{Z} v \, dz - v(Z) \frac{\partial Z}{\partial y} + v(Z_b) \frac{\partial Z_b}{\partial y} \tag{15-6}$$

If the function $Z(x, y, t)$ specifies the z-coordinate of the free water surface and if it is assumed that any particle on the surface does not leave it, then the vertical velocity of a particle on the water surface, $w(Z)$, is given by

$$w(Z) = \frac{DZ}{Dt} = \frac{\partial Z}{\partial t} + u(Z) \frac{\partial Z}{\partial x} + v(Z) \frac{\partial Z}{\partial y} \tag{15-7}$$

Similarly, if the bottom of the channel is rigid, then $F_b = Z_b(x, y) - z = 0$, in which $Z_b(x, y)$ gives the z-coordinate of the channel bottom. Hence,

$$w(Z_b) = \frac{DF_b}{Dt} = u(Z_b) \frac{\partial Z_b}{\partial x} + v(Z_b) \frac{\partial Z_b}{\partial y} \tag{15-8}$$

Substitution of Eqs. 15-6 through 15-8 into Eq. 15-5 leads to

$$\frac{\partial Z}{\partial t} + \frac{\partial (\bar{u} d)}{\partial x} + \frac{\partial (\bar{v} d)}{\partial y} = 0 \tag{15-9}$$

in which $\bar{u}$ and $\bar{v}$ are the mean values of u and v over the depth of the channel,

$$\bar{u} = \frac{1}{d} \int_{Z_b}^{Z} u \, dz \qquad \bar{v} = \frac{1}{d} \int_{Z_b}^{Z} v \, dz \tag{15-10}$$

in which $d = Z - Z_b$ is the water depth, measured perpendicular to the bottom of the channel:

Momentum equations

Since we are assuming the vertical acceleration to be negligible,

$$\frac{Dw}{Dt} \approx 0 \qquad \mu\nabla^2 w \approx 0 \tag{15-11}$$

Therefore, Eq. 15-4 reduces to

$$g_z - \frac{1}{\rho}\frac{\partial p}{\partial z} = 0 \tag{15-12}$$

Integrating Eq. 15-12 in the z-direction and considering the atmospheric pressure to be zero, we obtain

$$p = \rho g_z(z - Z) \tag{15-13}$$

Hence, it follows that

$$-\frac{1}{\rho}\frac{\partial p}{\partial x} = g_z \frac{\partial Z}{\partial x} \tag{15-14}$$

$$-\frac{1}{\rho}\frac{\partial p}{\partial y} = g_z \frac{\partial Z}{\partial y} \tag{15-15}$$

By multiplying Eq. 15-1 by u, adding to Eq. 15-2, substituting the expression for $-\frac{1}{\rho}\frac{\partial p}{\partial x}$ from Eq. 15-14, and rearranging the terms of the resulting equation, we obtain

$$\frac{\partial u}{\partial t} + \frac{\partial u^2}{\partial x} + \frac{\partial(uv)}{\partial y} + \frac{\partial(uw)}{\partial z} = g_x + g_z\frac{\partial Z}{\partial x} + \frac{\mu}{\rho}\nabla^2 u \tag{15-16}$$

Similarly, multiplying Eq. 15-1 by v, adding it to Eq. 15-3, substituting the expression for $-(1/\rho)\partial p/\partial y$ from Eq. 15-15, we obtain

$$\frac{\partial v}{\partial t} + \frac{\partial(uv)}{\partial x} + \frac{\partial v^2}{\partial y} + \frac{\partial(vw)}{\partial z} = g_y + g_z\frac{\partial Z}{\partial y} + \frac{\mu}{\rho}\nabla^2 v \tag{15-17}$$

Let us integrate Eqs. 15-16 and 15-17 in the z-direction. To simplify presentation, we consider the left- and right-hand sides of these equations separately. Integration of the left-hand side of Eq. 15-16 and application of the Leibnitz rule yield

$$\frac{\partial}{\partial t}\int_{Z_b}^{Z} u\,dz - u(Z)\frac{\partial Z}{\partial t} + \frac{\partial}{\partial x}\int_{Z_b}^{Z} u^2\,dz - u^2(Z)\frac{\partial Z}{\partial x}$$
$$+ u^2(Z_b)\frac{\partial Z_b}{\partial x} + \frac{\partial}{\partial y}\int_{Z_b}^{Z} uv\,dz - u(Z)v(Z)\frac{\partial Z}{\partial y}$$
$$+ u(Z_b)v(Z_b)\frac{\partial Z_b}{\partial y} + u(Z)w(Z) - u(Z_b)w(Z_b) \tag{15-18}$$

Based on the assumption of uniform velocity distribution (i.e., u and v are constants in the z-direction) and substituting Eqs. 15-7 and 15-8, Expression 15-18 simplifies as

$$\frac{\partial}{\partial t}(\bar{u}d) + \frac{\partial}{\partial x}(\bar{u}^2 d) + \frac{\partial}{\partial y}(\bar{u}\bar{v}d) \tag{15-19}$$

Similarly, the left-hand side of Eq. 15-17 becomes

$$\frac{\partial}{\partial t}(\bar{v}d) + \frac{\partial}{\partial x}(\bar{u}\bar{v}d) + \frac{\partial}{\partial y}(\bar{v}^2 d) \tag{15-20}$$

Integration of the right-hand side of Eqs. 15-16 and 15-17 yields

$$\left(g_x + g_z \frac{\partial Z}{\partial x}\right)d + \int_{Z_b}^{Z} \frac{\mu}{\rho}\nabla^2 u\,dz \tag{15-21}$$

$$\left(g_y + g_z \frac{\partial Z}{\partial y}\right)d + \int_{Z_b}^{Z} \frac{\mu}{\rho}\nabla^2 v\,dz \tag{15-22}$$

Since the x-y plane is parallel to the channel bottom, Z_b is constant. Therefore,

$$\frac{\partial Z}{\partial x} = \frac{\partial(Z_b + d)}{\partial x} = \frac{\partial d}{\partial x} \tag{15-23}$$

Similarly,

$$\frac{\partial Z}{\partial y} = \frac{\partial d}{\partial y} \tag{15-24}$$

Now, let us consider the shear stress terms. In turbulent flow, the dynamic viscosity is replaced by an eddy viscosity coefficient. Moreover, distinction is made between the stresses acting in the x-y plane and the stresses acting in the x-z and y-z planes. For example, the shear-stress term of the momentum equation in the x-direction may be written as

$$\epsilon_{xy}\left(\frac{\partial^2 u}{\partial x^2} + \frac{\partial^2 u}{\partial y^2}\right) + \epsilon_{zx}\frac{\partial^2 u}{\partial z^2} \tag{15-25}$$

in which ϵ_{xy} and ϵ_{zx} are the eddy-viscosity coefficients. In addition, it is assumed that the effective stresses are dominated by the bottom shear stresses. This means that the first term in Eq. 15-25 is negligible as compared to the second term. Therefore, the shear stress term of Eq. 15-25 reduces to $\epsilon_{zx}\partial^2 u/\partial z^2$. Integration of this expression with respect to z yields

$$\int_{Z_b}^{Z} \epsilon_{zx}\frac{\partial^2 u}{\partial z^2}dz = \epsilon_{zx}\left(\frac{\partial u}{\partial z}\right)_{z=Z} - \epsilon_{zx}\left(\frac{\partial u}{\partial z}\right)_{z=Z_b}$$

$$= \tau_{s_x} - \tau_{b_x} \tag{15-26}$$

in which τ_{s_x} and τ_{b_x} are the shear stresses at the water surface and at the channel bottom acting in the x-direction. Similarly, the shear stress term of Eq. 15-22 reduces to

$$\tau_{s_y} - \tau_{b_y} \qquad (15\text{-}27)$$

The shear stresses, τ_{s_x} and τ_{s_y}, due to wind velocity acting at the water surface are neglected and the shear stresses at the channel bottom, τ_{b_x} and τ_{b_y} are evaluated by using empirical formulas. For example, the Chezy equation gives

$$\tau_b = \frac{\rho g}{C^2} V^2 \qquad (15\text{-}28)$$

where V is the amplitude of flow velocity (i.e., $V = \sqrt{\bar{u}^2 + \bar{v}^2}$) and C is the Chezy coefficient. It follows from Eq. 15-28 that

$$\tau_{b_x} = \tau_b \cos \theta = \frac{\rho g}{C^2} \bar{u} V$$
$$\tau_{b_y} = \tau_b \sin \theta = \frac{\rho g}{C^2} \bar{v} V \qquad (15\text{-}29)$$

in which θ is the angle between the velocity vector and the x-axis.

Different terms of the depth-integrated momentum equations may now be assembled together. Substitution of Eqs. 15-19 to 15-24, 15-26, 15-27, and 15-29 into Eqs. 15-16 and 15-17 gives

$$\frac{\partial}{\partial t}(\bar{u}d) + \frac{\partial}{\partial x}(\bar{u}^2 d) + \frac{\partial}{\partial y}(\bar{u}\bar{v}d) = \left(g_x - g_z \frac{\partial d}{\partial x}\right) g d - \frac{g}{C^2} \bar{u} \sqrt{\bar{u}^2 + \bar{v}^2} \quad (15\text{-}30)$$

$$\frac{\partial}{\partial t}(\bar{v}d) + \frac{\partial}{\partial x}(\bar{u}\bar{v}d) + \frac{\partial}{\partial y}(\bar{v}^2 d) = \left(g_y - g_z \frac{\partial d}{\partial y}\right) g d - \frac{g}{C^2} \bar{v} \sqrt{\bar{u}^2 + \bar{v}^2} \quad (15\text{-}31)$$

Equations 15-30 and 15-31 are the momentum equations with respect to a coordinate system x-y parallel to the channel bottom. For the case of one-dimensional flow, the above equations reduce to those given by Yen [1973] for rectangular channels. For example, if $\bar{v} = 0$ and $\partial d / \partial y = 0$, Eq. 15-30 yields

$$\frac{\partial}{\partial t}(\bar{u}d) + \frac{\partial}{\partial x}(\bar{u}^2 d) + g d \cos \alpha_x \frac{\partial d}{\partial x} = g d (\sin \alpha_x - S_{f_x}) \qquad (15\text{-}32)$$

where α_x is the angle of inclination of the channel bottom.

Equations 15-9, 15-30 and 15-31 may be expressed in a horizontal system of coordinates, $\tilde{x}$-$\tilde{y}$-$\tilde{z}$ (Fig. 15-1). In this coordinate system, channels may have piecewise constant bottom slope. In order to transform from the inclined system x-y-z to the horizontal system $\tilde{x}$-$\tilde{y}$-$\tilde{z}$, it is required to rotate the former coordinate system. A rotation of this type is generally defined by using the direction cosines that give the angles between the axes in both systems. However, in this case it is better to express the rotation as a function of the angles between the bottom of the channel and the x and y axes (α_x and α_y in Fig. 15-1), since these angles are generally known. According to Fig. 15-1 and after some vectorial manipulations, the transformation between both systems of coordinates is given by

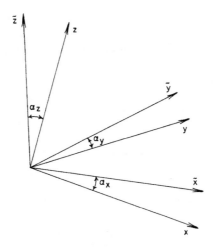

Fig. 15-1   Definition of α_x, α_y and α_z

$$\begin{pmatrix} \tilde{x} \\ \tilde{y} \\ \tilde{z} \end{pmatrix} = \begin{pmatrix} \cos\alpha_x & -\cos\varphi\cos\alpha_x/\sin\varphi & \tan\alpha_x\cos\alpha_z \\ 0 & \cos\alpha_y/\sin\varphi & \tan\alpha_y\cos\alpha_z \\ -\sin\alpha_x & -\sin\alpha_y\cos^2\alpha_x/\sin\varphi & \cos\alpha_z \end{pmatrix} \begin{pmatrix} x \\ y \\ z \end{pmatrix} \quad (15\text{-}33)$$

where $\cos\alpha_z = 1/\sqrt{1 + \tan^2\alpha_x + \tan^2\alpha_y}$; $\cos\varphi = \sin\alpha_x \cos\alpha_y$; and $\sin\varphi = \sqrt{1 - \sin^2\alpha_x \sin^2\alpha_y}$. The terms g_x, g_y and g_z may be computed from Eq. 15-33. For example,

$$g_x = \mathbf{g} \cdot \widehat{\mathbf{e}}_1 = -g\widetilde{\mathbf{e}}_3 \cdot \widehat{\mathbf{e}}_1 = g\sin\alpha_x \quad (15\text{-}34)$$

in which $\widehat{\mathbf{e}}_i$ are the unit vectors of the x-y-z system, and $\widetilde{\mathbf{e}}_i$ are the units vectors of the $\tilde{x}$-$\tilde{y}$-$\tilde{z}$ system. Note that the term $-\sin\alpha_x$ in Eq. 15-34 is the $(3, 1)$ element of the transformation matrix. Hence,

$$g_x = g\sin\alpha_x$$
$$g_y = g\frac{\sin\alpha_y \cos^2\alpha_x}{\sin\varphi}$$
$$g_z = -g\cos\alpha_z \quad (15\text{-}35)$$

where α_x, α_y and α_z are as defined in Fig. 15-1. Some simplification is needed before carrying out the transformation; otherwise the algebra becomes unwieldy. Let us assume that although $\sin\alpha_x$ and $\sin\alpha_y$ may not be small, their product is small, i. e.,

$$\sin \alpha_x \sin \alpha_y \approx \sin^2 \alpha_x \approx \sin^2 \alpha_y \approx 0 \qquad (15\text{-}36)$$

This approximation introduces a small error (< 3 percent) if $|\alpha_x|, |\alpha_y| < 10°$. It follows from Eq. 15-36 that

$$\sin \varphi = 1; \qquad \cos \varphi = 0 \qquad (15\text{-}37)$$

Based on these approximations and Eq. 15-33, we may write

$$\tilde{x} = x \cos \alpha_x + z \tan \alpha_x \cos \alpha_z$$
$$\tilde{y} = y \cos \alpha_y + z \tan \alpha_y \cos \alpha_z \qquad (15\text{-}38)$$

The transformed dependent variables become

$$h = \frac{d}{\cos \alpha_z}$$
$$\tilde{u} = \bar{u} \cos \alpha_x$$
$$\tilde{v} = \bar{v} \cos \alpha_y \qquad (15\text{-}39)$$

in which h is the flow depth measured *vertically* and $\tilde{u}$ and $\tilde{v}$ are the velocity components along the $\tilde{x}$- and $\tilde{y}$-directions respectively. Also, note that according to Eq. 15-38

$$\frac{\partial}{\partial x} = \cos \alpha_x \frac{\partial}{\partial \tilde{x}} + \tan \alpha_x \cos \alpha_z \frac{\partial}{\partial \tilde{z}}$$

$$\frac{\partial}{\partial y} = \cos \alpha_y \frac{\partial}{\partial \tilde{y}} + \tan \alpha_y \cos \alpha_z \frac{\partial}{\partial \tilde{z}} \qquad (15\text{-}40)$$

The presence of derivatives along the $\tilde{z}$-direction is undesirable, since the basic idea is to eliminate a spatial dimension out of the problem. However, it can be shown that terms like $\tan \alpha_x \cos \alpha_z \partial / \partial \tilde{z}$ are of the order of $\sin^2 \alpha_x$ and, therefore, are negligible. Introduction of Eqs. 15-35, 15-38, 15-39 and 15-40 into Eqs. 15-9, 15-30 and 15-31, simplifying, and dropping the $\sim$ symbol yield

$$\frac{\partial h}{\partial t} + \frac{\partial}{\partial x}(uh) + \frac{\partial}{\partial y}(vh) = 0$$
$$\frac{\partial}{\partial t}(uh) + \frac{\partial}{\partial x}(u^2 h) + \frac{\partial}{\partial y}(uvh)$$
$$= gh\Big[\cos \alpha_x S_{o_x} - (\cos \alpha_x \cos \alpha_z)^2 \frac{\partial h}{\partial x} - S_{f_x}\Big]$$
$$\frac{\partial}{\partial t}(vh) + \frac{\partial}{\partial x}(uvh) + \frac{\partial}{\partial y}(v^2 h)$$
$$= gh\Big[\cos \alpha_y S_{o_y} - (\cos \alpha_y \cos \alpha_z)^2 \frac{\partial h}{\partial y} - S_{f_y}\Big] \qquad (15\text{-}41)$$

in which

$$S_{o_x} = \sin \alpha_x; \qquad S_{o_y} = \sin \alpha_y$$

$$S_{f_x} = \frac{u\sqrt{u^2 + v^2}}{(C\cos\alpha_z)^2 h}; \qquad S_{f_y} = \frac{v\sqrt{u^2 + v^2}}{(C\cos\alpha_z)^2 h} \qquad (15\text{-}42)$$

For small channel bottom slope, Eqs. 15-41 may be written as

$$\mathbf{U}_t + \mathbf{E}_x + \mathbf{F}_y + \mathbf{S} = 0 \qquad (15\text{-}43)$$

in which

$$\mathbf{U} = \begin{pmatrix} h \\ uh \\ vh \end{pmatrix}; \quad \mathbf{E} = \begin{pmatrix} uh \\ u^2 h + \frac{1}{2}gh^2 \\ uvh \end{pmatrix}$$

$$\mathbf{F} = \begin{pmatrix} vh \\ uvh \\ v^2 h + \frac{1}{2}gh^2 \end{pmatrix}; \quad \mathbf{S} = \begin{pmatrix} 0 \\ -gh(S_{o_x} - S_{f_x}) \\ -gh(S_{o_y} - S_{f_y}) \end{pmatrix} \qquad (15\text{-}44)$$

and (uh) and (vh) are momenta convected in the x- and y-directions. If the Manning equation is used to compute the friction terms instead of the Chezy equation, then

$$S_{f_x} = \frac{n^2 u\sqrt{u^2 + v^2}}{C_o^2 h^{1.33}}; \qquad S_{f_y} = \frac{n^2 v\sqrt{u^2 + v^2}}{C_o^2 h^{1.33}} \qquad (15\text{-}45)$$

in which n = Manning coefficient and C_o = a dimensional constant ($C_o = 1$ for SI units and $C_o = 1.49$ for Customary units).

In terms of the primitive flow variables h, u, and v, the governing equations may also be written as

$$\mathbf{V}_t + \mathbf{P}_x + \mathbf{R}_y + \mathbf{T} = 0 \qquad (15\text{-}46)$$

in which

$$\mathbf{V} = \begin{pmatrix} h \\ u \\ v \end{pmatrix}; \quad \mathbf{P} = \begin{pmatrix} uh \\ \frac{1}{2}u^2 + gh \\ uv \end{pmatrix}$$

$$\mathbf{R} = \begin{pmatrix} vh \\ uv \\ \frac{1}{2}v^2 + gh \end{pmatrix}; \quad \mathbf{T} = \begin{pmatrix} 0 \\ -g(S_{o_x} - S_{f_x}) \\ -g(S_{o_y} - S_{f_y}) \end{pmatrix} \qquad (15\text{-}47)$$

It is necessary for certain numerical schemes that the equations be in nonconservation form. The nonconservation form of Eqs. 15-43 is

$$\mathbf{U}_t + \mathbf{A}\mathbf{U}_x + \mathbf{B}\mathbf{U}_y + \mathbf{S} = 0 \tag{15-48}$$

in which $\mathbf{A}$ and $\mathbf{B}$ are the Jacobians of $\mathbf{E}$ and $\mathbf{F}$

$$\mathbf{A} = \begin{pmatrix} 0 & 1 & 0 \\ -u^2 + gh & 2u & 0 \\ -uv & v & u \end{pmatrix} \qquad \mathbf{B} = \begin{pmatrix} 0 & 0 & 1 \\ -uv & v & u \\ -v^2 + gh & 0 & 2v \end{pmatrix} \tag{15-49}$$

Similarly, the nonconservation form of Eqs. 15-47 is

$$\mathbf{V}_t + \mathbf{G}\mathbf{V}_x + \mathbf{H}\mathbf{V}_y + \mathbf{T} = 0 \tag{15-50}$$

in which

$$\mathbf{G} = \begin{pmatrix} u & h & 0 \\ g & u & 0 \\ 0 & 0 & u \end{pmatrix}; \qquad \mathbf{H} = \begin{pmatrix} v & 0 & h \\ 0 & v & 0 \\ g & 0 & v \end{pmatrix} \tag{15-51}$$

Matrices $\mathbf{A}$ and $\mathbf{B}$, and $\mathbf{G}$ and $\mathbf{H}$ of Eqs. 15-49 and 15-51 have the important property that their eigenvalues or characteristic directions are identical and are given by

$$\mathbf{A} \text{ and } \mathbf{G} \begin{cases} \lambda_1 &= u \\ \lambda_2 &= u+c \\ \lambda_3 &= u-c \end{cases} \qquad \mathbf{B} \text{ and } \mathbf{H} \begin{cases} \omega_1 &= v \\ \omega_2 &= v+c \\ \omega_3 &= v-c \end{cases} \tag{15-52}$$

where c is the wave celerity $(c = \sqrt{gh})$.

The conservation form of the equations, Eqs. 15-43 and 15-46, has the advantage of being superior in conserving the flow variables as compared to the case when the equations are not in the conservation form. However, note that Eqs. 15-43 and 15-46 are not in full conservation form because of the presence of the vectors $\mathbf{S}$ and $\mathbf{T}$. When the source terms are not equal to zero, they act as sources or sinks. Since the contribution of these terms is usually small, the conservative properties are not significantly impaired.

15-3 Numerical Solution

The governing equations, Eqs. 15-43 and 15-46, are nonlinear first-order, hyperbolic partial differential equations for which analytical solutions are not available, except for very simplified one-dimensional cases. Therefore, they are solved numerically.

For gradually varied two-dimensional, unsteady flows, characteristics, explicit and implicit finite-difference methods have been used by a number of investigators [Leendertse, 1967; Katopodes and Strelkoff, 1978; Benque et al., 1982; Lai, 1986; Fennema and Chaudhry, 1986, 1987, 1989]. The method of characteristics was used by Katopodes and Strelkoff [1978] to simulate the two-dimensional propagation of dam-break flood waves. The bore was isolated and tracked explicitly along the downstream channel. Matsutomi [1983] used the leap-frog scheme to compute dam-break flow profiles over a dry bed. He employed shock-fitting to track the bore, as done by Sakkas and Strelkoff [1973] and used Ritter solution for the first few time steps. Katopodes [1984a] used a finite-element technique based on the Petrov-Galerkin formulation to solve several discontinuous flow situations in two- dimensions. Most of these computational procedures fail if subcritical and supercritical flows are present either simultaneously in different parts of the channel or if they occur in sequence at different times.

The finite-volume methods [Godunov, 1959] have been widely employed to solve 2-D depth-averaged equations. They conserve mass and momentum in each cell and fluxes can be evaluated at the cell faces by solving Riemann problem. Hirsch [1990] proposed computationally efficient Riemann solvers in combination with nonlinear flux limiters to prevent numerical oscillations where flow is highly sheared. Tamamidis and Assanis [1993] discussed the effect of including flux limiters in the higher-order schemes. Zhao et al. [1994] developed a finite-volume method with first-order spatial accuracy on unstructured grid. Toro [1992 and 1999] used a higher-order total variation diminishing versions of Godunov-type methods. Anastasiou and Chan [1997] and Sleigh et al. [1998] used second-order methods on triangular meshes. Mingham and Causon [1998] used a second-order accurate scheme that employs van Leer's Monotonic Upstream Schemes for the conservation laws and a robust approximation for Riemann solver. The artificial neural networks have been used by Dibike and Abbott [1999]and Benning et al. [2001] in conjunction with numerical calculations by identifying the nodes, and subsequently arranging them by the data obtained from numerical calculations to predict the entire flow field. Yoon and Kang [2004] used an upwind finite-volume method on unstructured triangular grids employing a cell-centered formulation. Chua and Holz [2005] proposed an artificial neural network numerical hybrid approach that is capable of producing higher resolution for the flow field at the specified areas.

The multiquadric method was first introduced by Hardy [1971] to approximate topographic surfaces from scattered data points and has received much attention for solving physical problems in the form of differential equations, e.g. Hon et al. [1997], Franke and Schaback [1997], Wong et al. [1999], among others. Hon et al. [1999] applied the multiquadric method to linear and nonlinear shallow water equations; however, this method depends on choosing an optimal shape parameter that depends on the properties of the numerical solution and on the number of collation points and no rigorous complete guidelines are available for choosing it.

The discontinuous Galerkin methods [Cockburn et al., 2000] couple a discontinuous spatial discretization involving flux balance across the interface of the elements with a higher-order polynomial approximation in every element. Therefore, these methods are good candidates for solving complex situations, such as dam break problems.

Fagherazzi et al. [2004] utilized the discontinuous Galerkin method for dam-break problem. They used an efficient Roe approximate Riemann solver to capture steep waves and bores and a projection limiter to eliminates oscillations near the discontinuities. Schwanenberg and Harms [2004] followed a similar approach and solved the shallow water equations using a discontinuous Galerkin finite element method for spatial discretization. The notation used here for the finite-difference mesh in x, y and t space is shown in Fig. 15-2. The x direction is designated by the subscript i, the y direction by the subscript j, and the t direction by the superscript k. The time level where the flow variables are known is denoted by superscript k and the unknown time level is shown by the superscript $k + 1$.

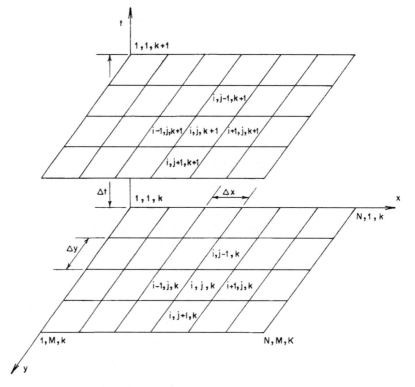

Fig. 15-2   Finite-difference grid

Finite difference methods involving the splitting of flux matrices $\mathbf{A}, \mathbf{B}, \mathbf{G}$ and $\mathbf{H}$ (Eqs. 15-49 and 15-51) have recently been introduced. Algorithms using this concept write finite-difference approximations along the characteristics directions and transmit information only in the correct directions, i.e., from the upwind direction in supercritical flow and from opposite sides in subcritical flow. Then, the compatibility equations are written in finite-difference form and integrated in the characteristic directions. These methods have the advantage over other methods when both subcritical and supercritical flows are present.

It is natural to assume that a good algorithm will be obtained if knowledge of this information is incorporated into the scheme. Basically this requires that the spatial difference operators should be one-sided and obtain information only from the upwind side in supercritical flow and from both sides in subcritical flow. In situations where both types of flows are present, the spatial operators should be switched appropriately to account for the correct signal direction. The method of characteristics has the capability of preserving directional information, but it is quite cumbersome to develop and program in two or more dimensions. In comparison, fixed grid capabilities of finite-difference schemes are attractive. By incorporating information on signal propagation in the finite-difference schemes, algorithms may be developed that are simpler to program, but emulate the method of characteristics.

Explicit and implicit finite-difference schemes are introduced in this chapter for the solution of the governing equations. These schemes are second-order accurate in both space and time, predict the location and height of the bores without the use of shock-fitting, and allow initial conditions with discontinuities. The steep wave fronts are usually spread only over a few mesh points. Fennema and Chaudhry used these schemes for the solution of one-dimensional [1986; 1987] and two-dimensional [1989] open-channel flows. Some of these schemes are capable of simulating both subcritical and supercritical flows.

15-4 MacCormack Scheme

Difference methods investigated by Lax and Wendroff [1960] have become very popular for solving hyperbolic systems. These methods, known as two-step schemes, are based on second-order Taylor series expansions in time. An interesting and simpler variation of the Lax-Wendroff scheme was introduced by MacCormacK [1969] and has been widely used in computational fluid dynamics. For linear problems, the scheme is identical to the original Lax-Wendroff method. For the analysis of two-dimensional open-channel flows, the scheme has been used by Fennema and Chaudhry [1989, 1990] and by Garcia and Kahawita [1986].

General formulation

This scheme consists of a two-step predictor-corrector sequence. Flow variables are known at time level k and their values are to be determined at time level $k + 1$. Then for grid points i, j, the following finite difference equations may be written from Eq. 15-43:

Predictor

$$\mathbf{U}^*_{i,j} = \mathbf{U}^k_{i,j} - \frac{\Delta t}{\Delta x}\nabla_x\mathbf{E}^k_{i,j} - \frac{\Delta t}{\Delta y}\nabla_y\mathbf{F}^k_{i,j} - \Delta t\,\mathbf{S}^k_{i,j} \quad \begin{cases} 2 \leq i \leq N \\ 2 \leq j \leq M \end{cases} \tag{15-53}$$

Corrector

$$\mathbf{U}^{**}_{i,j} = \mathbf{U}^k_{i,j} - \frac{\Delta t}{\Delta x}\Delta_x\mathbf{E}^*_{i,j} - \frac{\Delta t}{\Delta y}\Delta_y\mathbf{F}^*_{i,j} - \Delta t\,\mathbf{S}^*_{i,j} \quad \begin{cases} 1 \geq i \geq N - 1 \\ 1 \geq j \geq M - 1 \end{cases} \tag{15-54}$$

in which $\mathbf{U}^*$ and $\mathbf{U}^{**}$ are intermediate values for $\mathbf{U}$. The new values for the vector $\mathbf{U}$ at time $k + 1$ are obtained from

$$\mathbf{U}^{k+1}_{i,j} = \frac{1}{2}(\mathbf{U}^*_{i,j} + U^{**}_{i,j}) \tag{15-55}$$

The grid points i, j and k are as defined in Fig. 15-2. The scheme first uses backward space differences (∇_x and ∇_y) to predict an intermediate solution from known information at time-level k. The forward space differences (Δ_x and Δ_y) are used in the second step to correct the predicted values. The forward (Δ) and backward (∇) difference operators are defined as

$$\Delta_x\mathbf{U}_{i,j} = \mathbf{U}_{i+1,j} - \mathbf{U}_{i,j}$$
$$\nabla_x\mathbf{U}_{i,j} = \mathbf{U}_{i,j} - \mathbf{U}_{i-1,j} \tag{15-56}$$

where the subscript indicates the direction of differencing. The corrector step always uses one-sided differences opposite to the ones used in the predictor part; in this case forward finite-differences are used.

The differencing may be reversed or it may be alternated each time step. An example of a sequence that repeats itself every fourth time step is given in Fig. 15-3. This sequence removes most of the directional bias of this scheme that would otherwise be present if steps were not alternated. The differencing shown in the first sketch, the k time step, is equivalent to the differencing used in Eqs. 15-53 and 15-54.

The values of primitive variables are determined from the computed value of $\mathbf{U}$ at each step as follows:

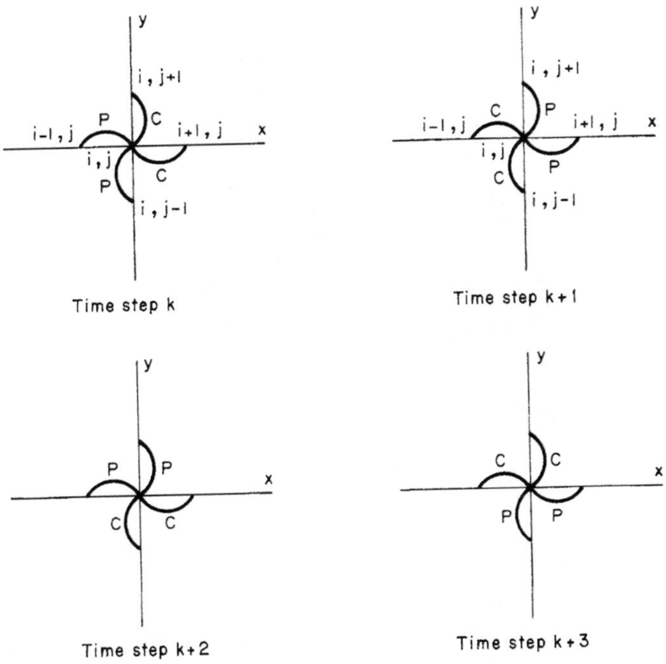
y
i , j+1
i-1 , j
P
C
i+1 , j
x
i , j
C
P
i , j-1
Time step k
y
i , j+1
i-1 , j
C
P
i+1 , j
x
i , j
P
C
i , j-1
Time step k+1
y
P
P
x
C
C
Time step k+2
y
C
C
x
P
P
Time step k+3

Fig. 15-3 Differencing sequence

$$
\begin{aligned}
h^{k+1} &= h^{k+1} \\
u^{k+1} &= \frac{(uh)^{k+1}}{h^{k+1}} \\
v^{k+1} &= \frac{(vh)^{k+1}}{h^{k+1}}
\end{aligned}
\tag{15-57}
$$

in which $k + 1$ refers to an intermediate value obtained during a current predictor or corrector sequence.

Boundary conditions

Reflection boundaries may be easily incorporated in the MacCormack scheme. In this procedure, fictitious points in the solid wall are replaced by immediate interior points. Antisymmetric reflection is incorporated by changing the sign of the normal component of the velocity. To illustrate, consider a solid boundary along the y-direction with the computational domain in the positive x-direction (Fig. 15-4). For the predictor step, the difference equations using the conservation form become

$$h_{i,j}^* = h_{i,j}^k - \frac{\Delta t}{\Delta x}\left[(uh)_{i,j}^k + (uh)_{i+1,j}^k\right] - \frac{\Delta t}{\Delta y}\left[(vh)_{i,j}^k - (vh)_{i,j-1}^k\right]$$

$$(uh)_{i,j}^* = (uh)_{i,j}^k - \frac{\Delta t}{\Delta x}\left[\left(u^2h + \frac{1}{2}gh^2\right)_{i,j}^k - \left(u^2h + \frac{1}{2}gh^2\right)_{i+1,j}^k\right]$$

$$- \frac{\Delta t}{\Delta y}\left[(uvh)_{i,j}^k - (uvh)_{i,j-1}^k\right] + gh_{i,j}^k \Delta t \left(S_{o_x} - S_{f_x}\right)_{i,j}^k$$

$$(vh)_{i,j}^* = (vh)_{i,j}^k - \frac{\Delta t}{\Delta x}\left[(uvh)_{i,j}^k + (uvh)_{i+1,j}^k\right]$$

$$- \frac{\Delta t}{\Delta y}\left[\left(v^2h + \frac{1}{2}gh^2\right)_{i,j}^k - \left(v^2h + \frac{1}{2}gh^2\right)_{i,j-1}^k\right]$$

$$+ gh_{i,j}^k \Delta t \left(S_{o_y} - S_{f_y}\right)_{i,j}^k \tag{15-58}$$

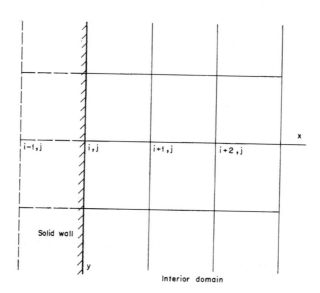

Fig. 15-4   Reflection boundary

In these equations, the values at all the $(i-1, j)$ points have been replaced by values at $(i+1, j)$ points and the sign of the normal velocity component, u, has been switched. No difficulty arises during the corrector step because none of the differences are from the interior of the solid boundary.

15-5 Gabutti Scheme

An explicit scheme based on the characteristic relations was introduced by Morreti [1979]; he called it the λ-scheme. The algorithm is based on the non-conservative equations (Eqs. 15-50) and relies on the theory of characteristics. Gabutti [1983] analyzed the numerical properties and presented an improved scheme. The resulting algorithm is an explicit scheme which consists of three sequential steps. This class of finite-difference schemes is also referred to as the class of split-coefficient matrix (SCM) methods. The objective is to write difference equations that transmit information from the relevant characteristic directions. This is accomplished by diagonalizing and splitting the coefficient matrices so that each part incorporates only positive or negative contributions of the eigenvalues.

General formulation

The algorithm is developed in the following manner. The Jacobians Eqs. 15-51 may be diagonalized by transformation matrices (see, for example, Anton, 1981) which are made up of the right eigenvectors of $\mathbf{G}$ and $\mathbf{H}$. Eqs. 15-50 may then be written as

$$\mathbf{V}_t + \mathbf{M}\mathbf{D}_G\mathbf{M}^{-1}\mathbf{V}_x + \mathbf{N}\mathbf{D}_H\mathbf{N}^{-1}\mathbf{V}_y + \mathbf{T} = 0 \tag{15-59}$$

in which $\mathbf{M}$ and $\mathbf{N}$ are defined by

$$\mathbf{M} = \begin{pmatrix} 0 & \frac{h}{2c} & \frac{h}{2c} \\ 0 & \frac{1}{2} & -\frac{1}{2} \\ 1 & 0 & 0 \end{pmatrix}; \quad \mathbf{N} = \begin{pmatrix} 0 & \frac{h}{2c} & \frac{h}{2c} \\ 1 & 0 & 0 \\ 0 & \frac{1}{2} & -\frac{1}{2} \end{pmatrix} \tag{15-60}$$

in which $\mathbf{D}_G$ and $\mathbf{D}_H$ are diagonal matrices of the eigenvalues presented in Eqs. 15-52. Diagonalization of the flux matrices leads to

$$\mathbf{G} = \mathbf{M}\mathbf{D}_G\mathbf{M}^{-1} = \begin{pmatrix} \frac{1}{2}(\lambda_2 + \lambda_3) & \frac{h}{2c}(\lambda_2 - \lambda_3) & 0 \\ \frac{c}{2h}(\lambda_2 - \lambda_3) & \frac{1}{2}(\lambda_2 + \lambda_3) & 0 \\ 0 & 0 & \lambda_1 \end{pmatrix} \tag{15-61}$$

$$\mathbf{H} = \mathbf{N}\mathbf{D}_H\mathbf{N}^{-1} = \begin{pmatrix} \frac{1}{2}(\omega_2 + \omega_3) & 0 & \frac{h}{2c}(\omega_2 - \omega_3) \\ 0 & \omega_1 & 0 \\ \frac{c}{2h}(\omega_2 - \omega_3) & 0 & \frac{1}{2}(\omega_2 + \omega_3) \end{pmatrix} \tag{15-62}$$

The diagonal matrices $\mathbf{D}_G$ and $\mathbf{D}_H$ may be split into positive and negative parts, $\mathbf{D}_G = \mathbf{D}_G^+ + \mathbf{D}_G^-$ and $\mathbf{D}_H = \mathbf{D}_H^+ + \mathbf{D}_H^-$. This can be accomplished in a number of ways. For example, one alternative is to test each eigenvalue by

$$\lambda_l^+ = \max(\lambda_l, 0); \quad \omega_l^+ = \max(\omega_l, 0)$$
$$\lambda_l^- = \min(\lambda_l, 0); \quad \omega_l^- = \min(\omega_l, 0) \tag{15-63}$$

In this manner, matrices are obtained which contain only positive or negative parts. The flux terms are thus split into

$$\mathbf{G} = \mathbf{G}^+ + \mathbf{G}^- = \mathbf{M}\mathbf{D}_G^+\mathbf{M}^{-1} + \mathbf{M}\mathbf{D}_G^-\mathbf{M}^{-1}$$
$$\mathbf{H} = \mathbf{H}^+ + \mathbf{H}^- = \mathbf{N}\mathbf{D}_H^+\mathbf{N}^{-1} + \mathbf{N}\mathbf{D}_H^-\mathbf{N}^{-1} \tag{15-64}$$

Incorporating these split flux terms in the system leads to the equation

$$\mathbf{V}_t + \mathbf{G}^+\mathbf{V}_x + \mathbf{G}^-\mathbf{V}_x + \mathbf{H}^+\mathbf{V}_y + \mathbf{H}^-\mathbf{V}_y + \mathbf{T} = 0 \tag{15-65}$$

Each split Jacobian matrix multiplies a spatial derivative, $\mathbf{V}_x$ or $\mathbf{V}_y$; thus with the established direction of the fluxes, appropriate differencing can be taken. With a positive coefficient matrix, a backward difference is used, e.g., $\mathbf{G}^+\mathbf{V}_x \approx \mathbf{G}_i^+ (\mathbf{V}_i - \mathbf{V}_{i-1})/\Delta x$ and with a negative coefficient matrix, a forward difference is used, e.g., $\mathbf{G}^-\mathbf{V}_x \approx \mathbf{G}_i^- (\mathbf{V}_{i+1} - \mathbf{V}_i)/\Delta x$. The propagation of the signal is now correctly applied along the characteristics. In subcritical flow, both positive and negative characteristics are used whereas in supercritical flow the information is carried only along the characteristics from the direction of flow. The spatial and temporal differencing can be implemented in many ways. In the Gabutti scheme the algorithm takes the following form

Predictor

Part A

$$\widetilde{\mathbf{V}}_{i,j} = \mathbf{V}_{i,j}^k - \frac{\Delta t}{\Delta x}\left(\mathbf{G}^+\nabla_x\mathbf{V}_{i,j}^k + \mathbf{G}^-\Delta_x\mathbf{V}_{i,j}^k\right)$$
$$- \frac{\Delta t}{\Delta y}\left(\mathbf{H}^+\nabla_y\mathbf{V}_{i,j}^k + \mathbf{H}^-\Delta_y\mathbf{V}_{i,j}^k\right)$$
$$- \Delta t\,\mathbf{T}_{i,j}^k = 0 \tag{15-66}$$

Predictor

Part B

$$\widehat{\mathbf{V}}_{i,j} = \mathbf{V}_{i,j}^k - \frac{\Delta t}{\Delta x}\left[\mathbf{G}^+(1 + \nabla_x)\nabla_x\mathbf{V}_{i,j}^k + \mathbf{G}^-(1 - \Delta_x)\Delta_x\mathbf{V}_{i,j}^k\right]$$
$$- \frac{\Delta t}{\Delta y}\left[\mathbf{H}^+(1 + \nabla_y)\nabla_y\mathbf{V}_{i,j}^k + \mathbf{H}^-(1 - \Delta_y)\Delta_y\mathbf{V}_{i,j}^k\right]$$
$$- \Delta t\,\mathbf{T}_{i,j}^k = 0 \tag{15-67}$$

Corrector

$$\overline{\mathbf{V}}_{i,j} = \widetilde{\mathbf{V}}_{i,j} - \frac{\Delta t}{\Delta x}\left(\widetilde{\mathbf{G}}^+ \nabla_x \widetilde{\mathbf{V}}_{i,j} + \widetilde{\mathbf{G}}^- \Delta_x \widetilde{\mathbf{V}}_{i,j}\right)$$
$$- \frac{\Delta t}{\Delta y}\left(\widetilde{\mathbf{H}}^+ \nabla_y \widetilde{\mathbf{V}}_{i,j} + \widetilde{\mathbf{H}}^- \Delta_y \widetilde{\mathbf{V}}_{i,j}\right)$$
$$- \Delta t\, \widetilde{\mathbf{T}}_{i,j} = 0 \qquad (15\text{-}68)$$

where $\widetilde{\mathbf{V}}$, $\widehat{\mathbf{V}}$ and $\overline{\mathbf{V}}$ are intermediate values for $\mathbf{V}$. The combination of the operators in the predictor step, part B are the difference approximations given by

$$(1 + \nabla_x)\nabla_x \mathbf{V}_{i,j} = 2\mathbf{V}_{i,j} - 3\mathbf{V}_{i-1,j} + \mathbf{V}_{i-2,j}$$
$$(1 - \Delta_x)\Delta_x \mathbf{V}_{i,j} = -2\mathbf{V}_{i,j} + 3\mathbf{V}_{i+1,j} - \mathbf{V}_{i+2,j}$$
$$(15\text{-}69)$$

At the end of each sequence the new values for $\mathbf{V}$ are obtained from
$$\mathbf{V}_{i,j}^{k+1} = \frac{1}{2}(\mathbf{V}_{i,j}^k + \widehat{\mathbf{V}}_{i,j} + \overline{\mathbf{V}}_{i,j} - \widetilde{\mathbf{V}}_{i,j}) \qquad (15\text{-}70)$$

Boundary conditions

Boundaries based on characteristic principles may be included in the Gabutti scheme. The compatibility equations valid along each characteristic aligned with the axes of the computational domain are obtained first, and the appropriate one is replaced by a specified boundary condition. The system is multiplied by the eigenvectors associated with the respective characteristic directions. In the case of a solid boundary shown in Fig. 15-4, the relevant eigenvectors corresponding to the eigenvalues given by Eqs. 15-52 are contained in $\mathbf{M}^{-1}$. Premultiplying each term of Eq. 15-59 by $\mathbf{M}^{-1}$ gives

$$\mathbf{M}^{-1}\mathbf{V}_t + \mathbf{D}_G\mathbf{M}^{-1}\mathbf{V}_x + \mathbf{M}^{-1}\mathbf{N}\mathbf{D}_H\mathbf{N}^{-1}\mathbf{V}_y + \mathbf{M}^{-1}\mathbf{T} = 0 \qquad (15\text{-}71)$$

The incoming characteristic, $\lambda_2 = u + c$, is associated with the second eigenvector $\mathbf{M}^{-1}$. The information along λ_2 comes from the wall and is replaced by the boundary condition $u = 0$ or $u_t = 0$. Thus the equations at the wall, after simplification, may be written as follows:

$$h_t + \lambda_3 h_x^- - \lambda_3 \frac{h}{c}u_x^- + \frac{1}{2}(\omega_2 + \omega_3)h_y - \frac{h}{c}\omega_1 u_y$$
$$+ \frac{h}{2c}(\omega_2 - \omega_3)v_y + \frac{gh}{c}(S_{o_x} - S_{f_x}) = 0$$

$$u_t = 0$$

$$v_t + \frac{c}{2h}(\omega_2 - \omega_3)h_y + \frac{1}{2}(\omega_2 + \omega_3)v_y - g(S_{o_y} - S_{f_y}) = 0$$
$$(15\text{-}72)$$

where the superscript indicates that a forward difference should be used. The first equation is solved for the head h^{k+1} using information along the backward characteristic $\lambda_1 = u - c$. The third equation is solved for v^{k+1} by using information along the path or world line $\lambda_3 = u$. These are the equations for solving the flow variables h and v at solid boundaries oriented parallel to the y-axis with the computational domain facing the positive x-direction. Similar equations may be written for boundaries with different orientations.

15-6 Artificial Viscosity

A characteristic of many second-order finite-difference schemes is that they produce numerical oscillations near discontinuities for Courant numbers less than one. These oscillations are due to truncation errors and are part of the diffusive properties of the scheme. Generally, if the leading term of the truncation error has odd derivatives, dispersive errors occur in the form of wiggles near the bore. It may be necessary to add an explicit damping term to smooth these oscillations. The procedure used here was developed by Jameson et al. [1981] and has the advantage of smoothing regions of sharp gradients while leaving relatively smooth areas undisturbed. These artificial dissipative terms are added to Eq. 15-43 in the form

$$\mathbf{U}_t + \mathbf{E}_x + \mathbf{F}_y + \mathbf{S} - \mathbf{D}\mathbf{U} = 0 \qquad (15\text{-}73)$$

where $\mathbf{D}$ is a dissipative operator along the principal axes defined by $\mathbf{D}\mathbf{U} = \mathbf{D}_x\mathbf{U} + \mathbf{D}_y\mathbf{U}$. Using second-order differences, the operator in the x-direction becomes

$$\mathbf{D}_x\mathbf{U} = \left[\epsilon_{x_{i+\frac{1}{2},j}}(\mathbf{U}_{i+1,j} - \mathbf{U}_{i,j}) - \epsilon_{x_{i-\frac{1}{2},j}}(\mathbf{U}_{i,j} - \mathbf{U}_{i-1,j})\right] \qquad (15\text{-}74)$$

in which ϵ_x is a parameter defined from a normalized form of the gradients of one variable (e.g., h) as

$$\nu_{x_{i,j}} = \frac{|h_{i+1,j} - 2h_{i,j} + h_{i-1,j}|}{|h_{i+1,j}| + |2h_{i,j}| + |h_{i-1,j}|}$$

$$\epsilon_{x_{i-\frac{1}{2},j}} = \kappa\frac{\Delta x}{\Delta t}\max(\nu_{x_{i-1,j}}, \nu_{x_{i,j}}) \qquad (15\text{-}75)$$

where κ is used to regulate the amount of dissipation. The computed variables are then modified by

$$\mathbf{U}_{i,j}^{k+1} = \mathbf{U}_{i,j}^{k+1} + \mathbf{D}_x\mathbf{U}_{i,j}^{k+1} + \mathbf{D}_y\mathbf{U}_{i,j}^{k+1} \qquad (15\text{-}76)$$

This statement should be viewed as a replacement statement. The terms are added after a predetermined number of timesteps, using the latest values of h, u, and v.

15-7 Beam and Warming Schemes

In this section, implicit finite-difference schemes are presented for the solution of the governing equations. Beam and Warming [1976, 1978] developed them for the solution of hyperbolic systems in conservation form and they have been used in computational fluid dynamics. The schemes are noniterative, which results in a considerable saving in computer time, especially in multi-dimensional problems. Most formulations of this scheme, discussed in the following paragraphs, are second-order accurate in time and can be made second- or fourth-order accurate in space.

The general formulation of the time differencing is presented first, followed by an efficient solution algorithm using an alternating direction implicit (ADI) procedure. Switching techniques, incorporated in the schemes, allow the analysis of flows if both subcritical and supercritical flows are present simultaneously. The inclusion of the boundary conditions of the numerical schemes is discussed.

General formulation

The system of Eqs. 15-43 may be solved by using time-difference approximations of the general form

$$
\mathbf{U}^{k+1} = \mathbf{U}^k + \Delta t \left[\frac{\theta}{1+\xi} \left(\frac{\partial \mathbf{U}}{\partial t} \right)^{k+1} + \frac{1-\theta}{1+\xi} \left(\frac{\partial \mathbf{U}}{\partial t} \right)^{k} + \frac{\xi}{1+\xi} \left(\frac{\partial \mathbf{U}}{\partial t} \right)^{k-1} \right]
$$

$$(15\text{-}77)$$

in which θ and ξ are parameters leading to a variety of schemes. Some of the common formulations [Richtmyer and Morton, 1967] are listed in Table 15-1.

Table 15-1 **Different formulations of Beam and Warming Schemes**

Scheme	θ	ξ
Euler Implicit (Backward Euler)	1	0
Three-Point Backward	1	$\frac{1}{2}$
Trapezoidal Formula (Crank Nicolson)	$\frac{1}{2}$	0

Substitution for $\partial \mathbf{U}/\partial t$ from Eq. 15-43 in terms of the flux and source terms $\mathbf{E}$, $\mathbf{F}$, and $\mathbf{S}$ yields

$$\mathbf{U}^{k+1} = \mathbf{U}^k - \Delta t \left[\frac{\theta}{1+\xi} \left(\frac{\partial \mathbf{E}}{\partial x} + \frac{\partial \mathbf{F}}{\partial y} + \mathbf{S} \right)^{k+1} \right.$$
$$\left. + \frac{1-\theta}{1+\xi} \left(\frac{\partial \mathbf{E}}{\partial x} + \frac{\partial \mathbf{F}}{\partial y} + \mathbf{S} \right)^k \right] + \frac{\xi \Delta t}{1+\xi} \left(\frac{\partial \mathbf{U}}{\partial t} \right)^{k-1}$$

$$(15\text{-}78)$$

Spatial finite-difference approximations will be substituted for the space derivatives later in the development, after a convenient form of the time discretization is obtained. The nonlinearity of the flux vectors, $\mathbf{E}^{k+1}, \mathbf{F}^{k+1}$ and $\mathbf{S}^{k+1}$, presents some difficulty, since they exist at the advanced time level. However, these are all functions of the flow variables, $\mathbf{U}$, for which solutions are to be obtained. These flux vectors may be linearized by using a local Taylor series expansion. For instance, the expansion of $\mathbf{E}^{k+1}$ yields

$$\mathbf{E}^{k+1} = \mathbf{E}^k + \Delta t \frac{\partial \mathbf{E}^k}{\partial t} + \frac{(\Delta t)^2}{2} \frac{\partial^2 \mathbf{E}^k}{\partial t^2} + \cdots \qquad (15\text{-}79)$$

Using the chain rule, $\partial \mathbf{E}^k / \partial t = (\partial \mathbf{E}^k / \partial \mathbf{U}) (\partial \mathbf{U}^k / \partial t)$. Since $\partial \mathbf{E}^k / \partial \mathbf{U}$ is simply the Jacobian $\mathbf{A}^k$, we may write $\partial \mathbf{E}^k / \partial t = \mathbf{A}^k \partial \mathbf{U}^k / \partial t$. Second-order accuracy is obtained if only the first two terms on the right-hand side of Eq. 15-79 are retained. Thus, retaining these two terms, substituting $\partial \mathbf{E}^k / \partial t = \mathbf{A}^k \partial \mathbf{U}^k / \partial t$ and writing $\partial \mathbf{U}^k / \partial t$ in difference form, we obtain

$$\mathbf{E}^{k+1} = \mathbf{E}^k + \mathbf{A}^k (\mathbf{U}^{k+1} - \mathbf{U}^k) \qquad (15\text{-}80)$$

Similarly, expansions for $\mathbf{F}^{k+1}$ and $\mathbf{S}^{k+1}$ may be written as

$$\mathbf{F}^{k+1} = \mathbf{F}^k + \mathbf{B}^k (\mathbf{U}^{k+1} - \mathbf{U}^k)$$
$$\mathbf{S}^{k+1} = \mathbf{S}^k + \mathbf{Q}^k (\mathbf{U}^{k+1} - \mathbf{U}^k) \qquad (15\text{-}81)$$

where $\mathbf{B}$ and $\mathbf{Q}$ are the Jacobian of $\mathbf{F}$ and $\mathbf{S}$, respectively. Substitution of Eqs. 15-80 and 15-81 into Eqs. 15-78 and combining terms of the same time level transforms the system into the following set of equations which contain $\mathbf{U}$ terms at the $(k+1)$ and k time levels

$$\mathbf{U}^{k+1} = \mathbf{U}^k - \Delta t \left[\frac{\theta}{1+\xi} \left(\frac{\partial}{\partial x} \mathbf{A}^k \mathbf{U}^{k+1} + \frac{\partial}{\partial y} \mathbf{B}^k \mathbf{U}^{k+1} + \mathbf{Q}^k \mathbf{U}^{k+1} \right) \right.$$
$$- \frac{\theta}{1+\xi} \left(\frac{\partial}{\partial x} \mathbf{A}^k \mathbf{U}^k + \frac{\partial}{\partial y} \mathbf{B}^k \mathbf{U}^k + \mathbf{Q}^k \mathbf{U}^k \right)$$
$$\left. + \frac{1}{1+\xi} \left(\frac{\partial \mathbf{E}}{\partial x} + \frac{\partial \mathbf{F}}{\partial y} + \mathbf{S} \right)^k \right] + \Delta t \frac{\xi}{1+\xi} \left(\frac{\partial \mathbf{U}}{\partial t} \right)^{k-1} \qquad (15\text{-}82)$$

Transposing the dependent variables at the advanced time level to the left-hand side of this equation yields a linear system for $\mathbf{U}^{k+1}$:

$$\left[\mathbf{I} + \Delta t \frac{\theta}{1+\xi} \left(\frac{\partial}{\partial x} \mathbf{A}^k + \frac{\partial}{\partial y} \mathbf{B}^k + \mathbf{Q}^k \right) \right] \mathbf{U}^{k+1}$$

$$= \left[\mathbf{I} + \Delta t \frac{\theta}{1+\xi} \left(\frac{\partial}{\partial x} \mathbf{A}^k + \frac{\partial}{\partial y} \mathbf{B}^k + \mathbf{Q}^k \right) \right] \mathbf{U}^k$$

$$- \Delta t \frac{1}{1+\xi} \left(\frac{\partial \mathbf{E}}{\partial x} + \frac{\partial \mathbf{F}}{\partial y} + \mathbf{S} \right)^k + \Delta t \frac{\xi}{1+\xi} \left(\frac{\partial \mathbf{U}}{\partial t} \right)^{k-1} \qquad (15\text{-}83)$$

in which $\mathbf{I}$ is the unit matrix and $(\partial/\partial x \, \mathbf{A}^k + \partial/\partial x \, \mathbf{B}^k) \mathbf{U}^{k+1}$ is to be interpreted as $\partial/\partial x \, (\mathbf{A}^k \mathbf{U}^{k+1}) + \partial/\partial x \, (\mathbf{B}^k \mathbf{U}^{k+1})$, i.e., the vector $\mathbf{U}$ is to be evaluated inside the derivatives. Note that the terms inside the brackets on both the left-hand and right-hand side of Eqs. 15-83 are identical. By using a forward difference operator, $\Delta_t \mathbf{U}^{k+1} = \mathbf{U}^{k+1} - \mathbf{U}^k$, and replacing the last terms of Eqs. 15-83 by a forward difference operator, Eqs. 15-83 may be written as

$$\left[\mathbf{I} + \Delta t \frac{\theta}{1+\xi} \left(\frac{\partial}{\partial x} \mathbf{A}^k + \frac{\partial}{\partial y} \mathbf{B}^k + \mathbf{Q}^k \right) \right] \Delta_t \mathbf{U}^{k+1}$$

$$= - \Delta t \frac{1}{1+\xi} \left(\frac{\partial \mathbf{E}}{\partial x} + \frac{\partial \mathbf{F}}{\partial y} + \mathbf{S} \right)^k + \frac{\xi}{1+\xi} \Delta_t \mathbf{U}^k$$

$$(15\text{-}84)$$

The above algorithm is said to be in delta form; the flow variables, $\mathbf{U}$, exist only in increments of $\mathbf{U}$ between two time levels. The principal advantage of this formulation is the computational efficiency due to a reduction in number of terms.

The order of the spatial differencing may be different between the right- and left-hand side of the equations. The preceding schemes lead to an unwieldy inversion (solution of a linear system) problem. The coefficients in the brackets have a band width of $2N$ in the matrix due to the addition of the components of $\mathbf{B}$. For example, an inversion of an 80-band matrix is needed for 40 mesh points in the x-direction A more efficient solution algorithm is obtained by factoring the left-hand side of Eqs. 15-84, as discussed in the following paragraphs.

Factored schemes

A matrix with a small band width can be obtained by reducing the two-dimensional problem to two one-dimensional problems. By the method of approximate factorization (AF), the left-hand side of the implicit schemes ($\theta \neq 0$) described by Eqs. 15-84 can be rewritten as a product of two components, each containing the terms of a specific direction:

$$\left[\mathbf{I} + \Delta t \frac{\theta}{1+\xi} \frac{\partial}{\partial x} \mathbf{A}^k\right] \left[\mathbf{I} + \Delta t \frac{\theta}{1+\xi} \left(\frac{\partial}{\partial y}\mathbf{B} + \mathbf{Q}\right)^k\right] \Delta_t \mathbf{U}^{k+1}$$
$$= -\Delta t \frac{1}{1+\xi} \left(\frac{\partial \mathbf{E}}{\partial x} + \frac{\partial \mathbf{F}}{\partial y} + \mathbf{S}\right)^k + \frac{\xi}{1+\xi} \Delta_t \mathbf{U}^k$$

$$(15\text{-}85)$$

This formulation contains two additional terms obtained by the multiplication on the left-hand side. Since the lowest order of accuracy of the scheme is $O(\Delta t)^2$, and since the additional terms are of the same order, the formal accuracy of the schemes is not affected by the new terms. Thus, without compromising the accuracy of the solution, the equations are factored in a series of steps which are directionally dependent. These steps are

$$\left[\mathbf{I} + \Delta t \frac{\theta}{1+\xi} \frac{\partial}{\partial x} \mathbf{A}^k\right] \Delta_t \widehat{\mathbf{U}} = -\Delta t \frac{1}{1+\xi} \left(\frac{\partial \mathbf{E}}{\partial x} + \frac{\partial \mathbf{F}}{\partial y} + \mathbf{S}\right)^k + \frac{\xi}{1+\xi} \Delta_t \mathbf{U}^k$$

$$\left[\mathbf{I} + \Delta t \frac{\theta}{1+\xi} \left(\frac{\partial}{\partial y}\mathbf{B} + \mathbf{Q}\right)^k\right] \Delta_t \mathbf{U}^{k+1} = \Delta_t \widehat{\mathbf{U}}$$

$$\mathbf{U}^{k+1} = \mathbf{U}^k + \Delta_t \mathbf{U}^{k+1}$$

$$(15\text{-}86)$$

where $\Delta_t \widehat{\mathbf{U}}$ is an intermediate value obtained by solving the system first along the rows (x-direction). This is an alternating-direction implicit (ADI) procedure. With these sequential steps, the inversion is reduced to solving a small band-width matrix along each row and column, which is more efficient. To allow for the presence of both subcritical and supercritical flows, a switching technique is used to obtain an appropriate spatial differencing. This switching technique is incorporated into the scheme by splitting the flux matrices into components containing either the positive or negative parts.

Implicit split-flux factoring

Flux splitting may be incorporated in the approximate-factored algorithm, Eqs. 15-86 as follows. The submatrices are used to operate on the flow variables with appropriate space differences. Substitution of the diagonalized matrices into the algorithm gives an implicit split-flux factored scheme, which may be implemented by the sequence

$$\left[\mathbf{I} + \Delta t \frac{\theta}{1+\xi} \frac{\partial}{\partial x} \left(\mathbf{A}^+ + \mathbf{A}^-\right)^k\right] \Delta_t \widehat{\mathbf{U}}$$

$$= -\Delta t \frac{1}{1+\xi} \left(\frac{\partial \mathbf{E}^+}{\partial x} + \frac{\partial \mathbf{E}^-}{\partial x} + \frac{\partial \mathbf{F}^+}{\partial y} + \frac{\partial \mathbf{F}^-}{\partial y} + \mathbf{S}\right)^k$$

$$+ \frac{\xi}{1+\xi} \Delta_t \mathbf{U}^k$$

$$\left[\mathbf{I} + \Delta t \frac{\theta}{1+\xi} \left(\frac{\partial}{\partial y}\mathbf{B}^+ + \frac{\partial}{\partial y}\mathbf{B}^- + \mathbf{Q}\right)^k\right] \Delta_t \mathbf{U}^{k+1} = \Delta_t \widehat{\mathbf{U}}$$

$$\mathbf{U}^{k+1} = \mathbf{U}^k + \Delta_t \mathbf{U}^{k+1} \tag{15-87}$$

where the sign of the flux components indicates the use of the split form – e.g., $\mathbf{A}^+ = \mathbf{M}\mathbf{D}^+\mathbf{M}^{-1}$;

$$\mathbf{M} = \begin{pmatrix} 0 & \frac{h}{2c} & \frac{h}{2c} \\ 0 & \frac{1}{2} & -\frac{1}{2} \\ 1 & 0 & 0 \end{pmatrix} \tag{15-88}$$

and $\mathbf{D}$ is the diagonal matrix of the eigenvalues of $\mathbf{A}$. The right-hand side of Eqs. 15-87 are evaluated as $\mathbf{E}_x = \mathbf{A}\mathbf{U}_x$ and $\mathbf{F}_y = \mathbf{B}\mathbf{U}_y$.

Further factorization of Eqs. 15-87 is possible since the left-hand side of the first step can be split into terms containing $\partial \mathbf{A}^+/\partial x$ and $\partial \mathbf{A}^-/\partial x$. This leads to a further reduction in the band-width of the coefficient matrix. When using second- or third-order accurate space differencing, some advantage may be gained by this additional factoring. However, when using first-order accurate spatial differencing, the nonfactored coefficient matrix is block-tridiagonal for which efficient solution algorithms are available. The complete finite-difference equations with first order spatial differencing (Fig. 15-2) become

$$\left[\mathbf{I} + \frac{\Delta t}{\Delta x} \frac{\theta}{1+\xi} \left(\nabla_x \mathbf{A}_{i,j}^+ + \Delta_x \mathbf{A}_{i,j}^- \right)^k \right] \Delta_t \widehat{\mathbf{U}}_{i,j} =$$

$$- \frac{\Delta t}{\Delta x} \frac{1}{1+\xi} \left(\mathbf{A}_{i,j}^+ \nabla_x \mathbf{U}_{i,j} + \mathbf{A}_{i,j}^- \Delta_x \mathbf{U}_{i,j} \right)^k$$

$$- \frac{\Delta t}{\Delta y} \frac{1}{1+\xi} \left(\mathbf{B}_{i,j}^+ \nabla_y \mathbf{U}_{i,j} + \mathbf{B}_{i,j}^- \Delta_y \mathbf{U}_{i,j} \right)^k$$

$$- \Delta t \frac{1}{1+\xi} \mathbf{S}_{i,j}^k + \frac{\xi}{1+\xi} \Delta_t \mathbf{U}_{i,j}^k$$

$$\left[\mathbf{I} + \Delta t \frac{\theta}{1+\xi} \left(\frac{1}{\Delta y} \left(\nabla_y \mathbf{B}_{i,j}^+ + \Delta_y \mathbf{B}_{i,j}^- \right) \right. \right.$$

$$\left. \left. + \mathbf{Q}_{i,j} \right)^k \right] \Delta_t \mathbf{U}_{i,j}^{k+1} = \Delta_t \widehat{\mathbf{U}}_{i,j}$$

$$\mathbf{U}_{i,j}^{k+1} = \mathbf{U}_{i,j}^k + \Delta_t \mathbf{U}_{i,j}^{k+1} \qquad (15\text{-}89)$$

Each step in the algorithm leads to a system of equations which have tridiagonal coefficient matrices. The first set of equations gives intermediate values of the flow variables and is a set of simultaneous equations along each row (x-direction). The coefficient matrix for each row has the structure

$$\begin{pmatrix} \mathbf{b}_1 & \mathbf{c}_1 & 0 & 0 & \cdots & 0 & 0 & 0 \\ \mathbf{a}_2 & \mathbf{b}_2 & \mathbf{c}_2 & 0 & \cdots & 0 & 0 & 0 \\ 0 & \mathbf{a}_3 & \mathbf{b}_3 & \mathbf{c}_3 & \cdots & 0 & 0 & 0 \\ \vdots & \vdots & \vdots & \vdots & \ddots & \vdots & \vdots & \vdots \\ \vdots & \vdots & \vdots & \vdots & \ddots & \vdots & \vdots & \vdots \\ 0 & 0 & 0 & 0 & \cdots & \mathbf{a}_{N-1} & \mathbf{b}_{N-1} & \mathbf{c}_{N-1} \\ 0 & 0 & 0 & 0 & \cdots & 0 & \mathbf{a}_N & \mathbf{b}_N \end{pmatrix} \times \begin{pmatrix} \Delta_t \widehat{\mathbf{U}}_i \\ \vdots \\ \vdots \\ \vdots \\ \vdots \\ \vdots \\ \Delta_t \widehat{\mathbf{U}}_N \end{pmatrix} = \begin{pmatrix} \text{RHS}_1 \\ \vdots \\ \vdots \\ \vdots \\ \vdots \\ \vdots \\ \text{RHS}_N \end{pmatrix}$$

$$(15\text{-}90)$$

where $\mathbf{a}_i, \mathbf{b}_i$ and $\mathbf{c}_i$ are 3 x 3 matrices of coefficients, $\Delta_t \widehat{\mathbf{U}}$ are three-element vectors of flow variables, h, uh and vh, in delta form and RHS are three-element vectors containing the terms of the right-hand side of Eqs. 15-89. Solution of this system may be obtained efficiently by special block tridiagonal solvers.

After a solution is obtained along the rows, the next set of equations solves a similar system of block-tridiagonal equations along the columns (y-direction). Finally, in the last calculation of Eqs. 15-89 the flow variables,

$\mathbf{U}_{i,j}^{k+1}$, are obtained by adding $\Delta_t\mathbf{U}^{k+1}$ to $\mathbf{U}_{i,j}^k$ values for the last time step values. The flow variables obtained from this sequence are in the form, h, uh, and vh. The primitive flow variables h, u and v are then obtained by solving the following equations

$$h_{i,j}^{k+1} = h_{i,j}^k + \Delta_t h_{i,j}^{k+1}$$

$$u_{i,j}^{k+1} = \frac{(uh)_{i,j}^k + \Delta_t(uh)_{i,j}^{k+1}}{h_{i,j}^{k+1}}$$

$$v_{i,j}^{k+1} = \frac{(vh)_{i,j}^k + \Delta_t(vh)_{i,j}^{k+1}}{h_{i,j}^{k+1}} \tag{15-91}$$

The solution at the next time step may now be obtained by repeating the entire process. Second-order spatial differencing may be used in Eqs. 15-89, but the coefficient matrix structure will be less efficient. However, this type of differencing may be used on the right-hand side without affecting the formal accuracy of the system.

Boundary conditions

Boundary conditions in an implicit formulation may be used by directly including physical boundary specifications [Anderson et al., 1984]. Extrapolation techniques can be used to calculate values at boundary nodes for which explicit boundary conditions are not given. For solid walls the normal velocity is zero and, depending on the orientation, either $\Delta_t(uh)$ or $\Delta_t(vh)$ is zero. The remaining two equations use either the positive part of $\mathbf{A}$ and $\mathbf{B}$ for boundaries facing the positive x- or y-direction or the negative part of $\mathbf{A}$ or $\mathbf{B}$ for boundaries facing the negative x- or y-direction. For example, the solid boundary as shown in Fig. 15-4 is evaluated using the equations

$$[\mathbf{I} - \frac{\Delta t}{\Delta x}\frac{\theta}{1+\xi}\mathbf{A}_{i,j}^-]\Delta_t\widehat{\mathbf{U}}_{i,j} + \frac{\Delta t}{\Delta x}\frac{\theta}{1+\xi}\mathbf{A}_{i+1,j}^-\Delta_t\widehat{\mathbf{U}}_{i+1,j} =$$

$$- \Delta t\frac{1}{1+\xi}\left[\frac{1}{\Delta x}\mathbf{A}_{i,j}^-(\mathbf{U}_{i+1,j} - \mathbf{U}_{i,j}) + \frac{1}{\Delta y}\mathbf{B}_{i,j}^+(\mathbf{U}_{i,j} - \mathbf{U}_{i,j-1})\right.$$

$$\left. + \frac{1}{\Delta y}\mathbf{B}_{i,j}^-(\mathbf{U}_{i,j+1} - \mathbf{U}_{i,j}) + \mathbf{S}_{i,j}\right]^k + \frac{\xi}{1+\xi}\Delta_t\mathbf{U}_{i,j}^k \tag{15-92}$$

The second equation is replaced by the boundary condition $\Delta_t\mathbf{U} = 0$. Matrix $\mathbf{A}_{i,j}^-$ becomes

$$\mathbf{A}_{i,j}^- = \frac{1}{2c}\begin{pmatrix} (u+c)\lambda_3 & 0 & 0 \\ 0 & 1 & 0 \\ (u+c)v\lambda_3 & 0 & 0 \end{pmatrix} \tag{15-93}$$

The spatial differencing for this example is in the positive x-direction, and normal switching techniques are used along the boundary in the y-direction. Similar equations may be written for boundaries facing other directions. In-flow and outflow and boundaries may also be handled in this manner if an appropriate boundary condition is specified.

15-8 Finite-Volume Scheme

Finite-volume schemes can be applied to irregular flow domains and at the same time they are as easy to implement as finite-difference schemes. The main advantages of this scheme are its simplicity and ease of implementation. It can also handle sharp gradients in the water surface profile if they are present. A finite-volume scheme [Singh, 1996], which is an explicit scheme and is second order accurate in space and time, is presented below. It is a two-step predictor-corrector scheme. Figure 15-5 shows the finite volume grid of the two-dimensional flow domain. The flow domain is divided into a set of cells (i, j), each of which is identified by the corresponding center point. The finite-volume grid need not be orthogonal as in the case of a finite-difference grid.

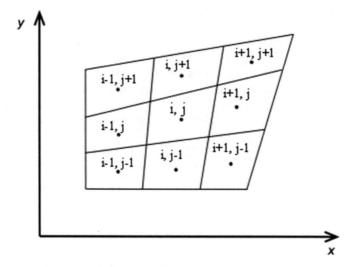

Fig. 15-5   Two-dimensional cells for an arbitrary shape flow domain (After Singh [1996])

The governing equations are integrated by a finite-volume technique on each of these cells covering the entire domain. Equation 15-43 may be written

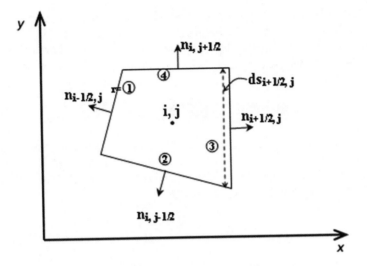

Fig. 15-6   **Normal vectors at the cell faces** (After Singh [1996])

in an integral form after applying the Gauss-divergence theorem, as shown in the following paragraphs

$$\int_{\forall} \frac{\partial \mathbf{U}}{\partial t} dv + \oint_{s} (Hn) ds = Sdv \tag{15-94}$$

where, H is the flux term at the control surface. The volume integral in the first term actually represents the integral of the time evolution of the function over the area of the cell. The surface integral in the second term is the total normal flux through the cell boundaries. The normal unit vectors to the cell walls are defined as shown in Fig. 15-6. The scalar product Hn in the second term of Eq. 15-94 can be expressed in terms of the Cartesian components as

$$Hn = En_x + Gn_y \tag{15-95}$$

where, n_x and n_y are the x and y components of the unit vector n. Assuming the vector U to be uniform over the cell, Eq. 15-94 may be written as

$$\frac{\partial U}{\partial t} \Delta A + \oint_{s} (Hn) ds = S \Delta A \tag{15-96}$$

where ΔA = area of the finite volume cell. The surface integral in Eq. 15-96 is approximated by a sum over the four walls of the finite-volume as follows.

$$\oint_{s} (Hn) ds \approx \sum_{r=1}^{4} (H_r n_r) dS_r \tag{15-97}$$

where, ds_r = lengths of the four walls which contour the cell (i,j), H_r = the numerical flux through the cell faces r which contour the cell (i,j).

Evaluation of the numerical flux at a cell face is explained here for the cell face between the nodes $(i+1,j)$ and (i,j). Similar procedure is adopted for other cell faces.

$$(H_3 n_3)ds = (Hn)_{i+\frac{1}{2},j} = \frac{1}{2}\left[H_R + H_L - \alpha(U_R - U_L)\right]n_{i+\frac{1}{2},j} \qquad (15\text{-}98)$$

where α = a positive coefficient, $HR = f(UR)$ = the flux computed using the information from the right side of the cell face and $HL = f(UL)$ = the flux computed using the information from the left side of the cell face. U_R and U_L are obtained using the following equations.

$$(U_L)_{i+\frac{1}{2},j} = U_{i,j} + \frac{1}{2}\delta U_{i,j} \qquad (15\text{-}99)$$

$$(U_R)_{i+\frac{1}{2},j} = U_{i+1,j} - \frac{1}{2}\delta U_{i+1,j} \qquad (15\text{-}100)$$

where subscript (i,j) refers to the value at node (i,j). The subscript $(i+\frac{1}{2},j)$ refers to the value at the interface between the nodes (i,j) and $(i+1,j)$. There are several ways to determine $\delta U_{i,j}$ and $\delta U_{i+1,j}$ using different slope limiter procedures [Yee, 1989; Alcrudo et al., 1992]. The integration of Eq. 15-96 in the time domain is done by using a predictor-corrector approach.

Predictor part

The predicted value of the vector U at the unknown time level $t + \Delta t$ is determined using the following discretization of Eq. 15-96.

$$U_{i,j}^* = U_{i,j}^k - \frac{\Delta t}{\Delta A_{i,j}}\left[H_{i+\frac{1}{2},j}^k ds_{i+\frac{1}{2},j} + H_{i,j+\frac{1}{2}}^k ds_{i,j+\frac{1}{2}} + H_{i-\frac{1}{2},j}^k ds_{i-\frac{1}{2},j}\right.$$
$$\left. + H_{i,j-\frac{1}{2}}^k ds_{i,j-\frac{1}{2}}\right] + \Delta t S_{i,j}^k$$

$$(15\text{-}101)$$

in which the superscripts k and $*$ refer to the value at the known time level, t and the predicted value at the unknown time level, $t + \Delta t$, respectively and Δt is the computational time step. Equation 15-101 should be interpreted component wise for the vector U. Equation 15-101 gives the predicted values of h, u and v at the time $t + \Delta t$ at node (i,j).

Corrector part

The vector U at the unknown time level $k + 1$ and at node (i,j) is computed by using the predicted values and the values at the time level k.

$$U_{i,j}^* = 0.5 \left[U_{i,j}^k + U_{i,j}^* - \frac{\Delta t}{\Delta A_{i,j}} \left(H_{i+\frac{1}{2},j}^* ds_{i+\frac{1}{2},j} + H_{i,j+\frac{1}{2}}^* ds_{i,j+\frac{1}{2}} \right. \right.$$

$$\left. \left. + H_{i-\frac{1}{2},j}^* ds_{i-\frac{1}{2},j} + H_{i,j-\frac{1}{2}}^* ds_{i,j-\frac{1}{2}} \right) + \Delta t S_{i,j}^* \right] \qquad (15\text{-}102)$$

The initial and boundary conditions and stability criteria should be satisfied, as discussed in the earlier sections.

15-9 Applications

To demonstrate the application of MacCormack and Gabutti schemes in hydraulic engineering, two typical problems are solved. In one of these problems, a strong bore is formed; the second problem deals with gradually varied flows. This is done intentionally to demonstrate robustness of the schemes to solve diversified type of problems. The governing equations may not be valid in the vicinity of the bore, where there are sharp curvatures. However, the computed results, such as maximum water levels and arrival time of a wave, may be used with confidence for typical engineering applications even though the details of the bore itself are not simulated in a rigorous manner. The inclusion of boundary conditions and the addition of artificial viscosity to smooth oscillations caused by dispersive errors of the finite-difference schemes are investigated.

The first example analyzes a partial dam breach, structural failure or instantaneous opening of sluice gates. The breach is nonsymmetrical to demonstrate analysis of a general case. The second problem is the passage of a flood wave through a channel contraction. Such flow conditions occur for flow through a bridge opening, cofferdams, or structures occupying partial channel widths, etc. The boundaries in these examples are taken parallel to the coordinate axes, which would usually be the case in actual problems of this kind – e.g., flow through a bridge opening or sudden opening of sluices. The 90°-degree corner imposes a rather severe test on the schemes and on the inclusion of boundaries. Other boundary shapes may be approximated by a staircase-type of configuration. Since the grid size is usually small, this approximation should not introduce significant errors in the computed results.

Application of these schemes yield flow velocities and flow depths throughout the computational domain at specified grid points. Such information may be utilized to determine the potential for scour or erosion, to assess the effectiveness of preventive measures to reduce scour and erosion, to determine the height of walls and dykes, and to determine the size of rocks needed for cofferdam closure during river diversion. In addition, these applications show that these schemes may be employed to determine flow conditions in the immediate vicinity of hydraulic structures, or they may be used to generate data for input to one-dimensional flow models away from the regions where flow may be assumed as one dimensional. With some ingenuity, any typical problem

involving two dimensional flows may be analyzed using the experience gained
from the present applications.

Although only unsteady flows are investigated, steady flows may be analyzed with the MacCormack and Gabutti schemes by letting the computations converge to a steady state subject to the specified end conditions. It
becomes necessary to use this procedure if both sub- and supercritical flows
are present in the flow field simultaneously. This is because the governing
equations are hyperbolic if the flow is subcritical; but they are not hyperbolic if the flow is supercritical. For this purpose, application of a scheme like
Gabutti scheme becomes imperative, since most other finite-difference schemes
presently used in hydraulic engineering either fail or give incorrect results. Because of the shock-capturing capabilities of the schemes presented herein, they
become attractive for solving flow situations where hydraulic jump is formed.
This procedure not only directly yields the location of the jump but gives an
approximate length of the jump as well.

Partial breach or opening of sluice gates

In this problem, the dam is assumed to fail instantaneously or the sluice gates
are assumed to be opened instantly. The discontinous initial conditions impose
severe difficulties in starting the computations, and most of the presently used
numerical schemes fail under such conditions. In the simulations included
herein, the channel downstream of the dam or the gates is assumed to have
some finite flow depth. This is quite normal for usual applications, where a
downstream control keeps the downstream channel in a "wet" condition. To
simulate a dry channel, however, a very small flow depth may be assumed in
the analysis. This procedure is much easier than to track the bore propagation
explicitly and should give results that are of the same, if not better, accuracy
than that of the other input variables.

The computational domain comprises of a 200-m long and 200-m wide
channel. The nonsymmetrical breach or sluice gates are 75m wide and the
structure or dam is 10m thick in the direction of flow. The grid is 41 by
41 points, which results in an individual mesh size of 5m by 5m. Additional
details are shown in Fig. 15-7. To prevent any damping by the source terms, a
frictionless, horizontal channel was used, and initial conditions (Fig. 15-8) had
a tailwater/reservoir ratio $h_t/h_r = 0.5$ in the initial few runs. Flow conditions
were analyzed for a wide variation of flow parameters, such as including the
friction losses (Manning n from 0 to 0.15), assuming a sloping channel (bottom
slope from 0 to 0.07), different ratios of the tailwater to reservoir depths (as low
as 0.2), symmetrical and unsymmetrical breach. However, only typical results
are included here to conserve space. The flow conditions were computed for
7.1 s after the dam failure or opening of the sluices. The results presented are
for $7.1 \le t < 7.1 + \Delta t$ seconds. At this time, the bore is well developed in
the central portion of the downstream channel and the wave front has reached
one bank of the channel.

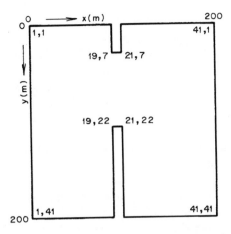

Fig. 15-7 Definition sketch for partial dam breach

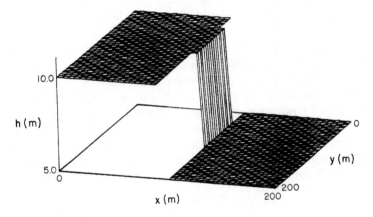

Fig. 15-8 Initial conditions

Two types of figures are used to present the computed results. The first figure is a perspective plot of the water surface. Remnants of the dam are represented by gaps near the middle of the plot. The vertical scale is exaggerated with respect to the horizontal scales. The second is a velocity vector plot. At each node the velocity is indicated by an arrow, with the magnitude represented by the length of the arrow. For esthetic reasons, velocities below a specified tolerance are not drawn (magnitude with a length less than 20 percent of the mesh size). The velocity vectors on the boundaries, parallel to the solid boundaries, and at right angles to inflow and outflow and boundaries are also not drawn.

To illustrate the difference between symmetric and anti-symmetric boundary conditions, both were incorporated in the MacCormack scheme. Figure 15-9 shows the perspective views of the water surface. The profile near boundaries, particularly in the reservoir area, illustrates the difference between the boundary conditions. Dispersion errors of this scheme, in the form of oscillations, are noticeable on the backside of the bore. Without affecting the quality of the profile, the solution may be smoothed by the addition of artificial viscosity. This is done in the results shown in Fig. 15-10(a). Antisymmetric boundaries are used in this run, and the computed velocity vectors are plotted in Fig. 15-10(b). In addition to eliminating the wiggles near the bore, the artificial-viscosity term also reduces undershoot near sharp corners. These steep depressions in the water surface are especially noticeable downstream of the breach. In these runs, the sharp corners are modeled by assuming they are boundaries parallel with the x-direction.

The solution of this problem by using the Gabutti scheme and with no artificial viscosity added is shown in Fig. 15-11(a). Since the diffusive properties of this scheme are different than those of the MacCormack scheme, the oscillations occur at different locations. The addition of artificial viscosity leads to the profile shown in Fig. 15-11(b).

Figures 15-8 to 15-11 are pictorial view of the computed water-surface profiles. These results show a qualitative comparison of these schemes, comparison of the procedures used for the inclusion of boundary conditions, and the effectiveness of addition of artificial viscosity to smooth the oscillations. For a quantitative comparison of the MacCormack and Gabutti schemes, computed results are shown in Figs. 15-12 and 15-13. Figure 15-12 shows the comparison of computed transverse water-surface profiles at $i = 16$ (upstream of the breach), at $i = 20$ (inside the breach), and at $i = 24$ (downstream of the breach). Figure 15-13 shows the time variation of flow depth at grid point (20, 15), i.e., inside the breach, and at grid point (24, 15), i.e., downstream of the breach. It is clear from these comparisons that both schemes give comparable results.

Propagation of a flood wave through channel contraction

The second example involves the propagation of a flood wave through a channel contraction (Fig. 15-14). The flood wave is shown in Fig. 15-15. The channel is 2 km wide and the flow conditions are computed in a 2 km length of the channel. The flow domain is divided into 50 x 50 grid. The unsymmetrical channel contraction is located near the midlength of the channel. The flow depth at all grid points at $t = 0$ is 5 m and the flow velocity is zero. The flood wave of Fig. 15-15 is introduced at the upstream end over the full channel width. The flow depth at the downstream end is kept constant and equal to the initial flow depth.

The flow depths computed by the MacCormack scheme at node (20, 21) and at node (25, 21) are shown in Fig. 15-16. Node (10, 21) is located upstream of the contraction and node (25, 21) is located downstream of the contraction. It is clear from this figure how the wave height and wave shape are modified as it travels through the computational flow domain. Figure 15-17 shows the

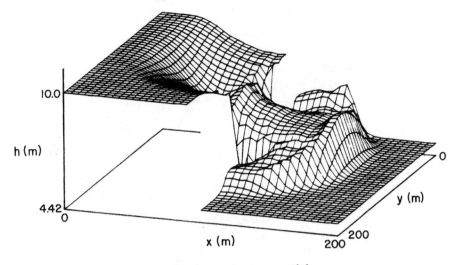

(a) Anti-symmetric boundary conditions

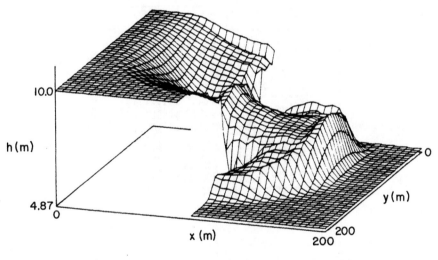

(b) Symmetric boundary conditions

Fig. 15-9 Water surface profile computed by MacCormack scheme with no artificial viscosity

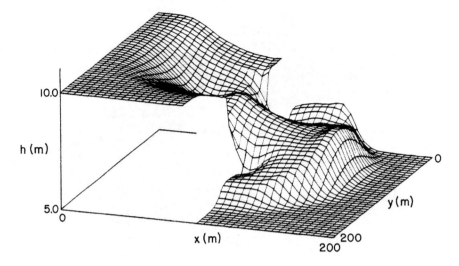

(a) Water surface profile

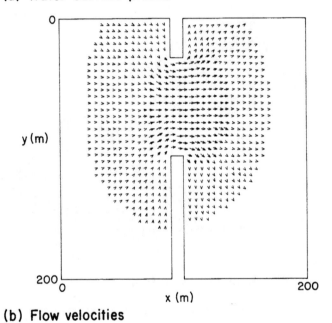

(b) Flow velocities

Fig. 15-10 Results computed by MacCormack scheme; artificial viscosity added

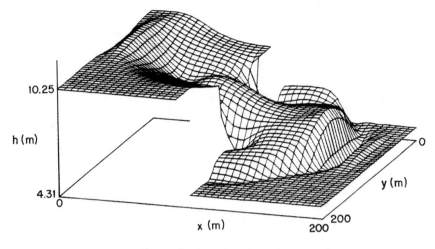

(a) No artifical viscosity

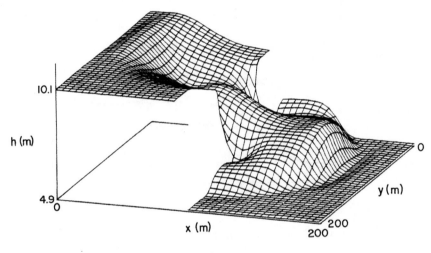

(b) With artificial viscosity

Fig. 15-11   Water surface profiles computed by Gabutti scheme

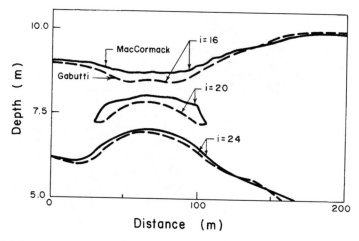

Fig. 15-12   Comparison of transverse profiles

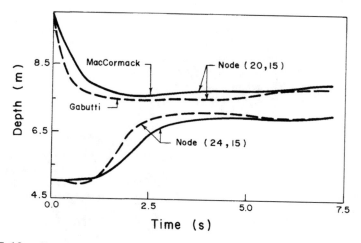

Fig. 15-13   Comparison of variation of flow depth with time

transverse water-surface profiles computed by the MacCormack scheme at $t = 220$ s at three locations. These locations are at $i = 20$ (upstream of the contraction), at $i = 25$ (inside contraction), and at $i = 35$ (downstream of contraction). It is clear from this figure how the wave is modified as it passes through the contraction. Since the contraction is unsymmetrical, the water surface is not symmetrical with respect to the centerline of the channel.

Comparison with other methods

Figure 15-18 compares the longitudinal water-surface profile at $t = 7.1$ s computed by the two explicit schemes presented here and the Beam and Warming

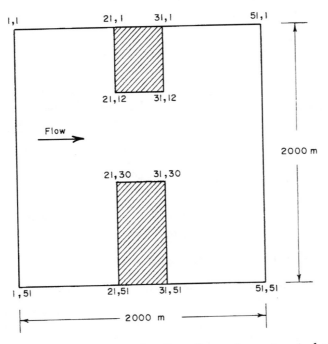

Fig. 15-14   Definition sketch for flow through contracted opening

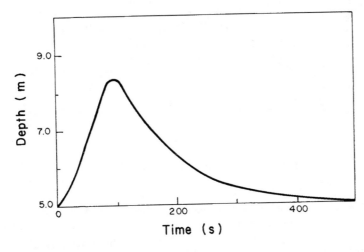

Fig. 15-15   Flood wave

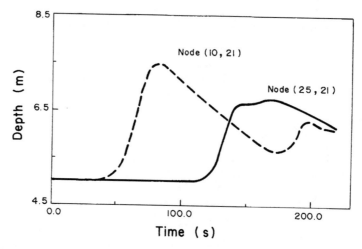

Fig. 15-16   Computed variation of flow depth

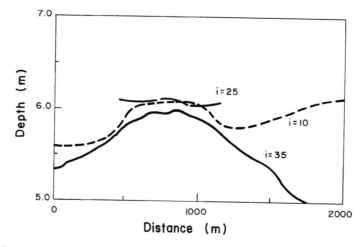

Fig. 15-17   Computed transverse flow profile

implicit schemes [Fennema and Chaudhry, 1989]. These water-surface profiles are for the case of partial failure of a dam or sudden opening of sluice gates (Fig. 15-7). A 41-by-41 grid was used to simulate the flow conditions in the computational domain. In the MacCormack and Gabutti schemes, artificial viscosity was added ($\kappa = 0.25$) to smooth the high-frequency oscillations in the computed flow depth. Anti-symmetric boundary conditions were used in the analysis. Flow was supercritical at several grid points downstream of the breach after a short time following the opening of the breach. The MacCormack scheme failed for h_t/h_r ratios less than 0.25, the Gabutti scheme for this ratio less than 0.2, and the Beam and Warming schemes for this ratio less than 0.001. This figure shows that the agreement between the results computed by these schemes is satisfactory.

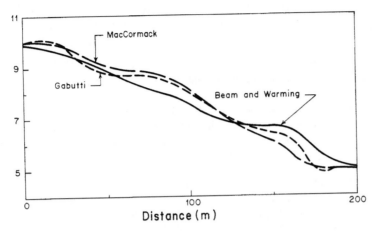

Fig. 15-18 **Comparison of computed water-surface profiles for dambreach of Fig. 15-7**

15-10 Summary

Depth-averaged equations describing two-dimensional, unsteady, free-surface flow were derived, starting with the Navier-Stokes equations. The friction losses were included using empirical relationships.

Explicit and implicit finite-difference schemes were presented for the analysis of two-dimensional, unsteady, free-surface flows. Numerical oscillations occurring near the sharp-fronted waves may be controlled by adding artificial viscosity. In these schemes, boundary conditions may be easily incorporated, initial conditions may have discontinuities, and explicit tracking of the bore is not necessary. In addition, these schemes are relatively easy to program and give good results.

The split-flux schemes allow the analysis of both subcritical and supercritical flows, with the Beam and Warming schemes especially able to handle flows with large shocks and bores.

The schemes described in this chapter may be used to provide reliable solutions to problems associated with events, such as dam breaks, dike breaches, and surge waves.

Problems

15-1 Develop a computer program for the computation of flows in the river channel and in the flood-plain from a ruptured levee by using MacCormack and Beam and Warming schemes.

15-2 By using the programs of Prob. 15-1, investigate the effect of different artificial viscosity on the computed results.

15-3 By using the programs of Prob. 15-1, compute the steady state flows in a channel expansion where the channel width varies from 2 m to 4 m in a distance of 5 m. Assume the flow depth at the channel entrance is 3 m and the flow velocity is 1 m/s.

15-4 Investigate the effect of the source terms (i.e., right-hand sides of Eq. 15-41) on the stability of MacCormack scheme by varying the bottom slope and the friction factors in flood plain. Select the dimensions of the channel and the flood plain and the rate of discharge so that the flow is subcritical.

References

Alcrudo, F., Garcia-Navarro, P., and Saviron, J. M., 1992, "Flux difference sphtting for 1D open channel flow equatione," *Inter. Jour. for Numerical Methods in Fluids,* vol. 14, pp. 1009–1018.

Anastasiou, K., and Chan, C. T. 1997, "Solution of the 2D Shallow Water Equations Using the Finite Volume Method on Unstructured Triangular Meshes," *Inter. Jour. Numer. Methods Fluids,* vol. 24, pp. 1225–1245.

Anderson, D. A., Tannehill J. D. and Pletcher, R.H., 1984, *Computational Fluid Mechanics and Heat Transfer.* McGraw-Hill, New York.

Anton, H., 1981, *Elementary Linear Algebra.* Wiley and Sons, New York.

Beam, R. M., and Warming, R. F., 1976, "An Implicit Finite-Difference Algorithm for Hyperbolic Systems in Conservation Form." *Jour. Comp. Phys.,* Vol. 22, pp. 87–110.

Benning, R. M., Becker, T. M., and Delgado, A., 2001, "Initial Studies of Predicting Flow Fields With an ANN Hybrid," *Adv. Eng. Software,* vol. 32, pp. 895–901.

Benque, J. P., Hauguel, A., and Viollet, P. L., 1982, *Engineering Applications of Computational Hydraulics,* Pitman Advanced Publishing Program, London, England.

Chaudhry, M. H., 2014, *Applied Hydraulic Transients.* 3rd edition, Chapter 3, Springer, New York, NY.

Chua, L. H. C., and Holz, K. P., 2005, "Hybrid Neural Network-Finite Element River Flow Model," *Jour. Hyd. Engineering,* vol. 131, no. 1, pp. 52–59.

Cockburn, B., Karniadakis, G., Shu, C. W., and Griebel, M., (eds.), 2000, "Discontinuous Galerkin Methods: Theory, Computation and Applications," *Lecture notes in computational science and engineering,* Springer, Berlin.

Courant, R., 1936, *Differential and Integral Calculus.* vol. II, Interscience, New York, NY.

Cunge J. A., Holly Jr., F. M., and Verwey, A., l980, *Practical Aspects of Computational River Hydraulics,* Pitman, London.

Dibike, Y. B., and Abbott, M. B., 1999, "Application of Artificial Neural Networks to the Simulation of a Two-Dimensional Flow," *Jour. Hyd. Research,* vol. 37, no. 4, pp. 435–446.

Fagherazzi, S., Rasetarinera, P., Hussaini, M. Y., and Furbish, D. J., 2004, "Numerical Solution of the Dam-Break Problem With a Discontinuous Galerkin Method," *Jour. Hyd. Engineering,* vol. 130, no. 6, pp. 532–539.

Fennema, R. J., 1985, "Numerical Solution of Two-Dimensional Transient Free-Surface Flows," Ph. D. Disseration, Washington State University, Pullman, WA.

Fennema, R. J., and Chaudhry, M. H., 1986, "Second-Order Numerical Schemes for Unsteady Free-Surface Flows with Shocks," *Water Resources Research,* vol. 22, no. 13, pp. 1923–1930.

Fennema, R. J., and Chaudhry, M. H., 1987, "Simulation of One-Dimensional Dam-Break Flows." *Jour. Hydraulic Research,* International Association for Hydraulic Research, vol. 25, no 1, pp. 41–51.

Fennema, R. J., and Chaudhry, M. H., 1989, "Implicit Methods for Two-dimensional Unsteady Free-Surface Flows," *Jour. Hyd. Research,"* Inter. Assoc. for Hydraulic Research, vol. 27, no. 3, pp. 321–332.

Fennema, R. J., and Chaudhry, M. H., 1990, "Explicit Methods for Two-dimensional Unsteady Free-Surface Flows," *Jour. Hyd. Engineering,"* Amer. Soc. of Civ. Engrs., vol. 116, no. 8, pp. 1013–1034.

Franke, C., and Schaback, R., 1997, "Convergence Orders of Meshless Collocation Methods and Radial Basis Functions," *Technical Report,* Dept. of Mathematics, University of Gottingen, Gottingen, Germany.

Gabutti, B., 1983, "On Two Upwind Finite-Difference Schemes for Hyperbolic Equations in Non-Conservative Form," *Computers and Fluids.* vol. 11, No. 3, pp. 207–230.

Garcia, R. and Kahawita, R. A., 1986, "Numerical Solution of the St. Venant Equations with MacCormack Finite-Difference Scheme," *International Jour. for Numerical Methods in Fluids,* vol. 6, pp. 259–274.

Godunov, S. K., 1959, "A Finite Difference Method for the Computation of Discontinuous Solutions of the Equations of Fluid Dynamics," *Matematicheski/u/i Sbornik. Novaya Seriya (Mathematics of the USSR-Sbornik),*vol. 47, pp. 357–393.

Hardy, R. L., 1971, "Multiquadric Equations of Topography and Other Irregular Surfaces," *Jour. Geophys. Res.,* vol. 76, no. 26, pp. 1905–1915.

Hirsch, H., 1990, *Numerical Computation of Internal and External Flows. Vol.2: Computational Methods for Inviscid and Viscous Flows,* Wiley, New York, NY.

Hon, Y. C., Lu, M. W., Xue, W. M., and Zhu, Y. M., 1997, "Multiquadric Method for The Numerical Solution of a Biphasic Mixture Model," *Appl. Math. Comput.,* vol. 88, no. 2, pp. 153–175.

Hon, Y. C., Cheung, K. F., Mao, X. Z., and Kansa, E. J., 1999, "Multiquadric Solution for Shallow Water Equations," *Jour. Hydraulic Engineering,* vol. 125, no. 5, pp. 524–533.

Jameson, A., Schmidt, W., and Turkel, E., (1981). "Numerical Solutions of the Euler equations by Finite Volume Methods Using Runge-Kutta Time-Stepping Schemes," *Proc.,* AIAA 14th Fluid And Plasma Dynamics Conference, Palo Alto, CA, AIAA–81–1259.

Jimenez, O., 1987, Personal communications with M. H. Chaudhry.

Katopodes, N., 1984a, "Two-Dimensional Surges and Shocks in Open Channels," *Jour. Hydraulic Engineering,* Amer. Soc. Civil Engrs., vol. 110, no. 6, pp. 794–812.

Katopodes, N. D., 1984b, "A Dissipative Galerkin Scheme for Open-Channel Flow." *Jour. Hyd. Div.,* Amer. Soc. Civ. Engrs., Vol. 110, No. HY6, pp. 450–466.

Katopodes, N. D., and Strelkoff, T., 1978, "Computing Two- Dimensional Dam-Break Flood Waves." *Jour. Hyd. Div.,* Amer. Soc. Civ. Engrs., vol. 104, no. HY9, pp. 1269–1288.

Lax, P. D. and Wendroff, B., 1960, "Systems of Conservation Laws." *Com. Pure Appl. Math.,* vol. 13, pp. 217–237.

Lai, C., 1986, "Numerical Modeling of Unsteady Open-Channel Flows," in *Advances in Hydroscience,* vol. 14, Academic Press, New York, NY., pp. 161–333.

Lax, P. D. and Wendroff, B., 1960, "Systems of Conservation Laws." *Com. Pure Appl. Math.,* vol. 13, pp. 217–237.

Leendertse, J. J., 1967, "Aspects of a Computational Model for Long Period Water-Wave Propagation," *Memo* RM-5294-PR, Rand Corporation, Santa Monica, CA, May.

MacCormacK, R. W., 1969, "The Effect of Viscosity in Hypervelocity Impact Cratering." *Amer. Inst. Aero. Astro.,* Paper 69–354, Cincinnati, Ohio.

Matsutomi, H., 1983, "Numerical Computations of Two-Dimensional Inundation of Rapidly Varied Flows due to Breaking of Dams." *Proc., XX Congress,* Inter. Assoc. Hyd. Research, Moscow, USSR, Subject A, vol. II, Sept. pp. 479–488.

Mingham, C. G., and Causon, D. M., 1998, "High-Resolution Finite-Volume Method for Shallow Water Flows," *Jour. Hydraulic Engineering,* vol. 124, no. 6, pp. 605–614.

Morreti, G., 1979, "The λ-Scheme," *Computer and Fluids,* vol. 7, pp. 191–205.

Richtmyer, R. D., and Morton, K. W., 1967, *Difference Methods for Initial-Value Problems,* John Wiley and Sons, New York, 2nd Edition.

Sakkas, J. G., and Strelkoff, T., 1973, "Dam-Break Flood in a Prismatic Dry Channel." *Jour. Hyd. Div.,* Amer. Soc. Civ. Engrs., vol. 99, no. HY12, pp.2195–2216.

Schwanenberg, D., and Harms, M., 2004, "Discontinuous Galerkin Finite-Element Method for Transcritical Two-Dimensional Shallow Water Flows," *Jour. Hyd. Engineering,* vol. 130, no. 5, pp. 412–421.

Singh, V. 1996, "Computation of shallow water flow over a porous medium," Ph.D. thesis, Indian Institute of Technology, Kanpur, India.

Sleigh, P. A., Gaskell, P. H., Berzins, M., and Wright, N. G., 1998, "An Unstructured Finite-Volume Algorithm for Predicting Flow in Rivers and Estuaries," *Comput. Fluids,* vol. 27, no. 4, 479–508.

Tamamidis, P., and Assanis, D. N. 1993, "Evaluation of Various High-Order-Accuracy Schemes With and Without Flux Limiters," *Int. Jour. Numer. Methods Fluids,* vol. 16, pp. 931–948.

Toro, E. F. 1992, "Riemann Problems and the WAF Method for Solving the Two-Dimensional Shallow Water Equations," *Philos. Trans. Royal Soc.,* London, 338, 43–68.

Toro, E. F., 1999, *Riemann Solvers and Numerical Methods for Fluid Dynamics,* 2nd Ed., Springer, Berlin.

Yen, B. C., 1973, "Open-Channel Flow Equations Revisited," *Jour. Engineering Mechanics Div.,* Amer. Soc. Civil Engrs., vol. 99, no. 5, pp. 979–1009.

Yee, H. C., 1989, "A class of high-resolution explicit and implicit shockcapturing methods," *NASA Technical Memorandum 101088,* NASA Ames Research Center, CA.

Yoon, T. H., and Kang, S. K., 2004, "Finite Volume Model for Two-Dimensional Shallow Water Flows on Unstructured Grids," *Jour. Hyd. Engineering,* vol. 130, no. 7, pp. 678–688.

Warming, R. F., and Beam, R. M., 1978, "On the Construction and Application of Implicit Factored Schemes for Conservation Laws." *Proc., Symposium on Computational Fluid Dynamics,* SIAM-AMS, vol. 11, NY, pp. 85–129.

Wong, S. M., Hon, Y. C., Li, T. S., Chung, S. L., and Kansa, E. J., 1999, "Multi-Zone Decomposition for Simulation of Time-Dependent Problems Using the Multiquadric Scheme," *Comput. Math. Appl.*

Zhao, D. H., Shen, H. W., Tabios, G. Q., Lai, J. S., and Tan, W. Y., 1994, "Finite-Volume Two-Dimensional Unsteady-Flow Model for River Basins," *Jour. Hyd. Engineering,* vol. 120, no. 7, pp. 863–883.

16

LEVEE BREACH MODELING

Flow through the breach in the Elm Point levee in St. Charles, MO. Breach occurred on June 23, 2008; with most of the flooded area being agricultural land (http://chl.erdc.usace.army.mil)

© Springer Nature Switzerland AG 2022
M. H. Chaudhry, *Open-Channel Flow*,
https://doi.org/10.1007/978-3-030-96447-4_16

16-1 Introduction

Levees are built along rivers, streams, and channels for flood protection or around water storage impoundments, typically using erodible materials. These earthen structures may fail due to overtopping, piping, or slope instability, with overtopping being the most common cause of failures. Levee failure may result in catastrophic property damage and loss of life, e.g., the damage in the City of New Orleans following Hurricane Katrina in 2005, US East Coast due to Hurricane Sandy in 2012, and in Columbia, South Carolina due to Hurricane Joaquin in 2015. An accurate prediction of the development of a breach, breach outflow, and the resulting flow field is needed for emergency preparedness and for flood-damage mitigation. The levee failure process is complex involving interaction between the water flow, sediment transport and the corresponding geomorphological changes.

The breaching of a dam differs from the breach in a levees due to the direction of flow in the main channel and in the breach. There are fewer studies on levee breach (main flow direction parallel to the embankment crest) than those on the dam breach (main flow direction perpendicular to the embankment crest). The simulation of dam-break flows is discussed in Chapter 15; the simulation of levee breach flows is presented in this chapter.

Flows through breached dams, embankments, levees and dykes are of considerable interest to researchers, modelers and practitioners for determining the extent of flooding, for investigating various options for flood-damage mitigation and for preparing emergency plans. A levee breach may be accidental or pre-planned; the latter is referred to as an *engineered breach*. For accurate flood simulation, the shape and size of a breach are needed. These are known a priori for an engineered breach. However, for an accidental or unplanned breach, the breach shape and size and breach development with time have to be determined simultaneously with the calculation of flows because of their inter-dependence. Most of the flood simulation models presently available require the breach size as an input and that is specified by the user. For proper dynamic modeling, however, the breach development has to be considered as a time-evolution process that depends upon the flows in and around the breach, levee material and levee cross section. The procedures for the breach evolution presented in this chapter may be utilized for this purpose.

In this chapter, the estimation of breach flows through a levee breach with a constant cross section and the modeling of the temporal development of a levee breach caused by overtopping are discussed. The presentation follows Elafy, Tabrizi, and Chaudhry [2018] and Elalfy [2015]. The breach in a non-cohesive earthen levee due to surface erosion is discussed herein; the failure of a cohesive levee is much more complex and requires the modeling of surface erosion as well as erosion due to head-cuts. A numerical model is developed based on the numerical solution of the two-dimensional shallow water equations simultaneously with the sediment-mass-conservation equation in which a new source term is included to account for the slumping failure. The numerical model is validated by comparing the computed results with the laboratory test data.

16-2 Estimation of Breach Flow

The lateral outflow through a levee breach may be estimated by numerically solving the one-dimensional, spatially varied flow equations or the two-dimensional shallow-water equations. As compared to the one-dimensional models, two-dimensional models require significantly more topographical information, more discretization effort and in addition require the inclusion of the breach as an internal boundary. Such topographical information is usually limited and/or not available for real-life projects. Overall, one-dimensional models are much easier to apply than the 2-D models.

For levee-breach flows, Han et al. [1998] developed a combined one- and two-dimensional hydrodynamic model using finite-difference method. The one-dimensional model solves the dynamic wave equation for the main channel and the two-dimensional model solves the diffusion wave equation for the floodplain. They applied their model to an actual levee-breach in the downstream reaches of the Han river and compared the simulated results with the observed data, including inundated depth, flood arrival time and the inundated areas. Roger et al. [2009] developed two models of levee-breach flows. The first model solves the shallow-water equations by using a total variation, diminishing Runge-Kutta discontinuous Galerkin finite-element method, and the second model solves the same equations by using a finite-volume scheme involving a flux-vector-splitting approach. Both models produced satisfactory agreement with the experimental results. Van Emelen et al. [2012] numerically solved the two-dimensional shallow-water equations and their computed results compared satisfactorily with those measured on a 1:50 scale model of the 17th street Canal breach in New Orleans, Louisiana. They concluded that the depth-averaged models give an acceptable flow prediction in the complex domains at reasonable computational cost.

Cheong [1991] utilized the energy and momentum equations and experimental results to estimate lateral outflow from a prismatic, trapezoidal channel. He recommended that the downstream depth should not be used to estimate the discharge coefficient of a side weir because of large variation in the downstream water surface level which may significantly affect the value of the discharge coefficient, especially at low Froude numbers. Experiments by Singh et al. [1994] on a rectangular, side weir in a prismatic rectangular channel with subcritical flow showed that the discharge coefficient of the side weir is a function of the upstream Froude number and the ratio of the sill height to the upstream depth. Most of the previous studies predicted the flow over side weirs by applying the energy principle in the longitudinal direction of the main channel and most of the developed equations are function of the local flow variables near the side weir [Hager, 1987; Singh et al., 1994; and Borghei et al., 1999]. Elafy et al. [2018] considered the lateral outflow through a levee breach as lateral outflow over a broad-crested side weir in rectangular and trapezoidal main channels, and solved the spatially varied flow equation utilizing the relationships for the lateral outflow, presented by Hager [1987]. They compared the computed results with the laboratory measurements and with those obtained by solving the two-dimensional depth-averaged flow equations numerically and presented correction factors so that simple one-dimensional models may be utilized. These studies are summarized in this section.

The equation for one-dimensional, spatially varied flow with lateral outflow may be written as [Henderson, 1966]

$$\frac{dh}{dx} = \frac{S_o - S_f - \frac{Q q_b}{g A^2}}{1 - Q^2(g A^2 D)} \tag{16-1}$$

in which dh/dx is the gradient of flow depth along the breach width (i.e., in the flow direction), S_o is the channel bottom slope, S_f is the slope of the energy grade line, Q is the main channel discharge (which is decreasing along the breach width), q_b is the lateral outflow intensity per unit width of the breach, A and D are the flow area and the hydraulic mean depth across the breach, respectively. For a broad-crested side weir on a horizontal prismatic channel, q_b may be calculated utilizing the data presented by Hager [1987].

To correct the discharge through a levee breach calculated by using a one-dimensional model, Elafy et al. [2018] introduced a correction factor,

$$C_f = \frac{Q_{bs} - Q_{b2}}{Q_i} \tag{16-2}$$

where Q_{bs} is the breach discharge calculated by using the spatially varied flow model and Q_{b2} is the breach discharge calculated by using the two-dimensional model. By comparing the results of the one- and two-dimensional models and by using multi-regression analysis, they developed the following expressions for C_f:

Rectangular main channel

$$C_f = 0.48 S_r^{6.2} b_r^{0.72} \mathbf{F}_{app}^{-0.15} \mathbf{F}_{(out)}^{-0.13} \tag{16-3}$$

Trapezoidal main channel

$$C_f = 0.3 S_r^{0.8} b_r^{0.72} \mathbf{F}_{app}^{-0.15} \mathbf{F}_{out}^{-0.13} \tag{16-4}$$

in which $S_r = (h_d - h_b)/h_u$; h_u and h_d are the water depths at the upstream and downstream ends of the breach on the main channel, h_b is the breach crest height above the bottom of the main channel, and $b_r = b_b/b_c$ is defined as the ratio of the breach width, b_b to the channel bed width, b_c. The ranges of S_r and b_r for the above equations are 0.71 to 0.90 and 0.12 to 0.86, respectively. However the ranges of $\mathbf{F}_{r_{app}}$ (Froude number in the main channel upstream of the breach) and $\mathbf{F}_{out}$ (Froude number at the model outlet) are based on the spatially varied flow model equal to 0.46 to 0.9 and 0.01 to 0.62, respectively. It is clear from the equations that the correction factor is affected more by the submergence ratio than by the relative breach width. Also, there is an inverse relationship between the correction factor and the approaching Froude number and the Froude number at the outlet.

16-3 Levee Breach due to Overtopping

In this section, we discuss the overtopping failure of non-cohesive earthen levees.

Faeh [2007] developed a two-dimensional numerical model using a finite-volume approach to solve the shallow-water equations along with a sediment mass conservation equation. The model was applied to the erosion-based dam breach cases and a levee breach of Elbe River. Faeh [2007] reported that the angle of the breach side-slope is the most sensitive parameter of an earthen dam breach which affects the lateral erosion. Kakinuma and Shimizu [2014] conducted four large-scale experiments on riverine levee breach with variation of inflow rate, levee material, and levee shape. They divided the levee failure process into four stages: The first two stages include the breach formation in the vertical direction, followed by the stage on breach widening in the upstream and downstream directions of the main channel, and finally the stage on deceleration in breach widening. They developed a two-dimensional, depth-averaged flow model utilizing a sediment transport equation based on their experiments. The model successfully predicted the breached volume during the breach widening stage.

A two-dimensional, depth-averaged model utilizing modified sediment mass conservation equation is developed [Elafy et al., 2018]. A new source term that accounts for the lateral erosion of breach sides is added to the classical sediment mass conservation equation. The model is validated by comparing the computed results with the laboratory measurements of steady and unsteady flows through a levee breach. Then the model is applied to the overtopping failure of a non-cohesive earthen levee to predict the breach evolution, including deepening and widening stages, and breach hydrograph.

16-4 Numerical Model

The governing equations are first presented, followed by a discussion of their solution numerically.

Governing Equations

The following governing equations describe the flow in the main channel and in the floodplain. Both vertical and lateral erosion are included in the model to compute the levee-breach evolution and breach characteristics.

Hydrodynamic equations

The depth-averaged, shallow water equations may be written in vector form as

$$\mathbf{U}_t + \mathbf{E}_x + \mathbf{F}_y + \mathbf{S} = 0 \tag{16-5}$$

in which

$$\mathbf{U} = \begin{pmatrix} h \\ q_x \\ q_y \end{pmatrix} ; \qquad \mathbf{E} = \begin{pmatrix} q_x \\ \frac{q_x^2}{h} + \frac{1}{2}gh^2 \\ \frac{q_x q_y}{h} \end{pmatrix}$$

$$\mathbf{F} = \begin{pmatrix} q_y \\ \frac{q_x q_y}{h} \\ \frac{q_y^2}{h} + \frac{1}{2}gh^2 \end{pmatrix} ; \qquad \mathbf{S} = \begin{pmatrix} 0 \\ -gh(S_{o_x} - S_{f_x}) \\ -gh(S_{o_y} - S_{f_y}) \end{pmatrix}$$

where, $\mathbf{U}$ = vector of conserved variables; $\mathbf{E}$ and $\mathbf{F}$ = flux vector functions; $\mathbf{S}$ = vector of source terms; t = time, x and y = horizontal Cartesian coordinates, h = flow depth, q_x and q_y = discharge per unit width in the x- and y-directions, respectively, g = gravitational acceleration, S_{o_x} and S_{o_y} = bed slope in the x- and y-directions, respectively, S_{f_x} and S_{f_y} = friction slope in the x- and y-directions, respectively.

Based on the Manning equation, the friction slope S_{f_x} and S_{f_y} may be written as

$$S_{f(x;y)} = \frac{n^2 q_{(x;y)} \sqrt{q_x^2 + q_y^2}}{h^{\frac{10}{3}}} \tag{16-6}$$

where n = Manning roughness coefficient.

Sediment equations

The two-dimensional form of the sediment-mass-conservation equation including the new source term to account for the lateral erosion may be written as

$$(1 - \lambda_p)\frac{\partial z}{\partial t} = -\frac{\partial q_{bx}}{\partial x} - \frac{\partial q_{by}}{\partial y} - \frac{\partial q_{sf}}{\partial x} \tag{16-7}$$

where the bed porosity, $\lambda_p = 0.43$, $z =$ bed elevation, $q_{sf} =$ sediment inflow rate induced by slope failure, q_{bx} and $q_{by} =$ volume bed-load transport rate per unit width in the x- and y-directions, respectively, and q_{bx} and q_{by} are calculated as follows.

Because the flow through levee breach includes flow over sloped beds in the x- and y-directions, the critical shear stress obtained based on the flat-bed assumption has to be corrected. According to Van Rijn [1993], the correction coefficients may be written as

$$k_1 = cos\beta_1 \left[1 - \left(\frac{tan^2\beta_1}{tan^2\phi} \right) \right]^{0.5} \tag{16-8}$$

$$k_2 = \begin{cases} \dfrac{sin(\phi - \beta_2)}{sin\phi} & \text{for flow on downward slope} \\ \dfrac{sin(\phi + \beta_2)}{sin\phi} & \text{for flow on upward slope} \end{cases} \tag{16-9}$$

$$\overline{\tau_c} = k_1 \quad k_2 \quad \overline{\tau_{c,f}} \tag{16-10}$$

in which, k_1 and $k_2 =$ correction coefficients in the x- and y-directions, respectively, β_1 and $\beta_2 =$ bed slope angle in the x- and y-directions, respectively, $\phi =$ repose angle of bed material, $\overline{\tau_c} =$ dimensionless corrected critical shear stress, and $\tau_{c,f} =$ dimensionless critical shear stress for a flat bed. The value of q_{bx} and q_{by} are affected by both the total shear stress and the bed slope components in the x- and y-directions. These are calculated as follows:

The shear stress components may be calculated as

$$\tau_{(x;y)} = \rho g h S_{f(x;y)} \tag{16-11}$$

in which $\rho =$ mass density of water.

The total shear stress

$$\tau = \sqrt{\tau_x^2 + \tau_y^2} \tag{16-12}$$

The dimensionless total shear stress is expressed as

$$\overline{\tau} = \frac{\tau}{\rho g R D} \tag{16-13}$$

where $R =$ submerged specific gravity of sediment, and $D =$ median grain size.

The dimensionless components of the rate of volume of bedload transport per unit width may be expressed as

$$\overline{q_{bx}} = \overline{q_b} cos\theta \tag{16-14}$$

$$\overline{q_{by}} = \overline{q_b} sin\theta \tag{16-15}$$

in which, $\theta =$ sediment transport angle which includes the effects of the main flow shear and bed slope in the x- and y-directions, and $\bar{q}_b =$ dimensionless total volume bedload transport rate per unit width and by using the Meyer-Peter and Müller [1948] (MPM) equation may be expressed as

$$\overline{q_b} = \alpha(\overline{\tau} - \overline{\tau_c})^\beta \tag{16-16}$$

in which α and β = coefficient and exponent of MPM formula, respectively. The sediment transport angle, θ, may be calculated according to Van Bendegom's formula [1947] after Talmon et al. [1995] as

$$\theta = tan^{-1}\left(\frac{sin\delta - \frac{1}{1.7\sqrt{\overline{\tau}}}\frac{\partial z}{\partial y}}{cos\delta - \frac{1}{1.7\sqrt{\overline{\tau}}}\frac{\partial z}{\partial x}}\right) \tag{16-17}$$

in which δ = direction angle of bed shear stress and is expressed as

$$\delta = tan^{-1}\left(\frac{V}{U}\right) \tag{16-18}$$

in which U and V = components of the depth-averaged flow velocity in the streamwise (x) and transverse (y) directions, respectively.

The volume bedload transport rate per unit width in the x- and y-directions may be calculated as

$$q_{b(x;y)} = \sqrt{RgD}D \quad \overline{q_{b(x;y)}} \tag{16-19}$$

Lateral sediment load due to slumping failure

Due to the vertical erosion caused by the flowing water through the breach, the breach side slopes become steep [Faeh, 2007; Kakinuma and Shimizu, 2014]. If the slope exceeds the soil repose angle, the unstable soil segment may slump downward, as shown in Fig. 16-1.

Spinewine et al. [2002] differentiated between the soil repose angle of the submerged (under the water surface) and emerged (above the water surface) regions. They stated that the emerged repose angle, ϕ_e is usually greater than submerged repose angle, ϕ_s, due to the apparent cohesion of the partially saturated soil above the water surface. They measured these angles using a simple experiment in a small water tank. The soil used in their experiment was coarse sand of uniform grain size of 1.8 mm. The estimated values of these angles were 35° and 87° for ϕ_s and ϕ_e, respectively. Wu et al. [2009] developed a two-dimensional, depth-averaged model which includes an avalanching algorithm to simulate breaching of a non-cohesive earthen dam due to overtopping. The model was validated with experimental measurements of the breach width and the breach hydrograph. The median grain size of the embankment material was 0.25 mm. A good agreement between the numerical

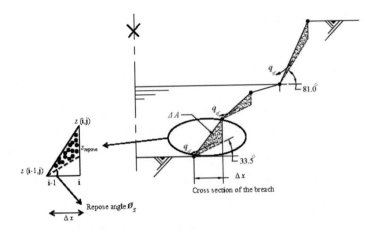

Fig. 16-1   Slumping failure of the breach sides

and experimental measurements was achieved when ϕ_s and ϕ_e are set as $33°$ and $79°$, respectively.

A cross-section of the right side of the breach with the submerged and emerged regions, separated by the water surface is shown in Fig. 16-1. On the left-hand side of the figure, the unstable soil section is shown by dotted area which is prone to fail because its slope angle is steeper than the repose angle. The sediment inflow induced by this slope failure, q_{sf}, may be calculated as

$$q_{sf} = \frac{\Delta A \zeta}{\Delta t} \tag{16-20}$$

in which ΔA = unstable area which falls downward, and ζ = relaxation co-efficient. The unstable area for the submerged region of the right side of the breach may be estimated as

$$\Delta A = 0.5\Delta x \left(z_{(i,j)} - z_{(i-1,j)} \right) - 0.5(\Delta x)^2 \tan \phi_s \tag{16-21}$$

in which Δx = the grid spacing in the x-direction, i indicates the node number in the x-direction, and j indicates the node number in the y-direction. The relaxation coefficient may be expressed as [Spinewine et al., 2002]

$$\zeta = \left(1 - e^{(-\Delta t/T_b)} \right) \tag{16-22}$$

in which Δt = the computational time step, and T_b = the time for the failed material to slump downward.

Numerical Solution

The governing equations are solved numerically by a finite-difference scheme as discussed in the following paragraphs.

Finite-difference Scheme

For steady flow through a levee breach, the hydrodynamic equations (Eq. 16-5) are solved by using MacCormack explicit finite-difference scheme [MacCormack, 1969]. For the details of this scheme, see Chapter 14. The scheme is second-order accurate, both in time and space and allows the solution of both gradually and rapidly varied flows [Tingsanchali and Chinnarasri, 2001] without requiring any special treatment for the steep wavefronts. The solution consists of a two-step predictor-corrector sequence. The spatial derivatives in the predictor step are replaced by forward-finite differences, and by backward finite-differences, in the corrector step. This sequence may be alternated every other time step.

For the simulation of the overtopping failure of a non-cohesive earthen levee, the sediment-mass-conservation equation (Eq. 16-7) is solved along with the hydrodynamic equations (Eq. 16-5) in a coupled manner, as outlined by Bhallamudi and Chaudhry [1991]. For the earthen levee and for the floodplain area, the slope stability is checked at the beginning of the predictor and corrector steps in order to calculate q_{sf}. The predicted values of bed elevation, z^*, are used in the corrector step to check the slope stability, and to evaluate the term, $\partial z / \partial x$. Also, the predicted values of the flow variables h^*, q_x^*, q_y^*, are used to evaluate the sediment transport capacity, $q_{b(x,y)}$, which are used in the corrector step. Thus, the final value of each variable for each time step considers the changes in all the other remaining variables.

The numerical scheme is used to calculate the flow variables at the interior nodes. However, at the boundaries, the flow variables are computed according to the flow conditions at each boundary, as discussed in the following paragraphs.

Initial and Boundary Conditions

For unsteady flow, the initial conditions are specified by a constant discharge equal to the inlet discharge and corresponding constant water depth in the main channel. However, for steady flow, the initial conditions are arbitrary since only the final steady state solution is of interest.

The boundaries for the computational domain may be classified as open or solid boundaries. The open boundaries include the upstream and the downstream ends of the main channel, and the floodplain exits; and the solid boundaries are the walls of the main channel [Kassem, 1996]. Since the flow at the upstream boundary is subcritical, q_x and q_y at all the time steps are specified and h is extrapolated from the interior nodes. The flow at the downstream boundary of the main channel is subcritical. Therefore, q_x and q_y are extrapolated from the interior nodes and h is specified equal to a constant value for steady flow and equal to stage hydrograph measured at Sensor 2 in unsteady flow (Fig. 16-3a). The boundary condition at the exit from the floodplain is a free-fall. Thus, it is governed by the type of flow: For subcritical flow, h

is specified equal to the critical depth and q_x, and q_y are extrapolated from the interior nodes; and for supercritical flow, h, q_x, and q_y are all extrapolated from the interior nodes. At the floodplain exit, z is specified equal to the original bed level. For the solid boundary, a free-slip boundary condition is imposed; h and z are extrapolated from the adjacent nodes and the flux adjacent to the wall is extrapolated from the interior nodes, while the flux normal to the wall is set equal to zero.

Stability condition

The computational time step is selected to satisfy the following Courant-Friedrichs-Lewy stability condition [Sanders et al., 2008].

$$\Delta t = \frac{C_n}{\dfrac{\left|\dfrac{q_x}{h}\right| + \sqrt{gh}}{\Delta x} + \dfrac{\left|\dfrac{q_y}{h}\right| + \sqrt{gh}}{\Delta y}} \tag{16-23}$$

where C_n = Courant number and is chosen to be 0.95 in this study.

Artificial viscosity

Due to the diffusive properties of the scheme, the truncation errors appear in the form of wiggles near the steep wave fronts. The technique developed by Jameson et al. [1981] is used herein to smooth the spurious oscillations near the discontinuities while leaving the smooth area undisturbed. The artificial dissipation terms are added to each of the conserved variables at the end of each time step, as discussed in Section 15-6.

The dissipation coefficient κ is used to regulate the amount of dissipation. It is desirable to use its lowest possible value, while still smoothing the high-frequency oscillations [Gharangik and Chaudhry, 1991]. Trials with different values of this coefficient indicated that the minimum value for the present simulations is 0.04 s/m and that the computed results are almost unchanged for values greater than 0.04 s/m.

16-5 Experimental Investigations

A brief description of the experimental set-up is first presented in this section. Then the experimental procedures are outlined and results are summarized.

Experimental Setup

Figure 16-2 shows the plan view of the experimental setup in the Hydraulics Laboratory of the University of South Carolina. The setup consists of a wooden flume and a floodplain on its left side. The setup is built on a raised, horizontal platform to allow a free fall at the end of the flume and at the exit of the floodplain. The flume is constructed with a vertical wall on the right side and a 2H : 1V sloped wall on the left side. The breach is located near the middle of the left wall, as shown in Fig. 16-2. The earthen levee cross section is trapezoidal, 0.20 m high, 2H : 1V side slopes, and 0.10 m crest width. A honeycomb, a flow straightener, and a wave suppressor are used at the intake of the flume to produce smooth flow conditions. An axial pump supplies a constant discharge of water to the main channel. The discharge is measured on the delivery side of the pump using an orifice plate. The water depth at the end of the flume is adjusted using a calibrated sharp-crested weir.

Experimental Procedures

Both steady and unsteady flow tests are conducted. The inflow discharge to the flume is 0.047 m^3/s and the downstream water depth is 0.1 m. The bed and water surface elevations near the breach are measured by using a Baumer ultrasonic distance measuring sensor (frequency = 290 kHz, accuracy ±0.3 mm) with a scanning range between 0.06 and 0.4 m. The Baumer is mounted on a movable bridge in order to measure the bed and water surface elevations along multiple sections. The water surface changes are measured along 19 sections in the vicinity of the breach area (Fig. 16-2). The first section is located 1 m upstream of the breach and the distance between the two other consecutive sections is 0.152 m (Fig. 16-2). The water surface velocity is measured using Particle Image Velocimetry (PIV). A large number of small floating particles are dropped at the upstream section of the flume and high-definition video of the flow is recorded from the top. The PIVLab MatLAB tool, developed by Thielicke and Stamhuis [2014a,b], is used for the post-test analysis. The water depth and velocity measured at 10 sections, Y1 through Y10 (Fig. 16-2), are compared with the computed results in the model applications section.

The unsteady experiment involves overtopping failure of a non-cohesive, non-compacted earthen levee. The inflow discharge to the flume is 0.055 m^3/s. The initial downstream water depth is 0.15 m. The earthen levee is built at the open section of the left wall of the flume, as shown in Fig. 16-2. The levee is constructed with a non-compacted, medium sand of uniform grain size of 0.6 mm. To initiate the overtopping, a pilot channel, 10 cm wide and 5 cm deep, is carved at the levee crest. The pilot channel is located at the first third of the levee crest since it is expected that the breach develops faster in the downstream direction due to the nature of the flow in the breached area. During the breaching of the levee, the changes in the water surface elevation are measured at two locations in the main channel using fixed Baumers, as shown in Fig. 16-2. The first location is along the centerline of the pilot channel and the second location is at the downstream end of the main channel. To measure the breach evolution, four rows of sliding rods, with seven rods in each row and the top part of the rods in each row of a specific color to represent

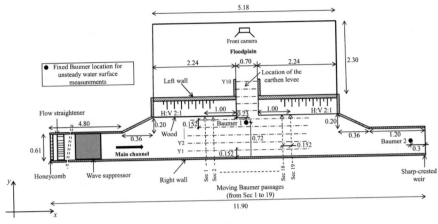

(a) Plan view of experimental setup

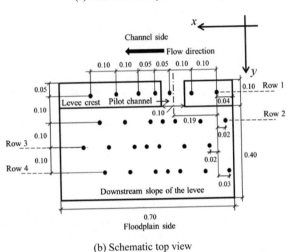

(b) Schematic top view

Fig. 16-2 (a) **Plan view of experimental setup and** (b) **Schematic top view (not to scale, all dimensions in meters)**

its location along the levee, are used. The rods are placed on the downstream slope of the levee with the first row aligned with the centerline of levee crest. In order to minimize the obstruction of flow through the breach, the rod diameter is 1 mm. The rod base is flat and square to prevent the tip of the rod from penetrating the soil surface, and only to allow the rod to drop when the soil underneath is eroded. The rods are placed on a staggered grid to measure changes in the bed elevation at different locations on the earthen levee. The movement of the rods is recorded by a high-definition video camera, recording at 60 frames/s. The camera is placed on the floodplain side and facing the downstream slope of the levee (Fig. 16-2). The top of each rod is digitized using GetData Graph Digitizer software to extract the breach shape every five seconds from the start of overtopping.

Experimental Results

Some salient features of the experimental results are discussed in this section.

Breach Hydrograph

Figure 16-3 shows the time variation of the water surface elevation, inflow, outflow, and breach hydrographs. The discharge through the breach is the difference between the main channel inflows and outflows and taking into consideration the change in the water storage in the main channel, upstream and downstream of the breach using the measured water surface elevation at Sensor 1 and at Sensor 2. At the beginning, the breach discharge is minimal, then it increases at a nearly fixed rate and then it decreases and becomes almost constant during the final stage of the breaching process. Kakinuma and Shimizu [2014] reported similar observations. The water levels in the upstream and downstream reservoirs are recorded using Baumer 1 and Baumer 2, respectively (Fig. 16-2). The unsteady discharge at the end of the main channel is computed from the formula for calibrated sharp-crested weir using the water depths measured by Baumer 2.

Breach-shape

Figure 16-4 show the breach evolution through the downstream slope of the levee. The flow is from left to right. To avoid the effects of the fixed walls in the breach area, only the first 35 s of the failure is reported. The breach growth has the same trend throughout the whole domain. First, the breach develops in the vertical direction. Then, the breach starts to widen due to the failure of breach side slopes. The breach widening is not symmetrical along the centerline of the pilot channel; it is faster at the right side than that at the left side. This may be due to the nature of flow through the breached levees. The flow depth and flow velocity are always high on the right side as compared to that on the left side.

16-6 Model Applications

In this section, an application of the numerical model to simulate steady flow through a levee breach and to an overtopping failure of a non-cohesive earthen levee is presented. Then a sensitivity analysis is introduced to study the effects of different model parameters on the breach shape. To insure that

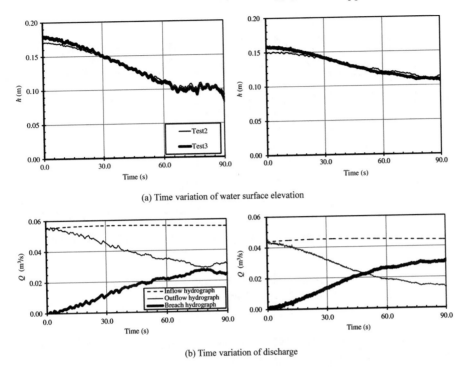

(a) Time variation of water surface elevation

(b) Time variation of discharge

Fig. 16-3 (a) **Time variation of water surface elevation for Tests 2 and 3: (Left) Sensor 1; (right) Sensor 2, (b) Time variation of discharge during (left) Test 2; (right) Test 3**

the numerical results are grid independent, three different grid sizes are tested for the case of steady flow through a levee breach. The average RMSE is also calculated between the measured and simulated velocity for ten sections; the difference between the results for three grid sizes is small.

Steady Flow through a Levee Breach

The developed hydrodynamic model is validated by simulating the steady flow through a levee breach. Computations are continued for the specified boundary conditions until the solution converges to the steady state. The simulated and the measured water depths are compared in Fig. 16-5. For sections Y1 to Y5 in the main channel (Fig. 16-2), the water surface upstream and downstream of the breach is nearly horizontal while the water surface in the main channel adjacent to the breach slope upwards in the downstream direction since the flow in the main channel is subcritical. This is consistent with the observations made by Borghei et al. [1999]. Also, it is observed that

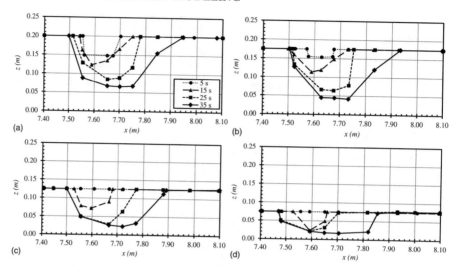

Fig. 16-4   Levee breach evolution during Test 2 along different rows:
(a) Row 1; (b) Row 2; (c) Row 3; (d) Row 4

the curvature of the water surface at the inner side of the breach increases moving towards the breach (sections Y4 and Y5) due to the formation of a wake zone on the upstream side of the breach. For sections Y6 to Y10 in the breach area, the water depth on the right side of the breach is higher than that on the left side due to the effect of the momentum flux of the main channel flow. Roger et al. [2009] reported the same observation. This effect decreases towards the end of the breach and the slope of the water surface decreases until it becomes nearly horizontal at the end of the breach (sections Y9 and Y10).

The simulated depth-averaged flow velocity and the measured surface velocity along sections Y1 to Y10 in the streamwise and transverse directions are compared in Figs. 16-6 and 16-7, respectively. The streamwise velocity in the main channel upstream of the breach is higher than that downstream of the breach (sections Y1 through Y5 in Fig. 16-6). This difference in the velocity is caused by the decrease in discharge in the downstream direction due to diversion of flow through the breach. This difference in velocity decreases in the breach area until it is almost zero (sections Y8 through Y10) since most of the flow is in the transverse direction. As shown in Fig. 16-7, the transverse velocity increases from section Y1 to Y5, with the peak velocity occurring almost at the middle of the cross section. However, in the breach area (sections Y6 to Y9), the peak velocity is shifted towards the right side of the breach. As mentioned earlier, this may be due the formation of a wake zone upstream of the breach. At the end of the breach area (section Y10), the effect of the wake zone vanishes and the velocity becomes almost uniform across the breach. Generally, the numerical model successfully captures the pattern of the water depth and flow velocity in the main channel and in the

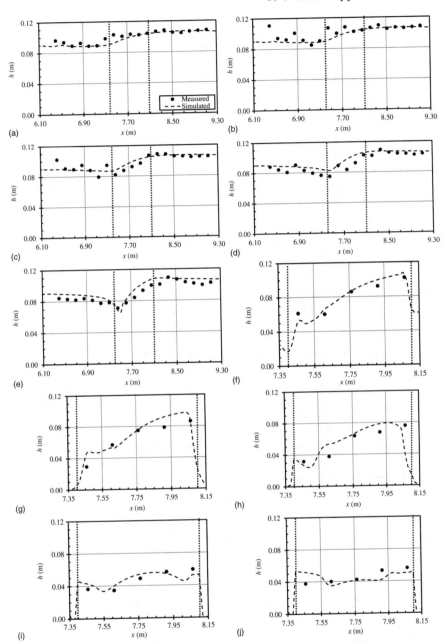

Fig. 16-5 Comparison between simulated and measured water depth for Test 1: (a) Y1; (b) Y2; (c) Y3; (d) Y4; (e) Y5; (f) Y6; (g) Y7; (h) Y8; (i) Y9; (j) Y10

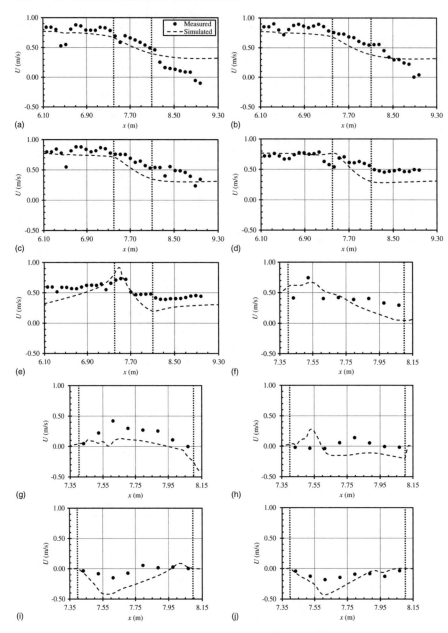

Fig. 16-6 Comparison between simulated and measured velocity in streamwise direction for Test 1: (a) Y1; (b) Y2; (c) Y3; (d) Y4; (e) Y5; (f) Y6; (g) Y7; (h) Y8; (i) Y9; (j) Y10

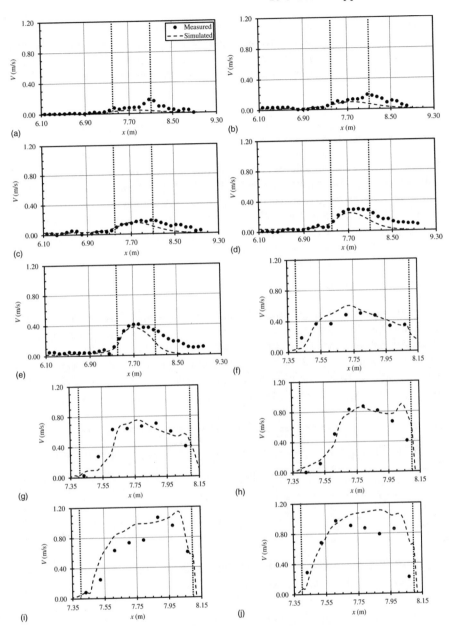

Fig. 16-7 Comparison between simulated and measured velocity in transverse direction for Test 1: (a) Y1; (b) Y2; (c) Y3; (d) Y4; (e) Y5; (f) Y6; (g) Y7; (h) Y8; (i) Y9; (j) Y10

breach area. The average RMSE for the flow velocity and depth at the ten sections in the main channel are less than that for the sections in the breach area. This may be due to the presence of vertical velocity components within the breach area which are neglected in the depth-averaged flow model [Van Emelen et al., 2012].

Levee Failure due to Overtopping

The hydrodynamic equations (Eq. 16-5) are solved along with the modified sediment-mass-conservation equation (Eq. 16-7) to simulate the overtopping failure of a non-cohesive earthen levee. Both vertical and lateral erosion are included in the numerical model to predict the breach evolution and breach hydrograph. The parameters used for this simulation are $n = 0.019$, $\phi_s = 33.5°$, $\phi_e = 81°$, $T_b = 1s$, $\alpha = 18$, and $\beta = 1.5$.

Breach Hydrograph

Figure 16-8 compares the simulated and measured time variation of the water depth for the unsteady flow test at Sensor 1. As shown in the figure, the numerical model successfully predicts the general trend of the water depth. Also, the breach hydrograph is successfully reproduced by the numerical model as shown in Fig. 16-9. The breach outflow reaches its peak value around 80 s and then becomes almost constant.

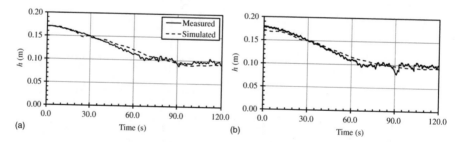

Fig. 16-8   Comparison between simulated and measured water depth at Sensor 1: (a) Test 2; (b) Test 3

Breach evolution

The measured and the simulated breach shapes along Row 2 are compared in Figure 16-10. The flow is from left to right. The location of the pilot channel is represented by two vertical dashed lines. As shown in Fig. 16-10, the numerical model successfully captures the unsymmetrical widening pattern of the breach

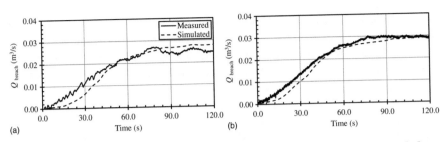

(a)   (b)

Fig. 16-9 **Comparison between simulated and measured breach hydrograph: (a) Test 2; (b) Test 3**

and also captures the maximum breach depth. The agreement between the numerical and the experimental results along Row 1 is better as compared to Row 2, Row 3, and Row 4. The average RMSE along Row 1 and Row 2 for the first 35 s of the failure is 0.0169 m and 0.0195 m, respectively. The increase of the RMSE along Row 2 as compared to that at Row 1 may be due to the complex flow pattern on the downstream face of the levee which includes vertical velocity components. The remaining model inaccuracies are due to the seepage through the levee which is not included in the present model.

Figure 16-11 shows the simulated time series of breach evolution through the

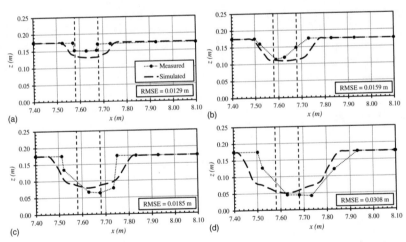

(a)   (b)
(c)   (d)

Fig. 16-10 **Comparison between simulated and measured breach evolution along Row 2 during Test 2: (a) time = 5 s; (b) time = 15 s; (c) time = 25 s; (d) time = 35 s**

downstream slope of the levee. At the beginning of overtopping, the flowing water through the pilot channel erodes vertically resulting in the pilot channel walls to collapse due to slope instability. The width of pilot channel increases allowing more flow through the breach which results in rapid deepening until reaching the fixed bed. Then the breach widens mainly in the downstream direction.

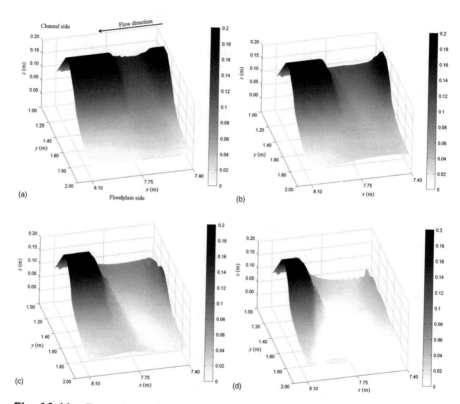

Fig. 16-11   Breach evolution for Test 2: (a) time = 15 s; (b) time = 30 s; (c) time = 45 s; (d) time = 60 s

Transient Velocity Field

Figure 16-12 shows the time variation of simulated velocity. At the early stages of the failure, the velocity is almost symmetric along the width of the breach, as shown in Fig. 16-12a. Then with time, the high velocity band shifts towards the downstream side of the breach, as shown in Fig.16-12b, c, and d. This is consistent with the observations of Kakinuma and Shimizu [2014]. This asymmetric velocity distribution may explain asymmetric breach widening during levee breach.

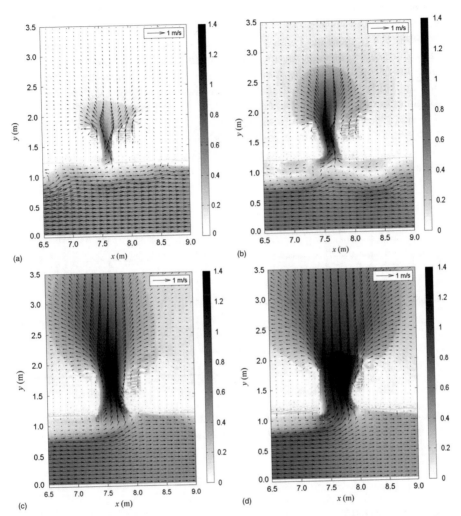

Fig. 16-12 Simulated velocity fields (m/s) at different times for Test 2: (a) time = 15 s; (b) time = 30 s; (c) time = 45 s; (d) time = 60 s

Sensitivity Analysis

A sensitivity analysis is presented to study the effects of different parameters on the levee breach shape. The unsteady flow is analyzed by changing the value of one parameter at a time while keeping the other parameters at their base value. A total of sixteen cases are tested. Figure 16-13 shows the relationship between the percentage change of each parameter and percentage change of breach top width and maximum breach depth. The breach dimensions of these cases are taken along the crest centerline 15 s after the failure starts. As shown in Fig. 16-13, the breach dimensions are directly proportional to Manning roughness coefficient and the coefficient of Meyer-Peter and Müller formula. It is also observed that, as the soil repose angle increases, the breach top width decreases and the maximum breach depth increases. This may be due to the fact that less material is expected to fail from the breach sides when the soil repose angle is high. Also, for the simulated cases, no major effect is found of T_b on the breach shape.

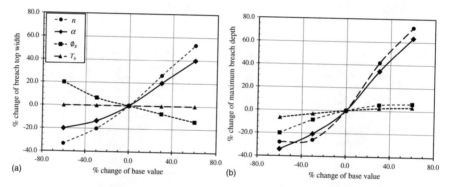

Fig. 16-13   Effect of different model variables on breach dimensions: (a) breach top width; (b) maximum breach depth

16-7 Summary

Two generalized cases of levee breach with constant cross section through rectangular and trapezoidal channels are studied. The resulting flow field is investigated by numerically solving the one-dimensional, spatially varied flow equation and the two-dimensional shallow-water equations. The flow field resulting from a levee breach is better predicted by the two-dimensional model than that by the one-dimensional model and the two-dimensional model computes higher lateral breach outflow than that by the one-dimensional model.

Non-dimensional equations are presented to correct the calculated breach outflow using the spatially varied flow equation for rectangular and trapezoidal channels.

To model the failure of non-cohesive earthen levees, new source term to account for lateral erosion due to the slope failure is added to the sediment-mass-conservation equation. The modified equation is solved along with the depth-averaged flow equations using a two-dimensional, finite-difference numerical model. The model is validated against a steady flow through a levee breach and for overtopping failure of a non-cohesive earthen levee. The numerical model successfully simulates the failure process of non-compacted and non-cohesive earthen levees. Additional investigations are needed to simulate the breaching process of cohesive levees which includes both the surface erosion as well as the headcut erosion.

Problems

16-1 Compute the flow through a 5-m breach in a 50-m wide drainage channel with flow of 400 m^3/s at a depth of 8 m. Assume the crest of the breach is 3-m above the channel bottom and the flood plain is 2-m above the channel bottom.

   i. Compute breach flow assuming the breach as broad-crested weir;
  ii. Using one-dimensional gradually varied flow with lateral outflow;
 iii. Using the methodology of Section 16-2; and
  iv. Compare the computed results of i through iv.

16-2 Compute breach outflow of Problem 16-1 for different flow depths in the flood plain. Select an appropriate value for any coefficient.

16-3 Write a computer code to compute the breach outflows using the equation for the one-dimensional gradually varied flow with lateral outflow. Compare the results for Problem 16-1.

16-4 Develop a computer code to compute the formation and growth of a levee breach for a flood increasing in flow from 400 m^3/s to 600 m^3/s in two hours and then remaking at this flow for several days.

  i. Neglect slumping failure; and
 ii. Include slumping failure in the breach development.

16-5 The flow in a 10-m wide canal with side slopes of 2H : 1V and canal bottom slope of 1 in 10,000 is suddenly increased due to sudden opening of the intake gates from 2 m to 4 m. Assume the reservoir is large with water level elevation 5 m above the canal bottom at the intake. Develop a computer code to compute the breach formation due to overtopping. Assume the breach is formed in a length of 10 m, and the canal freeboard is 0.5 m. Select appropriate values for the missing parameters.

16-6 Develop a computer code for the growth of a breach in a levee using the two dimensional St. Venant equations and the breach as an internal boundary condition. The river flow doubles in three hours on the upstream of the breach and remains constant afterwards at this value. Include the slumping failure mechanism in the growth of the breach.

References

ASCE/EWRI Task Committee on Dam/Levee Breaching, 2011 "Earthen embankment breaching," *Jour. Hyd. Engineering*, Amer. Soc. of Civil Engrs., vol. 137, no. 12, pp. 1549–1564.

Bhallamudi, S. M., and Chaudhry, M. H., 1991, "Numerical Modeling of Aggradation and Degradation in Alluvial Channels," *Jour. Hyd. Engineering.*, Amer. Soc. of Civil Engrs., vol. 117, no. 9, pp. 1145–1164.

Borghei, S., Jalili, M., and Ghodsian, M., 1999. "Discharge Coefficient for Sharp-Crested Side Weir in Subcritical Flow," *Jour. Hyd. Engineering*, vol. 125, no. 10, pp. 1051–1056.

Cheong, H. F., 1991, "Discharge Coefficient of Lateral Diversion From Trapezoidal Channel," *Jour. Irrig. Drain. Engineering*, vol. 117, no. 4, pp. 461–475.

Coleman, S. E., Andrews, D. P., and Webby, M. G., 2002, "Overtopping Breaching of Noncohesive Homogeneous Embankments," *Jour. Hyd. Engineering*, Amer. Soc. of Civil Engrs., vol 128, no 9, pp. 829–838.

Elalfy, E. Y. (2015), "Numerical and Experimental Investigations of Dam and Levee Failure," PhD dissertation, University of South Carolina, Columbia, SC, USA.

Elalfy, E. Y., Tabrizi, A. A., and Chaudhry, M. H., (2018), "Numerical and Experimental Modeling of Levee Breach Including Slumping Failure of Breach Ssides," *Jour. Hydraul. Eng.*, American Society of Civil Engrs., vol 144, no 2, pp. 04017066-1 to 18.

Faeh, R., 2007, "Numerical Modeling of Breach Erosion of River Embankments," *Jour. Hyd. Engineering*, vol. 133, no. 9, pp. 1000–1009.

Froehlich, David C, 2008, "Embankment Dam Breach Parameters and Their Uncertainties," *Jour. Hyd. Engineering*, vol. 134, no. 12, pp. 1708–1721

Gharangik, A. M. and Chaudhry, M. H., 1991, "Numerical Simulation of Hydraulic jump," *Jour. of Hyd. Engineering*, Amer. Soc. of Civil Engrs., vol. 117, no. 9, pp 1195–1211.

Hager, W. H., 1987, "Lateral Outflow Over Side Weirs," *Jour. Hydraul. Engineering*, vol. 113, no. 4, pp. 491–504.

Han, K. Y., Lee, J. T., and Park, J. H., 1998, "Flood Inundation Analysis Resulting from Levee-Break," *Jour. Hyd. Res.,* vol. 36, no. 5, pp. 747–759.

Henderson, F. M., 1966, *Open Channel Flow,* Macmillan, NY., New York.

Jameson, A., Schmidt, W., and Turkel, E., 1981, "Numerical solution of the Euler equations by finite volume methods using Runge Kutta time stepping schemes," In *14th fluid and plasma dynamics conference,* p. 1259

Kakinuma, T., and Shimizu, Y., 2014, "Large-scale Experiment and Numerical Modeling of a Rivirine Levee Breach," *Jour. of Hyd. Engineering,* Amer. Soc. of Civil Engrs., vol. 140, no. 9, 04014039.

Kassem, A. A., 1996, "Two-dimensional numerical modeling of sediment transport in unsteady open-channel flows," PhD dissertation, Washington State University, USA.

MacCormack, R. W., 1969, "The Effect of Viscosity in Hypervelocity Impact Cratering," Paper 69–354, American Institute Aeronautics Astronautics, Cincinnati, Ohio.

Riahi-Nezhad, C. K., 2013, "Experimental Investigation of Steady Flows at a Breached Levee," PhD dissertation, University of South Carolina, Columbia, SC, USA.

Roger, S., Dewals, B. J., Erpicum, S., Schwanenberg, D., Schüttrumpf, H., Köngeter, J., and Pirotton, M., 2009, "Experimental and Numerical Investigations of Dike-Break Induced Flows," *Jour. Hyd. Res.,* vol. 47, no. 3, pp. 349–359.

Singh, R. M., Manivannan, D., and Satyanarayana, T., 1994, "Discharge Coefficient of Rectangular Side Weirs," *Jour. Irrig. Drain. Engineering,* vol. 120, no. 4, pp. 814–819.

Van Emelen, S., Soares-Frazão, S., Riahi-Nezhad, C., Chaudhry, M. H., Imran, J., and Zech, Y., 2012, "Experimental and Numerical Simulations of the 17th Street Canal Breach," *Jour. Hyd. Research,* vol. 50, no. 1, 2012, pp. 1–12.

Schmocker, Lukas and Hager, W. H., 2009, "Modelling Dike Breaching due to Overtopping," *Jour. Hyd. Research,* vol. 47, no. 5, pp. 585–597.

Meyer-Peter, E., and Müller, R., 1948, "Formulas for Bed-Load Transport." *Proc., 2nd IAHR Congress,* International Association for Hydraulic Research, Stockholm, Sweden, pp. 39–64.

Pontillo, M. and Schmocker, L. and Greco, M. and Hager, W. H., 2010, "1D numerical Evaluation of Dike Erosion due to Overtopping," *Jour. of Hyd. Research,* vol. 48, no. 5, pp. 573–582.

Roger, S., et al., 2009, "Experimental and Numerical Investigations of Dike-Break Induced Flows." *Jour. Hyd. Res.* vol. 47, no. 3, pp. 349–359.

Sanders, B. F., Schubert, J. E., and Gallegos, H. A., 2008, "Integral Formulation of Shallow-Water Equations with Anisotropic Porosity for Urban Flood Modeling," Journal of Hydrology, vol. 362, no. 1–2, 19–38.

Spinewine, B., Capart, H., Le Grelle, N., Soares Frazao, S., and Zech, Y., 2002, "Experiments and Computations of Bankline Retreat Due to Geomorphic Dam-Break Floods," *Proc., 1st Int. Conf. on Fluvial Hydraulics River Flow,* A. Balkema, Rotterdam, Netherlands, pp. 651–661.

Talmon, A., Struiksma, N., and Van Mierlo, M., 1995, "Laboratory Measurements of the Direction of Sediment Transport on Tranverse Alluvial-Bed Slopes," *Jour. Hyd. Res.,* vol. 33, no. 4, pp. 495–517.

Thielicke, W., and Stamhuis, E. J., 2014a, "PIVlab: Time-Resolved Digital Particle Image Velocimetry Tool for MATLAB (Version: 1.35)." ¡https://doi.org/10.6084/m9.figshare.1092508.v6¿ (Nov. 22, 2017).

Thielicke, W., and Stamhuis, E. J., 2014b, "PIVlab: Towards User-Friendly, Affordable and Accurate Digital Particle Imagery Velocimetry in MATLAB," *Jour. Open Res. Software,* vol. 2, no. 1, e30.

Tingsanchali, T., and Chinnarasri, C., 2001, "Numerical Modelling of Dam Failure Due to Flow Overtopping," *Hydrol. Sci. Jour.,* vol. 46, no. 1, pp. 113–130.

Van Bendegom, L., 1947, Eenige Beschouwingen over Riviermorphologie en Rivierverbetering, De Ingenieur, vol. 59, no. 4, pp. 1–11 (in Dutch).

Van Rijn, L. C., 1993, Principles of Sediment Transport in Rivers, Estuaries and Coastal Seas, Aqua Publications, Amsterdam, Netherlands.

Van Emelen, S., Soares-Frazão, S., Riahi-Nezhad, C. K., Chaudhry, M. H., Imran, J., and Zech, Y., 2012 "Simulations of the New Orleans 17th street canal breach flood." *Journal of Hydraulic Research,* vol. 50, no. 1, pp. 70–81.

Visser, P. J., 1998, "Breach Growth in Sand-Dikes." Rep. No. 98-1, Delft Univ. of Technology, Delft, Netherlands.

Wang, Zhengang and Bowles, David S., 2006, "Three-Dimensional Non-Cohesive Earthen Dam Breach Model. Part 1: Theory and methodology, Part 2: Validation and Applications," *Advances in Water Resources,* vol. 29, no. 10, pp. 1528–1545

Wu, W., He, Z., and Wang, S. S., 2009, "A Depth-Averaged 2D Model of Non-Cohesive Dam/Levee Breach Processes," *Proc., World Environmental and Water Resources Congress Great Rivers,* S. Starrett, ed., Amer. Soc. Civ. Engrs., Reston, VA.

17

SEDIMENT TRANSPORT

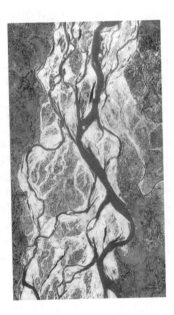

Satellite image of the braided Jamuna River in Bangladesh; the river carries one of the world's highest sediment loads (Courtesy, Institute of Water Modeling, Bangladesh)

This chapter is authored by **Jasim Imran**; some material from Prof. G. Parker's unpublished notes is used with permission.

17-1 Introduction

Sediment transport is essentially a two-phase flow problem in which the fluid phase is air or water and the solid phase is sediment particle. The processes of erosion, transport, and deposition of sediment, collectively termed as sedimentation, are natural processes and have been occurring throughout the geologic time. The landscape as well as the continental margin that includes the shelf, slope and canyons are continuously shaped by the process of sedimentation. Sediment transport occurs due to water, wind, and gravity. Interest in sediment transport stems from practical engineering importance of flood control, erosion control, and river basin management as well as economic interest associated with the extraction of petroleum and other mineral resources. The study of the movement of sediment particles under the influence of gravity and fluid drag constitutes a fascinating field. Let us treat river as a container. The typical container of fluid-sediment mixture, i.e. river, is constructed and deformed by its own content. Depending on the flow conditions and sediment size distribution, bedforms of various scale and shape can appear. These bedforms cause extra resistance to the flow and thus can alter the flow depth significantly.

Based on the environment in which sediment transport occurs, the transport process can be termed as aeolian, fluvial, marine, and submarine. In this chapter, focus primarily is given to sediment transport in rivers, i.e., fluvial transport. Sediment transport in rivers are classified in two modes based on the driving mechanism. These are bedload and suspended load. As bedload, sediment particles saltate, roll, and slide, but always staying close to the bed. As suspend load, sediment is carried by the fluid turbulence up in the water column. In the case of river, the volume concentration of solids in the water column tends to be rather dilute even during large floods. It is, therefore, possible to treat the sediment and fluid phase separately. It will take an entire textbook to cover various aspects of sediment transport. Here, the following important topics are presented in a condensed form: sediment property, sand-bed and gravel-bed rivers, threshold conditions for sediment movement and significant suspension, Shields diagram, sediment mass conservation in the river bed, resistance relations, and transport of sediment as bedload and suspended load.

17-2 Sediment Properties

Sediment in nature consists of particles originating from fragmented rocks. Sediment properties that play a role in the transport process include size, shape, and specific gravity of individual particles and porosity and size distribution of particles as a group. Quartz is the most common rock type encountered in the river. Other common rock types include basalt, granite, limestone, and magnetite. The specific gravity of quartz is in the range of $2.6 \sim 2.7$; an average value of 2.65 is most commonly used.

Sediment Size

The size distribution of sediment transported by rivers can be very wide. The most commonly used unit to describe sediment size, D, is mm. Natural sediment particles are irregular in shape and, therefore, the definition of size by a single dimension can be incomplete. For coarser particles, D represents the intermediate axis of the particle idealized as an ellipsoid. For sediment size ranging from 0.0625 mm to 16 mm, D denotes the smallest sieve size through which the particle will barely pass. For particles finer than 0.0625 mm, D represents an equivalent of sedimentation diameter obtained from the settling or fall velocity. As a matter of convenience, sedimentologists use the ϕ scale to describe sediment size. This scale allows description and plotting of grain size distribution within a narrow range and simplifies the statistical analysis of size distribution. The following equation describes the relationship between ϕ and D (mm)

$$\phi = -\log_2(D) = -\frac{\ln(D)}{\ln(2)} \qquad (17\text{-}1)$$

One disadvantage of the ϕ scale is that the value of ϕ decreases as D increases. With this in mind, Parker and Andrews [1985] introduced the ψ scale such that $\psi = -\phi$, i.e.,

$$\psi = \log_2(D) = \frac{\ln(D)}{\ln(2)} \qquad (17\text{-}2)$$

Based on size, sediment can be classified as clay, silt, sand, gravel, cobble, and boulder. Table 17-1 shows a commonly used classification of sediment into different size classes.

Table 17-1   **Classification of sediment based on size.**

Sediment type	D (mm)	ϕ	ψ
Boulder	> 256	< −8	> 8
Cobble	64 to 256	−8 to −6	6 to 8
Gravel	2 to 64	−6 to −1	1 to 6
Silt	0.0625 to 2	−1 to 4	−4 to 1
Clay	< 0.00195	> 9	< −9

Size Distribution

Once the size distribution of a sediment sample is determined, it is conveniently presented by plotting a cumulative frequency curve of percent by weight finer than a certain size against the logarithm of that size. From such a plot, the median size and other statistical properties can be easily determined. Figure 17-1 shows a typical size distribution curve. Plotting percent finer versus ψ also gives a similar curve. If $p(\psi)$ denotes the probability density of a sample associated with grain-size, ψ, then by definition

$$p(\psi) = \frac{dp_f}{d\psi} \tag{17-3}$$

where p_f indicates percent finer. The density $p(\psi)$ can be quite useful for extracting statistical information on the grain-size distribution. For example,

$$\psi_m = \int \psi p(\psi) d\psi; \quad \sigma^2 = \int (\psi - \psi_m)^2 p(\psi) d\psi \tag{17-4}$$

The geometric mean diameter, D_g, and the geometric standard deviation, σ_g, are given as

$$D_g = 2^{\psi_m}; \quad \sigma_g = 2^{\sigma} \tag{17-5}$$

Since the grain-size distribution is not a continuous function, the statistical properties are obtained from the following summation form rather than the integral form given by Eq. 17-4.

$$\psi_m = \sum_{i=0}^{N} \psi_i f_i; \quad \sigma^2 = \sum_{i=1}^{N} (\psi - \psi_m)^2 f_i \tag{17-6}$$

The grain-size distribution of a sample is characterized by $N + 1$ sizes of $D_{f,i}$ with corresponding fraction finer $f_{f,i}$. Now, consider the grain-size distribution of Fig. 17-1 presented in Table 17-2.

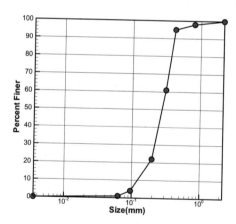

Fig. 17-1   Typical grain-size distribution

Table 17-2 **Fraction of individual size classes in a sediment sample**

i	$D_{f,i}$	$\psi_{f,i}$	$f_{f,i}$	D_i	ψ_i	f_i
1	0.0035	−8.158	0.00	0.015	−6.079	0.01
2	0.0625	−4.000	0.01	0.077	−3.690	0.03
3	0.0960	−3.381	0.04	0.137	−2.870	0.18
4	0.1950	−2.358	0.22	0.246	−2.024	0.39
5	0.3100	−1.690	0.61	0.361	−1.471	0.34
6	0.4200	−1.252	0.95	0.580	−0.787	0.03
7	0.8000	−0.322	0.98	1.327	0.408	0.02
8	2.2000	1.138	1.00			

The fraction of each size class D_i for $i=1$ to N can be calculated as follows:

$$\psi_i = \frac{1}{2}\left(\psi_{f,i+1} + \psi_{f,i}\right) \tag{17-7}$$

$$D_i = \left(D_{f,i}D_{f,i+1}\right)^{\frac{1}{2}} \tag{17-8}$$

and

$$f_i = f_{f,i+1} - f_{f,i} \tag{17-9}$$

For the distribution of Table 17-2, using Eqs. 17-5 and 17-6, D_g and σ_g have value of 0.251 mm and 1.74, respectively. If σ_g is less than 1.3, the sediment mixture is termed well-sorted and can be treated as uniform material; when σ_g exceeds 1.6, the mixture is considered to be poorly sorted.

17-3 Sand-bed and Gravel-bed Streams

Alluvial rivers can be classified broadly as sand-bed and gravel-bed streams based on grain-size distribution of the river bed. The median size, D_{50}, of the bed material of sand-bed streams typically range between 0.1 to 1 mm. The grain size distribution tends to be well sorted with σ_g varying from 1.1 to 1.5. The bed material of gravel-bed rivers often displays a wide range of grain-size distribution. In many gravel-bed rivers, the bed is vertically stratified with a coarse armor layer on the surface. In gravel-bed rivers, median size of the bed material varies between 15 mm to 200 mm or larger; the substrate is typically finer by a factor of 1.5 to 3. The value of σ_g for the substrate is quite large, with values in excess of 3.0. Armor layers are more typical of perennial streams with low sediment supply and moderate flood discharge. In ephemeral streams with violent flood and high sediment supply, the armor layer can disappear. The relationship between the existence of armor layer and flow discharge is evident in Fig. 17-2. At a relatively low discharge, an armor layer formed in a laboratory flume. At a moderately high discharge, the armor layer was partially destroyed. When the discharge was increased farther, the armor layer completely disappeared [Elhakeem, 2004].

(a)   (b)   (c)

Fig. 17-2   **The effect of discharge on the composition of bed surface (a) armor layer; (b) partially destroyed armor layer; (c) completely destroyed armor layer** (After Elhakeem [2004])

17-4 Threshold of Sediment Motion

Fluid drag exerted on sediment particles on the river bed is responsible for initiating sediment motion. For sediment motion to occur, the applied fluid drag must exceed a threshold value. The resistance to motion is due to cohesion in the case of consolidated clay-rich sediment and from the Coulomb friction in the case of non-cohesive sediment. We focus mainly on non-cohesive sediment. Consider the simplified case described in Fig. 17-3. The flow is over a granular bed with sediment size, D. The mean bed slope is small, i.e. $S \ll 1$. Assume that the roughness height, $k_s = n_k D$, where n_k is a dimensionless, $O(1)$ number (e.g., 2). Consider an "exposed" particle the centroid of which protrudes up from the mean bed by an amount $n_e D$, where n_e is again dimensionless and $O(1)$. The flow over the bed is assumed to be turbulent rough, and the drag on the grain is assumed to be in the inertial range. Fluid drag tends to move the particle and Coulomb resistance hinders motion. The impelling drag force, F_D, and the submerged weight of the particle, F_g, is given by

$$F_D = \frac{1}{2}\pi \left(\frac{D}{2}\right)^2 \rho C_D u_f^2 \qquad (17\text{-}10)$$

and

$$F_g = \frac{4}{3}\pi \left(\frac{D}{2}\right)^2 \rho R g \qquad (17\text{-}11)$$

where ρ denotes fluid density, $R = (\rho_s - \rho)/\rho$ is the submerged specific gravity of sediment particle, g is the gravitational acceleration, D is the sediment diameter, u_f is the fluid velocity at the level of the particle, and c_D is the drag coefficient related to the Reynolds number, $R_f = u_f D/\nu$. For spherical particle, C_D may be estimated from the standard drag curve. The Coulomb

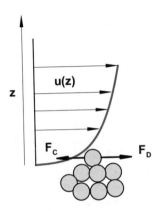

Fig. 17-3   Threshold condition for the entrainment of a sediment particle by water

resistive force F_C is given by

$$F_C = \mu_c F_g \tag{17-12}$$

where μ_c is the Coulomb friction coefficient. At the threshold of motion, $F_D = F_c$; therefore,

$$\frac{u_f^2}{RgD} = \frac{4}{3}\frac{\mu_c}{C_D} \tag{17-13}$$

For hydraulically rough flow, in which the roughness height, k_s, is much larger than the thickness of the viscous sublayer, the log-law may be written as

$$\frac{\overline{u}}{u_*} = \frac{1}{\kappa}\ln\left(\frac{z}{k_s}\right) + 8.5 \tag{17-14}$$

in which $\overline{u}$ is the Reynolds-averaged streamwise velocity, u_* is the shear velocity, κ is the von Kármán constant having a value of 0.41, k_s is the roughness height, and z is the distance from the bed. For the exposed particle of Fig. 17-3, Eq. 17-14 may be written as

$$\frac{u_f}{u_*} = \frac{1}{\kappa}\ln\left(\frac{n_e}{n_k}\right) + 8.5 \tag{17-15}$$

By using Eq. 17-15, Eq. 17-13 may be expressed as

$$\frac{u_*^2}{RgD} = \tau_c^* = \frac{4\mu_c}{3C_D}\left[\kappa \ln\left(\frac{n_e}{n_k}\right) + 8.5\right]^{-2} \tag{17-16}$$

The term τ_c^* in Eq. 17-16 is known as the *critical Shields stress*. For a sediment particle to be entrained from the bed and start to move, the non-dimensional bed shear stress or the Shields stress must exceed τ_c^*. The critical Shields stress can be estimated from Eq. 17-16 by making some reasonable assumptions on the values of μ_c, and n_e/n_k. The drag coefficient c_D can be estimated from the standard drag curve as a function of $u_f D/\nu$.

Critical Shields Stress for Sediment Mixture

River beds often comprise of a mixture of diverse sediment sizes. The larger grains in the mixture are heavier, and thus harder to move. However, the larger grains also protrude more into the flow and the smaller grains tend to hide in between them, so making the larger grains to move easier than the smaller grains. The net result is a mild tendency for coarser grains to be harder to move than finer grains, a fact first demonstrated by Egiazaroff [1965]. Let τ_{ci}^* and τ_{c50}^* denote the critical Shields stress for size D_i and the median size D_{50} of the surface sediment of a river bed. The following equation describes a general relationship between τ_{ci}^* and τ_{c50}^*

$$\frac{\tau_{ci}^*}{\tau_{c50}^*} = \left(\frac{D_i}{D_{50}}\right)^{-\gamma} \tag{17-17}$$

where γ ranges between 0.65 to 0.90 [Parker, 2008]. Noting the following relation between the Shields and dimensional bed shear stress

$$\tau_{ci}^* = \frac{\tau_{bci}}{\rho RgD_i}; \quad \tau_{c50}^* = \frac{\tau_{bc50}}{\rho RgD_{50}} \tag{17-18}$$

Equation 17-17 may be expressed as

$$\frac{\tau_{bci}}{\tau_{bc50}} = \left(\frac{D_i}{D_{50}}\right)^{1-\gamma} \tag{17-19}$$

If $\gamma = 1$, then all surface grains move at the same value of bed shear stress, i.e., at equal threshold. If on the other hand, $\gamma = 0$, then all particles move independently, i.e., they do not feel the effect of the neighboring particles. In most gravel-bed rivers, coarser surface grains are harder to move than finer surface grains, but only mildly, i.e., γ is closer to 1 than 0, but still < 1.

17-5 Condition for Significant Suspension

Sediment suspension and turbulence are closely-linked. One important measure of fluid turbulence is the rms or root mean square velocity, defined as

$$u_{\mathrm{rms}}{}^2 = \frac{1}{3}\left(\overline{u'^2} + \overline{v'^2} + \overline{w'^2}\right) \tag{17-20}$$

where prime represents the fluctuating component of velocity. For significant suspension to occur, u_{rms} near the bed must exceed the sediment fall velocity, v_s. Since turbulence is closely correlated [Tennekes and Lumley, 1972], the following approximate relationship between the rms velocity and the Reynolds stress may be written as

$$u_{\mathrm{rms}}{}^2 \sim \frac{\tau_{xz}}{\rho} = -\overline{u'w'} \tag{17-21}$$

For rough turbulent flow, the shear velocity can be estimated as

$$u_*^2 = -\overline{u'w'}|_{z=b} \tag{17-22}$$

in which b represents a near bed elevation such that $b \ll H$, where H is the flow depth. Between Eqs. 17-21 and 17-22

$$u_{\mathrm{rms}} \sim u_* \tag{17-23}$$

Bagnold [1966] proposed the following criterion for the onset of significant suspension

$$u_{*\mathrm{sus}} = v_s \tag{17-24}$$

Dividing both sides by $\sqrt{RgD}$, and taking square of Eq. 17-24, the following relationship for Shields stress required for sediment suspension is obtained

$$\tau_{*\mathrm{sus}} = \frac{v_s^2}{Rgd} \tag{17-25}$$

The fall velocity is a function of particle Reynolds number, $R_{ep} = (\sqrt{RgD}D)/\nu$. Dietrich [1982] developed the following relationships for estimating the fall velocity, v_s, of natural particles

$$D_* = \log(R_{ep}^2) = \log\left(\frac{RgD^3}{\nu}\right) \tag{17-26}$$

$$log(W_*) = -3.76 + 1.93D_* - 0.098D_*{}^2 - 0.00575D_*^3 + 0.00056D_*^4 \tag{17-27}$$

$$v_s = (Rg\nu W^*)^{1/3} \tag{17-28}$$

17-6 Shields Diagram

Albert Frank Shields' [1936a; 1963b; 1963c] work on incipient motion and bed-load transport is a benchmark study that has inspired numerous investigations and is widely applied in the fields of hydraulic engineering, fluvial geomorphology, and physical oceanography [Buffington, 1999]. Despite some inconsistencies and misconceptions as pointed out by Buffington [1999], Shields' work on the initiation of sediment motion has become somewhat of a legend because the principles of similarity are so eloquently presented with a very simple (yet based on physics) dimensionless diagram [García, 2000]. Shields diagram shows a relationship between the non-dimensional critical stress, τ_c^*, and the Shear Reynolds number, u_*D/ν. Shields diagram is not readily useful in its original form because in order to find τ_c, one must know $u_* = \sqrt{\tau_c/\rho}$. The relationship can be cast in an explicit form by plotting τ_c^* versus R_{ep}, noting the relationship

$$\frac{u_*D}{\nu} = \frac{u_*}{\sqrt{RgD}}\frac{\sqrt{RgD}D}{\nu} = \sqrt{\tau^*}R_{ep} \qquad (17\text{-}29)$$

Brownlie [1981] provided the following useful fit of Shields' data

$$\tau_c^* = 0.22R_{ep}^{-0.6} + 0.06\exp(-17.77R_{ep}^{-0.6}) \qquad (17\text{-}30)$$

Based on numerous field observations, it has been found that the Shields diagram leads to the overprediction of τ_c^* by a factor of 2 for fully rough flow in gravel-bed rivers. Therefore, Parker et al. [2003] amended Eq. 17-30 as

$$\tau_c^* = 0.5\left[0.22R_{ep}^{-0.6} + 0.06\exp(-17.77R_{ep}^{-0.6})\right] \qquad (17\text{-}31)$$

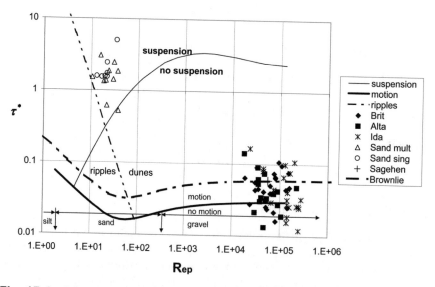

Fig. 17-4 **Modified Shields diagram** (Courtesy of G. Parker)

Shields' work can be extended to obtain a "regime" diagram for rivers [García, 2000]. Consider Fig. 17-4. The figure includes the fit of Shields' data given by Brownlie (Eq. 17-30), modified Brownlie equation (Eq. 17-31) for in-cipient motion, and the plot of Eq. 17-24 describing the condition for the onset of significant suspension. Figure 17-4 also shows the plot of τ^* versus R_{ep} for six different rivers under the bank-full condition. Here, both τ^* and R_{ep} are functions of the median grain size, i.e., D_{50}. The following observations can be made from this diagram: (i) sand-bed and gravel-bed rivers plot in distinctly different groups; (ii) in gravel-bed rivers, the Shields stress under the bank-full condition tends to be rather low; (iii) dunes do not usually develop in gravel-bed rivers, even under the bank-full condition; (iv) Shields stresses can be two orders of magnitude larger than τ_c^* in sand-bed streams during flood flows. This results in intense sediment transport; (v) in sand bed streams, most of the sediment transport occurs as suspended load while bedload transport is the norm in gravel-bed rivers.

17-7 The Exner Equation of Bed Sediment Conservation

Exner, an Austrian scientist, is the first to describe a morphodynamic prob-lem in quantitative terms [Exner, 1920, 1925] . The conservation equation of sediment mass at the river bed is commonly known as the Exner equation. Figure 17-5 shows a river segment with unit width. In this figure, the coor-dinate x denotes the streamwise direction, η denotes the bed elevation, q_b denotes the volume transport rate of bedload sediment per unit width per unit time, D_s is the deposition rate and E_s is the erosion rate of suspended sediment both expressed as volume per unit area per unit time. The bed of the river exchanges sediment with the flow at the upper surface of the control volume indicated by the shaded area. The following equation expresses the conservation of sediment with solid density of ρ_s and bed porosity of λ_p in the control volume

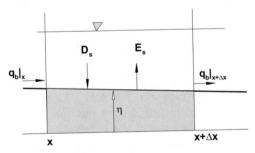

Fig. 17-5   **Mass conservation at the river bed**

$$\frac{\partial}{\partial t}\left[\rho_s(1-\lambda_p)\eta\right]\Delta x = \rho_s\left(q_b|_x - q_b|_{x+\Delta x}\right) + (D_s - E_s)\,\Delta x \qquad (17\text{-}32)$$

If the bed porosity is assumed to be independent of time, then for the limit $\Delta x \to 0$, Eq. 17-32 simplifies to

$$(1 - \lambda_p)\frac{\partial \eta}{\partial t} = -\frac{\partial q_b}{\partial x} + D_s - E_s \tag{17-33}$$

Equation 17-33 can be solved along with the Saint Venant Equations or the gradually varied flow equations to predict aggradation or degradation of channel bed. Equation 17-33 may be generalized to the following 2-D form to account for the lateral transport of sediment

$$(1 - \lambda_p)\frac{\partial \eta}{\partial t} = -\nabla . \overrightarrow{q_b} + D_s - E_s \tag{17-34}$$

where $\overrightarrow{q_b} = q_{bx}\mathbf{i} + q_{by}\mathbf{j}$ with q_{by} indicating the lateral transport rate of bedload. Numerous relationships have been developed for the prediction of the bedload transport rate in the streamwise direction as a function of the bed shear stress. Expression for q_{by} can be found in Sekine and Parker [1992]. In the case of dilute suspension of non-cohesive sediment

$$D_s = v_s \bar{c}_b; \quad E_s = v_s E \tag{17-35}$$

where $\bar{c}_b$ is the near-bed turbulence-averaged volume concentration of suspended sediment and $E = f(R_{ep}, \tau^*)$ is a non-dimensional entrainment rate. The near-bed concentration can be approximated as $\bar{c}_b = 2\overline{C}$ where $\overline{C}$ is the depth-averaged concentration [e.g., Parker et al., 1986].

Exner Equation for Multiple Size Fraction

Under most circumstances, sediment exchange between the flow and the river bed remains limited to a thin layer near the water-sediment interface. It is possible to define a rather thin active or exchange layer with thickness L_a which is typically taken to correspond to some multiple of characteristic surface grain size or dune height. The active layer concept was proposed by Hirano [1971] and later advanced and utilized by others [Parker, 1990a,b; Cui et al., 1996; Toro-Escobar et al., 1996] for modeling the morphodynamics of gravel-bed rivers, including downstream fining and armoring. Consider Fig. 17-6 showing an illustration of bedload transport, the active layer, and the substrate in 1-D. Let f_{bi}, F_i, and f_i respectively denote the fraction of the i-th size class in the bedload, the active layer, and the substrate. According to the active layer concept, F_i does not have a vertical structure while f_i does not generally vary with time. Under these conditions, considering bedload transport alone, the mass conservation equation for an individual size fraction in the river bed may be written as

$$(1 - \lambda_p)\left[f_{Ii}\frac{\partial \eta_b}{\partial t} + \frac{\partial}{\partial t}(L_a F_i)\right] = -\frac{\partial q_{bi}}{\partial x} \tag{17-36}$$

where f_{Ii} denotes the fraction of the i-th size class at the interface between the substrate and the active layer and q_{bi} is the volume transport rate of the same size class in the bedload. The derivation of Eq. 17-36 is straightforward and can be found in Parker and Sutherland [1990] and Parker et al. [2000]. Summing Eq. 17-35 over all grain-size classes, the following equation is obtained

$$(1 - \lambda_p)\frac{\partial \eta}{\partial t} = -\frac{\partial}{\partial x}\sum_{i=1}^{N} q_{bi} \tag{17-37}$$

With the aid of Eq. 17-37, η_b can be eliminated from Eq. 17-36 leading to

$$(1 - \lambda_p)\left[\frac{\partial}{\partial t}(L_a F_i) - f_{Ii}\frac{\partial L_a}{\partial t}\right] = -\frac{\partial q_{bi}}{\partial x} + f_{Ii}\sum_{i=1}^{N}\frac{\partial q_{bi}}{\partial x} \tag{17-38}$$

The generalized 2-D forms of Eqs. 17-38 and 17-37 including the contribution of suspended load may be written as

$$(1 - \lambda_p)\left[\frac{\partial}{\partial t}(L_a F_i) - f_{Ii}\frac{\partial L_a}{\partial t}\right] = -\nabla.\vec{q}_{bi} + f_{Ii}\sum_{i=1}^{N}\nabla.\vec{q}_{bi} + v_{si}(c_{bi} - E_i F_i) \tag{17-39}$$

and

$$(1 - \lambda_p)\frac{\partial \eta}{\partial t} = \sum_{i=1}^{N}\left[-\nabla.\vec{q}_{bi} + v_{si}(c_{bi} - E_i F_i)\right] \tag{17-40}$$

Along with appropriate flow equations, Eqs. 17-38 and 17-37 for 1-D or Eqs. 17-39 and 17-40 for 2-D cases can be solved to predict the change in the composition of the active layer and bed level changes. The interfacial size fraction f_{Ii} is defined as

$$f_{Ii} = \begin{cases} f_i|_{z=\eta-L_a}, & \frac{\partial \eta}{\partial t} < 0 \\ \alpha F_i + (1 - \alpha)f_{bi}, & \frac{\partial \eta}{\partial t} > 0 \end{cases} \tag{17-41}$$

where $0 \leq \alpha \leq 1$ [Toro-Escobar et al., 1996].

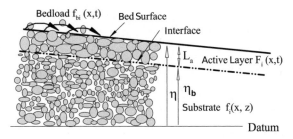

Fig. 17-6   Active layer concept

17-8 Bed-load Transport Relations

Scientists and engineers working in the field of morphodynamics have developed numerous relationships for estimating the rate of bedload transport. A common and useful approach to the quantification of bedload transport is to empirically relate the non-dimensional Einstein number, $q_b^* = q_b/(\sqrt{RgD}D)$ with either the Shields stress, τ^*, or the excess of the Shields stress, τ^*, above some appropriately defined "critical" Shields stress, τ_c^*, so as to fit the experimental data from which the relationship has been derived and to provide a useful demarcation of a range below which the bedload-transport rate is too low to be of interest. The functional relationship of interest in this form is

$$q_b^* = q_b^*(\tau^*) \tag{17-42}$$

or

$$q_b^* = q_b^*(\tau^* - \tau_c^*) \tag{17-43}$$

Some of the popular bedload transport relations for well-sorted sediments are presented below. These relationships apply to "plane-bed" conditions in the absence of bedforms.

1. Meyer-Peter and Müller [1948]:

$$q_b^* = 8(\tau^* - \tau_c^*)^{3/2}; \quad \tau_C^* = 0.047 \tag{17-44}$$

2. Wong and Parker [2006] correction of the Meyer-Peter and Müller relation:

$$q_b^* = 4.93(\tau^* - \tau_c^*)^{8/5}; \quad \tau_C^* = 0.047 \tag{17-45}$$

3. Einstein [1950]:

$$1 - \frac{1}{\sqrt{\pi}} \int_{-(0.143/\tau^*)-2}^{+(0.143/\tau^*)-2} \exp(-t^2)dt = \frac{43.5q_b^*}{1 + 43.5q_b^*} \tag{17-46}$$

4. Parker [1979] fit to Einstein [1950] equation:

$$q_b^* = 11.2\tau^{*3/2} \left(1 - \frac{\tau_c^*}{\tau^*}\right)^{9/2}; \quad \tau_c^* = 0.03 \tag{17-47}$$

5. Ashida and Michiue [1972]:

$$q_b^* = 17(\tau^* - \tau_c^*)\left(\tau^{*1/2} - \tau_c^{*1/2}\right); \quad \tau_c^* = 0.05 \tag{17-48}$$

6. Engelund and Fredsoe [1976]:

$$q_b^* = 18.74(\tau^* - \tau_c^*)\left(\tau^{*1/2} - 0.7\tau_c^{*1/2}\right); \quad \tau_c^* = 0.05 \tag{17-49}$$

Bed Load Transport Relations for Poorly Sorted Sediment

The relationships described previously cannot be applied to estimate the bed-load transport rate if the sediment is poorly sorted. As discussed earlier in this chapter, the threshold condition for the initiation of motion involving grain size with a wide range of size distribution is controlled by both the weight and exposure of the particle. The transport rate also depends on the availability of a given size fraction in the active layer. Some bedload models exclude sand-size particles from the formulation and consider the transport of sand as throughput (e.g., Parker [1982, 1990a; 1990b], Powell et al., 2001). The models of Ashida and Michiue [1972] and Wilcock and Crowe [2003] consider both sand and gravel transport in the formulation. Parker [2008] provides a comprehensive description of various bedload models of sediment mixture. To conserve space, only the Wilcock-Crowe model is presented here. Let us consider a sediment mixture with a total of N size classes. We may define the following non-dimensional relationships with reference to the size class i that varies from 1 to N.

$$\tau_i^* = \frac{\tau_b}{\rho R g D_i} \tag{17-50}$$

$$q_{bi}^* = \frac{q_{bi}}{\sqrt{R g D_i} D_i F_i} \tag{17-51}$$

and

$$W_i^* = \frac{q_{bi}^*}{\tau_i^{*3/2}} \tag{17-52}$$

in which τ_i^*, q_{bi}^* and W_i^* denote the grain-specific Shields stress, the Einstein number, and a dimensionless transport rate, respectively and F_i is the fraction of the size class in the surface layer. The Wilcock-Crowe model may be expressed by the following functional relation

$$W_i^* = G(\chi_i) \tag{17-53}$$

where

$$G(\chi) = \begin{cases} 0.002 \chi^{7.5} & \text{for } \chi < 1.35 \\ 14 \left(1 - \frac{0.894}{\chi^{0.5}}\right)^{4.5} & \text{for } \chi \geq 1.35 \end{cases} \tag{17-54}$$

The variable χ_i is defined as

$$\chi_i = \frac{\tau_{sg}^*}{\tau_{ssrg}^*} \left(\frac{D_i}{D_{sg}}\right)^{-b} \tag{17-55}$$

where

$$\tau_{sg}^* = \frac{\tau_b}{\rho R g D_{sg}};$$

$$\tau_{sg}^* = \frac{\tau_b}{\rho R g D_{sg}};$$

$$b = 0.67/[1 + \exp(1.5 - D_i/D_{sg})] \tag{17-56}$$

In Eq. 17-56, F_s represents the fraction of sand in the surface layer and D_{sg} is the geometric mean size of the surface material.

Recently, Elhakeem and Imran [2016] proposed a threshold-style semi-empirical formula for the transport of non-uniform sediment. The formula blends feedback from the composition of the surface, subsurface, and bedload material. The formulation was developed by describing sediment movement in a two-dimensional bedform. Surface and subsurface material were both incorporated to describe the interaction between the two, the formation and disintegration of the armor layer, and the hiding effects of larger grain sizes within the surface [Hinton et al., 2018]. The Elhakeem-Imran formula for the transport of the i-the fraction of sediment with grain size, D_i, can be expressed as

$$\Phi_{Bi} = \frac{q_{Bi}}{c_{Mi} f_{Oi} \sqrt{RgD_i} D_i} = aD_{*i}^{1.5} (\tau_{*g} - \tau_{*cg})^b \tag{17-57}$$

where $D_{*i} = D_i/D_{sg}$, D_{sg} being the geometric mean of the grain size; $f_{Oi} =$ the i-th fraction of the subsurface material (also known as parent, substrate, or original); $R =$ the reduced specific gravity of sediment; $\tau_{*j}' =$ Shields stress of the geometric mean size of the sediment mixture due to grain resistance only; $\tau_{*cg} =$ critical Shields stress of the geometric mean size of the sediment mixture, $c_u =$ Kramer coefficient of uniformity defined as

$$c_u = \frac{\sum_0^{50} f_{Oi} D_i}{\sum_0^{100} f_{Oi} D_i}; \tag{17-58}$$

the coefficients a and b are given as

$$a = 102 \exp[-3c_u]; \tag{17-59}$$

$$b = 2 - 0.33 \tan[0.9\pi(c_u - 0.5)]; \tag{17-60}$$

and c_{Mi} is a mobility coefficient given as

$$c_{Mi} = 0.8 c_u^{-0.45} \left(\frac{\tau_{*g}'}{\tau_{*cg}}\right)^{1.33(c_u - 0.5)} \exp[-0.66 c_u^{-0.3} \left(\frac{\tau_{*g}'}{\tau_{*cg}}\right)^{-2.1} (\ln D_{*i} + 0.15)] \tag{17-61}$$

Hinton et al. [2018] studied the performance of seven sediment transport models (three empirical and four semi-empirical) using 2,600 field measurements and found that the calibrated Elhakeem-Imran model performed best among the four semi-empirical models.

17-9 Suspended-load Transport

Suspended load constitutes a significant percent of the total sediment transported by a river. Once entrained from the river or introduced from the watershed, suspended load is carried by fluid turbulence. The Reynolds- or turbulence-averaged mass conservation equation of suspended sediment in open-channel flow may be written as

$$\frac{\partial \bar{c}}{\partial t} + \frac{\partial \bar{u}\bar{c}}{\partial x} + \frac{\partial \bar{v}\bar{c}}{\partial y} + \frac{\partial \bar{w}\bar{c}}{\partial z} - v_s \frac{\partial \bar{c}}{\partial z} = -\frac{\partial \overline{u'c'}}{\partial x} - \frac{\partial \overline{v'c'}}{\partial y} - \frac{\partial \overline{w'c'}}{\partial z} \qquad (17\text{-}62)$$

in which, u, v, and w are the velocity components in the x, y, and z direction, respectively and c is the volume concentration of suspended sediment, and v_s is the fall velocity. The over-bar and the prime denote the turbulence-averaged and the fluctuating component of a variable. For steady uniform flows in x (streamwise) and y (lateral) directions, Eq. 17-62 reduces to the following

$$-v_s \frac{d\bar{c}}{dz} = \frac{d}{dz}\left(-\overline{w'c'}\right) \qquad (17\text{-}63)$$

Similarly, the x-component of the momentum equation may be reduced to

$$0 = \frac{d}{dz}\left(-\overline{u'w'}\right) + gS \qquad (17\text{-}64)$$

where S is the longitudinal slope of the river bed and g is the gravitational acceleration. Under the condition of vanishing net sediment flux and shear stress at the water surface, i.e., at $z = H$, Eqs. 17-63 and 17-64 may be integrated to yield, respectively

$$F = \rho \overline{w'c'} = v_s \bar{c} \qquad (17\text{-}65)$$

and

$$\tau = -\rho \overline{u'w'} = \tau_b \left(1 - \frac{z}{H}\right) \qquad (17\text{-}66)$$

For steady uniform flow, the bed shear stress τ_b may be related to the flow field as

$$\tau_b = \rho u_*^2 = \rho g H S \qquad (17\text{-}67)$$

where u_* is the shear velocity. The eddy viscosity model may be utilized now to relate the Reynolds stress in Eq. 17-66 to the vertical gradient of the streamwise velocity as

$$\tau = -\rho \overline{u'w'} = \nu_t \frac{d\bar{u}}{dz} \qquad (17\text{-}68)$$

where ν_t is the eddy diffusivity. The log-law of Eq. 17-14 provides a good approximation of the velocity profile in developed steady open-channel flow. It follows from Eqs. 17-14 and 17-68

$$-\overline{u'w'} = \nu_t \frac{u_*}{\kappa z} \qquad (17\text{-}69)$$

Utilizing Eq. 17-66 and 17-67, Eq. 17-68 can be solved to obtain the following parabolic distribution of ν_t

$$\nu_t = \kappa u_* z \left(1 - \frac{z}{H}\right) \tag{17-70}$$

Equation 17-70 may be expressed in a non-dimensional form as

$$\frac{\nu_t}{\kappa u_* h} = \zeta\left(1 - \zeta\right); \quad \zeta = \frac{z}{H} \tag{17-71}$$

Similar to the Reynolds stress, the Reynolds mass flux $-\rho\overline{w'c'}$ may be expressed in terms of eddy diffusivity, ν_{st}, and the vertical gradient of the turbulence-averaged concentration as

$$-F = -\rho\overline{w'c'} = \nu_{st}\frac{d\bar{c}}{dz} \tag{17-72}$$

It is reasonable to make the approximation

$$\nu_{st} = \nu_t = \kappa u_* z \left(1 - \frac{z}{H}\right) \tag{17-73}$$

The balance Eq. 17-65 now becomes

$$\frac{d\bar{c}}{dz} + \frac{v_s}{\kappa u_* z \left(1 - \frac{z}{H}\right)}\bar{c} = 0 \tag{17-74}$$

The required boundary condition for the solution of Eq. 17-74 is a specified upward flux near the bed or the entrainment rate of suspended sediment

$$F|_{z=b} = v_s E \tag{17-75}$$

where E is a non-dimensional sediment entrainment rate. Rouse [1939] solved Eq. 17-74 to obtain the following profile of suspended sediment concentration in open channel flow under the equilibrium condition

$$\frac{\bar{c}}{\bar{c}_b} = \left[\frac{(1-\zeta)/\zeta}{(1-\zeta_b)/\zeta_b}\right]^Z \tag{17-76}$$

where $\bar{c}_b = E$, $\zeta = z/H$, and $\zeta_b = b/H$. The reference level, b, is typically chosen to be much smaller than H. The exponent $Z = v_s/\kappa u_*$ on the right hand of Eq. 17-76 is called the *Rouse Number*. Figure 17-7 shows the plot of $\bar{c}/\bar{c}_b$ for several values of Z and $b = 0.05H$.

Entrainment Relations

Several different relationships for estimating the rate of sediment entrainment into suspension are available in the literature (e.g., Smith and McLean, 1977; Van Rijn, 1984; García and Parker, 1991). For uniform sediment, the Van Rijn [1984] relationship may be written as

$$E = 0.015\frac{D_{50}}{b}\left(\frac{\tau_s^*}{\tau_c^*} - 1\right)^{1.5} R_{ep}^{-0.2} \tag{17-77}$$

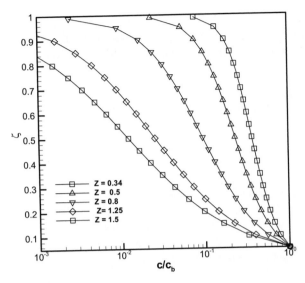

Fig. 17-7   **Vertical distribution of suspended sediment concentra-**
tion

in which D_{50} is the median grain size of the bed surface, τ_s^* is the Shields stress
due to skin friction, τ_c^* is the critical Shields stress that can be determined
from Eq. 17-30, and R_{ep} is the particle Reynolds number, defined by Eq. 17-29.
Bedforms often develop in sand-bed rivers during flood discharge. Van Rijn
[1984] defined b as 0.5 times the average bedform height if bedform is present
and larger of the Nikuradse roughness height k_s and $0.01H$ if bedform is not
present. García and Parker [1991] developed the following relationship for a
reference height, $b = 0.05H$

$$E = \frac{AZ_u^5}{1 + (A/0.3)Z_u^5}; \quad Z_u = \frac{u_{*s}}{v_s}R_{ep}^{0.6}; \quad A = 1.3 \times 10^{-7} \qquad (17\text{-}78)$$

The shear velocity, u_{*s} is related to the shear stress due to skin friction or
grain resistance as $u_{*s} = \sqrt{\tau_{bs}/\rho}$. Wright and Parker [2004] amended the
García-Parker relationship to the following form in order to generalize it for
large sand bed streams

$$E = \frac{AZ_u^5}{1 + (A/0.3)Z_u^5}; \quad Z_u = \frac{u_{*s}}{v_s}R_{ep}^{0.6}S^{0.07}; \quad A = 5.7 \times 10^{-7} \qquad (17\text{-}79)$$

in which S is the slope of the river bed. García and Parker [1991] also presented the following relationship for the i-th size class of poorly sorted sediment

$$E_i = \frac{AZ_{ui}^5}{1 + (A/0.3)Z_{ui}^5}; \quad Z_{ui} = \lambda_m\frac{u_{*s}}{v_{si}}R_{epi}^{0.6}\left(\frac{D_i}{D_{50}}\right)^{0.2}; \quad A = 1.3 \times 10^{-7} \qquad (17\text{-}80)$$

In Eq. 17-80, $\lambda_m = 1 - 0.298\sigma$, F_i denotes the fraction in the surface layer, $R_{epi} = \sqrt{RgD_i}D_i/\nu$ is the particle Reynolds number, and σ is defined by Eq. 17-6. Wright and Parker [2004] proposed the following modification to Eq. 17-80

$$E_i = \frac{AZ_{ui}^5}{1 + (A/0.3)Z_{ui}^5}; \quad Z_{ui} = \lambda_m \frac{u_{*s}}{v_{si}} R_{epi}^{0.6} \left(\frac{D_i}{D_{50}}\right)^{0.2} S^{0.08}; \quad A = 7.8 \times 10^{-7}$$

$$(17\text{-}81)$$

17-10 Resistance Relations

It is essential to estimate the bed shear stress to carry out any calculation involving sediment transport. The bed shear stress or boundary resistance in an open-channel flow may be expressed as

$$\tau_b = \rho u_*^2 = \rho C_f U^2 \tag{17-82}$$

The non-dimensional form of Eq. 17-82 is

$$\tau^* = \frac{u_*^2}{RgD} = \frac{C_f U^2}{RgD} \tag{17-83}$$

where u_* is the shear velocity, τ^* is the Shields stress, U is the depth-averaged velocity and C_f a friction factor that needs to be estimated from the available resistance relations. In the absence of bedforms, one of the following relations can be used directly for estimating the friction factor for fully rough flow

1. Keulegan law [1938]

$$C_f^{-1/2} = \frac{U}{u_*} = \frac{1}{\kappa} \ln\left(\frac{11H}{k_s}\right) \tag{17-84}$$

2. Manning-Strickler equation

$$C_f^{-1/2} = \frac{U}{u_*} = \alpha_r \left(\frac{H}{k_s}\right)^{1/6} \tag{17-85}$$

The dimensionless coefficient α_r in Eq. 17-85 varies between 8 and 9. For gravel-bed streams, Parker [1991] suggests a value of 8.1 for α_r. Keulegan's law is essentially the vertically integrated form of the log-law, given by Eq. 17-14 for the limit of k_s to H and the Manning-Strickler equation is a parabolic fit to the Keulegan law. In the absence of bedforms, the roughness height can be approximated as $k_s \cong 2D_{s90}$ with D_{s90} denoting a characteristic grain size such that 90% of the bed material is finer than this size. Under the uniform flow condition,

$$\tau_b = \rho C_f \frac{q_w^2}{H^2} = \rho g H S \qquad (17\text{-}86)$$

where q_w is flow discharge per unit width. The Chezy equation can be written as

$$U = \sqrt{\frac{g}{C_f}} H^{1/2} S^{1/2} \qquad (17\text{-}87)$$

Considering the resistance relation Eq. 17-85, Eq. 17-87 can be expressed as

$$U = \alpha_r \frac{\sqrt{g}}{C_f^{1/6}} H^{2/3} S^{1/2} \qquad (17\text{-}88)$$

Equations 17-83, 17-86 and 17-88 may be used to obtain the following relation for the Shields stress under the uniform-flow condition

$$\tau^* = \left(\frac{k_s^{1/3} q_w^2}{\alpha_r^2 g} \right)^{3/10} \frac{S^{7/10}}{RD}$$

Separation of Form Drag

Bedforms are often associated with the sand-bed rivers. The most common bedform observed in nature is dune. Other bedforms include ripples, and antidunes. Engelund and Hansen [1967] suggested that ripples can form if the sediment size on the bed is less than the thickness of the viscous sublayer, i.e., $\delta_v = 11.6\nu/u_* \leq D$. Ripples do not interact with the water surface. Dunes are bedforms that are approximately out of phase with the water surface, migrate downstream and occur in subcritical flows. Anti-dunes are approximately in phase with the water surface, typically occur in supercritical flows, and can migrate both upstream and downstream. When bedforms, such as dunes, are present, part of the drag is form drag associated with flow separation behind the dunes. Since this form drag is composed of stress that acts normal to the bed surface, it does not contribute directly to the motion of bed grains. As a result, it is usually subtracted out in performing sediment transport calculations. Einstein [1950] and Einstein and Barbarossa [1952] first described a method for separating form drag from the total resistance. Consider a uniform flow condition as illustrated in Fig. 17-8. In the upper panel, the bed is movable, i.e., dune is allowed to form and in the lower panel, the bed is rigid. In both cases, the flow velocity is the same and the bed has the same grain roughness. For the cases of Fig. 17-8(a), and 17-8(b), the balance between the gravitational and the drag force may be written as

$$\tau_b = \rho C_f U^2 = \rho g H S \qquad (17\text{-}89)$$

and

$$\tau_s = \rho C_{fs} U^2 = \rho g H_s S \qquad (17\text{-}90)$$

where C_f and C_{fs} are friction factors associated with the total drag and skin friction. We can now define a friction factor associated with form drag as

$$\tau_f = \rho C_{ff} U^2 = \rho g H_f S \tag{17-91}$$

The subscript f denotes variable associated with the form drag. From Eqs. 17-89, 17-90 and 17-91, we obtain the following relationships

$$\tau_f = \tau_b - \tau_s; \quad C_{ff} = C_f - C_{fs}; \quad H_f = H - H_s \tag{17-92}$$

Einstein and Barbarossa [1952] developed an empirical relationship between

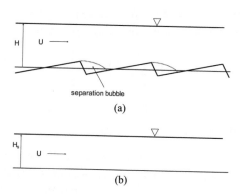

Fig. 17-8   Effect of bedform on flow depth
(a) With bedform; (b) Without bedform

C_{ff} and τ_{s35}^* with τ_{s35} being the Shields stress associated with the D_{35} of the bed material. The relationships developed by Engelund and Hansen [1967] for the separation of form drag from the total resistance is used widely. The relationship performs well for laboratory and small to medium scale field channels. However, it performs rather poorly for large, low slope sand-bed rivers. Wright and Parker corrected the Engelund-Hansen relationship to cover the entire range. The modified Engelund-Hansen relationship proposed by Wright and Parker [2004] is

$$\tau_s^* = 0.05 + 0.7 \left(\tau^* \mathbf{Fr}^{0.7}\right)^{0.8} \tag{17-93}$$

where $\mathbf{Fr} = U/\sqrt{gH}$ is the Froude number. Therefore,

$$\tau^* = \left(\frac{\tau_s^* - 0.05}{0.7}\right)^{5/4} \mathbf{Fr}^{-7/10} \tag{17-94}$$

In using the Wright-Parker model described above, the Shields stress due to skin friction τ_s^* is calculated using Manning-Strickler resistance relation Eq. 17-85 by setting $\alpha_r = 8.32$ and $k_s = 3D_{90}$.

17-11 Summary

In this chapter, the sediment transport process and common sediment properties were described. Procedure for analyzing sediment size distribution was presented. Conditions for incipient sediment motion and significant suspension were derived and the Shields diagram was presented. Several relationships for bed load transport were presented. The vertical distribution of sediment concentration under the equilibrium condition was derived. Relations for the entrainment of sediment into suspension were presented. Finally, resistance relations for calculating the bed shear stress were presented.

Problems

17-1 The size distribution of a sediment sample from the surface of a river bed is given below. Plot the distribution. Compute D_{50}, D_{90}, D_{sg} and σ_g for this sample. Classify the river as gravel-bed or sand-bed stream.

D (mm)	256	128	64	32	16	8	4	2	1	0.5	0.25	0.125
% finer	100	92	70	50	32	28	26	25	18	6	1	0

17-2 Using the Brownlie equation for critical Shields stress, compute τ_c^* for sediment sizes of 0.25 mm, 0.5 mm, 2 mm, and 16 mm. Consider $R = 1.65$ for all cases.

17-3 Compute and plot q_b^* versus τ^* in log-log scale for τ^* ranging from 0.05 to 1.0 using the following relationships:

  i. Meyer-Peter and Müller [1948];
 ii. Wong and Parker [2006] correction of the Meyer-Peter and Müller (MPM) relation; and
iii. Ashida and Michiue [1972].

17-4 Using the Manning-Strickler resistance relation (Eq. 17-85), show that under the uniform flow condition, flow depth in an open channel can be expressed as

$$H = \left(\frac{k_s^{1/3} q_w^2}{\alpha_r^2 gS} \right)^{3/10}$$

17-5 What is the fraction of sand F_s in the surface material of the river bed considered in Problem 17-1? Using the Wilcok-Crowe model, compute the total sediment transport rate if $S = 0.012$, and $q_w = 4$ m^2/s.

17-6 A sand bed river is characterized by a bed slope of 5.0×10^{-5}, surface D_{50} of 0.2 mm and D_{90} of 0.425 mm. Calculate the depth-discharge relationship for this river using the Wright-Parker version of Engelund-Hansen model. Consider H_s ranging from 0.9 m to 2.5 m.

References

Ashida, K. and M. Michiue, 1972, "Study on hydraulic resistance and bedload transport rate in alluvial streams." *Transactions*, Japan Society of Civil Engineering, 206: 59–69 (in Japanese).

Bagnold, R. A., 1966, "An approach to the sediment transport problem from general physics." *US Geol. Survey Prof. Paper* 422-I, Washington, D.C.

Brownlie, W. R., 1981, "Prediction of flow depth and sediment discharge in open channels." *Report No. KH-R-43A, W. M. Keck Laboratory of Hydraulics and Water Resources*, California Institute of Technology, Pasadena, California, USA, 232 p.

Buffington, J.M., 1999, "The Legend of A. F. Shields." *Journal of Hydraulic Engineering*, Vol. 125, No. 4, April 1999, pp. 376–387 , (doi 10.1061/(ASCE)0733-9429(1999)125:4(376))

Cui, Y., Parker, G. and C. Paola, 1996, "Numerical simulation of aggradation and downstream fining." *Jour. Hydraul. Res.*, 34, 185–203, 1996.

Dietrich, E. W., 1982, "Settling velocity of natural particles." *Water Resources Research* 18 (6), 1626–1982.

Egiazaroff, I. V., 1965, "Calculation of nonuniform sediment concentrations." *Journal of Hydraulic Engineering*, 91(4), 225–247.

Elhakeem, M., 2004, "A probabilistic approach to the modeling of entrainment, deposition and transport of bed load sediment." *PhD Thesis*, University of South Carolina, Columbia, USA.

Elhakeem M. and Imran, J., 2016, "Bedload model for nonuniform sediment." *Journal of Hydraulic Engineering*, 142(6), 04018038(1–11).

Einstein, H. A., 1950, "The Bed-load Function for Sediment Transportation in Open Channel Flows." *Technical Bulletin* 1026, U.S. Dept. of the Army, Soil Conservation Service.

Einstein H. A., and Barbarossa, N. L., 1952, "River Channel Roughness." *Journal of Hydraulic Engineering*, 117.

Engelund, F. and Hansen, E., 1967, "Hydraulic resistance in alluvial streams." *Acta Polytechnica Scandanavica*, V. Ci-35.

Engelund, F. and J. Fredsoe, 1976, "A sediment transport model for straight alluvial channels." *Nordic Hydrology*, 7 293–306.

Exner, F. M., 1920, "Zur Physik der Dunen." *Sitzber Akad. Wiss Wien*, Part IIa, Bd. 129 (in German).

Exner, F. M., 1925, "Uber die Wechselwirkung zwischen Wasser und Geschiebe in Flussen." *Sitzber. Akad. Wiss Wien*, Part IIa, Bd. 134 (in German).

García, M. and G. Parker, 1991, "Entrainment of bed sediment into suspension." *Journal of Hydraulic Engineering*, 117(4): 414–435.

Hinton, D., Hotchkiss, R. H., and Cope, W., 2018, "Comparison of calibrated mepirical and semi-Empirical methods for bedload transport rate prediction in gravel bed streams." *Journal of Hydraulic Engineering*, 144(7), 04018038 (1–17)

Hirano, M., 1971, "On riverbed variation with armoring." *Proceedings*, Japan Society of Civil Engineering, 195: 55–65 (in Japanese).

Keulegan, G. H., 1938, "Laws of turbulent flow in open channels." *National Bureau of Standards Research Paper* RP 1151, USA.

Meyer-Peter, E. and Müller, R., 1948, "Formulas for Bed-Load Transport." *Proceedings*, 2nd Congress, International Association of Hydraulic Research, Stockholm: 39–64.

Parker, G., 1979, "Hydraulic geometry of active gravel rivers." *Journal of Hydraulic Engineering*, 105(9), 1185 1201.

Parker, G., 1982, "Conditions for the ignition of catas trophically erosive turbidity currents." *Mar. Geol.* 46:307–27.

Parker, G., and Andrews, E. D., 1985, "Sorting of Bed Load Sediment by Flow in Meander Bends." *Water Resources Research* Vol. 21, No. 9, p 1361–1373.

Parker, G., Fukushima, Y., and Pantin, H. M., 1986, "Self-accelerating turbidity currents." *Jour. Fluid Mechanics*, 171, 145–181.

Parker, G., 1990a, "Surface-based bedload transport relation for gravel rivers." *Journal of Hydraulic Research*, 28(4): 417–436.

Parker, G., 1990b, "The ACRONYM Series of PASCAL Programs for Computing Bedload Transport in Gravel Rivers." *External Memorandum* M-200, St. Anthony Falls Laboratory, University of Minnesota, Minneapolis, Minnesota USA.

Parker, G. , and Sutherland, A. J., 1990, "Fluvial Armor." *J. Hydraul. Res.*, 28(5), 529–544.

Parker, G., 1991, "Selective sorting and abrasion of river gravel. I: Theory and II: Applications ." *Journal of Hydraulic Engineering*, 117(2): 131–171.

Parker, G., C. Paola, and Leclair, S., 2000, "Probabilistic form of Exner equation of sediment continuity for mixtures with no active layer." *Journal of Hydraulic Engineering*, 126(11): 818–826.

Parker, G. , Carlos M. Toro-Escobar, Michael Ramey, and Stuart Beck, 2003, "Effect of Floodwater Extraction on Mountain Stream Morphology." *Journal of Hydraulic Engineering*, Vol. 129, No. 11, pp. 885–895 , (doi 10.1061/(ASCE)0733-9429(2003)129:11(885))

Parker, G., 2008, "Sedimentation Engineering." *ASCE Manual 54*, Chapter 3 in "Sedimentation engineering: processes, measurements, modeling, and practice," *American Society of Civil Engineers*, by Garcia. M.

Powell, D. M., 1998, "Patterns and Processes of Sediment Sorting in Gravel-Bed Rivers." *Progress in Physical Geography*, Vol. 22, No. 1, 1–32.

Powell, D. M. , Reid, I. and Laronne, J. B. 2001, "Evolution of bedload grain-size distribution with increasing flow strength and the effect of flow duration on the caliber of bedload sediment yield in ephemeral gravel-bed rivers." *Water Resources Research*, 37(5), 1463–1474.

Rouse, H., 1939, "Experiments on the mechanics of sediment suspension." *Proceedings 5th International Congress on Applied Mechanics*, Cambridge, Mass., 550–554.

Sekine M. and Parker, G., 1992, "Bed-Load Transport on Transverse Slope. I." *Journal of Hydraulic Engineering*, Vol. 118, No. 4, pp. 513–535.

Shields, A., 1936a, "Anwendung der Aehnlichkeitsmechanik und der Turbulenzforschung auf die Geschiebebewegung." *Doktor-Ingenieurs dissertation* Technischen Hochschule, Berlin (in German).

Shields, A., 1936b, "Anwendung der Aehnlichkeitsmechanik und der Turbulenzforschung auf die Geschiebebewegung." *Mitteilungen der Preussischen Versuchsanstalt für Wasserbau und Schiffbau* Heft 26, Berlin (in German).

Shields, A., 1936c, "Application of similarity principles and turbulence research to bed-load movement." *Hydrodynamics Laboratory Publ. No. 167* W. P. Ott, and J. C. van Uchelen, trans., U.S. Dept. of Agr., Soil Conservation Service Cooperative Laboratory, California Institute of Technology, Pasadena, Calif.

Smith, J. D. and S. R. McLean 1977, "Spatially averaged flow over a wavy surface." *Journal of Geophysical Research*, 82(12): 1735–1746.

Tennekes, H. and Lumley, J. L., 1972, "A First Course in Turbulence." *MIT Press*, Cambridge, USA, 300 p.

Toro-Escobar, C. M. Parker, G. and C. Paola, 1996, "Transfer function for the deposition of poorly sorted gravel in response to streambed aggradation." *Journal of Hydraulic Research*, 34(1): 35–53.

Van Rijn, L., 1984, "Sediment transport, Part II: Suspended load transport." *Journal of Hydraulic Engineering*, 110(11), 1613–1641.

Wilcock, P. R., and Crowe, J. C., 2003, "Surface-based transport model for mixed-size sediment." *Journal of Hydraulic Engineering*, 129(2), 120–128.

Wong, M. and Parker, G., 2006, "The bedload transport relation of Meyer-Peter and Müller overpredicts by a factor of two." *Journal of Hydraulic Engineering*,132, 1159–1168.

Wright, S., and G. Parker, 2004, "Flow resistance and suspended load in sand-bed rivers: simplified stratification model." *Journal of Hydraulic Engineering*, 130(8), 796–805.

18

UNSTEADY FLOW SPECIAL TOPICS

Kunhar River Pakistan. Typical scour damage in Pakistan during major floods (approximately 1 in 10,000), July 2010 (Courtesy, Eric J Lesleighter)

© Springer Nature Switzerland AG 2022
M. H. Chaudhry, *Open-Channel Flow*,
https://doi.org/10.1007/978-3-030-96447-4_18

18-1 Introduction

In this chapter, we discuss a number of special topics, to which we apply concepts presented in the previous chapters. First, we discuss rating curve at a channel cross section during steady and unsteady flow conditions. Then, we describe different methods for flood routing. This is followed by a discussion of the aggradation and degradation of channel bottom due to imbalance between the actual amount of sediment in the flow and the carrying capacity of flow in a channel.

18-2 Rating Curve

A rating curve describes a relationship between the water level or stage at a channel cross section with the rate of discharge at that section. In Chapter 4, we presented several empirical resistance equations for steady-uniform flow that relate the channel discharge with the parameters of the channel. We may express these equations in a general form as

$$Q_n = kAR^m \sqrt{S_o}$$
(18-1)

in which Q_n = rate of discharge if the flow is uniform; k = a coefficient; A = flow area; R = hydraulic radius; S_o = channel-bottom slope, and m = an exponent which depends upon the equation used. For example, for the Manning equation, $k = C_o/n$, and $m = 2/3$; and for the Chezy equation, $k =$ Chezy C and $m = 1/2$. Similar to Eq. 18-1, a resistance equation for unsteady, nonuniform flow may be written as

$$Q = kAR^m \sqrt{S_f}$$
(18-2)

in which S_f = slope of the energy-grade line for discharge Q. If AR^m is a monotonically increasing function of flow depth y (this is the case for any regular channel cross section), then Eq. 18-1 plots as a dotted line, as shown in Fig. 18-1. This is a single-valued relationship between Q and y. Let us now discuss how unsteadiness with respect to time and nonuniformity with respect to distance modify this relationship.

It follows from Eqs. 18-1 and 18-2 that

$$Q = Q_n \sqrt{\frac{S_f}{S_o}}$$
(18-3)

Substitution of expression for S_f from the momentum equation (Eq. 12-18) into this equation yields

$$Q = Q_n \sqrt{1 - \frac{1}{S_o} \left(\frac{\partial y}{\partial x} - \frac{V}{g} \frac{\partial V}{\partial x} - \frac{1}{g} \frac{\partial V}{\partial t} \right)}$$
(18-4)

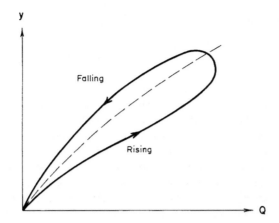

Fig. 18-1   Rating curve

Let us consider a channel reach in which inflow and stage are increasing with time. Thus, the flow depth and velocity at the upper end of the reach are higher than that at the lower end, i.e., the flow depth and flow velocity decrease with distance and the flow velocity increases with time. In other words, $\partial y/\partial x$ and $\partial V/\partial x$ are both negative and $\partial V/\partial t$ is positive. Normally, the last term of Eq. 18-4 representing the local acceleration is small as compared to the other two terms representing the convective acceleration. Therefore, it follows from Eq. 18-4 that Q during a rising stage is greater than Q_n for a given value of y. By proceeding similarly we can show that Q is less than Q_n for the same flow depth for a falling stage. Thus the rating curve has hysteresis, as shown in Fig. 18-1. Note that the difference between the discharges during rising and falling stages is due to unsteadiness and nonuniformity of the flow depth. The larger the hysteresis, the more pronounced is the effect of these terms. In such situations it becomes necessary to include these terms in the analysis.

18-3 Flood Routing

The computation of the height and velocity of a flood wave as it propagates in a body of water is referred to as *flood routing*. The body of water may be a lake, reservoir, channel, stream etc. For this purpose, we may solve the continuity and momentum equations derived in Chapter 11 by using various numerical schemes presented in Chapters 12 through 14. For two-dimensional flows, the governing equations and the numerical methods outlined in Chapter 15 may be utilized. Such computations have been referred to as *hydraulic routing*, and the models developed for this purpose are called *dynamic models*. In many situations, however, some terms of the governing equations are smaller than the other terms. For example, Henderson [1966] listed the following values in m/km for a fast-rising flood in a river in which the discharge increased in a

24-hour period from 280 m^3/s to 4,250 m^3/s and then decreased to 280 m^3/s: $S_o = 13.2; \partial y/\partial x = 0.25; (V/g)\partial V/\partial x = 0.06$ to $0.12; (1/g)\partial V/\partial t = 0.025$. In other words, the inertial terms are almost negligible as compared to the other terms and may be dropped without introducing significant errors. We will call such analysis procedures as *approximate methods.*

Depending upon the terms included in the analysis, the approximate procedures for flood routing have been given different names. For example, the continuity equation solved simultaneously with a simplified form of the momentum equation is called the *hydrologic routing.* If the simplified momentum equation is the steady-uniform equation, the routing procedure is called *kinematic routing,* and if an additional term for the slope of the water surface is included, the routing is called *diffusion routing.* By approximating the complex relationships between the storage capacity of a channel length and the inflow and outflow, several coefficient methods have been developed.

An approximate flood-routing procedure yields satisfactory results if the simplifying assumptions on which it is based are valid; i.e., the terms of the governing equations excluded in its development are negligible. These methods have the advantage in that they are simple to apply and that they do not require detailed data for the channel geometry. However, an improper application may yield totally incorrect results. We briefly discuss some of these approximate procedures in this chapter.

The shape and the magnitude of a flood wave may change as it travels in a body of water due to friction and due to storage effects. The storage may be produced by pondage or by cross-sectional changes in the natural channels. The wave is elongated and its magnitude is reduced due to storage. Such a reduction many times may be much more than that due to friction. In such situations, it is possible to simplify the analysis by disregarding the frictional effects.

18-4 Reservoir Routing

Lakes, reservoirs, ponds, and detention basins act as storage facilities. We will refer to each of them as reservoirs. The water level in such a storage facility may be considered horizontal. This simplifies the analysis significantly since dynamic effects are neglected and we need to consider only the continuity equation. According to this equation, the difference between the inflow and outflow is equal to the rate of change of volume of water stored in the reservoir. The stored volume is called *storage, S.*

We may write the continuity equation relating the inflow, I, outflow, O, and the rate of change of storage, S, in the reservoir (Fig. 18-2) as

$$\frac{dS}{dt} = I - O \tag{18-5}$$

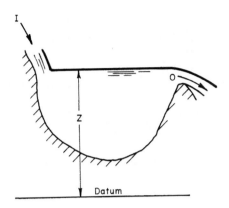

Fig. 18-2 Definition sketch

The inflow at different times is given as an inflow hydrograph; and the storage and outflow are specified as functions of the water level in the reservoir. To route a flood wave through a reservoir, we integrate this equation. For this purpose, any of the numerical schemes presented in Chapter 6 may be used. However, we will discuss in the following paragraphs a procedure initially developed for hand calculations.

A finite-difference approximation of Eq. 18-5 may be written as

$$\frac{\Delta S}{\Delta t} = \bar{I} - \bar{O} \tag{18-6}$$

in which $\bar{I}$ and $\bar{O}$ indicate the mean values during the time interval Δt. This interval is referred to as the *routing interval*. Let us designate the variables at the beginning of a routing interval by superscript k and the values at the end of the interval by superscript $k+1$. In addition, let us assume the variation of different variables is linear during the time interval. Then Eq. 18-6 becomes

$$\frac{S^{k+1} - S^k}{\Delta t} = \frac{1}{2}(I^k + I^{k+1}) - \frac{1}{2}(O^k + O^{k+1}) \tag{18-7}$$

By rearranging the known and unknown terms of this equation, we obtain

$$I^k + I^{k+1} + \frac{2S^k}{\Delta t} - O^k = O^{k+1} + \frac{2S^{k+1}}{\Delta t} \tag{18-8}$$

Now, if O and S are functions of the water level, z, in the reservoir, then we may say that O is a function of $2S/\Delta t$ as well. We can utilize this fact to solve Eq. 18-8 as follows.

1. We select a value for the routing interval and plot a curve between O and $(2S/\Delta t) + O$.
2. At the beginning of any routing interval, we know the values of water level in the reservoir, inflow I^k and outflow O^k. When we start the calculations,

the values of these variables are specified; later, they are computed during the previous time interval. In addition, we know I^{k+1} from the inflow hydrograph. For known O^k, we first read $(2S/\Delta t) + O$ from the curve between $(2S/\Delta t) + O$ and O. Then by subtracting $2O^k$ from this value, we compute $2S/\Delta t) - O$. Now, we can determine the left-hand side of Eq. 18-8.

3. For the value computed in step 2, which is also equal to $(2S^{k+1}/\Delta t) + O^{k+1}$ (from Eq. 18-8), we read the value of O^{k+1} from the curve between $(2S/\Delta t) + O$ and O.

4. We repeat steps 1 to 3 for the next time interval and continue this process until the computations for the desired period are done.

18-5 Channel Routing

In the previous section, we assumed that the water surface in the reservoir always remains level although it may change if the inflow is not equal to the outflow. Also, the storage and outflow were assumed as functions of water level. Hence, we could say that the storage is function of outflow. These assumptions are valid for a channel, if the flow in the channel reach is uniform. However, for nonuniform flow, the storage depends upon the inflow and outflow. The storage in a channel reach may be divided into *prism storage* where S is proportional to O and *wedge storage* where S is proportional to the difference between inflow and outflow (Fig. 18-3). Based on these assumptions, a procedure was presented in 1938 to do flood-routing studies on the Muskingum river by U.S. Army Corps of Engineers. Nowadays, this procedure is commonly called *Muskingum routing*.

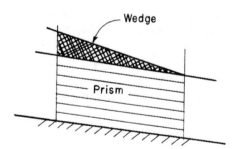

Fig. 18-3   Prism and wedge storage

In this method, the storage in a channel reach is expressed as a function of inflow and outflow as

$$S = KO + KX(I - O) \tag{18-9}$$

in which K and X are constants. Note that for dimensional reasons, K has units of time and X is dimensionless. We shall show later in Sect. 18-8 that K is the time for a wave to travel from one end of the reach to the other.

By using the notation of the previous section, we may write Eq. 18-9 for the storage at time k as

$$S^k = K[XI^k + (1 - X)O^k]$$ (18-10)

Similarly, the storage at time $k + 1$ may be written as

$$S^{k+1} = K[XI^{k+1} + (1 - X)O^{k+1}]$$ (18-11)

By substituting these equations into Eq. 18-8 and simplifying the resulting equation, we obtain

$$O^{k+1} = C_o I^{k+1} + C_1 I^k + C_2 O^k$$ (18-12)

in which

$$C_o = \frac{\Delta t - 2KX}{\Delta t + 2K(1 - X)}$$

$$C_1 = \frac{\Delta t + 2KX}{\Delta t + 2K(1 - X)}$$

$$C_2 = \frac{-\Delta t + 2K(1 - X)}{\Delta t + 2K(1 - X)}$$

We may use Eq. 18-12 for flood routing through a channel reach if we know the values of K and X. These may be determined from the observed flow records as discussed in the following paragraph; or their values may be computed from the expressions derived by a rigorous analysis in Sect. 18-8.

We may solve Eq. 18-10 [Roberson et al. 1998] for K as

$$K = \frac{0.5\Delta t[(I^k + I^{k+1}) - (O^k + O^{k+1})]}{X(I^{k+1} - I^k) + (1 - X)(O^{k+1} - O^k)}$$ (18-13)

For an observed hydrograph, we plot the numerator of this equation as the ordinate and the denominator as the abscissa for different time intervals and different assumed values of X, say 0.1, 0.2, 0.3, etc. The value of X that gives the plot close to a straight line is the X value to use and the slope of the graph is the value of K.

In some situations, the linear Muskingum model as discussed previously may be unsuitable for representing some reaches [Yoon and Padmanabhan, 1993]; moreover, the relationship between the weighted flow and the storage is not always linear [Geem, 2006]. Gill [1978] suggested two forms of nonlinear Muskingum models which account for the nonlinearity in storage and flow relationship.

$$S = K[XI + (1 - X)O]^m$$
$$S = K[XI^m + (1 - X)O^m] \tag{18-14}$$

where m is an exponent and is not known prior to the calculations; and, therefore, an additional unknown parameter is introduced. Several methods for the estimation of parameters for the nonlinear Muskingum model have been proposed, e.g., Gill [1978], Tung [1984], Yoon and Padmanabhan [1993], Mohan [1997], Kim et al. [2001], Das [2004], and Geem [2006].

18-6 Kinematic Routing

In kinematic routing, we solve the continuity equation, Eq. 12-4, simultaneously with an approximate form of the momentum equation. This approximate form is obtained by neglecting the local and convective acceleration terms of the momentum equation. The remaining terms represent the resistance equation for steady, uniform flow. In other words, we consider the flow to be steady for momentum conservation but take into consideration the effects of unsteadiness by an increase or decrease in the flow depth.

We may write the resistance equation in a form similar to Eq. 18-1, i.e.,

$$Q = f(A) \tag{18-15}$$

or

$$A = F(Q) \tag{18-16}$$

Hence, applying the chain rule we may write

$$\frac{\partial A}{\partial t} = \frac{\partial A}{\partial Q} \frac{\partial Q}{\partial t} \tag{18-17}$$

or

$$\frac{\partial A}{\partial t} = \frac{\partial Q}{\partial t} \frac{dA}{dQ}\bigg|_{x=x_o} \tag{18-18}$$

Substituting this equation into Eq. 12-4, assuming $q_l = 0$, and simplifying the resulting equation, we obtain

$$\frac{\partial Q}{\partial t} + a\frac{\partial Q}{\partial x} = 0 \tag{18-19}$$

in which $a = dQ/dA$. Since this is a kinematic model, the wave is called a *kinematic wave*.

It follows from the expression for a that it has dimensions of L/T. Thus, it represents a velocity. The following discussion will show that a is the velocity of a flood wave. Note that this expression is different from the one we derived in Chapters 12 and 13 for the absolute velocity of a disturbance as $V \pm c$. However, extensive field measurements of the propagation of the crest of flood waves confirm this relationship [Seddon, 1900].

Equation 18-19 is a first-order partial differential equation with Q as the dependent variable and x and t as the independent variables. It describes the movement of a flood wave in terms of the rate of discharge. If a is constant, then Eq. 18-19 is linear. D'Alembert presented a general solution of this linear equation as

$$Q = f(x - at) \qquad (18\text{-}20)$$

In this solution, we assume that the partial derivatives of f with respect to x and t exist. By taking the partial derivative of Eq. 18-20 with respect to x and t and substituting them into Eq. 18-19, we can prove that Eq. 18-20 represents the general solution of Eq. 18-19. At $t = 0$, $Q = f(x)$ represents the initial conditions. This curve describes the variation of discharge Q with distance x. The solution at time t_1 is $f(x - at_1)$, and at time t_2 it is $f(x - at_2)$. Let us assume that an observer is traveling at velocity a in the downstream direction. To this observer, this curve always appears as $f(x)$. We may draw the same conclusion by considering a moving coordinate system such that $\xi = x - at$. Then, $Q = f(\xi)$; the shape of the solution curve is always $Q = f(\xi)$, and it is independent of time t.

This discussion shows that a flood hydrograph in kinematic routing travels in the positive x-direction at velocity a; the shape of the hydrograph does not change and its peak does not attenuate (Fig. 18-4). However, note that these conclusions are based on the assumption that a is constant. If this is not the case – i.e., $dQ/dA \neq$ constant – then Eq. 18-19 becomes nonlinear and the wave shape may change due to nonlinear effects as it propagates in the channel. The change in the wave shape depends upon the variation of a with Q. The positive front of a wave steepens with distance if a increases with an increase in Q and the front flattens if a decreases with an increase in Q. Sometimes, the wave front may steepens so much as to form an almost vertical front; this is referred to as a *kinematic shock*.

A kinematic model is based on the solution of Eq. 18-19. The solution may be analytical or numerical. In a numerical solution the wave height and shape may be modified as it propagates. This modification is purely due to characteristics of the numerical method and does not represent simulation of the actual physical phenomenon. The wave modification may be in the form of reduction in the wave height, change in shape, or a combination of the two. The reduction in the height is called *dissipation,* the change in the shape is referred to as *dispersion,* and the combination of dissipation and dispersion is called *diffusion.*

The modification of the wave shape depends upon the numerical method employed and the value of the Courant number, $C_n = a\Delta t/\Delta x$. To illustrate this point, let us assume that the wave velocity, a, is constant, i. e., the slope of the curve between Q and A is constant. Let us now use the Lax scheme to numerically integrate Eq. 18-9. Then, by using the notation outlined in Chapter 14, we get

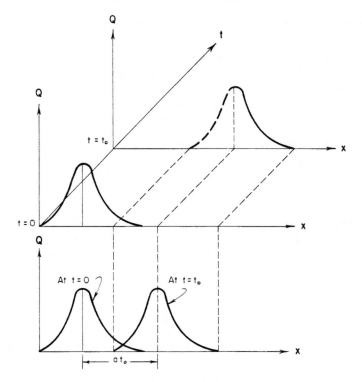

Fig. 18-4   Kinematic wave propagation

$$Q_i^{k+1} = \frac{1}{2}(Q_{i+1}^k + Q_{i-1}^k) - \frac{1}{2}a\frac{\Delta t}{\Delta x}(Q_{i+1}^k - Q_{i-1}^k) \qquad (18\text{-}21)$$

Rearrangement of the terms of this equation yields

$$Q_i^{k+1} = \frac{1}{2}(1 - C_n)Q_{i+1}^k + \frac{1}{2}(1 + C_n)Q_{i-1}^k \qquad (18\text{-}22)$$

It is clear from this equation that $Q_i^{k+1} = Q_{i-1}^k$ if $C_n = 1$. Thus, if the wave peak was at computational node $(i - 1)$ at time t_o, then it will be at node i at time $t_o + \Delta t$. Also note that the shape of the wave is not modified as it propagates from one node to the next. However, the wave shape is modified as it travels if $C_n \neq 1$. For example, if $C_n = 0.8$, then $Q_i^{k+1} = 0.1Q_{i+1}^k + 0.9Q_{i-1}^{k+1}$. In other words, the wave shape is modified and wave peak is attenuated while traveling from time k to time $k + 1$. According to the governing equation of a kinematic model, a flood wave travels without modifying its shape, and with no attenuation. However, numerical calculations with C_n different than 1 may result in attenuating the wave peaks and in modifying the wave shapes. This is mainly due to the limitations of the numerical solution and is not due to the simulation of actual dissipation.

Applicability criterion

For the applicability of the kinematic model to overland flow, Woolhiser and Liggett [1967] developed the following criterion:

$$K_f = \frac{S_o L_o}{y_n \mathbf{F}_r} \geq 20 \tag{18-23}$$

in which K_f = kinetic flow number; S_o = bottom slope; L_o = length of the overland flow plane; y_n = normal depth; and $\mathbf{F}_r$ = Froude number corresponding to uniform flow. Morris and Woolhiser [1980] state that it is also necessary for flows at low Froude number that $K_f \mathbf{F}_r^2 \geq 5$ in addition to the preceding criterion.

By using analytical solution of the linearized equations, Ponce et al. [1978] showed that the accuracy of the computed results for a sinusoidal perturbation of mean flow is within 95 percent accurate after one period of propagation if the dimensionless wave period

$$T_w = \frac{T S_o V_o}{y_o} > 171 \tag{18-24}$$

in which T = wave period; V_o = reference mean velocity; and y_o = reference flow depth. The wave period T may be taken as twice the time of rise of the flood wave.

18-7 Diffusion Routing

In the diffusion routing, we solve a simplified form of the momentum equation with the continuity equation. The simplified form of the momentum equation includes the convective acceleration term representing the spatial change in the flow depth as well as the source terms but neglects the temporal derivative term as well as the convective acceleration terms due to spatial change in the flow velocity. Thus, the momentum equation with these simplifications becomes

$$S_f = S_o - \frac{\partial y}{\partial x} \tag{18-25}$$

We showed in Chapter 4 that any resistance formula may be written in the form $Q = K S_f{}^2$, where K is the conveyance factor. Substitution of this expression into Eq. 18-25 yields

$$\frac{Q^2}{K^2} = S_o - \frac{\partial y}{\partial x} \tag{18-26}$$

Let us eliminate y from this equation and the continuity equation (Eq. 12-4) so that the resulting equation describes the variation of Q with respect to x and t. To do this, let us first differentiate Eq. 18-26 with respect to t and Eq. 12-4 with respect to x, assume $q_l = 0$, and then eliminate $\partial^2 y/\partial x \partial t$ from the resulting equations. Differentiation of Eq. 18-26 with respect to t gives

$$\frac{2Q}{K^2}\frac{\partial Q}{\partial t} - \frac{2Q^2}{K^3}\frac{\partial K}{\partial t} = -\frac{\partial^2 y}{\partial x \partial t} \qquad (18\text{-}27)$$

Differentiating Eq. 12-4 with respect to x, noting that $\partial A/\partial x = B\partial y/\partial x$, and dividing throughout by B yield

$$\frac{\partial^2 y}{\partial x \partial t} = -\frac{1}{B}\frac{\partial^2 Q}{\partial x^2} \qquad (18\text{-}28)$$

By eliminating $\partial^2 y/\partial x \partial t$ from Eqs. 18-27 and 18-28, we obtain

$$\frac{1}{B}\frac{\partial^2 Q}{\partial x^2} - \frac{2Q}{K^2}\frac{\partial Q}{\partial t} + \frac{2Q^2}{K^3}\frac{\partial K}{\partial t} = 0 \qquad (18\text{-}29)$$

Based on the chain rule, we may write that $\partial K/\partial t = (\partial K/\partial A)(\partial A/\partial t)$. Substitution for $\partial A/\partial t$ from Eq. 12-4 into this expression gives

$$\frac{\partial K}{\partial t} = \frac{dK}{dA}\bigg|_{x=x_o}\left(-\frac{\partial Q}{\partial x}\right) \qquad (18\text{-}30)$$

To simplify the derivation, let us obtain the derivative dK/dA from the expression $Q = K\sqrt{S_o}$ for uniform flow instead of from the general expression $Q = K\sqrt{S_f}$. Then, it follows from Eq. 18-20 that

$$\frac{\partial K}{\partial t} = \frac{1}{\sqrt{S_o}}\frac{dQ}{dA}\left(-\frac{\partial Q}{\partial x}\right) \qquad (18\text{-}31)$$

By eliminating $\partial K/\partial t$ from Eqs. 18-29 and 18-31, noting that $Q = K\sqrt{S_o}$ and $a = dQ/dA$, multiplying throughout by $K^2/(2BQ)$ and simplifying the resulting equation, we obtain

$$D\frac{\partial^2 Q}{\partial x^2} - \left(\frac{\partial Q}{\partial t} + a\frac{\partial Q}{\partial x}\right) = 0 \qquad (18\text{-}32)$$

in which $D = Q/(2BS_o)$.

A comparison of Eqs. 18-19 and 18-32 shows that other than the first term, the remaining equation is the same as the equation for the kinematic model. The first term of this equation represents the diffusion of a flood wave as it travels in the channel. By using the coefficients D and a determined from the observed hydrographs, the attenuation of a flood wave due to storage and friction may be included in the analysis.

Applicability

The following criterion may be used [Ponce et al., 1978] for the applicability of the diffusion model:

$$K_w = \mathbf{F}_r T S_o \sqrt{\frac{g}{y_o}} \geq 30 \qquad (18\text{-}33)$$

in which $\mathbf{F}_r$ = reference flow Froude number and the other variables are as defined for the kinematic model.

18-8 Muskingum-Cunge Routing

In Section 18-5, we discussed the Muskingum routing and presented a procedure to determine the coefficients K and X from the observed flood hydrographs. Cunge [1969] derived expressions for these coefficients from a finite-difference approximation of the kinematic wave equation, Eq. 18-19. We outline this procedure in this section.

Let us substitute the following finite-difference approximations into Eq. 18-19

$$\frac{\partial Q}{\partial x} = \frac{(O^k + O^{k+1}) - (I^k + I^{k+1})}{2\Delta x}$$
$$\frac{\partial Q}{\partial t} = \frac{\alpha(I^{k+1} - I^k) + (1 - \alpha)(O^{k+1} - O^k)}{\Delta t} \qquad (18\text{-}34)$$

in which α is the weighting coefficient for the time derivative. The simplification of the resulting equation yields Eq. 18-12 except that the following expressions for different coefficients are obtained instead of those given in Eq. 18-13:

$$C_o = \frac{0.5\Delta t - \alpha \Delta x/a}{0.5\Delta t + (1 - \alpha)\Delta x/a}$$
$$C_1 = \frac{0.5\Delta t + (\alpha \Delta x/a)}{0.5\Delta t + (1 - \alpha)\Delta x/a}$$
$$C_2 = \frac{-0.5\Delta t + (1 - \alpha)\Delta x/a}{0.5\Delta t + (1 - \alpha)\Delta x/a} \qquad (18\text{-}35)$$

Note that for $\alpha = X$ and $\Delta x/a = K$, these expressions reduce to those given in Eq. 18-13. Thus, $K = \Delta x/a$, which is the travel time for a flood wave to propagate through the reach.

We can show that if $\alpha = 0.5$ and $a\Delta t/\Delta x = 1$ in the Muskingum-Cunge routing, then a wave does not attenuate and does not change shape as it propagates through a channel reach. This result is similar to that obtained by kinematic routing.

18-9 Aggradation and Degradation of Channel Bottom

In this section, we present a mathematical model to simulate the aggradation and degradation of the channel bottom due to imbalance between the actual amount of sediment transported by flow in the channel and the sediment-carrying capacity of the flow.

Introduction

The channel bottom may aggrade or degrade if the balance among water discharge, sediment flow and the channel shape is disturbed. Such a disturbance may be due to natural or other factors, such as the construction of a dam, change in the sediment supply rate, lowering of the channel bottom, migration of knickpoints etc. A reliable, quantitative estimates of the bed aggradation or degradation become necessary in river control engineering and water-management projects.

Several experimental studies have been conducted to investigate the short- and long-term bed-level changes in alluvial channels. Lane and Borland [1954] conducted experiments to study river bed scour during floods. Brush and Wolman [1960] measured the time variation of bed levels in a laboratory channel due to the migration of knickpoints or the points of abrupt change in the longitudinal profile. Newton [1951] obtained laboratory data for degradation due to sediment diminution and Soni et al. [1980] studied aggradation due to sediment overloading. Begin et al. [1981] experimentally studied degradation of alluvial channels in response to lowering of the channel bottom. Suryanarayana [1969] obtained laboratory data for the degradation of alluvial channels downstream of a dam.

A number of analytical solutions have been developed by simplifying the governing equations describing the aggradation and degradation processes. Soni et al. [1980] used a linear diffusion model to predict the transient-bed profiles due to sediment overloading. Jain [1981] pointed out an error in their boundary conditions and presented an analytical solution utilizing more appropriate boundary conditions. His computed results compared satisfactorily with the experimental data. Begin et al. [1981] used a diffusion model to compute longitudinal profiles produced by base-level lowering. Gill [1983a,b] solved the linear diffusion equation describing the aggradation and degradation process by the Fourier series and by the error-function methods. Jaramillo and Jain [1984] developed a nonlinear parabolic partial differential equation, solved it by the method of residuals, and compared their computed results with the available experimental data. Zhang and Kahawita [1987] and Gill [1987] presented nonlinear solutions for aggradation and degradation that compared better with the experimental data than the linear solutions.

The linear and nonlinear parabolic models are based on the assumption of quasi-steady water flow. This assumption may not be valid during floods or during other unsteady flow conditions. This may not also be valid even if the discharge is constant if the slope of the channel bottom changes during the period of interest. Therefore, the complete unsteady water flow equations along with the sediment continuity equation are solved by numerical techniques. Holly [1986], Dawdy and Vanoni [1986] and Cunge et al. [1980] reviewed the literature on the numerical simulation of alluvial hydraulics. Lu and Shen [1986] tested several numerical aggradation and degradation models by comparing the computed results with laboratory data [Suryanarayana, 1969].

One-dimensional, unsteady sediment transport models may be classified into two categories: the uncoupled models, in which the water-flow equations and sediment continuity equation are uncoupled during a given time step, and the quasi-steady flow models, in which the energy equation is solved along with the sediment continuity equation. Lyn [1987] used perturbation techniques to identify multiple time scales in the governing equations and suggested that complete coupling between the full unsteady water-flow equations and the sediment continuity equation is desirable in cases where the conditions are rapidly changing at the boundaries. He proposed to use the Preissmann linearized implicit scheme for a simultaneous solution of the governing equations. Park and Jain [1986] used this procedure in their unsteady, uncoupled model for the analysis of the aggradation due to sediment overloading. To avoid instability, they had to iterate the solution whenever the spatial gradient of change in bed level became too large.

In this section, a one-dimensional, unsteady, coupled deformable bed model [Bhallamudi and Chaudhry, 1991] is presented. The complete Saint-Venant equations for water flow and the sediment continuity equation are solved simultaneously by the MacCormack explicit scheme [MacCormack, 1969]. The computed results are compared with the experimental data to verify the model.

Governing equations

The following equations describe unsteady flow in a wide rectangular channel with deformable bed (Fig. 18-5):

Continuity equation for water

$$\frac{\partial h}{\partial t} + \frac{\partial q}{\partial x} = 0 \tag{18-36}$$

Momentum equation for water

$$\frac{\partial q}{\partial t} + \frac{\partial}{\partial x}\left(\frac{q^2}{h} + \frac{1}{2}gh^2\right) + gh\frac{\partial z}{\partial x} + ghS_f = 0 \tag{18-37}$$

Continuity equation for sediment

$$\frac{\partial}{\partial t}\left[(1-p)z + \frac{q_s h}{q}\right] + \frac{\partial q_s}{\partial x} = 0 \tag{18-38}$$

in which q = water discharge per unit width; h = flow depth; z = channel bottom elevation; q_s = unit sediment discharge, and p = porosity of the bed layer. The sediment discharge may be estimated by an empirical power function of the flow velocity

$$q_s = a(\frac{q}{h})^b \tag{18-39}$$

in which a and b are empirical constants whose values depend upon the sediment properties. Note that this relationship for the sediment discharge is used herein mainly for simplicity. More elaborate and complex relationships may easily be included in the model since the governing equations are solved by an explicit numerical scheme.

Experimental investigations of Soni et al. [1977] show that the Manning n under nonuniform conditions in aggrading channels is smaller than its value for uniform flow. The resistance equation for uniform flow may be used under nonuniform conditions, provided local friction slope is used instead of the bottom slope. The applicability of the steady-uniform sediment transport formula under nonuniform conditions is questionable and needs further investigations.

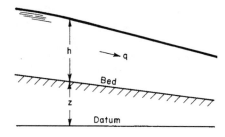

Fig. 18-5   Definition sketch

Numerical Scheme

Equations 18-36 through 18-38 are a set of nonlinear hyperbolic equations and closed form solutions are available only for idealized cases. We solve these equations numerically by using the MacCormack scheme. As we discussed in Chapters 14 and 15, this scheme is simple to implement and captures shocks without any special treatment, and the incorporation of general empirical equations for roughness and sediment discharge is easy.

We use the forward finite-differences for the spatial partial derivatives in the predictor part and the backward finite-differences in the corrector part. As discussed in Chapters 14 and 15, this sequence may be reversed every second time step.

Predictor

In the following equations, a superscript * indicates value of the variable computed at the end of the predictor part.

$$h_i^* = h_i^k - \frac{\Delta t}{\Delta x} \left(q_{i+1}^k - q_i^k \right)$$

$$q_i^* = q_i^k - \frac{\Delta t}{\Delta x} \left[\frac{\left(q_{i+1}^k \right)^2}{h_{i+1}^k} - \frac{\left(q_i^k \right)^2}{h_i^k} + \frac{g}{2} \left\{ \left(h_{i+1}^k \right)^2 - \left(h_i^k \right)^2 \right\} \right]$$

$$- gh_i^k \frac{\Delta t}{\Delta x} \left(z_{i+1}^k - z_i^k \right) - gh_i^k \Delta t \frac{\left(q_i^k n \right)^2}{\left(h_i^k \right)^{3.33}}$$

$$z_i^* = z_i^k + \frac{1}{1-p} \left[\left(\frac{q_s h}{q} \right)_i^k - \left(\frac{q_s h}{q} \right)_i^* \right]$$

$$- \frac{\Delta t}{(1-p)\,\Delta x} \left[(q_s)_{i+1}^k - (q_s)_i^k \right]$$

$$(q_s)_i^* = a \left(\frac{q_i^*}{h_i^*} \right)^b \tag{18-40}$$

Corrector

$$h_i^{**} = h_i^* - \frac{\Delta t}{\Delta x} \left[q_i^* - q_{i-1}^* \right]$$

$$q_i^{**} = q_i^* - \frac{\Delta t}{\Delta x} \left[\frac{\left(q_i^* \right)^2}{h_i^*} - \frac{\left(q_{i-1}^* \right)^2}{h_{i-1}^*} + \frac{g}{2} \left\{ \left(h_i^* \right)^2 - \left(h_{i-1}^* \right)^2 \right\} \right]$$

$$- gh_i^* \frac{\Delta t}{\Delta x} \left(z_i^* - z_{i-1}^* \right) - gh_i^* \Delta t \frac{\left(q_i^* n \right)^2}{\left(h_i^* \right)^{3.33}}$$

$$z_i^{**} = z_i^* + \frac{1}{1-p} \left[\left(\frac{q_s h}{q} \right)_i^* - \left(\frac{q_s h}{q} \right)_i^{**} \right] - \frac{\Delta t}{(1-p)\,\Delta x} \left[(q_s)_i^* - (q_s)_{i-1}^* \right]$$

$$(q_s)_i^{**} = a \left(\frac{q_i^{**}}{h_i^{**}} \right)^b \tag{18-41}$$

in which the superscript ∗∗ denotes the value of the variable after the corrector step.

Now, the values of the unknowns at $k + 1$ time level (i.e., at the end of time) are given by interval Δt are given by

$$h_i^{k+1} = \frac{1}{2} \left(h_i^k + h_i^{**} \right)$$

$$q_i^{k+1} = \frac{1}{2} \left(q_i^k + q_i^{**} \right)$$

$$z_i^{k+1} = \frac{1}{2} \left(z_i^k + z_i^{**} \right) \tag{18-42}$$

By using the preceding algorithm, the values of h, q, and z at the new time level $k + 1$ are determined at every interior node ($i = 2, ...N$). The values of the dependent variables h, q and z at the boundary nodes 1 and $N+1$ are determined by using the boundary conditions. For subcritical flow, it can be shown by using the characteristic theory that two boundary conditions at the upstream boundary and one condition at the downstream boundary have to be specified. The values of the dependent variables which are not specified through boundary conditions may be determined from the characteristic equations. Their values may also be determined by interpolation from the known values at the interior nodes [Roache, 1972]. The inclusion of these boundary conditions into the finite-difference scheme is problem specific and is discussed for each problem later.

Stability

For stability, the MacCormack scheme has to satisfy the Courant-Friedrichs-Lewy (CFL) condition. Since the water waves travel at a much higher velocity than the bed transients, this condition is given by the following equation:

$$C_n = \frac{\left(q/h + \sqrt{gh}\right)\Delta t}{\Delta x} \leq 1 \qquad (18\text{-}43)$$

in which C_n is the Courant number. Equation 18-43 has to be satisfied at every grid point for the scheme to be stable.

Artificial Viscosity

Numerical oscillations near the steep wave fronts may be dampened by introducing artificial viscosity. For this purpose, Jameson procedure discussed in Chapter 14 may be utilized. Of the several cases studied, smoothing was required only in the case of knickpoint migration.

Computational Procedure

Let us say the values of h_i^k, q_i^k and z_i^k are known at the known time level k at all grid points $(i = 1, \cdots, N+1)$ of a channel divided into N reaches and we want to determine their values at the unknown time level, $k+1$. The known values are the initial conditions at $t = 0$ if the computations are just starting; otherwise, they are the computed values during the previous time interval.

The values at time level $k+1$ may be computed as follows:

1. The values of h_i^*, q_i^*, and z_i^* at the interior nodes $(i = 2, ...N)$ are computed by using Eqs. 18-26 and their values at the boundaries $(i = 1$ and $i = N+1)$ are computed by using the appropriate boundary conditions.
2. Now, h_i^{**}, q_i^{**} and z_i^{**} at the interior nodes $(i = 2, ...N)$ are determined from Eqs. 18-27 and at the boundaries from the boundary conditions.
3. Then, h, q and z at the end of time interval Δt, i.e., h_i^{k+1}, q_i^{k+1} and z_i^{k+1} are determined by using Eqs. 18-28 to 18-30.

4. The values determined in step 3 are modified if necessary to dampen the high-frequency oscillations by using the Jameson procedure for artificial viscosity presented in Chapter 14.

5. The values of h_i^k, q_i^k and z_i^k for the next time interval are set equal to h_i^{k+1}, q_i^{k+1} and z_i^{k+1}; the time interval, Δt, for the next step is determined from Eq. 18-43; and the procedure is repeated until simulation for the specified time is done.

The flow equations and the sediment continuity equation are coupled in this procedure because it uses a two-level predictor-corrector approach. Strictly speaking, there is no coupling during the predictor part. However, the predicted values of the bed elevation are used to determine the correct values of discharge in Eq. 18-23. Similarly, the predicted values of h and q are used to determine q_s and to evaluate the spatial derivative term in Eq. 18-24. Therefore, each dependent variable computed at the end of the time step takes into account the changes in all the other variables. In this sense, this procedure may be called *coupled*. On the other hand, coupling is not achieved if Eq. 18-24 is solved after completely solving Eqs. 18-22 and 18-23, i.e., after corrector step for Eqs. 18-22 and 18-23. This later approach is similar to the *uncoupled implicit method*. According to Park and Jain [1986], this approach results in numerical instabilities, whenever the gradient of bed level becomes too large. They iterated the solution during each time step for numerical stability. In addition to its simplicity, the computational procedure presented herein does not require iterations.

Applications

To illustrate the application of this model, the computed results are compared with experimental results for two cases. For the comparisons for other cases, see Bhallamudi and Chaudhry [1991].

Aggradation due to sediment overloading

Test results on the aggradation process [Soni et al. 1980] in a laboratory channel were compared with the computed results. The channel was 0.2 m wide and 30 m long. The sand forming the bed and the injected material had a mean diameter of 0.32 mm. From the uniform-flow measurements, the following values were computed: $a = 1.45 \times 10^{-3}$; $b = 5.0$; Manning $n = 0.022$; and the porosity of the sediment bed layer, $p = 0.4$. For the results presented here, the initial water discharge, $q_o = 0.020$ m^2/s; the uniform flow depth, $h_o = 0.05$ m; the initial bed slope, $S_o = 0.00356$; and the equilibrium sediment discharge, $q_{so} = 4$.

In the mathematical model, uniform unit discharge, uniform flow depth, and the initial bed elevations, calculated from S_o, were specified at every node as the initial conditions. The transient state was initiated by increasing the rate of sediment discharge at the upstream end by Δq_s. As mentioned earlier, one boundary condition at the downstream end and two boundary conditions at the upstream end were specified. The upstream boundary representing constant discharge was implemented by specifying $q(0, t) = q_o$ for $t \geq 0$. However, the inclusion of the second boundary, $q_s(0, t) = q_{so} + \Delta q_s$ was not as straightforward as the former; and it had to be translated into an equation so that the bed elevation at the upstream end could be calculated. This was achieved by assuming a fictitious node upstream of node 1 and specifying the sediment discharge at that node equal to $q_{so} + \Delta q_s$. By using the sediment continuity equation and applying the backward finite-difference on the spatial differential term, we obtain

$$\left[(1 - p)z + \frac{q_s h}{q} \right]_1^{k+1} = \left[(1 - p)z + \frac{q_s h}{q} \right]_1^k + \frac{\Delta t}{\Delta x} \left[\left(q_{so} + \Delta q_s \right) - \left(q_s \right)_1^k \right]$$

(18-44)

The left-hand side of Eq. 18-44 and therefore, z, at the unknown time level $k + 1$ can be calculated, since the terms on the right-hand side are known for the time level k. The flow depth at node 1 is determined from the characteristic equation by using the known discharge at the boundary. The constant-depth downstream boundary condition was specified by $h(N\Delta x, t) = h_o$ for $t \geq 0$. The discharge and the bed elevation at the downstream end are determined by extrapolation from the values at the interior nodes. Note that the constant-depth boundary condition is valid for long channels in which bed transients do not reach the downstream end within the period for which conditions are computed. This boundary condition is not valid for flood flows and short channels. The boundary values in these cases may be evaluated by the characteristic method along with a rating curve between discharge and depth as the specified boundary condition.

The mathematical model was used on a 50-m long channel, which was divided into 50 reaches ($\Delta x = 1.0$ m). The computational time step, Δt, was selected so that the Courant condition for stability ($C_n = 0.9$) was satisfied. Figure 18-6 compares the computed transient bed and water-surface profiles at $t = 40$ minutes with the measured values. The "measured" points here correspond to the average transient profiles. Averaging of the actual data was required because of the presence of ripples and dunes on the bed. As can be seen, the mathematical model satisfactorily simulates the aggradation of the channel bed as well as the transient water-surface profile.

Knickpoint migration

A *knickpoint* is defined as an abrupt change in the longitudinal bottom profile of a channel (Fig. 18-7). In a channel with nonerodible bed, a knickpoint remains intact indefinitely. However, in channels flowing over an erodible bed, the knickpoints are obliterated as they migrate upstream. Brush and Wolman [1960] explained the migration of knickpoints as a result of erosion potential (hS_f) becoming maximum at the point where slope changes. Referring

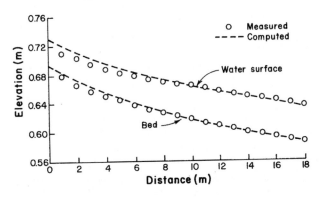

Fig. 18-6 **Comparison of computed and measured bed and water-surface profiles**

to Fig. 18-5, the flow is subcritical on the upstream side of the knickpoint and supercritical on the downstream side. The boundary shear ($\tau_o = \gamma h S_f$) is maximum at the break in the bottom slope, it decreases downstream as the flow depth decreases, and it also decreases on the upstream side of the knickpoint as the energy-grade line flattens even though flow depth is higher. Since the sediment transport is directly proportional to the shear stress, the higher shear stress at the knickpoint results in more material being carried away from this location than from the upstream or downstream reaches. Therefore, the knickpoint migrates upstream, the eroded material deposits downstream, and finally the oversteepened reach flattens.

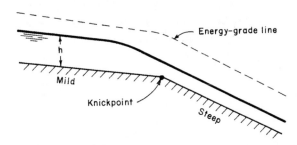

Fig. 18-7 **Definition sketch for knickpoint migration**

The experimental data obtained by Brush and Wolman [1960] in a 15.8-m-long and 1.2-m-wide flume was compared with the computed results. Before the experiment, a 0.21-m-wide, trapezoidal channel with rounded corners was molded in noncohesive sands with a median size of 0.67 mm. A fall of approximately 0.0305 m was provided at a distance of 10.8 m from the upstream end to simulate the oversteepened reach or the knickpoint. The slope of the channel upstream and downstream of this point was approximately equal to 0.00125. Water was then turned on ($q_o = 0.0028$ m^2/s, $h_o = 0.0305$ m) and the bed levels were recorded at successive times.

To simulate this experiment by the mathematical model, the channel was divided into 52 reaches ($\Delta x = 0.3048$ m). The initial and boundary conditions were included as in the previous case; Courant number was 0.85; and the Jameson procedure (see Chapter 14) was used to add artificial viscosity. The sediment discharge was determined from Eq. 18-39, with a value of $b = 4.2$ and coefficient a was assumed as a function of the energy grade line, S_e. At the start of the computations for each new time step, the slope of the energy grade line at every grid point was computed using the backward finite difference and the values of q, h and z at the known time level:

$$S_e = \frac{1}{\Delta x}\left(z_{i-1} + h_{i-1} + \frac{q_{i-1}^2}{2gh_{i-1}^2}\right) - \frac{1}{\Delta x}\left(z_i + h_i + \frac{q_i^2}{2gh_i^2}\right) \qquad (18\text{-}45)$$

Then the value of a was determined from $a = S_e^{1.71}$. The exponents in Eq. 18-39 were estimated from the sediment transport measurements [Brush and Wolman, 1960]. The conditions were identical in both the experiments except for the channel slope, and the difference in sediment transport rates could be related to the slope of the energy grade line by a power law.

Figure 18-8 compares the computed and measured results at $t = 2.67$ hours. The computed bed profile satisfactorily matches the measured bed profile despite the inherent uncertainty in the sediment transport equation. As can be seen from Fig. 18-8, the downstream bed level in the experiment is higher than that predicted by the model, although the volume of predicted upstream erosion is almost equal to the deposition on the downstream side. Channel widening on the downstream side of the knickpoint was observed in the experiments and the larger deposition in the experiments might be due to the extra sediment available from erosion of the bank.

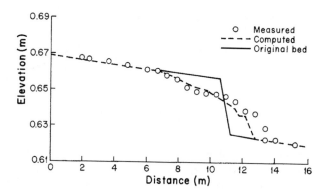

Fig. 18-8 Bed level variation due to knickpoint migration

18-10 Summary

A number of special topics were discussed in this chapter. the rating curve at a channel cross section during steady and unsteady flow was discussed. Different simplified flood routing methods were presented. A mathematical model for the simulation of aggredation and degredation due to imbalance between the actual and carrying-capacity of sediment transport was presented.

Problems

18-1 Compute the outflows from a reservoir by using a routing interval of 5 minutes for the following data:

  i. Spillway outflow $= 50\ H^{1.5}$, where $H =$ head above the spillway crest, in ft;
  ii. The reservoir has vertical sides and the surface area is 300,000 ft^2;
  iii. The inflow increases linearly from zero at $t = 0$ to 500 ft^3/s at $t = 15$ minutes and then linearly decreases to 100 ft^3/s in 10 minutes, after which the inflow remains constant at this value;
  iv. The reservoir is at the spillway crest level at time $t = 0$.

18-2 Write a computer program to route an inflow hydrograph through a reservoir. Assume the data for the inflow hydrograph are given at discrete times, and the reservoir surface area and outflow through the spillway at specified elevations are specified. Use a second-order accurate finite-difference scheme to integrate the governing equation and parabolic interpolation to compute values from the stored data.

18-3 For the following data, compute the outflow from a detention pond until time $t = 20$ minutes.

  i. Spillway crest level $=$ El. 10 ft;
  ii. Spillway discharge (in ft^3/s) $= 100(z - 10)^{1.5}$, $z =$ water level, in ft;
  iii. The pond surface area at El. 0 is 200,000 ft^2 and it linearly increases to 300,000 ft^2 at El. 40 ft;
  iv. Inflow for $t < 10$ min is $5t$ in which t is in seconds and inflow remains constant at 3000 ft^3/s for $t > 10$ min;
  v. The pond level at time $t = 0$ is at El. 8 ft.

18-4 Prove that if $\alpha = 0.5$ and $\alpha \Delta t / \Delta x = 1$ in the Muskingum-Cunge routing, a wave does not attenuate as it is routed through a channel reach. [*Hint:* Determine C_o, C_1, and C_2 for these values and then show that a flood wave is not attenuated].

18-5 By expanding Eq. 18-12 in a Taylor series and comparing it with the diffusion equation, show that

$$X = \frac{1 - Q_o}{2S_o aB\Delta x}$$

18-6 Route the flood hydrograph of Prob. 18-1 through a channel reach using kinematic routing for different values of C_n. Use Lax, MacCormack, and Preissmann schemes and compare their results with the exact solution.

18-7 By using the Lax and MacCormack schemes, study the effects of different values of the dispersion coefficient, D, in Eq. 18-32 on the computed results. Route the hydrograph of Prob. 18-1.

References

Anonymous, 1985, *HEC-1, Flood Hydrograph Package: Users' manual,* U.S. Army Corps of Engineers, Hydrologic Engineering Center, Davis, CA.

Bhallamudi, S. M., and Chaudhry, M. H., 1991, "Numerical modeling of aggradation and degradation in alluvial channels," *Jour. Hydraulic Engineering,* Amer. Soc. Civil Engrs., vol 117, no 9, pp. 1145–1164.

Begin, Z. B., Meyer, D. F. and Schumm, S. A., 1981, "Development of longitudinal profiles of alluvial channels in response to base-level lowering." *Earth Surface Processes and Land Forms,* vol 6, no 1, pp. 49–68.

Brush, L. M. and Wolman, M. G., 1960, "Knickpoint behavior in noncohesive material: A laboratory study." *Bulletin of the Geological Society of America,* vol 71, pp. 59–74.

Cunge, J. A., Holly Jr, F. M. and Verwey, A., 1980, *Practical Aspects of Computational River Hydraulics,* Pitman Advanced Publishing Program, London.

Cunge, J. A., 1969, "On the subject of a flood propagation computation method (Muskingum method," *Jour. Hydraulic Research,* Inter. Assoc. Hydraulic Research, vol 7, no 2, pp. 205–230.

Das, A., 2004, "Parameter Estimation for Muskingum Models," *Jour. Irrigation and Drainage Engineering,* vol. 130, no. 2, pp. 140–147.

Dawdy, D. R. and Vanoni, V. A., 1986, "Modeling alluvial channels," *Water Resources Research,* vol 22, no 9, pp. 71S-81S.

Fennema, R. J., and Chaudhry, M. H., 1990, "Numerical Solution of 2-D Free-Surface Flows: Explicit Methods," *Jour. Hydraulic Engineering,* Amer. Soc. Civil Engrs., vol 116, no 8, pp. 1013–1034.

Geem, Z. W., 2006, "Parameter Estimation for the Nonlinear Muskingum Model Using the BFGS Technique," *Jour. Irrigation and Drainage,* vol. 132, no. 5, pp. 474–478.

Gill, M. A., 1978, "Flood Routing by the Muskingum Method," *Jour. Hydrology,* Amsterdam, The Netherlands, vol. 36, pp. 353–363.

Gill, M. A., 1983a, "Diffusion model for aggrading channels," *Jour. Hyd. Research,* Inter. Assoc. Hydraulic Research, vol 21, no 5, pp. 355–367.

Gill, M. A., 1983b, "Diffusion model for degrading channels," *Jour. Hyd. Research,* Inter. Assoc. Hyd. Research, vol 21, no 5, pp. 369–378.

Gill, M. A., 1987, "Nonlinear solution of aggradation and degradation in channels," *Jour. Hyd. Research,* Inter. Assoc. Hyd. Research, vol 25, no 5, pp. 537–547.

Kim, J. H., Geem, Z. W., and Kim, E. S., 2001, "Parameter Estimation of the Nonlinear Muskingum Model Using Harmony Search," *Jour. Amer. Water Resources Assoc.,* vol. 37, no. 5, pp. 1131–1138.

Hayami, S., 1951, "On the propagation of flood waves," *Bulletin of the Disaster Prevention Research Institute*, Disaster Prevention Research Institute, Kyoto, Japan, vol 1, no 1, pp. 1–16.

Henderson, F. M. 1966, *Open-Channel Flow*, Macmillan, New York, NY.

Holly F. M., Jr, 1986, "Numerical simulation in alluvial hydraulics," *Proc. 5th Congress of the Asian and Pacific Regional Division,* Inter. Assoc. Hyd. Research, Aug. 18–21, Seoul, Korea.

Hromadka, T. V., and DeVries, J. J., 1988, "Kinematic wave and computational error," *Jour. Hydraulic Engineering,* Amer. Soc. Civil Engrs., vol 114, no 2, pp. 207–217, see also discussions and closure, vol 116, no 2, pp. 278–289.

Jain, S. C., C, "River bed aggradation due to overloading," *Jour. Hyd. Div.,* Amer. Soc. Civ. Engr., vol. 107, no. 1, pp. 120–124.

Jain, S. C. and Park, I., 1989, "Guide for estimating riverbed degradation," *Jour. Hyd. Engineering,* Amer. Soc. Civ. Engr., vol. 115, no. 3, pp. 356–366.

Jameson, A., Schmidt, W., and Turkel, E., 1981, "Numerical Solutions of the Euler equations by Finite Volume Methods Using Runge-Kutta Time-Stepping Schemes," *AIAA 14th Fluid And Plasma Dynamics Conference,* Palo Alto, California, AIAA-81-1259.

Jaramillo, W. F. and Jain, S. C., 1984, "Aggradation and degradation of alluvial-channel beds," *Jour. Hyd. Engineering,* Amer. Soc. Civ. Engrs., vol. 110, no. 8, pp. 1072–1085.

Katopodes, N.D., 1982, "On Zero-Inertia and Kinematic Waves," *Jour. Hydraulics Div.,* Amer. Soc. Civil Engrs., vol 108, no 11, pp. 1380–1387.

Lane, E.W. and Borland, W.M., 1954, "River-bed scour during floods," *Trans. Amer. Soc. Civ. Engrs.,* vol 119, pp.1069–1079.

Lee, J. K., and Froehlich, D. C., 1989, "Two-Dimensional Finite-Element Hydraulic Modeling of Bridge Crossings," Report FHWA-RD-88-146, U.S. Dept. of Transportation, McLean, VA.

Lu, J-y and Shen, H. W., 1986, "Analysis and comparisons of degradation models," *Jour. Hyd. Engineering,* Amer. Soc. Civ. Engrs., vol. 112, no. 4, pp. 281–299.

Lyn D. A., 1987, "Unsteady sediment transport modeling," *Jour. Hyd. Engineering,* Amer. Soc. Civ. Engr., vol. 113, no. 1, pp. 1–15.

Lighthill, M. J., and Whitham, G. B., 1955, "On kinematic waves. I: Flood movement in long rivers," *Proc., Royal Society,* London, UK, A229, pp. 281–316.

MacCormack, R. W., 1969, "The Effect of Viscosity in Hypervelocity Impact Cratering," *Amer. Inst. Aero. Astro.*, Paper 69–354, Cincinnati, OH.

Mohan, S., 1997, "Parameter Estimation of Nonlinear Muskingum Models Using Genetic Algorithm," *Jour. Hyd. Engineering,* vol. 13, no. 2, p. 142.

Morris, E. M., and Woolhiser, D. A., 1980, "Unsteady one-dimensional flow over a plane: Partial equilibrium and recession hydrograph," *Water Resources Research,* vol. 16, no. 2, pp. 355–360.

Newton, C. T., 1951, "An experimental investigation of bed degradation in an open channel," *Trans. Boston Soc. Civ. Engrs.,* pp. 28–60.

Park, I. and Jain, S. C., 1986, "River-bed profiles with imposed sediment load," *Jour. Hyd. Engineering,* Amer. Soc. Civ. Engrs., vol. 112, no. 4, pp. 267–279.

Ponce, V. M., Li, R. M., and Simons, D. B., 1978, "Applicability of kinematic and diffusion models," *Jour. Hydraulic Engineering,* Amer. Soc. Civil Engrs., vol. 104, no. 3, pp. 353–360.

Ponce, V. M., and Theurer, F. D., 1982, "Accuracy criteria in diffusion routing," *Jour. Hydraulic Engineering,* Amer. Soc. Civil Engrs., vol. 108, no. 6, pp. 747–757.

Ponce, V. M., 1990, *Engineering Hydrology,* Prentice Hall, Englewood Cliffs, NJ.

Ponce, V. M., 1991, "The kinematic wave controversy," *Jour. Hydraulic Engineering,* Amer. Soc. Civil Engrs., vol. 117, no. 4, pp. 511–525.

Roache, P. J, 1972, *Computational Fluid Dynamics,* Hermosa Publishers, Albuquerque, NM.

Roberson, J. A., Cassidy, J. J., and Chaudhry, M. H., 1997, *Hydraulic engineering,* Second ed., John Wiley & Sons, New York, NY.

Roberson, J.A., Cassidy, J.J., and Chaudhry, M.H. (1998), *Hydraulic Engineering,* Houghton Mifflin Company, Boston. pp. 222–225, 611–635.

Seddon, J. A., 1900, "River hydraulics," *Trans.,* Amer. Soc. Civ. Engrs., pp. 179–229.

Seed, R. B., Nicholson, P. G., Dalrymple, R. A., et al., 2005, "Preliminary Report on the Performance of the New Orleans Levee System in Hurricane Katrina on August 29, 2005," Report No. UCB/CITRIS–05/01, sponsored by NSF, Nov.

Soni, J. P., Garde, R. J. and Raju, K. G., 1980, "Aggradation in streams due to overloading, " *Jour. Hyd. Div.,* Proc. Amer. Soc. Civ. Engrs., vol. 106, no. 1, pp. 117–132.

Soni, J. P., Garde, R. J. Raju, K. G., and Kittur, G. J., 1977, "Nonuniform Flow in Aggrading Channels," *Jour. Waterways Port Coastal Ocean Div.,* Amer Soc. Civ. Engrs., vol 103, no 3, Aug., pp. 321–333.

Suryanarayana, B., 1969, "Mechanics of degradation and aggradation in a laboratory flume," *thesis* presented for the degree of Doctor of Philosophy, Colorado State University, Ft. Collins, CO.

Tung, Y.K., 1984, "River Flood Routing by Nonlinear Muskingum Method," *Jour. Hyd. Div.,* Amer. Soc. Civ. Engrs., vol. 111, no. 12, pp. 1447–1460.

Yoon, J., and Padmanabhan, G., 1993, "Parameter Estimation of Linear and Nonlinear Muskingum Models," *Jour. Water Resources Plannning,* Amer. Soc. Civ. Engrs., vol. 119, no. 5, pp. 600–610.

Woolhiser , D. A., and Liggett , J.A., 1967, "Unsteady one-dimensional flow over a plane – the rising hydrograph," *Water Resources Research,* vol. 3, no. 3, 753–771.

Zhang, H. and Kahawita, R., 1987, "Nonlinear model for aggradation in alluvial channels," *Jour. Hyd. Engineering,* Amer. Soc. Civ. Engrs., vol 113, no 3, pp. 353–369.

Authors Index

Guillén, N. F., 333

Abbett, M., 261, 262, 279
Abbott, D. E., 209, 244
Abbott, M. B., 59, 258, 273, 274,
    279, 281, 346, 361, 364, 369,
    378, 393, 411, 413, 457
Ackers, P., 217, 244, 318, 320, 331
Advani, R. M., 244
Ahmadi, R. M., 80, 90, 122
Ahuja, K. C., 247
Akan, A. O., 105, 120
Albertson, M. L., 299
Alcrudo, F., 444, 456
Ali, K. H. M., 244
Alsdorf, D. E., 332
Amein, M., 378, 393, 411
Amorocho, J., 27, 30
Anastasiou, K., 425, 456
Anderson, A. G., 312, 314, 333
Anderson, D. A., 257–260, 269,
    279, 378, 390, 397, 404, 411,
    441, 456
Anderson, E., 331
Anderson, V. M., 245
Andrews, E. D., 491, 513
Anton, H., 431, 456
Arnell, N. W., 332
Arumugam, K., 237, 238, 247

ASCE, 462
Ashida, K., 502, 503, 512
Assanis, 425
Assanis, D. N., 425, 459
Austin, L. H., 248

Babb, A. F., 27, 30
Bagge, G., 252, 279
Bagnold, 497, 512
Baker, J. A., 376, 378
Bakhmeteff, B. A., 39, 59, 159,
    198, 221, 245
Bala, S. K., 334
Banks, R. B., 112, 121
Barbarossa, N. L., 509, 510,
    512
Barbetta, S., 335
Barnes, H. H., 101, 103, 120
Basak, S., 248
Basco, D. R., 273, 279, 346, 348,
    359, 361
Bastola, S., 335
Beam, R. M., 397, 411, 435, 452,
    454–456, 459
Beck, 513
Becker, T. M., 456
Begin, Z. B., 528, 538
Bennel, R. S., 248
Benning, R. M., 425, 456

© Springer Nature Switzerland AG 2022
M. H. Chaudhry, *Open-Channel Flow*,
https://doi.org/10.1007/978-3-030-96447-4

Subject Index

Abrupt rise, 226–228

Absolute wave velocity, 66, 339, 346, 354

Acceleration, 4, 16, 17, 37, 38, 94, 126, 208, 355, 418
    centrifugal, 18
    convective, 6, 517, 522, 525
    local, 4, 206, 353, 517

Acoustic Doppler Current Profiler, 322

Adverse slope, 129

Aggradation, 527, 528
    sediment overloading, 533

Air entrainment, 220, 231, 252, 272, 306, 311–313, 335
    air-free zone, 312
    mixing zone, 312
    underlying zone, 312
    upper zone, 312

Alluvial channel, 299, 300, 528, 538

Alternate depths, 40, 86

Alternating direction implicit, 438

Amplification factor, 402, 404

Amplitude, 338, 404

Angle of repose, 294, 295

Artificial channels
    flume, 276, 535, 540

Artificial neural networks, 425

Artificial viscosity, 269, 272, 392, 434, 445, 448, 454, 455, 532, 533, 536

Baffle blocks, 226, 234

Basin, 518

Beam and Warming scheme, 435

Bedforms, 490, 502, 507–509

Bedload, 490, 502, 512–514

Bernoulli equation, 38, 39, 41

Best hydraulic section, 286, 287

Bisection method, 77, 80, 86, 109, 110, 167

Bore, 338, 340, 348, 350, 359, 368, 374, 376, 377, 390, 425–427, 434, 445, 446, 448, 455

Bottom slope
    mild, 306
    steep, 194, 310, 311

Boundary conditions, 260, 261, 266, 267, 274, 364, 370, 371, 377, 382, 384, 387, 392, 394–396, 405, 406, 429, 433, 435, 441, 445, 447, 448, 454, 455, 528, 532, 534, 536
    non-periodic, 376

Boundary element method, 376, 379

Boundary layer, 208, 218, 231, 311–314, 329

© Springer Nature Switzerland AG 2022
M. H. Chaudhry, *Open-Channel Flow*,
https://doi.org/10.1007/978-3-030-96447-4

Printed in the United States
by Baker & Taylor Publisher Services